Cognitive and Working Memory Training

Cognitive and Working Memory Training

Perspectives From Psychology, Neuroscience, and Human Development

Edited by

JARED M. NOVICK

MICHAEL F. BUNTING

MICHAEL R. DOUGHERTY

RANDALL W. ENGLE

OXFORD

UNIVERSITY PRESS

OXFORD
UNIVERSITY PRESS

Oxford University Press is a department of the University of Oxford. It furthers
the University's objective of excellence in research, scholarship, and education
by publishing worldwide. Oxford is a registered trade mark of Oxford University
Press in the UK and certain other countries.

Published in the United States of America by Oxford University Press
198 Madison Avenue, New York, NY 10016, United States of America.

Library of Congress Control Number: 2019949802
ISBN 978–0–19–997446–7

3 5 7 9 8 6 4 2

Printed by Integrated Books International, United States of America

To our families.

For our students.

Contents

I. COGNITIVE PERSPECTIVE

II. NEUROCOGNITIVE PERSPECTIVE

III. DEVELOPMENTAL PERSPECTIVE

Acknowledgments

We are indebted to the range of scholars who contributed their time and expertise to write the chapters for this collected volume, and who also provided ample work to peer-review each other's submissions. We thank them for these efforts and for slogging through our rounds of gentle edits and comments. A special thanks to Karly Schwarz and Claire Crossman, graduate assistants in the Department of Hearing and Speech Sciences at the University of Maryland, for their tireless endeavors to pull this entire project together in all the ways that matter. It is not an overstatement to say that, without them, this book would not have been completed.

Contributors

Jacky Au
School of Education
University of California, Irvine
Irvine, CA, USA

Claudia C. von Bastian
Department of Psychology
University of Sheffield
Sheffield, UK

Daphne Bavelier
Brain and Learning Lab
Psychology and Education
Sciences
University of Geneva
Geneva, Switzerland

Rossana De Beni
Department of General Psychology
University of Padova
Padova, Italy

Erika Borella
Department of General Psychology
University of Padova
Padova, Italy

Michael F. Bunting
Senior Research Scientist and
Director of Research Development
Applied Research Laboratory
for Intelligence and Security
(ARLIS)
University of Maryland
College Park, MD, USA

Martin Buschkuehl
Director Education Research
MIND Research Institute
Irvine, CA, USA

Barbara Carretti
Department of General Psychology
University of Padova
Padova, Italy

Cesare Cornoldi
Department of General Psychology
University of Padova
Padova, Italy

Adele Diamond
Department of Psychiatry
University of British Columbia
Vancouver, BC, Canada

Sandra Dörrenbächer
Department of Psychology
Saarland University
Saarbrücken, Germany

Michael R. Dougherty
Professor of Psychology
Department of Psychology
University of Maryland
College Park, MD, USA

Adam Eichenbaum
Helen Wills Neuroscience Institute
University of California, Berkeley
Berkeley, CA, USA

Randall W. Engle
Professor of Psychology
School of Psychology
Georgia Institute of Technology
Atlanta, GA, USA

Anders M. Fjell
Department of Psychology
University of Oslo
Oslo, Norway

C. Shawn Green
Department of Psychology
University of Wisconsin, Madison
Madison, WI, USA

Sabrina Guye
University Research Priority Program
"Dynamics of Healthy Aging"
University of Zurich
Zurich, Switzerland

Henk J. Haarmann
Applied Research Lab for Intelligence
and Security
University of Maryland
College Park, MD, USA

Kenny Hicks
School of Psychology
Georgia Institute of Technology
Atlanta, GA

Susanne M. Jaeggi
School of Education
University of California, Irvine
Irvine, CA

Benjamin Katz
Assistant Professor
Human Development and Family
Science
Virginia Tech
Blacksburg, VA

Jutta Kray
Department of Psychology
Saarland University
Saarbrücken, Germany

Stefanie E. Kuchinsky
Applied Research Lab for Intelligence
and Security
University of Maryland
College Park, MD, USA

Ulman Lindenberger
Max Planck Institute for Human
Development
Center for Lifespan Psychology
Berlin, Germany

Daphne S. Ling
Department of Psychiatry
University of British Columbia
Vancouver, BC, Canada

Martin Lövdén
Department of Neurobiology, Care
Sciences and Society (NVS), H1
Karolinska Institutet
Stockholm, Sweden

Jared M. Novick
Associate Professor
Department of Hearing and Speech
Sciences
Program in Neuroscience and Cognitive
Science
University of Maryland
College Park, MD, USA

Lars Nyberg
Department of Integrative Medical
Biology
Umeå University
Umeå, Sweden

Brooke M. Okada
Graduate Researcher
Department of Psychology
University of Maryland, College Park
College Park, MD, USA

Florian Schmiedek
Department of Education and Human
Development
German Institute for International
Educational Research (DIPF)
Frankfurt am Main, Germany

Priti Shah
Department of Psychology
University of Michigan
Ann Arbor, MI

Carla De Simoni
Department of Psychology,
University of Zurich
Zurich, Switzerland

L. Robert Slevc
Associate Professor
Department of Psychology
Program in Neuroscience and
Cognitive Science
University of Maryland
College Park, MD, USA

Kristine B. Walhovd
Department of Psychology
University of Oslo
Oslo, Norway

Prologue

"Brain training" exploded into the marketplace earlier this century with the publication of a few scientific studies that caught the attention of the media and, inevitably, excited the public at large. The findings offered hope to populations that no pill could provide: Could we offset the effects of dementia through consistent mental challenge? Could we close the achievement gap for at-risk children? Unsurprisingly, the global buzz promoted the advent of websites and video games that issued regimens for training intelligence. The promise of brain training (also known as cognitive or working memory training) was that persistent engagement of core cognitive functions like fluid reasoning, inhibition, or attention control could increase cognitive functioning in general, and that the effect endures over time.

However, a problem with many brain-training applications on the market then, and today, is the lack of scientific consensus in the field about whether they actually work. Certainly, the natural course of things is that new tests of popular ideas give way to data that offer different results and, thus, put a spotlight on different ways of interpreting original findings. This is good for science, as alternative viewpoints (theories, ideas, etc.) push a field forward through a conspiracy of evidence that attempts to converge on the truth. Sometimes it means reconsidering our initial excitement through the lens of nuance and reminding ourselves—scientists and civilians alike—that knowledge can only advance through steady interrogation of traditional hypotheses. While brain training remains popular, some of the initial excitement has faded under the glare of scientific scrutiny about the true efficacy of the programs and the unavoidable press that comes along with it.

Here, we bring a range of viewpoints into a single volume that represents the ideas (and empirical discoveries) of leading researchers in the arena of cognitive training. Critically, perspectives from both proponents and detractors are included in an accessible format so that students of psychology (of both the professional and armchair varieties) can assess the current state of the science. The chapters are highly interdisciplinary and cover a range of related issues, which are organized thematically around three "cluster" topics in working memory training: cognitive perspectives, cognitive neuroscience perspectives, and developmental perspectives across the lifespan. Each cluster is introduced by a "challenge" chapter in which the authors pose certain fundamental questions that are intended to identify key issues, spark debate, and yield an array of viewpoints within the clusters. The subsequent chapters within a cluster are organized around the authors' responses to the challenge questions and the defense of their responses. They offer best practices in the field regarding effective training methods, tools, designs, and uses for applied, theoretical, and computational goals, and what has been learned—and what remains to be learned—about the advantages and potential caveats of cognitive training.

The goal of this book is to offer an objective and balanced appraisal of the science of cognitive training by framing the book's theme within a broader context of cognitive psychology in terms of past, present, and the contours of future theoretical and empirical endeavors in the domain of cognitive and working memory training.

A Brief History of the Scientific Study of Human Abilities

Human intelligence—the mental traits and abilities that are uniquely human and the province of *Homo sapiens* alone—has permitted people to accomplish feats unmatched by any other creature on Earth. Societies are products of human intelligence, which enables our knowledge and understanding of the world, both the physical and the spiritual. Intelligence has taken humans to the moon, sent robots to Mars, and developed tools for looking across the cosmos and listening for decipherable sounds from space. Through their intelligence, humans possess the cognitive abilities to learn, understand, remember, and teach. Humans can think concretely or abstractly, and they can apply logic and reason, and solve novel problems. They can recognize patterns, plan, make decisions, and draw conclusions. Humans can use language to communicate thoughts, feelings, intentions, and deceptions in infinitely creative and productive ways.

For much of the history of the scientific study of intelligence, two related questions have driven the conversation: What are the dominant abilities involved in intelligent behavior, and what is the organization of the mechanisms that support it? Spearman (1904, 1927, 1946; Spearman & Wynn-Jones, 1950) insisted that, instead of an infinite number of unrelated mental abilities, a single common factor runs through all intellectual operations, generating commonly observed correlations among tests. Spearman's original contention was that performance associations across mental tests can be accounted for by the combination of two factors. One of these factors, g, is common to all the tests and is domain-free. The other factor, designated the s factor, is specific, and it captures what remains in the intercorrelations among the tests (except what is attributable to test errors). Spearman regarded g as a cognitive-general factor and argued that it is present to varying degrees in every intellectual operation.

Spearman's theory was informed by his observations of performance on mental ability tests. He observed that people who do well in one area also tend to do well in other areas. This lent credence to his notion that general intelligence influences performance on all cognitive tasks. A commonly invoked metaphor for understanding general intelligence is to compare it to athleticism. One probably would not expect a professional basketball player to be equally impressive on the softball or soccer fields; however, because of her all-around athleticism, she is likely to excel at those sports compared to your average couch potato. Yet, many things in the world are related but may reflect lurking influences of confounding variables. Take, for instance, the oft-cited correlation between a rise in ice cream sales and homicide rates. It is considerably unlikely that one has a causal effect on the other, as many factors (e.g., hot weather, more milling about) independently contribute to both. Thus, the same reasoning ought to be applied to cognitive training: rather than focusing on the products of interventions (does it work or not? does performance increase from pretest to posttest?), the effects of brain training (or lack thereof) may be better understood through the mechanisms that drive the process.

Important questions remain about how to measure and define intelligence and understand all the factors that comprise this squishy term; but contemporary study asks how to improve it nonetheless in hopes of preventing cognitive abilities from declining with age. Intelligence may not be static. Cognitive skills, memory, reasoning, motor skills, and the speed of thinking can and do increase or decrease over time and are subject to experience-induced plasticity for a multitude of reasons. Genetics and environmental conditions are contributing factors, both to the growth of intelligence and the delaying of its deterioration. Nutrition, pharmacological and psychological factors, and behavior can have positive or negative impacts. Understanding the relationship among these factors and how they affect cognitive performance is a true scientific challenge, let alone the challenge of isolating what is cause, and what is effect.

The majority of readily noticeable changes in intelligence occur at either the onset of development, during the critical period, or during old age. But is intelligence, or at least complex cognitive behavior, immutable between these two poles on the developmental spectrum? The predominant view for much of the twentieth century was that it was: intellectual capacity was shaped during childhood and fixed during adulthood until old age brought an inevitable decline. Recent research on the malleability of intelligence has begun to challenge this view, indicating that certain types of mental workouts, also known as "brain training," can actually improve core mental abilities and protect against such declines. As the metaphor goes, as exercising a muscle increases physical strength, exercising the mind can increase mental fitness in terms of how much information can be temporarily maintained and processed, including the ability to focus attention on a current task at hand.

There is little doubt that "brain training" is a hot and hotly contested topic in the interdisciplinary fields of psychology, cognitive neuroscience, and human development. This edited volume asks questions about the nature of intellect and cognitive abilities and explores evidence that these attributes are amenable to change from training, and what mechanisms that contribute to complex cognition may or may not be malleable via intervention (i.e., is there more promise in a process-specific approach to training, rather than an all-encompassing one subsumed under the broad heading of "intelligence"?). Importantly, one focus of the book is on the notion of transfer—namely, the extent to which cognitive training generalizes to learning and performance measures that were decidedly not part of the training regimen, but still tap into specific process that were part of the regimen despite ostensible differences in task characteristics. This edited volume is inspired by the outcome of a 2011 workshop on this topic and features a series of chapters by 12 leading scholars in the cognitive and neural sciences. Generally, the issues addressed are:

- What is the scientific evidence that cognitive training influences and benefits performance on a range of everyday tasks, including intelligence, memory, attention, vision, learning, creativity, and language processing—in both healthy and special populations (e.g., young children, aging adults, those diagnosed with Attention Deficit Disorder)?
- What best practices exist in the field regarding effective training methods, tools, designs, and uses for applied, theoretical, and computational goals?
- What has been learned—and what remains to be learned—about the advantages and caveats of cognitive training? What should, and should we not, buy into?

We have assembled an interdisciplinary group of distinguished authors—all experts in the field—who have been testing the efficacy of cognitive and working-memory training using a combination of behavioral, neuroimaging, meta-analytic, and computational modeling methods. As will become clear, there is a range of views on the extent to which cognitive training remains promising. As such, this edited volume will be a defining resource on the practicality, utility, and validity of the field of cognitive training research in general and working memory training in particular.

This book is the first of its kind and is therefore expected to appeal broadly to academics in the cognitive and neural sciences, to students of psychology, to clinical practitioners interested in cognitive remediation, and to government stakeholders whose principal concern is to increase the learning and performance capabilities of their workforce. We therefore anticipate that this book will play a key role in the field by integrating a host of research efforts on cognitive training and cognitive plasticity into a single, comprehensive volume accessible to a wide audience. In the end, we also hope that it will spring new research that addresses still-open questions through collaborators on both sides— namely, the enthusiasts and critics of current data—who set aside dogma, their scientific differences, and agree upon a study design in pursuit of scientific rigor and knowledge, which can only advance through constant questioning of conventional models, ideas, and approaches.

Michael F. Bunting and Jared M. Novick

References

Spearman, C. (1904). "General intelligence," objectively determined and measured. *American Journal of Psychology, 15,* 201–293.

Spearman, C. (1927). *The abilities of man.* New York, NY: Macmillan.

Spearman, C. (1946). Theory of general factor. *British Journal of Psychology, 36,* 117–131.

Spearman, C., & Wynn-Jones, L. L. (1950). *Human ability: A continuation of "the abilities of man."* London, England: Macmillan.

SECTION I
COGNITIVE PERSPECTIVE

1

Cognitive Perspectives of Working Memory Training

Current Challenges in Working Memory Training

Kenny Hicks and Randall W. Engle

Introduction

Working memory training is an emergent field aimed at improving general cognitive abilities through targeted brain exercises. The prospect of improving cognitive abilities like attention control, comprehension, and reasoning has piqued the interest of the scientific community and the general public alike. If cognitive abilities like working memory capacity can be improved, it is assumed that this improvement will result in benefits to a broad range of real-world abilities associated with working memory capacity, including reading comprehension, math performance, and attention control (Holmes et al., 2010; Jaeggi, Buschkeuhl, Jonides, & Perrig, 2008; Klingberg, 2005; Klingberg, Forssberg, & Westerberg, 2002). However, to date, there is no clear answer to the question of whether cognition will improve through interventions designed to enhance working memory capacity. One reason is an absence of discussion among researchers of various training paradigms, which has resulted in a lack of consensus on the basic underlying principles of the research, including differences in the operational definition of working memory training, inconsistent ways to measure increases in working memory, and little integration of findings into a larger literature on cognitive training or, more broadly, working memory capacity.

This line of research is theoretically important, but it is also unique because of its potential for real-world impact. Working memory training has far-reaching implications for many diverse stakeholders, including not only academics but also any group interested in cognitive improvements. Those concerned with such diverse topics as improving selection, job training, and cognitive remediation are interested in the efficacy of working memory training and its potential for future applications. Products that extoll the benefits of "brain training" and other targeted exercises aimed at increasing cognitive abilities have permeated the public sphere. Such widespread attention has led to an influx of working memory training literature in psychology and other disciplines, but our enthusiasm must be tempered by the evidence.

Preliminary studies on working memory training are not often referred to as pilot studies in media headlines or by the commercial programs that use them for advertisement (Holmes, Gathercole, & Dunning, 2009; Holmes et al., 2010). Simply put, although coverage of preliminary studies may be well intentioned, the results can be easily oversold. It is our hope that the reader will gain perspective on the outcomes of training, evaluate the strength of the current evidence, develop an understanding of the current debate on the most controversial findings, and understand how these findings might transfer to the real world.

Researchers of working memory training claim that cognitive training interventions result in transfer to a domain-general ability (Klingberg, 2005) above and beyond task-specific abilities, such as strategy use (e.g., chunking items into groups). Although perspectives on what constitutes successful transfer are varied, it is generally conceptualized as near and far transfer.

Near transfer refers to gains in tasks that tap the same construct that the intervention seeks to improve. In terms of working memory training, near transfer would be achieved by demonstrating improved performance on novel working memory tasks. Evidence of far transfer occurs when subjects demonstrate superior performance on tasks that require working memory but reflect a fundamentally different construct (e.g., fluid intelligence). The logic here is that working memory is a key component of other higher-order cognitive abilities, such as fluid intelligence, and that improving working memory should lead to an increase in any ability dependent on the working memory construct. Ideally, after working memory training, improvements would be observed in tasks designed to assess both near and far transfer. However, to claim that working memory training increases fluid intelligence, the results should also provide evidence that working memory was improved. That is to say, at a minimum, researchers should demonstrate near transfer before claiming to show evidence of far transfer.

To better understand the complexity of transfer, we can look to the 1992 case study of Rajan Mahadevan. Mahadevan had an exceptional memory for numbers: he could recall a record of more than 30,000 digits of pi. Investigators found that, rather than possessing an innate ability for memorization, he used a mnemonic strategy, namely grouping numbers into blocks of ten. Further evidence that Mahadevan used a memory strategy for numbers came when scientists found that his memory for spatial objects was merely average (Biederman, Cooper, Fox, & Mahadevan, 1992). What is important is that Mahadevan's superior memory performance for digits did not transfer to other domains of memory (Ericsson, Delaney, Weaver, & Mahadevan, 2004).

The goal of working memory training is to demonstrate broad transfer to tasks that involve the same components of working memory that were targeted during training. Therefore, improvements should be observed on a broad range of tasks that tap the ability being trained. This is measured by observing the difference between pre- and posttest performance on cognitive tasks that subjects have not practiced.

Investigating Transfer

Given our lab's substantial contributions to the theoretical aspects of working memory capacity and its relationship to fluid intelligence (Engle & Kane, 2004; Engle, Tuholski, Laughlin, & Conway, 1999; Kane et al., 2004; Shipstead, Redick, & Engle, 2012b), we were intrigued by the prospect of improving cognition and the potential theoretical and real-world implications. In particular, we were motivated to further our understanding of the mechanisms that drive the cognitive improvements.

The task of selecting which training regimen to investigate was difficult because no unified approach to the study of working memory training exists. Due to the highly influential claim that training on the dual N-back task led to improvements on matrix reasoning, one of the best indicators of fluid intelligence (Jaeggi et al., 2008), our first study implemented the dual N-back paradigm. While offering an exciting prospect for the field, Jaeggi and colleagues' article included methodological shortcomings, such as measuring fluid intelligence with a single test instead of multiple indicators. Further, the 2008 article actually represents the combination of four studies that differed significantly, including different measures of fluid intelligence across studies, differing deadlines for completing the test of fluid intelligence, and the use of no-contact control groups (see Redick et al., 2012, for a more in-depth discussion of the studies). In an attempt to replicate Jaeggi and colleagues' 2008 findings, we conducted a follow-up study that addressed the previous study's shortcomings by including a dual N-back training group, an active control group that performed an adaptive visual search task, as well as a no-contact control group. Practice on a visual search task was chosen as the active control condition because previous research found no relationship between working memory capacity and visual search performance across a number of studies that varied the difficulty of the search task (Kane, Poole, Tuholski, & Engle, 2006). The training study also included measures designed to simulate real-world performance (e.g., the ability to manage air traffic and perform complex multitasks). Despite adequate statistical power and the inclusion of multiple indicators of both fluid intelligence and multitasking, the results of the training study failed to demonstrate any behavioral improvements after dual N-back training (Redick et al., 2012).

After our failure to replicate far transfer to fluid intelligence after dual N-back training, our research became focused on investigating near transfer. Therefore, our next study explored the effects of training on measures of short-term memory and the complex span (Chein & Morrison, 2010; Unsworth, Heitz, Schrock, & Engle, 2005; Unsworth, Redick, Heitz, Broadway, & Engle, 2009). Subjects were randomly assigned to a short-term memory training condition, a complex span training condition, or an active control group that practiced a visual search task. All three were adaptive tasks (i.e., increased in difficulty with success). To investigate far transfer, we administered multiple measures of fluid intelligence at pre- and posttest, including Raven's Progressive Matrices, the Letter Sets task, and the Number Series task. We also included other measures related to working memory capacity, such as free recall. None of our training groups showed improvements in fluid reasoning (e.g., far transfer). We found partial evidence for moderate transfer (tasks representing moderate transfer that were different from the training paradigm but still relied heavily on many of the same memory processes involved in the training tasks). For instance, improvements in the secondary memory portion of immediate free recall were observed, while evidence for the Keep Track Task was less interpretable (Harrison et al., 2013). Overall, this study represented our second failure to demonstrate domain-general improvements to cognitive abilities after extensive working memory training.

Criticisms of Working Memory Training

Shipstead and colleagues (2010) published the first systematic review analyzing the methods and results used in the working memory training literature. Prior to 2010, the claims that working memory training led to vast improvements in intelligence and attention control had gone largely unchallenged. The Shipstead article was fittingly titled "Does Working Memory Training Generalize?" If working memory capacity can predict performance across a range of important cognitive outcomes, does working memory training lead to improvements across a wide range of cognitive tasks? Although the question posed is simple, it has received surprisingly little attention. The conclusion of Shipstead and colleagues was that most working memory training studies failed to control for threats to internal validity, failed to include adequate control groups for comparison, and consistently relied on a single test to represent constructs of interest—practices that continue in the working memory training literature today.

In the publication, "Cogmed Training: Let's Be Realistic About Intervention Research" by Gathercole and colleagues (2012), the researchers argued that many of the criticisms raised about working memory training are impractical. They justified their perspective with two arguments. First, the authors maintained that training studies are costly and time consuming. Second, they argued that too much rigor in the early stages of investigation could be wasteful because the outcome of a training study is uncertain. They advised initial studies on working memory training to focus on ways to minimize costs in order to maximize the chance for a successful outcome (e.g., no-contact and active control groups are expensive, so they shouldn't be required during early stages of research). The authors made their case by describing the results of two pilot studies they were able to complete successfully while reducing the cost of training. The first study (Holmes et al., 2009) did not include a control group, while the second study (Holmes et al., 2010) included a no-contact control group. For Gathercole and colleagues' logic to hold true, they must assume that the results of working memory training studies are not dependent on whether the study includes a passive or active control group. However, meta-analytic work investigating this issue has found the opposite. Researchers have found substantial differences in training outcomes that depend on the type of control group included in the study. Specifically, training studies including no-contact control groups have much higher rates of finding positive training effects, whereas studies including active control groups have found little to no reliable effects of training (for a more in-depth review, see Melby-Lervåg & Hulme, 2013; Melby-Lervåg, Redick, & Hulme, 2016). This result is further supported by Dougherty and colleagues (2016), who conducted a re-analysis of N-back training studies. In line with the broader literature on working memory training, the authors found significant evidence in support of working memory training when studies included a no-contact control group, but no evidence for training-related benefits for studies that included an active control group. In light of these findings, we agree with Shipstead and colleagues (2012b) that researchers should pursue more robust experimental designs. In addition to the inclusion of active control groups, researchers should also be cautious when interpreting their findings. An in-depth analysis of five studies by Redick (2016) showed that several studies supporting working memory training are the result of a general decline in the performance of the control group from pre- to posttest, which render the results uninterpretable from any theoretical perspective.

In consideration of this work, the aim of the current chapter is to pose a series of questions to researchers investigating the efficacy of working memory training. By adopting a common-question approach, we are following the example set by *Variation in Working Memory* (Conway, Jarrold, Kane, Miyake, & Towse, 2007), where each contributing research group is asked to address questions that motivate discussion about the commonalities and distinctions in the field of working memory training. Through answering these directed questions, each research group will discuss what motivates their work, cover the research results of their particular paradigm, and express their views on the current state of working memory training as it relates to various training interventions. The questions raised here touch on fundamental issues integral to working memory training research and will open a dialogue among researchers with various perspectives and methodologies.

Four Directed Questions

Question 1: Theory of Working Memory Training

What is your theory of working memory capacity? What theoretical framework guides your perspective on working memory capacity? How has this perspective influenced the way you design interventions aimed at improving it?

One of the first things proponents of working memory training need to present before claiming to improve cognitive abilities is a theory of working memory capacity that outlines the mechanisms responsible for its relationship to higher-order cognition. Before claims of cognitive improvements can be taken seriously by academic researchers and the general public, proponents of working memory training first need to describe a theory of working memory that outlines the mechanisms responsible for its relationship to higher-order cognition (see von Bastian & Oberauer, 2014, for a more in-depth discussion of this issue).

Although a number of studies have reported increases in reasoning (Jaeggi et al., 2008; Klingberg, 2005; Klingberg et al., 2002), attention control (Klingberg et al., 2005), and memory performance (Holmes et al., 2009), the question is: Why? So far, the literature lacks a concrete theoretical framework that can shed light on this question.

While we acknowledge that the assumptions regarding the underlying structure of working memory capacity may differ greatly between research groups, we encourage each contributor to provide evidence for the malleability of working memory capacity and other aspects of higher-order cognition. Since differences in perspectives influence how researchers investigate the effects of working memory training (e.g., guiding selection and creation of training paradigms), it is important for each research group to outline their concept of working memory capacity and the underlying mechanisms that govern it. Future work in this area will need to more clearly explain specific mechanisms responsible for individual differences in working memory capacity and the way in which working memory training paradigms target and improve the mechanisms.

Question 2: Major Claims of Working Memory Training

What specific claims do you make concerning the efficacy of your training program? What are the specific claims your research group makes about the improvements that should result from your training regimen? How do your claims differ from those of other research groups? Have your initial claims changed in response to newer evidence?

The research groups selected for this volume implement a diverse range of training interventions, which has resulted in substantial differences in the claims researchers have made regarding their findings. For example, the claims made by Eichenbaum, Bavelier, and Green (see Chapter 9) are specific to improvements in visual attention, while the claims about improving attention made by Jha, Krompinger, and Baime (2007) are much more general.

The questions in this section seek to clarify the discrepancies between the past and present claims of researchers of working memory training. Before we can fully consider the efficacy and utility of training, it is important to understand the current positions of each group and how their understanding has evolved based on developments in the field.

Question 3: Methodological Issues

What are the biggest methodological issues in your area? How do you address these issues presently, and how do you plan to address them in the future?

The purpose of this question is to require each research group to address common methodological pitfalls that inhibit researchers' ability to draw causal inferences about their interventions. Some researchers have tackled this issue head on. For example, Eichenbaum, Bavelier, and Green (see Chapter 9) discuss the importance of adequate sampling and control groups. Specifically, they claim that wait-list or no-contact controls are insufficient and that active control groups are becoming the gold standard in the field (see Green & Bavelier, 2012).

No Single Test Is Process Pure: The Importance of Measuring Constructs With Multiple Indicators

The strength behind a theoretical account of working memory depends on how well researchers can measure the underlying construct. Since all laboratory tasks are imperfect, they include variance related to the ability of interest (e.g., working memory capacity) and variance that is unrelated (error variance). The only way to tease the two apart is to administer several working memory tests to derive the common variance among them and to isolate the task-specific variance of each task. This way, the only commonality among the tasks will be working memory capacity. The point here is that observing performance on a single task does not afford an adequate measure of working memory. Instead, latent abilities like working memory capacity can be measured only by observing performance across multiple indicators and investigating the common variance shared among them.

Multiple indicators are rarely used in the working memory training literature. As noted by Shipstead and colleagues (2012b), near transfer can occur for a variety of reasons other than improvements in working memory capacity. The most parsimonious account of improvements on a single task of working memory capacity is that subjects simply learn a strategy that is specific to performing the training tasks and can, in turn, apply this strategy to other memory tasks (Shipstead et al., 2012b). This result should not necessarily lead to the conclusion that working memory capacity has improved. In fact, it could imply that performance on a particular task is artificially inflated after training due to practice effects and that the measure is no longer a valid estimate of the subject's ability.

Active Control Groups

To infer causality, researchers must ensure internal validity. Internal validity refers to the degree of certainty that an intervention is responsible for causing changes at posttest. Without a robust experimental design, any evidence supporting working memory training could be explained by extraneous factors unrelated to the training program. In order to assume that an intervention is responsible for changes in behavior, the experimenter must eliminate these threats (Campbell & Stanley, 1966).

To claim causation, experimenters must make certain that the treatment is the only viable explanation for their results. Protecting against threats to internal validity requires training studies to include an active control group. Subjects should be randomized into treatment and control conditions to control for sampling bias and the influence of experimenter expectations (Campbell & Stanley, 1966). A good active control group is one that equates to the treatment group in every way except for the critical manipulation thought to be responsible for the treatment effect.

The measurement of latent abilities with a single indicator, inclusion of no-contact control groups, and inadequate subject randomization have not been an exception; these practices have been the rule, and the widespread use of these methods has stymied researchers' ability to provide a convincing answer to the most basic questions related to working memory training.

Question 4: Contributions to the Field

How has your research contributed to understanding of the working memory construct? How do you integrate your findings into the context of a larger literature on working memory training?

A primary concern about the present state of working memory training studies is that there is little standardization and convergence of findings. Research groups have different working memory training approaches and assess varying cognitive skills and transfer measures, and minimal research has been done on the comparative efficacy of the various approaches. This question encourages each contributing research team to discuss the implications of their findings for other researchers of working memory capacity and their perspectives on various interventions designed to improve cognitive abilities.

Conclusion

In summary, the purpose of this chapter is to address the best way to integrate and to synthesize the current findings on working memory training. The strength of this volume is the inclusion of various techniques in, and perspectives on, working memory training. Consideration of different perspectives is critical for advancing understanding of the overall efficacy and effect size of working memory training. Identifying the commonalities in results and focusing on key differences in theoretical perspectives, methodologies, and outcomes will lead to more focused developments in the field.

References

Biederman, I., Cooper, E. E., Fox, P. W., & Mahadevan, R. S. (1992). Unexceptional spatial memory in an exceptional memorist. *Journal of Experimental Psychology: Learning, Memory, and Cognition, 18*(3), 654–657. doi:10.1037//0278-7393.18.3.654

Campbell, D. T., & Stanley. J. C. (1966). *Experimental and quasi-experimental designs for research.* Boston, MA: Houghton Mifflin.

Chein, J. M., & Morrison, A. B. (2010). Expanding the mind's workspace: Training and transfer effects with a complex working memory span task. *Psychonomic Bulletin & Review, 17*(2), 193–199. doi:10.3758/PBR.17.2.193

Conway, A. R., Jarrold, C. E., Kane, M. J., Miyake, A., & Towse, J. N. (2007). *Variation in working memory.* Oxford, UK: Oxford University Press.

Dougherty, M. R., Hamovitz, T., & Tidwell, J. W. (2016). Reevaluating the effectiveness of N-back training on transfer through the Bayesian lens: Support for the null. *Psychonomic Bulletin and Review, 23*(1), 306–316. doi:10.3758/s13423-015-0865-9

Engle, R. W., & Kane, M. J. (2004). Executive attention, working memory capacity, and a two-factor theory of cognitive control. In B. H. Ross (Ed.), *The psychology of learning and motivation: Advances in research and theory* (Vol. 44, pp. 145–199). New York, NY: Elsevier Science. doi:10.1016/S0079-7421(03)44005-X

Engle, R. W., Tuholski, S. W., Laughlin, J. E., & Conway, A. R. (1999). Working memory, short-term memory, and general fluid intelligence: A latent-variable approach. *Journal of Experimental Psychology: General, 128*(3), 309–331. doi:10.1037//0096-3445.128.3.309

Ericsson, K. A., Delaney, P. F., Weaver, G., & Mahadevan, R. (2004). Uncovering the structure of a memorist's superior "basic" memory capacity. *Cognitive Psychology, 49*(3), 191–237. doi:10.1016/j.cogpsych.2004.02.001

Gathercole, S. E., Dunning, D. L., & Holmes, J. (2012). Cogmed training: Let's be realistic about intervention research. *Journal of Applied Research in Memory and Cognition, 1*(3), 201–203. doi:10.1016/j.jarmac.2012.07.007

Green, C. S., & Bavelier, D. (2012). Learning, attentional control and action video games. *Current Biology, 22*(6), 197–206. doi:10.1016/j.cub.2012.02.012

Harrison, T. L., Shipstead, Z., Hicks, K. L., Hambrick, D. Z., Redick, T. S., & Engle, R. W. (2013). Working memory training may increase working memory capacity but not fluid intelligence. *Psychological Science, 24*(12), 2409–2419. doi:10.1177/0956797613492984

Holmes, J., Gathercole, S. E., & Dunning, D. L. (2009). Adaptive training leads to sustained enhancement of poor working memory in children. *Developmental Science, 12*(4), F9–F15. doi:10.1111/j.1467-7687.2009.00848.x

Holmes, J., Gathercole, S. E., Place, M., Dunning, D. L., Hilton, K. A., & Elliott, J. G. (2010). Working memory deficits can be overcome: Impacts of training and medication on working memory in children with ADHD. *Applied Cognitive Psychology, 24*(6), 827–836. doi:10.1002/acp.1589

Jaeggi, S. M., Buschkeuhl, M., Jonides, J., & Perrig, W. J. (2008). Improving fluid intelligence with training on working memory. *Proceedings of the National Academy of Sciences of the United States of America, 105*(19), 6829–6833. doi:10.1073/pnas.0801268105

Jha, A. P., Krompinger, J., & Baime, M. J. (2007). Mindfulness training modifies subsystems of attention. *Cognitive Affective and Behavioral Neuroscience, 7*(2), 109–119. doi:10.3758/cabn.7.2.109

Kane, M. J., Hambrick, D. Z., Tuholski, S. W., Wilhelm, O., Payne, T. W., & Engle, R. W. (2004). The generality of working memory capacity: A latent-variable approach to verbal and visuospatial memory span and reasoning. *Journal of Experimental Psychology: General, 133*(2), 189–217. doi:10.1037/0096-3445.133.2.189

Kane, M. J., Poole, B. J., Tuholski, S. W., & Engle, R. W. (2006). Working memory capacity and the top-down control of visual search: Exploring the boundaries of "executive attention." *Journal of Experimental Psychology: Learning Memory and Cognition, 32*(4), 749–777. doi:10.1037/0278-7393.32.4.749

Klingberg, T. (2005). Training and plasticity of working memory. *Trends in Cognitive Sciences, 14*(7), 317–324. doi:10.1016/j.tics.2010.05.002

Klingberg, T., Fernell, E., Olesen, P. J., Johnson, M., Gustafsson, P., Dahlström, K., . . . Westerberg, H. (2005). Computerized training of working memory in children with ADHD—A randomized, controlled trial. *Journal of the American Academy of Child and Adolescent Psychiatry, 44*(2), 177–186. doi:10.1097/00004583-200502000-00010

Klingberg, T., Forssberg, H., & Westerberg, H. (2002). Training of working memory in children with ADHD. *Journal of Clinical and Experimental Neuropsychology, 24*(6), 781–791. doi:10.1076/jcen.24.6.781.8395

Melby-Lervåg, M., & Hulme, C. (2013). Is working memory training effective? A meta-analytic review. *Developmental Psychology, 49*(2), 270–291. doi:10.1037/a0028228

Melby-Lervåg, M., Redick, T. S., & Hulme, C. (2016). Working memory training does not improve performance on measures of intelligence or other measures of "far transfer": Evidence from a meta-analytic review. *Perspectives on Psychological Science, 11*(4), 512–534. doi:10.1177/1745691616635612

Redick, T. S., Shipstead, Z., Harrison, T. L., Hicks, K. L., Fried, D. E., Hambrick, D. Z., . . . Engle, R. W. (2012). No evidence of intelligence improvement after working memory training: A randomized, placebo-controlled study. *Journal of Experimental Psychology: General, 142*(2), 359–379. doi:10.1037/a0029082

Redick, T. S., Shipstead, Z., Wiemers, E. A., Melby-Lervåg, M., & Hulme, C. (2016). What's working in working memory training? An educational perspective. *Educational Psychology Review, 27*(4), 617–633. https://doi.org/10.1007/s10648-015-9314-6.

Shipstead, Z., Hicks, K. L., & Engle, R. W. (2012a). Cogmed working memory training: Does the evidence support the claims? *Journal of Applied Research in Memory and Cognition, 1*(3), 185–193. doi:10.1016/j.jarmac.2012.06.003

Shipstead, Z., Redick, T. S., & Engle, R. W. (2010). Does working memory training generalize? *Psychologica Belgica, 50*(3–4), 245–276. doi:10.5334/pb-50-3-4-245

Shipstead, Z., Redick, T. S., & Engle, R. W. (2012b). Is working memory training effective? *Psychological Bulletin, 138*(4), 628–654. doi:10.1037/a0027473

Shipstead, Z., Redick, T. S., Hicks, K. L., & Engle, R. W. (2012). The scope and control of attention as separate aspects of working memory. *Memory, 20*(6), 608–628. doi:10.1080/09658211.2012.691519

Unsworth, N., Heitz, R. P., Schrock, J. C., & Engle, R. W. (2005). An automated version of the operation span task. *Behavior Research Methods, 37*(3), 498–505. doi:10.3758/BF03192720

Unsworth, N., Redick, T. S., Heitz, R. P., Broadway, J. M., & Engle, R. W. (2009). Complex working memory span tasks and higher-order cognition: A latent-variable analysis of the relationship between processing and storage. *Memory, 17*(6), 635–654. doi:10.1080/09658210902998047

von Bastian, C. C., & Oberauer, K. (2014). Effects and mechanisms of working memory training: A review. *Psychological Research, 78*(6), 803–820. doi:10.1007/s00426-013-0524-6

2

Working Memory Training From an Individual Differences Perspective

Efficacy in Older Adults

Erika Borella, Barbara Carretti, Cesare Cornoldi, and Rossana De Beni

Introduction

Working memory (WM) is the ability to retain and manipulate information for use in complex cognitive tasks, and it is a core mechanism involved in higher-order cognitive abilities like fluid intelligence, problem-solving, and reading comprehension (Borella, Ghisletta, & de Ribaupierre, 2011; de Ribaupierre, 2001). Although WM is characterized by a limited capacity, it is a crucial mechanism in cognition. It is also a cognitive process that suffers a clear and linear decline with aging (Borella, Carretti, & De Beni, 2008; Borella et al., 2017). WM is consequently one of the general processes targeted by the new generation of process-based cognitive training.

The assumption that WM is trainable is based on evidence of the plasticity of the cognitive system across the whole life span (Hertzog, Kramer, Wilson, & Linderberger 2009). Further, according to some WM models, such as the continuity model (see Cornoldi, 2010; Cornoldi & Vecchi, 2003), WM is characterized by different processes that depend on the type of content processed (verbal vs. spatial) and also on the involvement of executive control. Therefore, theoretically, improving WM can also enhance its related processes. The Cornoldi and Vecchi WM model distinguished between a "basic structure" (a sort of personal biological equipment), and a "used ability" determined by the way individuals use their WM. On this basis, the benefits of training may presumably concern not only the basic structure of WM, but also its usage.

This chapter presents and discusses the model of WM proposed by Cornoldi and Vecchi that is based on an analysis of individual and age-related differences. The model is used as a framework for discussing the efficacy of a WM training procedure developed for older adults. The model (a) assumes that different WM tasks (and underlying processes) may be located along two continua that describe the type of content to be processed and the degree of active control required by the task and (b) considers metacognitive/motivational aspects, which also have a role in determining WM performance. The training procedure discussed here takes into account not only the capacity to use WM resources and attentional control by adopting an adaptive procedure, but also the importance of planning challenging and engaging training activities that sustain motivation and favor the training's short- and long-term efficacy, at least in older adults. These aspects seem crucial in explaining the results obtained with this verbal WM training program in aging, which is in fact one of the few training programs to date that has shown promising and consistent results in terms of both short- and long-term training gains and transfer effects (see Borella, Carbone, Pastore, De Beni, & Carretti, 2017).

Question 1: Theory of Working Memory Training

Before proponents of WM training claim that WM training can improve cognitive abilities, they need to present a theory of WM that outlines its structure and the mechanisms responsible for its relationship with higher-order cognition.

The Cornoldi and Vecchi WM model (see Cornoldi, 1995, 2010; Cornoldi & Vecchi, 2003) considers differences between groups of individuals who were distinguished by age or specific cognitive impairments (in terms of both sensory and cognitive components). The model can be seen as the combination of a multicomponent (Baddeley, 1986, 2000) and unitary capacity-limited model (Engle, Cantor, & Carullo, 1992). It draws a distinction within WM between different processes, according to the type of content processed (verbal vs. spatial) and the involvement of executive control. Subtle differences identified between some groups (poor comprehenders, individuals with intellectual disabilities, etc.) suggested that a dichotomous distinction between mainly active and passive executive processes related to the domain-specific components was overly simple. For instance, poor comprehenders were found to have difficulties in active verbal WM tasks that were not seen in passive verbal tasks or in active visuospatial tasks (see meta-analysis by Carretti, Borella, Cornoldi, & De Beni, 2009). In other words, the pattern of results suggested that poor comprehenders' WM impairment depends on the involvement of attentional resources and the modality of the task, a finding that challenges the unitary view of WM (given that these individuals' visuospatial WM is not impaired) and the unique role of executive functions in explaining their deficit. This fine-tuned analysis could be applied to cases of intellectual disabilities, too, such as in genetic conditions like Down syndrome (e.g., Doerr, Carretti, & Lanfranchi, 2019 for a review) or Fragile-X Syndrome (e.g., Lanfranchi, Cornoldi, Drigo, & Vianello, 2009). For example, several findings highlighted different degrees of failure in tasks typically assumed to measure the central executive function, which is responsible for allocating attentional resources and managing controlled, intentional, conscious, or else effortful processes. Impairments could be mild in the backward span task, more severe in listening span tasks, and even more severe in updating or dual tasks (Lanfranchi et al., 2009).

The model was also extended to the case of cognitive aging and showed the importance of distinguishing WM tasks by the type of control required and the type of content (Bisiacchi, Borella, Bergamaschi, Carretti, & Mondini, 2008; Bopp & Verhaeghen, 2007; Borella et al., 2008; Borella, Pezzuti, De Beni, & Cornoldi, in press). It also showed how tasks are clearly distinguishable on the basis of their content, but only at lower control levels (Vecchi & Cornoldi, 1999).

Based on the evidence obtained by examining individual and age-related differences, we found more support for the hypothesis of a vertical continuum of active control along which each WM task (and the associated WM processes) could be found at a different level (see Figure 2.1).

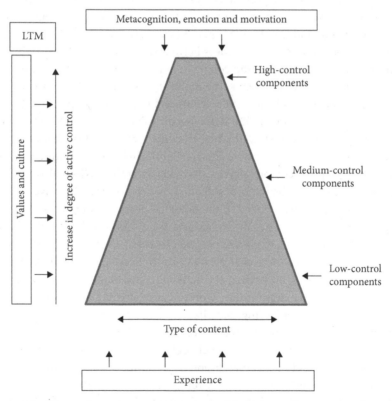

Figure 2.1. The working memory continua model: the vertical continuum describes the degree of active control required and the horizontal continuum describes the type of content processed.

Similarly, we found that the basic distinction between verbal and visuospatial processes could represent a very useful heuristic method, but it could not fully account for the wide range of dissociations and specific deficits found in WM, depending on the type of content. This was the case of individuals with Down syndrome, who had impaired verbal WM performance but a more adequate performance in the visuospatial WM domain when passive tasks were considered (Lanfranchi, Cornoldi, & Vianello, 2004). A deeper analysis of the visuospatial component revealed a more complex picture, however, with a specific deficit in the spatial simultaneous component, but not in the sequential one (Carretti, Lanfranchi, & Mammarella, 2013; Lanfranchi, Carretti, Spanò, & Cornoldi, 2009). Studies using an adult life-span approach even showed that visuospatial and verbal WM decline at the same rate (Borella et al., 2008; Park et al., 2002). Altogether, these results suggest that the rigid distinction between two modality-specific components cannot account for individual and age-related differences.

Such deficits in individuals with Down syndrome and the similar development of WM across the adult life span could not be related to separate, equidistant, different WM components; some of the components seem to be closer to, or farther away from, one another (see Figure 2.1). For example, nonverbal learning-disabled children are generally characterized by a weakness in visuospatial WM and a relatively strong verbal WM (Mammarella & Cornoldi, 2014), but they may have different specific degrees of weakness within the visuospatial domain (Mammarella et al., 2006; Mammarella, Pazzaglia, & Cornoldi, 2008). Consistent with this view, Mammarella, Borella, Pastore, and Pazzaglia (2013) found that visuospatial memory tasks in adulthood should also be differentiated in terms of the task presentation format (spatial simultaneous, spatial sequential, visual and visuospatial complex span), not just according to the degree of attentional control demanded by the task.

Group effects on more subtle differences within the visuospatial domain were also supported by evidence using a dual task paradigm (Pazzaglia & Cornoldi, 1999). Hence our assumption that, for the same degree of control, tasks can be divided according to the type of content, and different types of content can be closer or farther away from one another.

The result of these considerations was a continuity model defined by two main continua (i.e., a vertical continuum describing the degree of active control required and a horizontal continuum describing the type of content processed; see Figure 2.1). This model is therefore unitary, since there is no "interruption" between the different processes, and it can distinguish between the processes at work. The model is also represented as a cone to describe not only the variation in content in different possible directions, but also the declining variation coinciding with an increasing amount of control required and a loss of specificity for tasks demanding high levels of control.

The main assumptions of the model are:

- Continuity: Each task (and each underlying process) has a place in the WM system defined by its position along the vertical continuum (describing the degree of active control required by the task) and the horizontal continuum (describing the type of content processed). Tasks (and processes) are not completely dissociable, but their distinctiveness increases with the increasing distance between them.
- Domain specificity: Each task (process) is characterized by a domain specificity defined by the type of content to be processed. As the cone shape suggests, however, the smaller the distance between different points on the circumference as we approach the vertex, the higher the specificity for low-control tasks and the lower the specificity (down to nil) for high-control tasks.
- Resources: The WM system has a limited capacity, and the resources available can be used either locally (N) or globally (T). Each process subtracts N local resources shared with another process in a proportion inversely related to the distance between the two processes. The quantity of resources subtracted by a process increases in line with the increasing active control required. The overall system has a total quantity of resources (T) that cannot be exceeded by a single task or combination of them. Each individual is characterized by having different amounts of T and N resources available.
- Long-term memory: Each point in the system can be connected to long-term memory, which supports the encoding of information and its maintenance.

The model was used (Cornoldi, 2007, 2010) to examine the relationship between WM and human intelligence, based on the assumption that intellectual functioning is supported by WM and that its structure could somehow be considered isomorphic to the structure of intelligence. In other words, human intelligence can also be represented as a cone, with its higher points corresponding to the most-controlled intellectual functions (e.g., reasoning, reading comprehension), which are supported by controlled active WM processes, and its lower points, which coincide with less-controlled functions that depend more on the type of content to be processed and are supported by less-controlled WM processes. This model of intelligence was further articulated to explain why individuals' manifestations of intelligence (intelligence in use) do not always coincide with their potential abilities (basic structure of intelligence). This was done by including factors that explain why individuals with the same basic structure of intelligence may have different levels of intellectual production (Cornoldi, 2007, 2010).

Two main implications derive from the assumption that WM and intelligence share a similar organization. First, in both cases, we can distinguish between a basic structure deeply rooted in an individual's biological equipment and a used ability determined by the actual use the individual makes of his or her basic intelligence. Second, the factors affecting the actual use of a person's basic intelligence can also be considered for their influence on the actual use made of WM structure. The factors include (see Figure 2.1): (a) experience, which typically directly affects the low-control processes; (b) values and education, potentially influencing all the processes; and (c) metacognition (knowledge of how the mind works and its strategic control), along with emotions and motivational aspects, which typically directly affect the high-control processes. In this sense, taking action on different levels can predictably modify WM performance.

This further specification of the model seemed particularly appropriate for examining the case of older adults, individuals more than 64 years old (in particular for the so-called young-old, 64–74 years old, and old-old, who have more than 75 years old). In fact, the model may explain why adequate cognitive abilities may be preserved despite a poor performance, since a declining basic ability can presumably be compensated by higher levels' being achieved in the factors influencing the said ability in use. In particular, elderly people may take advantage of a wealth of experience and values and a high level of emotional control and regulation.

Question 2: Major Claims of Working Memory Training

As our WM model mainly originated from, and was supported by, evidence concerning WM weaknesses in specific groups of individuals, it seemed particularly appropriate to consider the issue of treatment for people with poor WM performance, such as older adults. In fact, there is an increasing interest in focusing on training schemes for people who need to enhance or sustain their WM to cope with a specific weakness in this crucial mechanism. There is evidence to suggest that WM training is of greater benefit to poor performers (Jaeggi, Buschkuehl, Jonides, & Perrig, 2008), especially in the developmental domain (Klingberg et al., 2005). Our WM model suggests that the benefits of training concern not only the basic structure of WM but also its usage. Since some individuals revealing WM weaknesses were also characterized by poor performance in relevant everyday cognitive activities (e.g., reading comprehension, problem-solving), our approach—like other WM models—suggests that there is some causal relationship between WM and everyday activities, but it also predicts that a change in WM could affect everyday activities. Therefore, improvements in

WM must concern not only its basic structure but also WM in use (i.e., usage in everyday activities).

Another important feature of our model is that different components can be trained and the generalizability of a training effect depends on the distance—in terms of the processes shared—between the components. Since high-control components referring to different domains are strictly related, we can predict a stronger transfer effect on another domain when training focuses on a high-control component rather than a low-control component.

Many WM training studies have been conducted in the developmental domain, on typically or atypically developing children, and some have focused on adults (generally students), but relatively few studies have investigated the potential role of WM training in aging. This is somewhat surprising, given the increasing life expectancy and the population's growing proportion of older adults, as well as the fact that older adults experience difficulties in WM tasks. It is therefore imperative to develop training procedures to preserve or enhance older adults' cognitive functioning, to ensure their independence and well-being. Hence, our decision is to focus on WM training procedures for older adults.

When we began to develop our WM program for older adults, our starting point was our WM model and the fact that cognitive plasticity, in the broad sense of the term, characterizes the whole life span, including aging. The first generation of training studies, in which strategies were taught to improve episodic memory, had shown that older adults' memory could be improved. This evidence enabled us to ascertain that, even in old age, despite the age-related decline in cognitive resources, there is still some potential for improving memory performance, acquiring new skills, and making a more efficient use of available cognitive resources (what is called cognitive plasticity) with training. One of the limits of the training programs tested, however, was that the procedure did not induce participants to develop a metacognitive attitude and to conduct a personal search for the best strategies to use depending on the nature of the training tasks used. The training only proposed specific strategies, which worked effectively within a given domain or for a particular task and context. Transfer effects in other tasks/situations could demand additional resources, which would explain why generalized effects were rarely found (but see Carretti, Borella, & De Beni, 2007), especially in older adults, who often did not spontaneously use the strategies they had learned after completing their training (Rebok, Carlson, & Langbaum, 2007; Verhaeghen & Marcoen, 1996; Verhaeghen, Marcoen, & Goosens, 1992).

WM has been considered a "crucial" process to train, because it is a basic mechanism of cognition that is involved in a wide variety of abilities of higher-order cognition (attention, reasoning, reading comprehension, etc.) with a bearing on everyday functioning (Borella et al., 2011); declines with aging (Park et al., 2002); relies on the frontoparietal network, which shows a more pronounced decline in aging than other brain networks (Birren & Schaie, 2006); and is crucially related to age and group differences and is therefore useful in the clinical or applied setting. Thus, if the ability to use WM can be improved, then—given its central role in cognition—it is also reasonable to expect such an improvement to facilitate cognitive plasticity or at least to contribute toward building a scaffold for sustaining cognitive functioning by means of compensation mechanisms (Park & Reuter-Lorenz, 2009). Our WM model suggested this as well.

The general procedure adopted for basic WM training involved asking participants (older adults) to practice repeatedly with a WM task without teaching them any context-specific strategies that would help them to complete only the task in question. The WM training thus focused on enabling changes in participants' underlying ability and inducing a process-based plasticity, with consequences for other WM-related skills engaging similar processes or brain regions (for a review, see Buschkuehl, Jaeggi, & Jonides, 2012). In fact, several studies of young adults (Takeuchi et al., 2010; Westerberg & Klingberg, 2007) and children (Klingberg, Forssberg, & Westerberg, 2002) have shown that WM training may succeed in inducing changes in cognitive performance associated with changes in the neural activity of the prefrontal cortex. Few studies have examined neural changes induced by training in aging, and it showed an increased activation of the striatum in older adults trained with an N-back task (e.g., Dahlin, Nyberg, Bäckman, & Neely, 2008; Salminen, Kühn, Frensch, & Schubert, 2016).

WM training for older people seems to be a promising way to postpone or compensate for the age-related decline in WM, and some evidence exists for immediate and longer-term improvements in the task involved in the training (specific effects) and also for benefits in very similar tasks (i.e., near-transfer effects; but see Dahlin et al., 2008). On the other hand, although WM training for older adults has shown potential generalizability of training gains in other, dissimilar tasks (i.e., far transfer), results have continued to be limited; in the rare cases when far transfer did occur, the effects did not last for long (see Karbach & Verhaeghen, 2014).

No doubt numerous factors could contribute to the success or failure of WM training, and they are probably not easy to identify, given that most studies have adopted different training procedures, tasks, and so on. When we developed our training procedure, we considered length of training, requirements of the training task, and the participant's point of view. This is because we had noticed that these aspects had not been seen as crucial and had not been assessed in other training studies. Concerning the length of the training, the training procedures tested in the literature (even in the latest studies) are typically long, ranging from 10 (2-hour) sessions to 45 daily sessions. We felt that so many sessions (especially for young or older adults) spent in front of a computer, alone or as a group, repeating the same exercises, would affect participants' motivation and engagement in the activities, with a negative fallout on the training's efficacy. Judging from our theoretical framework of WM, motivational and metacognitive aspects could affect WM performance (see Question 1). In addition, WM training uses clinical settings, where clinicians have to deal with time constraints, so the number of training sessions would need to be limited.

We therefore designed a training procedure that takes these considerations into account. Partly inspired by the first WM training studies by Jaeggi and collaborators (2008), we adopted an adaptive procedure (i.e., the difficulty of the training task was increased if participants were successful at a given level; if not, a less difficult task was presented). Unlike procedures involving a fixed level of difficulty, an adaptive procedure enables participants to be trained on a level of difficulty coming close to the limits of their capacity. From a metacognitive and motivational standpoint, adaptive procedures also give participants the opportunity to experience success in their recall during training sessions. This can positively influence their attitude, changing any misconceptions about the efficacy and control of memory with aging, making them much more flexible in their thinking, and enabling them to improve their control over the efficacy of their performance (Cavallini, Dunlosky, Bottiroli, Hertzog, & Vecchi, 2010; Vranic, Carretti, Španic, & Borella, 2013). Unlike all the other procedures used, however, our procedure combined the adaptive procedure with a constant variation in the maintenance and processing demands of the training task. Across and within training sessions (see Table 2.1) participants' activities always changed. The difficulty of the WM training task was manipulated by changing the amount of information to recall (first training session) or by varying the processing demand (second and third training sessions) and maintenance requirements (phase two of the first, second, and third sessions), obliging participants to adapt quickly to the new demands of the task.

This procedure (for more details, see Borella et al., 2010) ensures that the task is always challenging, cognitively demanding, and novel, thus helping to ensure that participants pay attention to the task. Schmidt and Bjork (1992) have argued that challenging training conditions often promote generalization. Our practice task engages multiple processes, demanding varying degrees of attentional control and requiring encoding, maintenance of information, inhibition of no-longer-relevant information, simultaneous management of two tasks, attention shifting, and the ability to control attention, all combined with an adaptive procedure and variations in the materials and demands. In this respect, our procedure is a hybrid (i.e., an adaptive procedure combined with variations in the demands of the task), distinct from others in the literature.

We believe that these features may stimulate participants' cognitive flexibility and plasticity as well as their interest, because the demands of the training task

Table 2.1. *Description of Training Sessions for Basic WM Training for Elderly People*

First training session	This session was divided into three phases.
	In each of the three phases, participants listened to sets of two, three, four, or five word lists (each list containing 5 words), and were asked to remember target words (the first or last word on each list, in serial order) and to tap their hand on the table whenever they heard an animal noun. They were presented with three sets of word lists for each level of difficulty (which depended on the number of lists in a set). If they recalled the words correctly for two of the three sets, the task's difficulty was increased (up to sets of five word lists). When they were unable to do so, they moved on to the next phase of the task, which started from the easiest level (sets of two word lists). The three phases of the task differed in that participants had to recall the last word on each list during the first phase, the first word on each list during the second, and the last word again during the third.
Second training session	Participants listened to sets of two, three, four, or five word lists (each list containing 5 words), and were asked to remember target words (the words followed by a sound, which could be anywhere on the list, in serial order) and to tap their hand on the table whenever they heard an animal noun. They were presented with four sets of word lists for each level of difficulty (which depended on the number of lists in a set). Sets with two word lists could contain 2–8 animal nouns, sets with three lists could contain 4–9 animal nouns, sets with four lists could contain 6–11 animal nouns, and sets with five lists could contain 8–17 animal nouns.
Third training session	Participants listened to sets of two, three, four, or five word lists (each list containing 5 words), and were asked to remember target words (the first or last word on each list, in serial order) and to tap their hand on the table whenever they heard an animal noun. They were presented with four sets of word lists for each level of difficulty (which depended on the number of lists in a set), and they had to recall (i) the last word on each list in the first set; (ii) the first word on each list in the second set; (iii) the last word on each list in the third set; and (iv) the first word on each list in the fourth.

Note. See Borella, Carretti, Riboldi, & De Beni (2010).

are always novel and challenging. All these aspects may also help to motivate older participants, who feel they can cope with the task and succeed in a task they might have believed difficult (overcoming stereotypes related to aging), and this also keeps them interested in the training activities (see also Hertzog & Hultsch, 2000; Schmidt & Bjork, 1992). During debriefing, participants told us they felt involved in the task and they enjoyed the training. They also reported feeling more engaged in everyday activities as they tried to put their training into practice.

Our training consists of just three 1-hour (1-hour and a half) sessions, instead of the numerous training sessions used in other studies, as mentioned previously. While the chances of obtaining transfer effects after such a brief training may appear counterintuitive, its intensive nature (concentrated in terms of the timescale) may facilitate the short-term transferability of WM training to other tasks, making participants stay interested in the training activities. The training schedule was also arranged so that the sessions had a fixed interval between them, giving participants sufficient time to consolidate the skills they acquired, while also reducing the risk of losing any beneficial effects from having practiced with the task (Cepeda, Pashler, Vul, Wixted, & Rohrer, 2006), thereby favoring generalization (Vlach & Sandhofer, 2014).

Finally, the experimenter attended all the training sessions and introduced each training task. This human–human interaction in the training setting, instead of a machine–human interaction, may have increased participants' interest in the training activities. The experimenter's presence also may have favored the success of the training because participants did not feel alone, and they were encouraged to stay on task and not to let their minds wander, an issue that can occur more easily when participants are left alone in front of the computer to complete lengthy, repetitive tasks. Our experimenter also attended the alternative activities conducted with the active control group, to match the amount of social interactions experienced by the training and control groups. Metacognitive aspects, including promoting participants' self-awareness and their monitoring of any progress in their performance, reducing their anxiety, and increasing their confidence in their cognitive abilities (Dunlosky et al., 2007) and motivation, may also play a part in explaining the success of training (Carretti, Borella, Zavagnin, & De Beni, 2013) because the experimenters are an active part of a trainee's progress. In some training groups, participants also reported feeling stimulated to "make more use of their memory" in everyday life. We attributed this to the above-mentioned aspects, and it may have led to the development of new skills outside the training environment as well as given rise to maintenance effects.

In sum, our training procedure, and its schedule and setting in particular (i.e., the experimenter's presence), seem to be crucial factors for inducing and maintaining the training benefits that we have found so far in both normal and pathological aging (amnestic mild cognitive impairment; Borella et al., 2010; Carretti, Borella, Zavagning, et al., 2013), and those stressful situations that can affect older adults' mental health such as major surgery (Carbone, Carretti, Vianello, & Borella, in press).

Meanwhile, we have also identified several variables, and individual characteristics (see for a synthesis Borella et al., 2017), that influence the success of our training procedure, including the age range of older adult participants, and the nature of the training task (Borella et al., 2017). As we hypothesized in our first study in 2010, the broad and promising transfer effects obtained in the short term, some of which were maintained, may have depended on the inclusion of only young-old participants (adults 64–74 years of age). Fluid abilities decline linearly with increasing age (Park et al., 2002), and crystallized abilities start to decline beyond the age of 74 (Schaie & Willis, 1996; Singer, Verhaeghen, Ghisletta, Lindenberger, & Baltes, 2003). Although cognitive and neural plasticity mechanisms can be stimulated by training and new learning even in old age (Greenwood & Parasuraman, 2010), these mechanisms appear to be less effective in very late adulthood (Schmiedek, Lövdén, & Lindenberger, 2010). Advanced age could therefore negatively affect the magnitude and pattern of any training gains, because adults age 75 and over (the old-old) experience a more accentuated cognitive decline than the young-old. We found that, although even the old-old could benefit from our training procedures, demonstrating that WM performance can be improved even in advanced old age, the related transfer effects seemed less robust and more limited than those seen in the young-old. This suggests that old-old participants acquire skills associated more with verbal WM than on an information-processing level (i.e., training does not generate changes in the efficacy of their cognitive mechanisms and, more generally, of their cognitive resources). It also suggests that more time (e.g., training sessions) is probably needed to modify performance in the very elderly (Borella et al., 2013).

There is debate about the greater age-related memory decline in visuospatial rather than in verbal tasks (Bopp & Verhaeghen, 2007; Jenkins, Myerson, Joerding, & Hale, 2000; Myerson, Emery, White, & Hale, 2003; Myerson, Hale, Rhee, & Jenkins, 1999) and older adults' limited familiarity with visuospatial material (Borella et al., 2014). Therefore, we also tested whether training task modality has a role in favoring any transfer and maintenance effects in older adults. In other training studies that identified few or no transfer effects, the training task was either spatial (Li et al., 2008), visual (Buschkuehl et al., 2008), or combinations of verbal and visuospatial WM tasks (Richmond, Morrison, Chein, & Olson, 2011; Zinke, Zeintl, Eschen, Herzog, & Kliegel, 2012). Therefore, the use of a verbal rather than a visual or spatial WM training task may explain the efficacy of our training procedure.

The Padua WM Training Procedure: The Expectations

Given the characteristics of our training, which engages multiple processes, and its hybrid procedure that favors engagement and motivation, we hypothesized that our WM training would produce transfer effects (for a categorization of transfer effects, see Noack, Lövdén, Schmiedek, & Lindenberger, 2009). The transfer effects include: (a) nearest-transfer effects to tasks that involve processes (elaboration and processing phase) similar to the one practiced (i.e., the same narrow ability but with material of a different nature and different secondary requirements from those of the trained task, such as a visuospatial WM task, since the criterion task is a verbal WM task); (b) near-transfer effects to tasks that represent the general memory factor but with different requirements of the task (i.e., the same broad ability, such as a short-term memory task); and (c) far-transfer effects to tasks that have some relationship with WM processes (Borella et al., 2010), such as reasoning abilities (the Cattell Test), inhibitory processes (the Stroop Color Test), processing speed (the Pattern Comparison test), reading comprehension (Nelson Denny test; see Carretti, Borella, Zavagnin, et al., 2013), and objective performance-based tasks (Cantarella, Borella, Carretti, De Beni, & Kliegel, 2017; Borella, Cantarella, Carretti, De Lucia, & De Beni, in press).

Across different studies, we systematically tested the different types of transfer effect using the same tasks or variants of them and predicted transfer gains. This enabled us to replicate and to confirm the efficacy of our training program. As previously mentioned, the fact that we found limited transfer effects in old-old groups suggests the need to modify the procedure (e.g., to plan more sessions for participants with lower levels of plasticity, as discussed further below). While transfer effects could always be expected in the short term depending on the age range of participants and the nature of the training tasks (with larger effects for young-old than for old-old, and for verbal rather than for visuospatial training tasks), we do not have any specific expectations regarding the maintenance of these effects. Maintenance effects are a crucial issue in training studies but, unfortunately, are not always examined thoroughly or even taken into account. Finding long-term training gains would shed light on whether training favors plasticity (if the effects are maintained) or flexibility (if the effects found soon after the training are short-lived). Note that, so far, our results point to a greater flexibility than plasticity, since the maintenance effects we found (even in old-old participants and using visuo-spatial material) are not generalizable to all the transfer measures considered. Our results nonetheless suggest that older adults are still capable of improving their WM performance and other more or less closely related abilities, and of maintaining these gains, which is an encouraging finding in the aging domain.

The Padua WM Training Studies: From the Laboratory Tasks to Skills Related to Everyday Functioning

We have the same goals as other groups researching WM training, which is to examine the feasibility of increasing WM, and we have similar expectations concerning the benefits of WM training. In particular, we have focused on examining whether training WM can improve performance in other tasks—those not trained directly and more or less closely related to WM—in different age groups (young-old, old-old) and populations (normally aging older adults or those with mild cognitive impairment [MCI]), and using different types of training material (e.g., verbal, visuospatial). Other differences in our research relate to the explicit inclusion in our procedure of features that might also sustain participants' motivation and engagement by using tasks that present challenging situations, having the experimenter attend the training sessions, and limiting the number of sessions. As mentioned, these aspects can explain the large transfer effects that we found using the verbal WM training procedure for at least the young-old and participants with MCI.

Another aspect that distinguishes our study from others is that we tested our procedure in different studies on different groups of older adults (young-old, old-old, and participants with MCI). Concerning the old-old, in light of our recent results, a more intensive training program that includes booster sessions (Ball et al., 2002) might be more appropriate for nurturing transfer gains, given the old-old group's likely greater cognitive decline (especially as concerns WM and brain regions associated with WM functioning). Our training procedure may show large and persistent transfer gains in the young-old but be insufficient for the old-old because of the more accentuated decline in their cognitive resources. To induce cognitive plasticity in the old-old adult's cognitive system may take a longer period of training. Because the young-old have more malleable cognitive skills than the old-old (Schmiedek et al., 2010) and cognitive plasticity declines with aging (Lövdén, Bäckman, Lindenberger, Schaefer, & Schmiedek, 2010), our training might prove more effective if more time is spent on the WM training sessions (Jaeggi et al., 2008). This possibility merits exploration in future studies.

Finally, given the role of WM in complex abilities, we also made an effort to examine its efficacy not only in laboratory tasks but also in skills related to everyday functioning, such as reading comprehension (Carretti, Borella, Zavagnin, et al., 2013) and everyday problem-solving using the Everyday Problem test (as adapted by Marsiske and Willis, 1995; see Cantarella et al., 2017; Borella et al., in press).

The Padua WM Training Efficacy

Overall, the results of our WM training studies have been consistent and support the short- and long-term efficacy of our hybrid training procedure (see Borella et al., 2017). Its efficacy has been proved also in other laboratories and cultural context (e.g., Brum Schimidt, Borella, Carretti, & Yassuda Sanches, 2018). Manipulating the nature of the training task and the age range of older adults involved in the studies prompted us not to change our initial claims, but to clarify some crucial variables that could mediate the efficacy of WM training.

Question 3: Methodological Issues

Study Design: The Importance of the Active Control Group

As mentioned, we started from the idea that individuals of all ages have some degree of cognitive plasticity. The implication of this statement is that specific training activities can enhance cognitive performance even in older adults, as some studies on fluid intelligence, for example, have demonstrated (Baltes, Kliegl, & Dittmann-Kohli, 1988). Our model of WM predicts the feasibility of modifying WM performance by improving what we have termed an individual's WM in use.

The efficacy of a training procedure is related to its internal validity (i.e., the ability to demonstrate that any changes are produced by the experimental procedure and not by other confounders). This is crucial, because there is a risk that an intervention's efficacy could be overestimated due to the hidden influence of other, uncontrolled variables. In the particular case of training studies, different problems can emerge, as the classical studies on experimenter bias (Rosenthal & Jacobson, 1968) and participants' expectations have clearly shown.

Intervention studies are open to several biases. For instance, simple pre- vs. posttest comparisons cannot support the conclusion that changes in participants' performance over time are necessarily due to the intervention, because participants change naturally over time and because of the well-known test–retest effect relating to the participants' greater familiarity with the tests, and for other statistical reasons (e.g., the regression toward the mean). It can be impossible to isolate changes due to the intervention from other changes that may have occurred while the program was underway.

Another example of methodological problems is associated with the "matched comparison" approach, in which researchers use statistical techniques to find participants who attend only pre- and posttest assessments but are comparable with participants completing the training, allowing for the creation of a control group that is usually described as a passive or no-contact control group. The posttest performance of participants in the training program is thus compared with that of nonparticipants, and any differences in scores are attributed to the training program. But, although the matched comparison approach enables the effects of statistical or measurement artifacts to be controlled, it can have limited validity depending on how much information researchers are able to use to match the groups, because the individuals are not grouped using a randomization procedure, and this exposes the results to a number of potential biases.

Another variable affecting the validity of the study design is the difference in social contact between the groups, which is greater for the experimental group than for the controls. As classically demonstrated, participants' performance can be influenced not only by the experimenter's expectations (Rosenthal & Fode, 1963) but also by their own perception of the experimenter's expectations (i.e., the well-known Hawthorne effect) or by other nonspecific factors. This means that failure to see any improvement in a passive control group could be partly because the procedure did not expose the group to the same "conditions," such as social exposure to the experimenter (due to absence of the experimenter) and the lack of the involvement of the participants of this group in any type of alternative activities.

In an effort to overcome these potential problems, in our training studies we adopted the classical "pre- and posttest control group design," with an active control group (Campbell & Stanley, 1963). In other words, we randomly assigned participants to the training or control arm, measured their initial pretest and final posttest performance using parallel versions of the tasks, and proposed different activities to the two groups (i.e., training sessions for the experimental group and alternative activities for the active control group). We based our randomization procedure on the creation of a list of participants, homogeneous in terms of age, education, and mental and physical health. We assigned the participants randomly: half of them to the training program condition and the other half to the active control condition. We also tested the two groups' initial and final (pre- and posttest) performances using parallel versions of the tasks to limit any re-test effects (something unfortunately not often done in training studies). Our training program also usually included a follow-up session to investigate maintenance effects. The inclusion of a follow-up session is an aspect to be seriously considered because of the so- called training sleeper effect, which is the presence of clear transfer effects only at follow-up (Borella et al., 2017). This effect may in fact indicate that certain abilities take longer to show a significant improvement in performance, at least when older adults are involved in training studies.

As mentioned, different activities were presented to the two groups (training tasks for the experimental group and alternative activities for the active control group). In all our studies, we always included an active control group who took part in activities related to memory but not expected to have any beneficial effect on WM performance. The participants in both groups met with the same experimenter and attended the same number of sessions, and each session lasted a similar amount of time, but the control group did not practice with the WM task.

Ethical Considerations

For all the participants, who in our studies were mainly active, healthy older adults, we seriously considered the type of activities proposed to the active control group. The alternative activities presented had to meet the control participants' expectations, and we had to control for the level of motivation, to exclude that the results obtained in the trained group were due to differences in motivation. Thus, we asked participants of the active control group to complete questionnaires about the vividness of their autobiographical memories, their feelings about their memories of the past, and their perceived well-being. Participants usually accepted these activities readily, and the completion of the questionnaires encouraged a motivating and engaging context for the control group, too.

A different approach should be taken when the trained and control groups consist of individuals with special needs, as was the case in our assessment of the efficacy of WM training in people with MCI (Carretti, Borella, Fostinelli, et al., 2013). In this case, the active control group was involved in educational activities about memory, which included giving them some explanation of how to use external memory aids to improve their memory performance. This was done to give those in the control group some potentially useful input (although it was not specifically designed to improve their WM performance) in the place of asking them to answer questionnaires, which could have been difficult given their particular cognitive profile (see Carretti Borella, Fostinelli, et al., 2013).

Another approach for overcoming the ethical problems of training vs control group design, would be to invite the participants who were part of the active control arm to take part in the training program as soon as the study was completed. Placing those in the control group at the top of the list for enrollment in training once the study was over could work as an incentive for their participation in the study and solve the ethical issue in the case of participants with special needs.

Another methodological issue relates to the type of training activities. Our typical intervention with elderly adults involved using a complex WM task, like the Categorization Working Memory Span (CWMS) task, which involves simultaneously retaining and processing information. This choice may have offered some advantages over tasks like the N-back task or updating tasks. The N-back task involves manipulating and retaining information as well as updating the temporal order, contextual information, and processes for binding stimuli with certain representations (Oberauer, 2005). Although theoretically the N-back should share common processing mechanisms with other complex span tasks like the CWMS task, its underlying mechanisms are not fully understood (Schmiedek, Hildebrandt, Lövodén, Wilhelm, & Lindenberger, 2009). The few studies that used the N-back task in combination with other complex WM span tasks have shown variable and often modest correlations (from very low or null in Roberts & Gibson, 2002, and Kane, Conway, Miura, & Colflesh, 2007, to large in Shamosh et al., 2008, and Schmiedek et al., 2009), suggesting that the processes involved in N-back task are—partially—different from those of classic complex WM span tasks. Evidences in this direction come from some training studies results with young adults, too: Linares and colleague, for instance, reported that the training benefit of an N-back training did not transfer to an updating task based on a semantic criteria, and viceversa, as well as to the Operation span task, a classic measure of WM (Linares, Borella, Lechuga, Carretti, & Pelegrina, 2018; 2019).

In contrast, the mechanisms or processes involved in complex WM span tasks have been studied more extensively (as in Miyake et al., 2000, or Friedman & Miyake, 2004, for example) and so has their association with complex—high order—cognitive tasks, such as intelligence (Ackerman, Beier, & Boyle, 2005), thus enabling more precise predictions regarding any effects of WM training on them (or on producing transfer gains on them).

Maintenance Effects

A related issue warranting further discussion is the importance of examining maintenance effects. The challenge of training studies is not only to produce skill changes that are evident immediately after completion of training, but also to create the right conditions for the benefits to be durable. As Hertzog, Kramer, Wilson, and Linderberger (2009) said in a review, "Without information on transfer and maintenance, it is difficult to discern whether intervention effects are restricted to the formation of a new skill, with little consequences for behavior unrelated to this skill, or whether cognitive mechanisms and capacities of general applicability have been enhanced" (p. 18).

The question is whether a WM training program can be judged effective if only short-term benefits are considered. In our opinion, a follow-up session is as crucial as an active control group. In the field of aging, for instance, some studies report transfer effects found at the posttest but not at a later follow-up, or vice versa, depending on the ability considered (i.e., Borella, Cantarella, Carretti et al., in press). In one of our studies (Borella et al., 2010), verbal WM training led to a better performance in a visuospatial WM task at posttest (and in other measures, such as the Cattell Test), but the improvement in visuospatial WM was no longer significant at follow-up.

Analyzing maintenance effects could be useful to clarify the processes involved in training gains and could shed more light on the structure of WM. As an example of the latter, we found that, after completing a verbal WM training program, participants performed better on a visuospatial WM task at posttest, but the effect disappeared at a later follow-up. This could be attributed to differences in the format of the information to process (verbal vs. spatial). That is why, although some models conceive WM as a unitary system regardless of the material processed, the findings of Borella et al. (2013) suggest that a distinction based on format is relevant in WM training studies (at least as far as older adults are concerned) and are in agreement with other studies not focusing on training (Kane et al., 2004; Miyake & Shah, 1999, Park et al., 2002). This pattern of results fits with our model, in which content's interaction with attentional control is crucial, and the transfer effects presumably decrease with the increasing distance (in terms of modality) between the two processes involved.

Effect Sizes

The reporting of training gains is also important, including whether the effect sizes are reported. Effect sizes are crucial for enabling comparisons between studies and for assessing the WM system's malleability. Effect-size indexes are not always discussed thoroughly, and a great variety of formulas and methods have been used to compute effect sizes in the area of WM training. For example, some authors report the classical Cohen's *d*, adjusting for within-group comparisons (although this is not always the case), while others use the standardized gains score (which adjusts for pretest variability) or prefer to consider the net effect (which adjusts the index on the dimension of the change obtained by the control group). All these methods and formulas are valuable, but it would be beneficial to discuss how best to report the effect sizes relating to training programs so as to facilitate comparisons between studies. We hold that it would be useful to report effect-size indexes because they compare both between- and within-group changes (i.e., the differences between groups at pre- and posttest and the size of the increase separately for each group). This yields information on the distance between the groups at the different stages (pretest, posttest, follow-up) and enables comparison of the dimension of the change in different groups across studies as well.

In this respect, Cohen's *d* (Cohen, 1988) has been used appropriately in the literature to represent training gains. Other options may be useful, too, such as relative difference scores computed as *(posttest score – pretest score)/pretest score* (see Carretti et al., 2007). Finally, another useful method is the standardized gain score (Jaeggi, Buschkuehl, Jonides, & Shah, 2011), which enables a direct comparison between gains, also adjusting for possible differences at the pretest stage. In the field of age and individual differences, the standardized gain score seems particularly appropriate, because it enables us to control for initial group differences.

Multiple Indicators

Another often-ignored issue in the field of training studies concerns the importance of having robust measures for assessing pre- and posttest gains. There are at least two ways to deal with this requirement: one option is to use tasks that have good psychometric properties, while the other (more time-consuming) option is to consider multiple indicators of the same process. The latter option has recently been strongly recommended by Shipstead, Readick, and Engle (2010). Although we agree with the points raised in their article, we have addressed the issue by using the first approach (i.e., by choosing tasks with adequate psychometric properties).

It is worth noting that the use of multiple indicators of the different abilities targeted can also pose a question about the length of training programs (see Brums et al., in press). Our training consisted of only three sessions, so the use of multiple indicators would risk involving more assessment sessions than training sessions, which would not only be paradoxical but also generate effects due to participants' practicing with these measures rather than with the training activity itself. We therefore recommend paying more attention to the accurate selection of relevant processes and mechanisms associated with improvements in WM rather than considering the problem of multiple indicators. This is particularly important when applying WM training programs in the clinical setting.

Question 4: Contributions to the Field

Our results on the effects of WM training support an articulated view of WM, as explained in response to Question 1.

In particular, our results show that focusing the training procedure on more active-controlled WM components can produce good results, which can be transferred to different types of content and also to complex cognitive abilities (as predicted by the hypothesis that these WM processes are also involved in complex skills, such as reading comprehension, reasoning, everyday problems etc.). At the same time, our results show that, where content distances exist between different tasks—as in the case of verbal and visuospatial tasks—then any transfer effects may be more limited, especially at the follow-up.

Our results also further support the hypothesis that WM is involved in a series of intellectual functions and that improving WM performance can have an effect on these functions.

Another crucial element emerging from our studies concerns the importance of motivational, emotional, and metacognitive factors affecting WM and intellectual performance. In our training programs, we concentrated on sustaining these aspects by proposing a procedure that was always challenging and novel in terms of the activities involved (i.e., we implicitly considered the metacognitive and motivational aspects). Measuring the extent of the effects of the training on either the basic structure of WM or its use—or even in both—is difficult, but the fact that the training was particularly effective in individuals whose WM performance was suboptimal could be because the individuals made poor use of their abilities. In other words, it may be that their poor WM performance was due, at least in part, to a poor use of WM, and training improved its usage.

References

Ackerman, P. L., Beier, M. E., & Boyle, M. O. (2005). Working memory and intelligence: The same or different constructs? *Psychological Bulletin, 131*(1), 30–60. doi:10.1037/0033-2909.131.1.30

Baddeley, A. D. (1986). *Working memory.* Oxford, U.K.: Oxford University Press.

Baddeley, A. D. (2000). The episodic buffer: A new component of working memory? *Trends in Cognitive Sciences, 4*(11), 417–423. doi:10.1016/S1364-6613(00)01538-2

Ball, K., Berch, D. B., Helmers, K. F., Jobe, J. B., Leveck, M. D., Marsiske, M., . . . Willis, S. L. (2002). Effects of cognitive training interventions with older adults: A randomized controlled trial. *Journal of the American Medical Association, 288*(18), 2271–2281. doi:10.1001/jama.288.18.2271

Baltes, P. B., Kliegl, R., & Dittmann-Kohli, F. (1988). On the locus of training gains in research on the plasticity of fluid intelligence in old age. *Journal of Educational Psychology, 80*(3), 392–400. doi:10.1037//0022-0663.80.3.392

Birren, J. E., & Schaie, K. W. (Eds.). (2006). *Handbook of the psychology of aging* (6th ed.). San Diego, CA: Elsevier.

Bisiacchi, P., Borella, E., Bergamaschi, S., Carretti, B., & Mondini, S. (2008). Interplay between memory and executive functions in normal and pathological aging. *Journal of Clinical and Experimental Neuropsychology. 30*(6), 723–733. doi:10.1080/13803390701689587

Bopp, K. L., & Verhaeghen, P. (2007). Age-related differences in control processes in verbal and visuospatial working memory: Storage, transformation, supervision, and coordination. *The Journals of Gerontology, Series B: Psychological Sciences, 62*(5), 239–246.

Borella, E., Cantarella, A., Carretti, C., De Lucia, & De Beni, R. (*in press*). Improving everyday functioning in the old-old with a working memory training. *The American Journal of Geriatric Psychiatry.*

Borella, E., Cantarella, A., Joly, E., Ghisletta, P., Carbone, E., Coraluppi, D., . . . De Beni, R. (2017). Performance-based everyday functional competence measures across the adult lifespan: The role of cognitive abilities. *International Psychogeriatrics, 29*(12), 2059–2069. doi:10.1017/S1041610217000680

Borella, E., Carbone, E., Pastore, M., De Beni, R., & Carretti, B. (2017). Working memory training for healthy older adults: The role of individual characteristics in explaining short-and long-term gains. *Frontiers in Human Neuroscience, 11*, 99. doi:10.3389/fnhum.2017.00099

Borella, E., Carretti, B., Cantarella, A., Riboldi, F., Zavagnin, M., & De Beni, R. (2014). Benefits of training visuospatial working memory in young-old and old-old. *Developmental Psychology, 50*(3), 714–727. doi:10.1037/a0034293

Borella, E., Carretti, B., & De Beni, R. (2008). Working memory and inhibition across the adult life span. *Acta Psychologica, 128*(1), 33–44. doi:10.1016/j.actpsy.2007.09.008

Borella, E., Carretti, B., Riboldi, F., & De Beni, R. (2010). Working memory training in older adults: Evidence of transfer and maintenance effects. *Psychology and Aging, 25*(4), 767–778. doi:10.1037/a0020683

Borella, E., Carretti, B., Zanoni, G., Zavagnin, M., & De Beni, R. (2013). Working memory training in old-age: An examination of transfer and maintenance effects. *Archives of Clinical Neuropsychology, 28*(4), 331–347. doi:10.1093/arclin/act020

Borella, E., Ghisletta, P., & de Ribaupierre, A. (2011). Age differences in text processing: The role of working memory, inhibition, and processing speed. *The Journals of Gerontology, Series B: Psychological Sciences, 66*(3), 311–320. doi:10.1093/geronb/gbr002

Borella, E., Pezzuti, L., De Beni, R., & Cornoldi, C. (*in press*). Intelligence and working memory: evidence from administering the WAIS-IV to Italian adults and elderly. *Psychological Research.* doi:10.1007/s00426-019-01173-7

Brum Schimidt, P., Borella, E., Carretti, B., & Yassuda Sanches, M. (2018). Verbal working memory training in older adults: an investigation of dose response. *Aging & Mental Health.* doi:10.1080/13607863.2018.1531372

Buschkuehl, M., Jaeggi, S. M., Hutchison, S., Perrig-Chiello, P., Däpp, C., Müller, M., . . . Perrig, W. J. (2008). Impact of working memory training on memory performance in old-old adults. *Psychology and Aging, 23*(4), 743–753. doi:10.1037/a0014342

Buschkuehl, M., Jaeggi, S., & Jonides, J. (2012). Neuronal effects following working memory training. *Developmental Cognitive Neuroscience, 2*(1)(Suppl.), 167–179. doi:10.1016/j.dcn.2011.10.001

Campbell, D., & Stanley, J. (1963). *Experimental and quasi-experimental designs for research.* Chicago, IL: Rand-McNally.

Cantarella, A., Borella, E., Carretti, B., De Beni, R., & Kliegel, M. (2017). Can working memory training enhance older adults performance in fluid intelligence and in everyday life tasks? *International Journal of Geriatric, 32*(1), 86–93.

Carbone, E., Vianello, E., Carretti, B., Borella, E. (in press). Working Memory Training for Older Adults After Major Surgery: Benefits to Cognitive and Emotional Functioning. *The American Journal of Geriatric Psychiatry.* doi:10.1016/j.jagp.2019.05.023. IF: 3.48

Carretti, B., Borella, E., Cornoldi, C., & De Beni, R. (2009). Role of working memory in explaining the performance of individuals with specific reading comprehension difficulties: A meta-analysis. *Learning and Individual Differences, 19*(2), 246–251. doi:10.1016/j.lindif.2008.10.002

Carretti, B., Borella, E., & De Beni, R. (2007). Does a strategic memory training improve the working memory capacity of younger and older adults? *Experimental Psychology, 54*(4), 311–320. doi:10.1027/1618-3169.54.4.311

Carretti, B., Borella, E., Fostinelli, S., & Zavagnin, M. (2013). Benefits of training working memory in amnestic mild cognitive impairment: Specific and transfer effects. *International Psychogeriatrics, 25*(4), 617–626. doi:10.1017/S1041610212002177

Carretti, B., Borella, E., Zavagnin, M., & De Beni, R. (2013). Gains in language comprehension relating to working memory training in healthy older adults. *International Journal of Geriatric Psychiatry, 28*(5), 539–546. doi:10.1002/gps.3859

Carretti, B., Lanfranchi, S., & Mammarella, I. C. (2013). Spatial-simultaneous and spatial-sequential working memory in individuals with Down syndrome: The effect of configuration. *Research in Developmental Disabilities, 34*(1), 669–675. doi:10.1016/j. ridd.2012.09.011

Cavallini, E., Dunlosky, J., Bottiroli, S., Hertzog, C., & Vecchi, T. (2010). Promoting transfer in memory training for older adults. *Aging Clinical and Experimental Research, 22*(4), 314–323. doi:10.3275/6704

Cepeda, N. J., Pashler, H., Vul, E., Wixted, J. T., & Rohrer, D. (2006). Distributed practice in verbal recall tasks: A review and quantitative synthesis. *Psychological Bulletin, 132*(3), 354–380. doi:10.1037/0033-2909.132.3.354

Cohen, J. (1988). *Statistical power analysis for the behavioral sciences* (2nd ed.). Hillsdale, NJ: Lawrence Erlbaum Associates.

Cornoldi, C. (1995). La memoria di lavoro visuo-spaziale. In F. Marucci (Ed.), *Le immagini mentali* (pp. 145–181). Rome, Italy: La Nuova Italia Scientifica.

Cornoldi, C. (2007). *L'intelligenza.* Bologna, Italy: Il Mulino.

Cornoldi, C. (2010). Metacognition, intelligence, and academic performance. In H. Salatas Waters & W. Schneider (Eds.), *Metacognition, strategy use, and instruction* (pp. 257–277). New York, NY: Guilford Press.

Cornoldi, C., & Vecchi, T. (2003). *Visuo-spatial working memory and individual differences*. Hove, U.K.: Psychology Press.

Dahlin, E., Nyberg, L., Bäckman, L., & Neely, A. S. (2008). Plasticity of executive functioning in young and old adults: Immediate training gains, transfer, and long-term maintenance. *Psychology and Aging, 23*(4), 720–730. doi:10.1037/a0014296

Doerr, E., Carretti, B., & Lanfranchi, S. (2019). The working memory of individuals with Down syndrome. In S. Lanfranchi (Eds.), *State of the Art of Research on Down Syndrome, 56,* 45–56.

de Ribaupierre, A. (2001). Working memory and attentional processes across the lifespan. In P. G. N. Otha (Ed.), *Lifespan development of human memory* (pp. 59–80). Cambridge, MA: MIT Press.

Dunlosky, J., Cavallini, E., Roth, H., McGuire, C. L., Vecchi, T., & Hertzog, C. (2007). Do self-monitoring interventions improve older adult learning? [Special issue 1]. *Journal of Gerontology: Psychological Sciences, 62,* 70–76. doi:10.1093/geronb/62.special_issue_1.70

Engle, R. W., Cantor, J., & Carullo, J. J. (1992). Individual differences in working memory and comprehension: A test of four hypotheses. *Journal of Experimental Psychology: Learning, Memory and Cognition, 18*(5), 972–992. doi:10.1037/0278-7393.18.5.972

Friedman, N. P., & Miyake, A. (2004). The relations among inhibition and interference control functions: A latent-variable analysis. *Journal of Experimental Psychology: General, 133*(1), 101–135. doi:10.1037/0096-3445.133.1.101

Gathercole, S. E., Dunning, D. L., & Holmes, J. (2012). Cogmed training: Let's be realistic about intervention research. *Journal of Applied Research in Memory and Cognition, 1*(3), 201–203. doi:10.1016/j.jarmac.2012.07.007

Greenwood, P. M., & Parasuraman, R. (2010). Neuronal and cognitive plasticity: A neurocognitive framework for ameliorating cognitive aging. *Frontiers in Aging Neuroscience, 2,* Article 150. doi:10.3389/fnagi.2010.00150

Hertzog, C., & Hultsch, D. F. (2000). Metacognition in adulthood and aging. In T. A. Salthouse & F. I. M. Craik (Eds.), *Handbook of aging and cognition* (2nd ed., pp. 417–466). Mahwah, NJ: Erlbaum.

Hertzog, C., Kramer, A. F., Wilson, R. S., & Linderberger, U. (2009). Enrichment effects on adult cognitive development: Can the functional capacity of older adults be preserved and enhanced? *Psychological Science in the Public Interest, 9*(1), 1–65. doi:10.1111/j.1539-6053.2009.01034.x

Jaeggi, S. M., Buschkuehl, M., Jonides, J., & Perrig, W. J. (2008). Improving fluid intelligence with training on working memory. *Proceedings of the National Academy of Sciences of the United States of America, 105*(19), 6829–6833. doi:10.1073/pnas.0801268105

Jaeggi, S. M., Buschkuehl, M., Jonides, J., & Shah, P. (2011). Short- and long-term benefits of cognitive training. *Proceedings of the National Academy of Sciences of the United States of America, 108*(25), 10081–10086. doi:10.1073/pnas.1103228108

Jenkins, L., Myerson, J., Joerding, J. A., & Hale, S. (2000). Converging evidence that visuospatial cognition is more age-sensitive than verbal cognition. *Psychology and Aging, 15*(1), 157–175. doi:10.1037//0882-7974.15.1.157

Kane, M. J., Conway, A. R. A., Miura, T. K., & Colflesh, G. J. H. (2007). Working memory, attention control, and the N-back task: A question of construct validity. *Journal* of *Experimental Psychology: Learning, Memory, and Cognition, 33*(3), 615–622. doi:10.1037/0278-7393.33.3.615

Kane, M. J., Hambrick, D. Z., Tuholski, S. W., Wilhelm, O., Payne, T. W., & Engle, R. W. (2004). The generality of working memory capacity: A latent-variable approach to verbal and visuospatial memory span and reasoning. *Journal of Experimental Psychology: General, 133*(2), 189–217. doi:10.1037/0096-3445.133.2.189

Karbach, J., & Verhaeghen, P. (2014). Making working memory work: A meta-analysis of executive-control and working memory training in older adults. *Psychological Science, 25*, 2027–2037. doi:10.1177/0956797614548725

Klingberg, T., Fernell, E., Olesen, P. J., Johnson, M., Gustafsson, P., Dahlström, K., . . . Westerberg, H. (2005). Computerized training of working memory in children with ADHD—A randomized, controlled, trial. *Journal of the American Academy of Child and Adolescent Psychiatry, 44*(2), 177–186. doi:10.1097/00004583-200502000-00010

Klingberg, T., Forssberg, H., & Westerberg, H. (2002). Training of working memory in children with ADHD. *Journal of Clinical and Experimental Neuropsychology, 24*(6), 781–791. doi:10.1076/jcen.24.6.781.8395

Lanfranchi, S., Carretti, B., Spanò, G., & Cornoldi, C. (2009). A specific deficit in visuospatial simultaneous working memory in Down syndrome. *Journal of Intellectual Disability Research, 53*(5), 474–483. doi:10.1111/j.1365-2788.2009.01165.x

Lanfranchi, S., Cornoldi, C., Drigo, S., & Vianello, R. (2009). Working memory in individuals with fragile X syndrome. *Child Neuropsychology, 15*(2), 105–119. doi:10.1080/09297040802112564

Lanfranchi, S., Cornoldi, C., & Vianello, R. (2004). Verbal and visuospatial working memory deficits in children with Down syndrome. *American Journal on Mental Retardation, 109*(6), 456–466. doi:10.1352/0895-8017(2004)1092.0.CO;2

Li, S. C., Schmiedek, F., Huxhold, O., Röcke, C., Smith, J., & Lindenberger, U. (2008). Working memory plasticity in old age: Practice gain, transfer, and maintenance. *Psychology and Aging, 23*(4), 731–742. doi:10.1037/a0014343

Linares, R., Borella, E., Lechuga, M. T., Carretti, B., & Pelegrina, S. (2018). Training working memory updating in young adults. *Psychological Research, 82*, 535–548. doi:10.1007/s00426-017-0843-0

Linares, R., Borella, E., Lechuga, M. T., Carretti, B., & Pelegrina, S. (2019). Nearest transfer effects of working memory training: A comparison of two programs focused on working memory updating. *PlosONE, 13;*14(2):e0211321. doi:10.1371/journal.pone.0211321

Lövdén, M., Bäckman, L., Lindenberger, U., Schaefer, S., & Schmiedek, F. (2010). A theoretical framework for the study of adult cognitive plasticity. *Psychological Bulletin, 136*(4), 659–676. doi:10.1037/a0020080

Mammarella, I. C., Borella, E., Pastore, M., & Pazzaglia, F. (2013). The structure of visuospatial memory in adulthood. *Learning and Individual Differences, 25*, 99–110. doi:10.1016/j.lindif.2013.01.014

Mammarella, I. C., & Cornoldi, C. (2014). An analysis of the criteria used to diagnose children with non-verbal learning disabilities (NLD). *Child Neuropsychology, 20*(3), 255–280. doi:10.1080/09297049.2013.796920

Mammarella, I. C., Cornoldi, C., Pazzaglia, F., Toso, C., Grimoldi, M., & Vio, C. (2006). Evidence for a double dissociation between spatial-simultaneous and spatial-sequential working memory in visuospatial (nonverbal) learning disabled children. *Brain and Cognition, 62*(1), 58–67. doi:10.1016/j.bandc.2006.03.007

Mammarella, I. C., Pazzaglia, F., & Cornoldi, C. (2008). Evidence for different components in children's visuospatial working memory. *British Journal of Developmental Psychology*, *26*(3), 337–355. doi:10.1348/026151007X236061

Marsiske, M., & Willis, S. L. (1995). Dimensionality of everyday problem solving in older adults. *Psychology and Aging*, *10*(2), 269–283. doi:10.1037/0882-7974.10.2.269

Miyake, A., Friedman, N. P., Emerson, M. J., Witzki, A. H., Howerter, A., & Wager, T. (2000). The unity and diversity of executive functions and their contributions to complex "frontal lobe" tasks: A latent variable analysis. *Cognitive Psychology*, *41*(1), 49–100. doi:10.1006/cogp.1999.0734

Miyake, A., & Shah, P. (1999). *Models of working memory: Mechanisms of active maintenance and executive control*. New York, NY: Cambridge University Press.

Myerson, J., Emery, L., White, D. A., & Hale, S. (2003). Effects of age, domain, and processing demands on memory span: Evidence for differential decline. *Aging, Neuropsychology, and Cognition*, *10*(1), 20–27. doi:10.1076/anec.10.1.20.13454

Myerson, J., Hale, S., Rhee, S. H., & Jenkins, L. (1999). Selective interference with verbal and spatial working memory in young and older adults. *The Journals of Gerontology, Series B: Psychological Sciences*, *54*(3), 161–164. doi:10.1093/geronb/54B.3.P161

Noack, H., Lövdén, M., Schmiedek, F., & Lindenberger, U. (2009). Cognitive plasticity in adulthood and old age: Gauging the generality of cognitive intervention effects. *Restorative Neurology and Neuroscience*, *27*(5), 435–453. doi:10.3233/RNN-2009-0496

Oberauer, K. (2005). Binding and inhibition in working memory: Individual and age differences in short-term recognition. *Journal of Experimental Psychology: General*, *134*(3), 368–387. doi:10.1037/0096-3445.134.3.368

Park, D. C., Lautenschlager, G., Hedden, T., Davidson, N. S., Smith, A. D., & Smith, P. K. (2002). Models of visuospatial and verbal memory across the adult life span. *Psychology and Aging*, *17*(2), 299–320. doi:10.1037//0882-7974.17.2.299

Park, D. C., & Reuter-Lorenz, P. A. (2009). The adaptive brain: Aging and neurocognitive scaffolding. *Annual Review of Psychology*, *60*, 173–196. doi:10.1146/annurev.psych.59.103006.093656

Pazzaglia, F., & Cornoldi, C. (1999). The role of distinct components of visuo-spatial working memory in the processing of texts. *Memory*, *7*(1), 19–41. doi:10.1080/096582199388057

Rebok, G. W., Carlson, M. C., & Langbaum, J. B. (2007). Training and maintaining memory abilities in healthy older adults: Traditional and novel approaches [Special issue 1]. *The Journals of Gerontology, Series B: Psychological Sciences*, *62*, 53–61. doi:10.1093/geronb/62.special_issue_1.53

Richmond, L. L., Morrison, A. B., Chein, J. M., & Olson, I. R. (2011). Working memory training and transfer in older adults. *Psychology and Aging*, *26*(4), 813–822. doi:10.1037/a0023631

Roberts, R., & Gibson, E. (2002). Individual differences in sentence memory. *Journal of Psycholinguistic Research*, *31*(6), 573–598. doi:10.1023/A:1021213004302

Rosenthal, R., & Fode, K. (1963). The effect of experimenter bias on the performance of the albino rat. *Systems Research and Behavioral Science*, *8*(3), 183–189. doi:10.1002/bs.3830080302

Rosenthal, R., & Jacobson, L. (1968). *Pygmalion in the classroom: Teacher expectations and pupils' intellectual development*. New York, NY: Holt, Rinehart and Winston.

Salminen, T., Kühn, S. Frensch, P. A., & Schubert T. (2016). Transfer after Dual n-Back Training Depends on Striatal Activation Change. *Journal of Neuroscience*, *36*(39), 10198–10213; doi:https://doi.org/10.1523/JNEUROSCI.2305-15.201

Schaie, K. W., & Willis, S. L. (1996). Psychometric intelligence and aging. In F. Blanchard-Fields & T. M. Hess (Eds.), *Perspectives on cognition in adulthood and aging* (pp. 293–322). New York, NY: McGraw-Hill.

Schmidt, R. A., & Bjork, R. A. (1992). New conceptualizations of practice: Common principles in three paradigms suggest new concepts for training. *Psychological Science, 3*(4), 207–217. doi:10.1111/j.1467-9280.1992.tb00029.x

Schmiedek, F., Hildebrandt, A., Lövdén, M., Wilhelm, O., & Lindenberger, U. (2009). Complex span versus updating tasks of working memory: The gap is not that deep. *Journal of Experimental Psychology: Learning, Memory, and Cognition, 35*(4), 1089–1096. doi:10.1037/a0015730

Schmiedek, F., Lövdén, M., & Lindenberger, U. (2010). Hundred days of cognitive training enhance broad abilities in adulthood: Findings from the COGITO study. *Frontiers in Aging Neuroscience, 2*, 1–10. doi:10.3389/fnagi.2010.00027

Shamosh, N. A., Deyoung, C. G., Green, A. E., Reis, D. L., Johnson, M. R., Conway, A. R., . . . Gray, J. R. (2008). Individual differences in delay discounting: Relation to intelligence, working memory, and anterior prefrontal cortex. *Psychological Science, 19*(9), 904–911. doi:10.1111/j.1467-9280.2008.02175.x

Shipstead, Z., Readick, T. S., & Engle R. W. (2010). Does working memory training generalize? *Psychologica Belgica, 50*(3–4), 245–276. doi:10.5334/pb-50-3-4-245

Singer, T., Verhaeghen, P., Ghisletta, P., Lindenberger, U., & Baltes, P. B. (2003). The fate of cognition in very old age: Six-year longitudinal findings in the Berlin Aging Study (BASE). *Psychology and Aging, 18*(2), 318–331. doi:10.1037/0882-7974.18.2.318

Takeuchi, H., Sekiguchi, A., Taki, Y., Yokoyama, S., Yomogida, Y., Komuro, N., . . . Kawashima R. (2010). Training of working memory impacts structural connectivity. *The Journal of Neuroscience, 30*(9), 3297–3303. doi:10.1523/JNEUROSCI.4611-09.2010

Vecchi, T., & Cornoldi, C. (1999). Passive storage and active manipulation in visuospatial working memory: Further evidence from the study of age differences. *European Journal of Cognitive Psychology, 11*(3), 391–406. doi:10.1080/713752324

Verhaeghen, P., & Marcoen, A. (1996). On the mechanisms of plasticity in young and older adults after instruction in the method of loci: Evidence for an amplification model. *Psychology and Aging, 11*(1), 164–178. doi:10.1037//0882-7974.11.1.164

Verhaeghen, P., Marcoen, A., & Goossens, L. (1992). Improving memory performance in the aged through mnemonic training: A meta-analytic study. *Psychology and Aging, 7*(2), 242–251. doi:10.1037//0882-7974.7.2.242

Vlach, H. A., & Sandhofer, C. M. (2014). Retrieval dynamics and retention in cross-situational statistical learning. *Cognitive Science, 38*, 757–774. doi:10.1111/cogs.12092

Vranic, A., Carretti, B., Španic, A. M., & Borella, E. (2013). The efficacy of a multifactorial memory training in institutionalized older adults. *International Psychogeriatrics, 25*(11), 1885–1897. doi:10.1017/S1041610213001154

Westerberg, H., & Klingberg, T. (2007). Changes in cortical activity after training of working memory—A single-subject analysis. *Physiology and Behavior, 92*(1–2), 186–192. doi:10.1016/j.physbeh.2007.05.041

Zinke, K., Zeintl, M., Eschen, A., Herzog, C., & Kliegel, M. (2012). Potentials and limits of plasticity induced by working memory training in old-old age. *Gerontology, 58*(1), 79–87. doi:10.1159/000324240

3

Training Working Memory for 100 Days

The COGITO Study

Florian Schmiedek, Martin Lövdén, and Ulman Lindenberger

This chapter is based on the theoretical framework for the study of adult cognitive plasticity by Lövdén, Bäckman, Lindenberger, Schaefer, and Schmiedek (2010) and on empirical findings from the COGITO Study (Schmiedek, Lövdén, & Lindenberger, 2010), conducted at the Max Planck Institute for Human Development. In the study, 101 younger and 103 older adults practiced a battery of working memory (WM), episodic memory, and perceptual speed tasks for 100 sessions. The design and analyses of the study include key features for producing and detecting transfer effects at the level of cognitive abilities. Among the features are: (a) an intensity and dosage of training that is likely to induce an enduring mismatch between functional supply and demand, which is conducive to plastic changes in cognitive abilities, and (b) a multivariate and heterogeneous battery of transfer tasks and sufficiently large samples to allow for the investigation of transfer of training at the level of latent factors. Younger adults showed short-term and long-term transfer effects for reasoning and episodic memory, whereas older adults showed only short-term transfer on a WM latent factor composed of tasks that resembled the practiced tasks, something that younger adults did as well. The chapter covers possible interpretations of the findings in terms of increases in WM capacity, improvements in the efficiency of material-independent or material-specific processes or strategies, and improvements in motivation and self-concept.

Question 1: Theory of Working Memory Training

In a nutshell, WM is a system for keeping all sorts of content active in memory and organizing the processing of the content. Rather than assuming that this system has a fixed (traitlike) capacity (which training then would have to try to increase), we think that the WM system, in addition to capacitylike aspects, involves mechanisms that can work more or less efficiently and reliably, with the efficiency being subject to both traitlike and statelike variability and potentially also being amenable to improvements through practice.

Our view on WM is guided by concentric models of its architecture (Cowan, 1995; Oberauer, 2002) and by a focus on mechanisms of creating, maintaining, updating, and releasing bindings of different kinds of information (e.g., objects to spatial positions; Oberauer, 2005; Treisman & Gelade, 1980; Zimmer, Mecklinger, & Lindenberger, 2006). Furthermore, we consider models from computational neuroscience that aim to explain the conflicting modes of stability (of contents to be held in WM) and flexibility (of replacing contents and task sets in WM) as important (Cools & D'Esposito, 2011; Durstewitz & Seamans, 2008). Neuroscience models of selective updating (Frank, Loughry, & O'Reilly, 2001) are also relevant in this regard. Regarding our theoretical view on the plasticity of WM, we apply the general framework for the study of adult cognitive plasticity proposed by Lövdén et al. (2010).

Concentric models, such as those of Cowan (1995) and Oberauer (2002), conceive of WM as an activated part of long-term memory, with activated elements being directly accessible for cognitive processing (i.e., the region of direct access). The processing of elements requires that they be put in the focus of attention. Whether this focus can hold more than one element (or chunk of elements) at a time has been questioned (Oberauer & Bialkova, 2009) and so has the claim that training can enlarge the size of the focus (see Oberauer, 2006 and Verhaeghen, Cerella, & Basak, 2004, for opposing evidence).

A fixed-size focus of attention as well as a (possibly) immutable region of direct access do not preclude the possibility, however, that WM processes can gain efficiency with practice, thereby leading to improved performance (von Bastian & Oberauer, 2014). On the one hand, those processes might include basic cognitive processes like switching the focus of attention (Dorbath, Hasselhorn, & Titz, 2011) and component processes of updating information in WM (e.g., transformation, substitution, and retrieval; see Ecker, Lewandowsy, & Oberauer, 2014; Ecker, Lewandowsky, Oberauer, & Chee, 2010) that are relatively independent of the specific material being processed. On the other hand, efficiency might also improve in ways that are more specific to the particular material of the practiced task. This might be due to the use of strategies that work only for specific content (e.g., visualization of nouns) or due to material-specific automatization.

We supplement this view of WM with models from cognitive and computational neuroscience. Here we consider an attention function (of the lateral prefrontal cortex) that deals with active maintenance of content-unspecific (e.g., goal- and context-related) variables that serve to bind together distributed, capacity-limited, and content-specific internal representations that are held active in sensory cortices (Luck & Vogel, 2013; Sreenivasan, Curtis, & D'Esposito, 2014; Unsworth, Fukuda, Awh, & Vogel, 2014, 2015). This function, which serves the focus of attention, must, when faced with internal and external distraction, balance the demands of maintaining the focus of attention (i.e., stability) while also allowing rapid shifting (i.e., flexibility) of focus when needed (Cools & D'Esposito, 2011; Durstewitz & Seamans, 2008). Selective updating (not necessarily complete shifting) of parts of the focus of attention may occur through gating mechanisms (in striatal–prefrontal cortex interactions; Frank et al., 2001). We view WM training as potentially affecting the efficiency of these processes, without necessarily affecting any WM capacity per se. For example, achieving optimal tuning of the balance between stability and flexibility may be more beneficial for cognitive performance (Cools & D'Esposito, 2011). Rapid selective updating may be a partly content-unspecific neural process (Dahlin, Neely, Larsson, Bäckman, & Nyberg, 2008; D'Ardenne et al., 2012). Finally, the binding and biasing aspect of focusing attention may be critically dependent on reliable functional and structural large-scale brain connectivity, which is modifiable by cognitive training (Anguera et al., 2013; Lövdén, Bodammer, et al., 2010).

Regarding gains in material-specific efficiency, ample evidence exists that practice can improve the speed and accuracy of any basic cognitive decision or transformation process that is not yet fully overlearned (i.e., asymptotic performance levels have been reached). For example, most people will show practice-related improvements on an alpha span task simply because, before training, they are not at the asymptotes of their learning curves regarding the skill of quickly sorting words alphabetically. While such practice effects may lead to increases in the observable capacity specific to the alpha span (i.e., the set size of words that can reliably be processed), they should have no effect on WM tasks that do not involve this particular skill.

Generally, the basis for our concept of cognitive plasticity, and thus one of the backgrounds to our design of training studies, is a distinction of plastic changes from changes based on behavioral flexibility (Lövdén, Bäckman, et al., 2010). The human cognitive system generally exhibits an impressive amount of flexibility in adapting to changing environmental demands. When confronted with new WM tasks, for example, existing skills and available strategies can be explored and adapted to the task. This includes strategic choices of goal settings (e.g., trying to remember only a part of a presented memory list; Shing, Schmiedek, Lövdén, & Lindenberger, 2012), prioritization of subtasks (e.g., the primary task at the cost of the secondary task in complex span paradigms), speed–accuracy trade-off settings, the employment of verbalization and visualization strategies (Hertzog, Lövdén, Lindenberger, & Schmiedek, 2017), and more.

Such exploitation of behavioral flexibility can potentially lead to considerable improvements in performance. We distinguish such changes from plastic changes of the cognitive system by defining *plasticity* as the capacity for reactive change in the presence of an enduring mismatch between the demands confronting the cognitive system and the supply it is able to offer (Lövdén, Bäckman, et al., 2010). The defining characteristic of plastic changes is a widening of the range of behavioral flexibility itself. That is, we consider training-related changes in cognitive performance to be indicative of plasticity only if an increased functional supply allows for new or more difficult tasks to be dealt with. Of crucial importance is the proposition that only a considerable mismatch of demand and supply that endures for an extended time should be able to lead to such plastic changes.

Several characteristics that we think are essential for successful training programs follow from this. First, training should contain several tasks that differ in paradigm and content, to reduce the likelihood of successfully working on them using a limited number of strategies contained in the toolbox of behavioral flexibility. For the same reason, we even see advantages in changing or adding tasks during the course of training. Second, task difficulty should be dynamically adapted to each individual's performance level to keep up the mismatch of supply and demand. Third, training duration needs to be extensive. We consider it unlikely that a cumulated training time of a few hours (e.g., one week of daily practice for one hour each day) will lead to the kind of changes at the neuronal level that constitute plasticity.

In sum, we think that extended practice on diverse and challenging WM tasks can lead to improvements in the efficiency of creating, maintaining, updating, and releasing bindings and the corresponding interplay of stability and flexibility of cognitive representations in WM. To exclude the possibility that observed improvements in performance can be interpreted as manifestations of behavioral flexibility, it is of great importance to demonstrate the emergence of plastic changes. This can be achieved by showing improvements on transfer tasks that minimize the likelihood of improvements based on the application of strategies and skills specific to the practiced tasks, or general improvements in motivation and/or self-concept.

Question 2: Major Claims of Working Memory Training

Earlier cognitive intervention work at the Max Planck Institute for Human Development has shown that instruction and practice of certain strategies (e.g., mnemonic techniques like the method of loci) can lead to considerable improvements of performance in episodic memory tasks in both younger and older adults (Baltes & Kliegl, 1992). Also, improvements in performance on fluid intelligence tasks due to practice were shown (Ball et al., 2002; Baltes, Dittmann-Kohli, & Kliegl, 1986). Common to these training studies, however, was the finding that the effects were highly task-specific, that is, the effects did not show transfer beyond the paradigms that were part of the training (for a review of this work, see Baltes & Lindenberger, 1988). Therefore, the improvements have to be considered to be largely manifestations of behavioral flexibility.

In a first attempt to evaluate to effectiveness of WM training (Li et al., 2008), 19 younger adults (age 20–30 years) and 21 older adults (age 70–80 years) practiced two versions of a spatial 2-back task in 45 practice sessions, which led to transfer to spatial 3-back and numerical 2-back tasks. The results indicated that specific updating mechanisms could be improved independently from the content of the tasks. Whether the improvements also led to improvements at the level of more general factors of WM or fluid intelligence, however, could not be investigated with the small samples in the study.

The COGITO Study, which was designed with a focus on investigating day-to-day fluctuations in cognitive performance (Schmiedek, Lövdén, & Lindenberger, 2013), implemented several of the features discussed above as being relevant for plastic changes of WM. Compared to the training in the study by Li et al. (2008), the training was more diverse, more extensive, and better tailored to individual performance levels. Diversity was ensured by including three WM tasks that differed in paradigm as well as in content (figural-spatial, 3-back; verbal, alpha span; and numerical, memory updating) together with three episodic memory tasks and a total of six perceptual speed tasks (three comparison and three two-choice reaction tasks). The inclusion of non-WM tasks potentially served two beneficial purposes in the service of improving the effectiveness of the WM training. First, diversity of abilities required by the tasks can make a training program more varied and motivating. Second, episodic memory tasks especially might also train binding mechanisms that are recruited by WM.

Training was extensive, with a total duration of 100 training sessions, each lasting for 45 minutes to one hour. Task difficulty was individualized by setting presentation times of the WM and episodic memory tasks and masking times of the three choice-reaction time tasks of perceptual speed to appropriate levels based on individual pretest performance (see Schmiedek et al., 2010). The levels were chosen in a way to keep participants' performance from both floor and ceiling across the 100 training sessions. This worked quite well for most of the participants. In fact, average learning curves for the WM tasks (as well as the episodic memory tasks) indicated that training gains did not reach an asymptote within 100 training sessions but continued to show small but steady improvements until the very end of the observation period.

Transfer effects were assessed at the latent factor level. To this end, comprehensive transfer batteries were included in extensive pretest and posttest sessions. The sessions contained three WM tasks based on the same paradigms as the trained ones but with different task content. This included numerical 3-back, spatial memory updating, and animal span (i.e., sorting animals according to size), three complex span WM tasks, nine tasks each of reasoning, episodic memory, perceptual speed from the Berlin Intelligence Structure Test (Jäger, Süß, & Beauducel, 1997), a paired associates test, and Raven's matrices.

This allowed us to create measurement models for factors of WM updating, WM complex span, and the broad abilities of reasoning, episodic memory, and perceptual speed. In addition, the sample sizes for both younger adults (101 in the training group and 44 in the no-training control group) and older adults (103 in the training group and 39 in the no-training control group) were large enough to allow for structural equation modeling and the investigation of transfer effects at the latent factor level using latent change score models (McArdle, 2009).

Results at the observed task level were mixed, with several tasks (or task parcels) of WM, reasoning, and episodic memory showing significant interactions of occasion and group in the younger and older groups (see Figure 3.1; Schmiedek et al., 2010). More importantly, results from latent change score models demonstrated significant transfer at the latent ability factor level for the near WM factor (i.e., of tasks based on the same paradigms as the trained ones) for both younger and older adults and also for reasoning and episodic memory

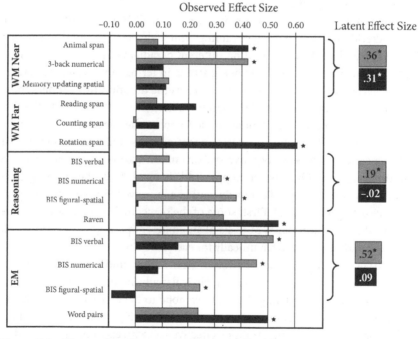

Figure 3.1. Observed and latent net effect sizes of performance gains from pretest for WM, reasoning, and episodic memory (EM). Bars show net effect sizes (standardized changes in the experimental group minus standardized changes in the control group), separately for younger adults (gray bars) and older adults (black bars). Statistically significant net effect sizes correspond to reliable interactions (*: $p < .05$) between group (experimental vs. control) and occasion (pretest vs. posttest). Reproduced from Schmiedek et al. (2010).

for the younger adults (Schmiedek et al., 2010). Furthermore, analyses of individual differences in training and transfer effects showed ability-specific relations of corresponding factors of practiced and transfer tasks for WM and episodic memory. This renders general motivational effects less likely and strengthens an interpretation of the effects being located at the level of broad abilities. This is important, especially considering that the control group was not active.

Using data from a two-year follow-up study, which repeated the posttest assessment, we could further investigate whether the effects for the latent ability factors of reasoning and episodic memory were maintained over an extended period of time for the group of younger adults. Latent change score models for the participants who returned for follow-up sessions (80 in the training group and 32 in the control group) indicated that there were reliable long-term effects for both broad abilities (see Figure 3.2; Schmiedek, Lövdén, & Lindenberger, 2014a). Comparisons of self-reported motivation to work on the tasks from both groups on the different occasions did not provide any evidence for motivational factors' being responsible for these effects.

In sum, the latent change score analyses of the COGITO Study data have provided evidence that extensive training with a challenging task battery that includes several WM tasks based on different paradigms can produce transfer effects that, while not necessarily being strong in terms of conventional evaluation of effect sizes, have the breadth as well as the duration that renders them potentially beneficial for everyday cognitive functioning. We attribute the fact that we found effects at the latent ability level not to the superiority of our particular choice of tasks (many other WM updating tasks could serve as well), but to the amount of time the cognitive systems of our participants were put in a condition of mismatch between supply and demand. Still, the training program implemented in COGITO likely could be improved if larger effectiveness and efficiency were the goal. First, some of the perceptual speed tasks could be substituted with additional WM tasks of still different paradigms. Second, task difficulty could be adapted dynamically for each participant.

We do not see our main contribution to the field of cognitive training research as the development of a task battery of particular advantage over other training that focuses on WM updating. Instead, we hope to have provided convincing arguments for the importance of implementing training programs with sufficient dosage and of evaluating training effectiveness with appropriate methods.

Question 3: Methodological Issues

We are not contesting the benefits of randomization in evaluation research; rather, we think that the question of which control conditions are used might be

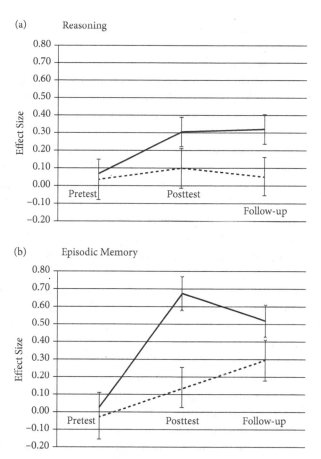

Figure 3.2. Latent means and associated standard errors for the training and control groups at pretest, posttest, and follow-up. Training group shown with solid lines, control with dashed lines. (a) latent factor of reasoning; (b) latent factor of episodic memory. As the indicator tasks of the latent factors were standardized by SDs at pretest, latent means are in effect size metric. Reproduced from Schmiedek et al. (2014).

more important than how participants get into the different conditions. When important aspects like motivation and subjective evaluation of the potential benefit of the training are equated across experimental conditions, the danger of participants' selecting themselves into treatment conditions in ways that confound treatment effects with pretraining ability may be minimized even in the absence of randomized assignment to groups. Conversely, perfect randomization of participants into groups may not allow answering the question about whether plastic changes have occurred if the control condition is less motivating or less able to foster a self-concept regarding cognitive ability than the condition of the target training.

Regarding the question of the mechanisms of training and transfer effects, it can be as important to show the limits of transfer effects as it can be to show their breadth. To exclude the possibility that, for example, increased motivation or self-concept is responsible for presumed improvements in cognitive resources, it might be helpful to define ability factors that should not show transfer effects. Ideally, theoretical models of the mechanisms of transfer allow for prediction of patterns of present and absent transfer to support the convergent and discriminant validity of the training effects. Such an approach has been used in an exemplary way by von Bastian and Oberauer (2013), who tested predictions of a cognitive process model by comparing transfer at the latent factor level for different target constructs across training conditions that focused on storage and processing, relational integration, and supervision as a theoretically defined functional category of WM.

Testing the predictions empirically can most convincingly be done based on latent factors for the constructs that define the convergent and discriminant relations (Noack, Lövdén, & Schmiedek, 2014). The proof of cognitive training's effectiveness is transfer to cognitive abilities relevant to competence in everyday life, including educational or job achievement. Regarding scientific approaches to demonstrate transfer, we hold critical views on two commonly applied practices, namely the use of single transfer tasks per ability and the attempt to classify those into ordinal categories of transfer distance (e.g., near versus far). First, we think that it does not suffice to demonstrate improvements attributable to training (e.g., by means of a control group design) in single tasks that are thought to measure the targeted ability. Abilities can never be directly measured by single tasks, because the latter always contain variance due to measurement error and task-specific processes and skills. Showing training-related improvements on a single task will always leave open the question of whether the improvements can really be attributed to improvements in the latent (i.e., not directly observable) ability or if they are due to improvements of task-specific skills, the acquisition of task-specific strategies, and the like.

Second, we think that the common practice of attempting to classify transfer tasks as indicating near or far transfer is not fruitful for research on cognitive interventions because it seems close to impossible to agree on definitions of those classifications. What one group of researchers considers far transfer might be near transfer from the theoretical perspective of another group. In our view, the problems associated with investigating transfer based on target abilities classified as near or far and operationalized by single tasks can and should be overcome by approaches that take theory and empirical knowledge regarding the structural relations among the trained and transfer tasks into account. Preferably, we should aim to demonstrate transfer at the level of latent factors, with the definition of factors and the choice of tasks determined by established hierarchical models of cognitive abilities (Lövdén, Bäckman, et al., 2010; Noack, Lövdén, Schmiedek, & Lindenberger, 2009).

With the aim of interpreting such structural relations among tasks within a theoretical frame, models like the three-stratum model of Carroll (1993) can serve to replace arbitrary classifications as near versus far with classifications of transfer "at the level of" narrow, broad, or general abilities, thereby providing a common ground for the communication and evaluation of transfer effects. Demonstrating transfer at the level of the broad ability of fluid intelligence, for example, would then require showing improvements at the level of a latent factor that is operationalized with several heterogeneous tasks (i.e., differing in paradigm and/or content).

The tasks should psychometrically be good indicators of fluid intelligence and be sampled from the construct space of this broad cognitive ability in a way that the space is broadly covered (Little, Lindenberger, & Nesselroade, 1999). If, for instance, only tasks of figural-spatial induction were used (e.g., figural analogies, figural series, and figural matrices), then possible improvements of the factor that represents their common variance cannot unambiguously be interpreted as improvements of fluid intelligence, because effects might be constrained to a narrower factor of figural inductive reasoning, which is nested in the fluid intelligence factor.

The correlational associations between training and transfer tasks also permit the formulation of expectations about the size of transfer effects (McArdle & Prindle, 2008). Specifically, based on the observed correlation between the trained task (or trained ability construct) and the transfer task (or transfer ability construct) at pretest and the effect size of gains on the trained task (or trained ability construct), one can compute the expected effect size of gains on the transfer task (or transfer ability construct) under the assumption that training gains reflect unbiased improvements in the ability targeted by the transfer task (or transfer ability construct; cf. Lawley, 1943; Pearson, 1903). If the observed transfer effect falls below this expectation, as Rode, Robson, Purviance, Geary, and Mayr (2014) found when training WM in school-age children, then this means that improvements on the trained tasks were biased toward factors that are unrelated to the cognitive ability targeted by the transfer task (or transfer ability construct). It would be desirable for researchers to routinely compare the magnitude of observed transfer effects against this expectation.

Some commercial cognitive training programs seem to promise improvements at the level of general cognitive ability, which comprises broad cognitive abilities like reasoning, episodic memory, and perceptual speed. Such effects, in our view, would have to be demonstrated by showing training-related changes of a latent factor at the top of a hierarchy of a considerable number of factors that are each based on a comprehensive assessment battery of tasks. Such general effects are not necessary for a training to be evaluated as effective, however, because improvements even in narrow abilities might be of sufficient practical relevance. It is important, though, that the true scope of WM training's effectiveness be determined by locating transfer in hierarchical models. Knowing about the validity of established cognitive ability factors regarding real-life outcomes, then, can help in the evaluation of potential training benefits. Certainly, one can also attempt to measure real-life outcomes directly. If this is done with objective tasks that cover a predefined domain of everyday functioning, again, latent factor approaches could be used to test whether the desired general improvements really have been achieved. If the scientific focus is instead on general resource constructs from cognitive psychology, like WM, rather than on psychometric abilities, latent factor approaches and hierarchical structure models can also be readily applied (Miyake et al., 2000; Miyake & Friedman, 2012; Schmiedek, Lövdén, & Lindenberger, 2014b).

Statistically, change at the level of latent ability factors can be investigated with latent change score models (McArdle, 2009; McArdle & Prindle, 2008). This requires measurement models with factor loadings, intercepts, and preferably also residual variance parameters being invariant across occasions (e.g., pretest, posttest, and follow-up in intervention studies). Not being able to show such measurement invariance can be indicative of the presence of task-specific effects (Noack et al., 2014). Careful investigation of this issue should, therefore, not be seen as a nuisance but as an enterprise that is informative about the levels at which transfer effects might take place. It should be noted, however, that the use of confirmatory measurement models in a structural equation modeling framework creates a need for larger samples than provided by many cognitive intervention studies. Because different experimental (e.g., training and control) groups can be analyzed simultaneously, using multigroup structural equation modeling with measurement model parameters constrained to be equal across groups, sample sizes do not necessarily have to be unrealistically large. A review of the training literature indicated that examples of empirical studies that meet the standards of latent change modeling (multiple indicators and a relatively large sample) exist but are the exception rather than the rule (Noack et al., 2014).

Question 4: Contributions to the Field

Conclusions regarding the understanding of WM based on our studies should be modest. Our studies have contributed a pattern of findings of transfer at the observed and latent factor level that fit well into theoretical perspectives that propose a central role for binding and updating mechanisms in WM and its relation to reasoning (Wilhelm, Hildebrandt, & Oberauer, 2013). The observed pattern is that transfer effects are present at the latent factor level for reasoning and for WM updating tasks of different content than the trained ones, but not for complex span tasks. One can ask whether and how this pattern of findings can be explained in terms of the possible effects of training: (a) increases in WM capacity, (b) improvements in the efficiency of material-independent basic WM processes, (c) improvements of material-specific processes, (d) improvements in more or less material-specific strategies, and (e) improvements in motivation and self-concept.

While we cannot exclude improvements in motivation and self-concept as contributing factors to the transfer effects based on our study design, which included a no-contact control group, the findings on self-reported motivation to work on the training tasks do not give any indication that (changes in) motivation differed between the experimental groups. Furthermore, the explanation of the pattern of transfer effects in terms of motivational factors would necessitate the assumption that motivational effects are less pronounced for complex span tasks.

Improvements in more or less material-specific strategies are likely to contribute to the transfer effects on WM updating tasks with different content. While the applicability of certain cognitive strategies (like visualization) is often tied to specific material and their efficient use requires task-specific adaptations, one can also think of less material-dependent strategies that might help to optimize the aspects of performance that were fed back to our participants and used as dependent measures in our analyses. Such strategies could include the adaptive setting of goals or speed–accuracy trade-offs, which might be applicable to the trained as well as the untrained WM updating tasks (cf. von Bastian & Oberauer, 2014). For example, participants who used a strategy of selectively trying to memorize only a subset of items in the numerical updating task (see Shing et al., 2012, for evidence indicating the presence of such a strategy) might transfer this strategy to the spatial updating task that was included in the transfer battery. It is very unlikely, however, that such strategies could help to improve performance on the reasoning transfer tasks. The potential creativity of people means they may come up with new strategies, and it is impossible to definitely exclude this possibility, but accounts based on strategies that were applied to— and themselves practiced with—the trained WM tasks do not seem to work well as an explanation for the overall pattern of findings.

If we wanted to explain the observed transfer effects on WM updating tasks and on reasoning in terms of an increased capacity of the WM system, we would have to provide an explanation for the lack of transfer to complex span tasks, which serve as well-established measures of this capacity. While ceiling effects of our versions of the complex span tasks for a number of our younger participants might serve as such an explanation, we think that it might also be that WM capacity did not improve in such a way that the WM system could hold more pieces of any kind of information ready for processing after training.

It is more likely that improvements in the efficiency of relatively material-independent basic WM processes underlie the transfer effects to WM updating as well as to reasoning. The processes include the creation, maintenance, updating, and dissolving of bindings that give rise to mental representations needed to do WM updating tasks and that can be argued to be required for successfully solving reasoning tasks of many kinds (Wilhelm et al., 2013). Evidence has been provided that updating processes can be improved by training, possibly due to increased striatal dopamine release (for a review, see Bäckman & Nyberg, 2013). In line with this notion, we have reported findings from the COGITO Study showing that carriers of the Val allele of the COMT polymorphism, who have less dopamine expressed in the prefrontal cortex, performed worse on the WM updating tasks at baseline but showed larger practice gains in these tasks from training than carriers of the Met allele (Bellander et al., 2015).

We relate this result to available evidence indicating that Met carriers perform better than Val carriers in WM tasks mainly taxing maintenance, whereas Val carriers perform better at updating (Bilder, Volavka, Lachman, & Grace, 2004; Colzato, Waszak, Nieuwnhuis, Posthuma, & Hommel, 2010). Val carriers may show larger training gains, because updating operations carry greater potential for plasticity than maintenance operations or because the task demands reward improvements in updating more than improvements in maintenance. Finally, we have reported alterations of the white-matter tracts that connect the left and right hemispheres of the frontal lobes from the COGITO intervention (Lövdén, Bodammer, et al., 2010). Training affected several metrics of white-matter microstructure, as probed with diffusion-tensor imaging, and increased the area of the anterior part of the corpus callosum (i.e., the genu). The alterations were of similar magnitude in younger and older adults. These findings are in line with animal evidence on the activity-dependent regulation of adult myelination (Fields, 2008; Zatorre, Fields, & Johansen-Berg, 2012). In line with evidence on functional connectivity from other groups (Anguera et al., 2013; Chapman et al., 2015), our findings on training-induced change in structural connectivity suggest that the binding and biasing aspect of focusing attention, which critically depends on reliable large-scale connectivity (Wang, 2010), may be malleable through experience.

If such improvements are at the heart of transfer effects to WM and rea-soning, why does improved efficiency of updating bindings not benefit indi-vidual performance on complex span tasks? This might seem particularly surprising given that latent factors of updating and complex span tasks have been found to be highly, or even perfectly, correlated in untrained samples (Schmiedek, Hildebrandt, Lövdén, Wilhelm, & Lindenberger, 2009; Schmiedek et al., 2014b; Wilhelm et al., 2013). Besides the already mentioned possibility that improvements on complex span tasks might have been constrained to some degree by ceiling effects, we would like to offer another interpretation. Generally, the fact that individual differences in performance on two kinds of paradigms correlate highly in a sample of untrained participants does not imply that improvements on one of the paradigms need to be matched by improvements in the other. It might be that the common variance in complex span and updating tasks before training was largely dominated by individual differences in the capacity of the WM system to reliably hold a limited number of elements active in WM. The extensive practice on WM updating tasks might have increased the efficiency of the WM system in quickly and reliably changing these elements and their associations. The resulting capacity to quickly estab-lish and manipulate complex mental representations might aid successful pro-cessing of complex reasoning tasks. It might not be beneficial for performance on complex span tasks, though, with their smaller demand on permanently and flexibly updating mental representations. Improvements on complex span tasks might depend more on the efficient employment of retrieval strategies from secondary (long-term) memory for items that exceed the capacity of primary (short-term) memory (Unsworth & Engle, 2007). Recent evidence suggests that the ability to establish and retrieve associations in secondary memory might contribute to reasoning performance independently of the role of pri-mary memory (Shipstead, Lindsey, Marshall, & Engle, 2014; Unsworth et al., 2014), so an interesting question for future research would be whether training modules that focus on this ability might produce transfer to complex span tasks and further increase the transfer effects to reasoning that can be produced by updating training alone.

In sum, a number of explanatory mechanisms might be involved in the total pattern of effects, involving increased use of more or less material-dependent strategies and adaptations to task demands. The core of the latent transfer effects to reasoning may be improvements in the general efficiency and reliability in building and updating bindings as a basis for complex mental representations and manipulations. To achieve such improvements, extensive and intensive training of several updating tasks with different contents, adaptive difficulty, and a motivating implementation seems to be the most promising approach.

References

Anguera, J. A., Boccanfuso, J., Rintoul, J. L., Al-Hashimi, O., Faraji, F., Janowich, J., . . . Gazzaley, A. (2013). Video game training enhances cognitive control in older adults. *Nature, 501*(7465), 97–101. doi:10.1038/nature12486

Bäckman, L., & Nyberg, L. (2013). Dopamine and training-related working-memory improvement. *Neuroscience and Biobehavioral Reviews, 37*(9), 2209–2219. doi:10.1016/j.neubiorev.2013.01.014

Ball, K. K., Berch, D. B., Helmers, K. F., Jobe, J. B., Leveck, M. D., Marsiske, M., . . . Willis, S. L. (2002). Effects of cognitive training interventions with older adults: A randomized controlled trial. *Journal of the American Medical Association, 288*(18), 2271–2281. doi:10.1001/jama.288.18.2271

Baltes, P. B., Dittmann-Kohli, F., & Kliegl, R. (1986). Reserve-capacity of the elderly in aging-sensitive tests of fluid intelligence: Replication and extension. *Psychology and Aging, 1*(2), 172–177. doi:10.1037//0882-7974.1.2.172

Baltes, P. B., & Kliegl, R. (1992). Testing-the-limits research suggests irreversible aging loss in memory based on mental imagination. *Developmental Psychology, 28*(1), 121–125. http://dx.doi.org/10.1037/0012-1649.28.1.121

Baltes, P. B., & Lindenberger, U. (1988). On the range of cognitive plasticity in old-age as a function of experience: 15 years of intervention research. *Behavioral Therapy, 19*(3), 283–300. doi:10.1016/S0005-7894(88)80003-0

Bellander, M., Bäckman, L., Liu, T., Schjeide, B.-M. M., Bertram, L., Schmiedek, F., . . . Lövdén, M. (2015). Lower baseline performance but greater plasticity of working memory for carriers of the Val allele in the COMT Val^{158}Met polymorphism. *Neuropsychology, 29*(2), 247–254. doi:10.1037/neu0000088

Bilder, R. M., Volavka, J., Lachman, H. M., & Grace, A. A. (2004). The catechol-*O*-methyltransferase polymorphism: Relations to the tonic-phasic dopamine hypothesis and neuropsychiatric phenotypes. *Neuropsychopharmacology, 29*(11), 1943–1961. doi:10.1038/sj.npp.1300542

Carroll, J. B. (1993). *Human cognitive abilities.* Cambridge, U.K.: Cambridge University Press.

Chapman, S. B., Aslan, S., Spence, J. S., Hart, J. J., Jr., Bartz, E. K., Didehbani, N., . . . Lu, H. (2015). Neural mechanisms of brain plasticity with complex cognitive training in healthy seniors. *Cerebral Cortex, 25*(2), 396–405. doi:10.1093/cercor/bht234

Colzato, L. S., Waszak, F., Nieuwenhuis, S., Posthuma, D., & Hommel, B. (2010). The flexible mind is associated with the catechol-*O*-methyltransferase (COMT) Val158Met polymorphism: Evidence for a role of dopamine in the control of task-switching. *Neuropsychologia, 48*(9), 2764–2768. doi:10.1016/j.neuropsychologia.2010.04.023

Cools, R., & D'Esposito, M. (2011). Inverted-U-shaped dopamine actions on human working memory and cognitive control. *Biological Psychiatry, 69*(12), e113–125. doi:10.1016/j.biopsych.2011.03.028

Cowan, N. (1995). *Attention and memory: An integrated framework.* New York, NY: Oxford University Press.

Dahlin, E., Neely, A. S., Larsson, A., Bäckman, L., & Nyberg, L. (2008). Transfer of learning after updating training mediated by the striatum. *Science, 320*(5882), 1510–1512. doi:10.1126/science.1155466

D'Ardenne, K., Eshel, N., Luka, J., Lenartowicz, A., Nystrom, L. E., & Cohen, J. D. (2012). Role of prefrontal cortex and the midbrain dopamine system in working memory updating. *Proceedings of the National Academy of Sciences, 109*(49), 19900–19909. doi:10.1073/pnas.1116727109

Dorbath, L., Hasselhorn, M., & Titz, C. (2011). Aging and executive functioning: A training study on focus switching. *Frontiers in Psychology, 2,* Article 257. doi:10.3389/fpsyg.2011.00257

Durstewitz, D., & Seamans, J. K. (2008). The dual-state theory of prefrontal cortex dopamine function with relevance to catechol-O-methyltransferase genotypes and schizophrenia. *Biological Psychiatry, 64*(9), 739–749. doi:10.1016/j.biopsych.2008.05.015

Ecker, U. K. H., Lewandowsky, S., & Oberauer, K. (2014). Removal of information from working memory: A specific updating process. *Journal of Memory and Language, 74,* 77–90. doi:10.1016/j.jml.2013.09.003

Ecker, U. K. H., Lewandowsky, S., Oberauer, K., & Chee, A. E. H. (2010). The components of working memory updating: An experimental decomposition and individual differences. *Journal of Experimental Psychology: Learning, Memory, and Cognition, 36*(1), 170–189. doi:10.1037/a0017891

Fields, R. D. (2008). White matter in learning, cognition, and psychiatric disorders. *Trends in Cognitive Psychology, 31*(7), 361–370. doi:10.1016/j.tins.2008.04.001

Frank, M. J., Loughry, B., & O'Reilly, R. C. (2001). Interactions between frontal cortex and basal ganglia in working memory: A computational model. *Cognitive and Affective Behavioral Neuroscience, 1*(2), 137–160. doi:10.3758/CABN.1.2.137

Hertzog, C., Lövdén, M., Lindenberger, U., & Schmiedek, F. (2017). Age differences in coupling of intraindividual variability in mnemonic strategies and practice-related associative recall improvements. *Psychology and Aging, 32*(6), 557–571. doi:10.1037/pag0000177

Jäger, A. O., Süß, H.-M., & Beauducel, A. (1997). *Berliner Intelligenzstruktur-Test, BIS-Test. Form 4. Handanweisung.* Göttingen, Germany: Hogrefe.

Lawley, D. N. (1943). A note on Karl Pearson's selection formulae. *Proceedings of the Royal Society of Edinburgh, 62*(1), 28–30. doi:10.1017/S0080454100006385

Li, S.-C., Schmiedek, F., Huxhold, O., Röcke, C., Smith, J., & Lindenberger, U. (2008). Working memory plasticity in old age: Practice gain, transfer, and maintenance. *Psychology and Aging, 23*(4), 731–742. doi:10.1037/a0014343

Little, T. D., Lindenberger, U., & Nesselroade, J. R. (1999). On selecting indicators for multivariate measurement and modeling with latent variables: When "good" indicators are bad and "bad" indicators are good. *Psychological Methods, 4*(2), 192–211. doi:10.1037//1082-989X.4.2.192

Lövdén, M., Bäckman, L., Lindenberger, U., Schaefer, S., & Schmiedek, F. (2010). A theoretical framework for the study of adult cognitive plasticity. *Psychological Bulletin, 136*(4), 659–676. doi:10.1037/a0020080

Lövdén, M., Bodammer, N. C., Kühn, S., Kaufmann, J., Schütze, H., Tempelmann, C., . . . Lindenberger, U. (2010). Experience-dependent plasticity of white-matter microstructure extends into old age. *Neuropsychologia, 48*(13), 3878–3883. doi:10.1016/j.neuropsychologia.2010.08.026

Luck, S. J., & Vogel, E. K. (2013). Visual working memory capacity: From psychophysics and neurobiology to individual differences. *Trends in Cognitive Sciences, 17*(8), 391–400. doi:10.1016/j.tics.2013.06.006

McArdle, J. J. (2009). Latent variable modelling of differences and changes with longitudinal data. *Annual Review of Psychology, 60,* 577–605. doi:10.1146/annurev.psych.60.110707.163612

McArdle, J. J., and Prindle, J. J. (2008). A latent change score analysis of a randomized clinical trial in reasoning training. *Psychology and Aging, 23*(4), 702–719. doi:10.1037/a0014349

Miyake, A., & Friedman, N. P. (2012). The nature and organization of individual differences in executive functions: Four general conclusions. *Current Directions in Psychological Science, 21*(1), 8–14. doi:10.1177/0963721411429458

Miyake, A., Friedman, N. P., Emerson, M. J., Witzki, A. H., Howerter, A., & Wager, T. D. (2000). The unity and diversity of executive functions and their contributions to complex "frontal lobe" tasks: A latent variable analysis. *Cognitive Psychology, 41*(1), 49–100. doi:10.1006/cogp.1999.0734

Noack, H., Lövdén, M., & Schmiedek, F. (2014). On the validity and generality of transfer effects in cognitive training research. *Psychological Research, 78*(6), 773–789. doi:10.1007/s00426-014-0564-6

Noack, H., Lövdén, M., Schmiedek, F., & Lindenberger, U. (2009). Cognitive plasticity in adulthood and old age: Gauging the generality of cognitive intervention effects. *Restorative Neurology and Neuroscience, 27*(5), 435–453. doi:10.3233/RNN-2009-0496

Oberauer, K. (2002). Access to information in working memory: Exploring the focus of attention. *Journal of Experimental Psychology: Learning, Memory, and Cognition, 28*(3), 411–421. doi:10.1037//0278-7393.28.3.411

Oberauer, K. (2005). Binding and inhibition in working memory: Individual and age differences in short-term recognition. *Journal of Experimental Psychology: General, 134*(3), 368–387. doi:10.1037/0096-3445.134.3.368

Oberauer, K. (2006). Is the focus of attention in working memory expanded through practice? *Journal of Experimental Psychology: Learning, Memory & Cognition, 32*(2), 197–214.

Oberauer, K., & Bialkova, S. (2009). Accessing information in working memory: Can the focus of attention grasp two elements at the same time? *Journal of Experimental Psychology: General, 138*(1), 64–87. doi:10.1037/0278-7393.32.2.197

Pearson, K. (1903). Mathematical contributions to the theory of evolution: XI. On the influence of natural selection on the variability and correlation of organs. *Philosophical Transactions of the Royal Society of London (Series A), 200*, 1–66. doi:10.1098/rsta.1903.0001

Rode, C., Robson, R., Purviance, A., Geary, D. C., & Mayr, U. (2014). Is working memory training effective? A study in school setting. *PLOS One, 9*, e104796. doi:10.1371/journal.pone.0104796

Schmiedek, F., Hildebrandt, A., Lövdén, M., Wilhelm, O., & Lindenberger, U. (2009). Complex span versus updating tasks of working memory: The gap is not that deep. *Journal of Experimental Psychology: Learning, Memory, and Cognition, 35*(4), 1089–1096. doi:10.1037/a0015730

Schmiedek, F., Lövdén, M., & Lindenberger, U. (2010). Hundred days of cognitive training enhance broad cognitive abilities in adulthood: Findings from the COGITO Study. *Frontiers in Aging Neuroscience, 2*, 1–10. doi:10.3389/fnagi.2010.00027

Schmiedek, F., Lövdén, M., & Lindenberger, U. (2013). Keeping it steady: Older adults perform more consistently on cognitive tasks than younger adults. *Psychological Science, 24*(9), 1747–1754. doi:10.1177/0956797613479611

Schmiedek, F., Lövdén, M., & Lindenberger, U. (2014a). Younger adults show long-term effects of cognitive training on broad cognitive abilities over two years. *Developmental Psychology, 50*(9), 2304–2310. doi:10.1037/a0037388

Schmiedek, F., Lövdén, M., & Lindenberger, U. (2014b). A task is a task is a task: Putting complex span, N-back, and other working memory paradigms in psychometric context. *Frontiers in Psychology, 5*, Article 1475. doi:10.3389/fpsyg.2014.01475

Shing, Y.-L., Schmiedek, F., Lövdén, M., & Lindenberger, U. (2012). Memory updating practice across 100 days in the COGITO Study. *Psychology and Aging, 27*(2), 451–461. doi:10.1037/a0025568

Shipstead, Z., Lindsey, D. R. B., Marshall, R. L., & Engle, R. W. (2014). The mechanisms of working memory capacity: Primary memory, secondary memory, and attention control. *Journal of Memory and Language, 72*(1), 116–141. doi:10.1016/j.jml.2014.01.004

Sreenivasan, K. K., Curtis, C. E., & D'Esposito, M. (2014). Revisiting the role of persistent neural activity during working memory. *Trends in Cognitive Science, 18*(2), 82–89. doi:10.1016/j.tics.2013.12.001

Treisman, A. M., & Gelade, G. (1980). A feature-integration theory of attention. *Cognitive Psychology, 12*(1), 97–136. doi:10.1016/0010-0285(80)90005-5

Unsworth, N., & Engle, R. W. (2007). The nature of individual differences in working memory capacity: Active maintenance in primary memory and controlled search from secondary memory. *Psychological Review, 114*(1), 104–132. doi:10.1037/0033-295X.114.1.104

Unsworth, N., Fukuda, K., Awh, E., & Vogel, E. K. (2014). Working memory and fluid intelligence: Capacity, attention control, and secondary memory. *Cognitive Psychology, 71*, 1–26. doi:10.1016/j.cogpsych.2014.01.003

Unsworth, N., Fukuda, K., Awh, E., & Vogel, E. K. (2015). Working memory delay activity predicts individual differences in cognitive abilities. *Journal of Cognitive Neurosciences, 27*(5), 853–865. doi:10.1162/jocn_a_00765

Verhaeghen, P., Cerella, J., & Basak, C. (2004). A working memory workout: How to change to size of the focus of attention from one to four in ten hours or less. *Journal of Experimental Psychology: Learning, Memory, and Cognition, 30*, 1322–1337. doi:10.1037/0278-7393.30.6.1322

von Bastian, C. C., & Oberauer, K. (2013). Distinct transfer effects of training different facets of working memory capacity. *Journal of Memory and Language, 69*(1), 36–58. doi:10.1016/j.jml.2013.02.002

von Bastian, C. C., & Oberauer, K. (2014). Effects and mechanisms of working memory training: A review. *Psychological Research, 78*(6), 803–820. doi:10.1007/s00426-013-0524-6

Wang, X.-J. (2010). Neurophysiological and computational principles of cortical rhythms in cognition. *Physiological Reviews, 90*(3), 1195–1268. doi:10.1152/physrev.00035.2008

Wilhelm, O., Hildebrandt, A., & Oberauer, K. (2013). What is working memory capacity, and how can we measure it? *Frontiers in Psychology, 4*, Article 433. doi:10.3389/fpsyg.2013.00433

Zatorre, R. J., Fields, R. D., & Johansen-Berg, H. (2012). Plasticity in grey and white: Neuroimaging changes in brain structure during learning. *Nature Neuroscience, 15*(4), 528–536. doi:10.1038/nn.3045

Zimmer, H. D., Mecklinger, A., & Lindenberger, U. (2006). Levels of binding: Types, mechanisms and functions of binding in remembering. In H. D. Zimmer, A. Mecklinger, & U. Lindenberger (Eds.), *Handbook of binding and memory: Perspectives from cognitive neuroscience* (pp. 3–24). Oxford, U.K.: Oxford University Press.

4

How Strong Is the Evidence for the Effectiveness of Working Memory Training?

Claudia C. von Bastian, Sabrina Guye, and Carla De Simoni

Introduction

The potential of working memory (WM) training to enhance cognitive abilities has been attracting strong academic and public interest. The effectiveness of WM training in terms of producing transfer to untrained WM tasks and to other related cognitive abilities, such as fluid intelligence, however, is still under debate, with several studies finding null effects and evoking growing skepticism. This chapter argues that the question of whether WM training can induce cognitive plasticity in terms of transfer effects is not a simple yes or no question and it cannot be conclusively answered due to persisting methodological issues across the literature. In particular, the chapter discusses the lack of theoretically motivated selection of training and transfer tasks, the lack of active control groups, and small sample sizes. Consequently, despite the rapidly growing number of training studies, the evidence produced is still inconclusive and favors neither the null nor the alternative hypothesis, as we demonstrate using Bayes factors (BFs). There is, therefore, a need for solidly designed studies based on theory-driven models of transfer tested in larger samples and compared to active controls. Finally, research materials and protocols should be shared and made more easily accessible for the research community in order to improve collaboration and the reproducibility of training research.

Question 1: Theory of Working Memory Training

WM has been established as a strong predictor of fluid intelligence (Conway, Kane, & Engle, 2003; Engle, Tuholski, Laughlin, & Conway, 1999; Kyllonen & Christal, 1990; Oberauer, Süß, Wilhelm, & Wittmann, 2008; Süß, Oberauer, Wittmann, Wilhelm, & Schulze, 2002) and is involved in a wide range of other cognitive abilities, such as reading comprehension (Barrett, Tugade, & Engle, 2004). Therefore, WM has become a popular target of interventions aimed at inducing generalized improvements in cognitive performance through extensive practice. Early process-based training studies in which participants repetitively practiced WM tasks indeed yielded promising transfer effects to fluid intelligence measures (Jaeggi, Buschkuehl, Jonides, & Perrig, 2008; Klingberg et al., 2005; Klingberg, Forssberg, & Westerberg, 2002). These seminal WM training interventions were intriguing because of their potential to enhance intelligence, to counteract cognitive decline in aging, and to improve WM impairments accompanying neurological disorders. Moreover, training interventions as a means of directly manipulating WM capacity would allow a deeper theoretical and causal (instead of a merely correlational) understanding of WM's relationship with fluid intelligence, as it becomes possible to investigate the relationship experimentally. From this theoretical perspective, it is particularly interesting to examine what specific WM mechanisms could underlie the transfer effects observed in some WM training studies. In turn, identifying the mechanisms that are most malleable allows for designing maximally effective interventions for applied contexts.

Generally, training-induced cognitive improvements can be caused by either an increase in capacity through structural changes or by a more efficient use of the capacity available (for a more detailed discussion, see von Bastian & Oberauer, 2014). Expanded capacity should affect performance in any task that draws on the capacity limit and, therefore, should result in improvements across a wide range of tasks. Conversely, enhanced efficiency can cause both narrow and relatively broad transfer effects. For example, whereas strategies developed during training may be useful if the training and transfer task have the same structure or material, they are difficult to apply in other contexts, and thus yield no transfer, or only near transfer to similar tasks. In contrast, however, enhancing efficiency through speeding up specific processes may result in transfer effects across a range of tasks that share these basic processes. Distinguishing improvements that are due to expanded capacity from improvements that are due to increased efficiency requires a WM model that differentiates both aspects. Therefore, our current work is based on the three-embedded-components model of WM (Oberauer, 2009; Oberauer & Hein, 2012).

Three-Embedded-Components Model of WM

Based on Cowan's WM model (Cowan, 1995), the three-embedded-components model (Oberauer, 2009; Oberauer & Hein, 2012) differentiates three functional levels of information selection: the activated part of long-term memory (aLTM), the region of direct access (RDA), and the focus of attention (FA). The aLTM provides all information that is relevant for a current task. Contents of long-term memory are activated through perceptual input or the spread of activation of other WM representations. For example, for solving an arithmetic problem (e.g., 3 + 5 = 8), digits and math operators are activated in long-term memory. Of the activated representations, a subset of approximately four items, or chunks, is held in the capacity-limited RDA, where they are temporarily bound into new structures. In regard to the arithmetic problem above, this means that the elements of the equation are bound to their roles (e.g., 3 is the first operand, + is the operator). The RDA is comparable to the FA in Cowan's model (1995). The FA in Oberauer's model, in contrast, contains just the item (or chunk) held in the RDA that is currently processed (e.g., the digit 3 of the equation). It is accessed through the binding to its context (e.g., first operand), which serves as a retrieval cue.

Using this model, it is possible to evaluate training and transfer effects in regard to the mechanisms underlying training-induced changes in WM performance. Improvements in WM driven by expanded capacity (i.e., an increased number of bindings that can be maintained at any given time) would be reflected by broad transfer effects to other cognitive abilities that also draw on the ability to temporarily bind elements to a cognitive coordinate system (e.g., deductive and inductive reasoning; Oberauer, 2009). Better WM performance could, however, also be caused by more efficient information processing. For example, practice could yield reduced time costs for switching between items in WM (i.e., faster focus shifts; Oberauer, 2006; Verhaeghen, Cerella, & Basak, 2004) or faster removal of no longer relevant information from WM (Ecker, Lewandowsky, & Oberauer, 2014).

Question 2: Major Claims of Working Memory Training

The existing literature (including our own studies) does not yield sufficiently convincing evidence to support the claim that training induces changes in WM capacity, as cognitive performance is rarely affected that broadly (see also Question 3 below). However, given the relatively consistent transfer effects we observed on other, nontrained WM measures (Licini, 2014; von Bastian & Eschen, 2016; von Bastian, Langer, Jäncke, & Oberauer, 2013; von Bastian & Oberauer, 2013), it appears that training may induce changes in WM efficiency. More specifically, despite the lack of consistent far-transfer effects, we frequently found that WM training affected performance in tasks demanding item-context binding (von Bastian, Langer, et al., 2013; von Bastian & Oberauer, 2013). Therefore, our current and future work is directed toward examining training effects on the efficiency of specific processes and exploring whether such changes affect other cognitive performance despite the lack of changes in capacity.

Question 3: Methodological Issues

Despite the extensive discussion of methodological issues in the past (Bogg & Lasecki, 2015; Moody, 2009; Shipstead, Redick, & Engle, 2012), multiple methodological issues still persist in the training literature. The three most pressing issues from our perspective are the lack of theory-driven selection of multiple indicators of cognitive abilities, the lack of adequate control groups, and small sample sizes (see also Guye, Röcke, Mérillat, von Bastian, & Martin, 2016).

Measuring Cognitive Abilities, Not Task-Specific Performance

As discussed in more detail by Hicks and Engle (see Chapter 1), the only way to avoid misinterpreting task-specific performance changes as ability-related improvements is to measure the theoretical constructs of interest with multiple indicators. To tackle this issue in our work, we made it a standard practice to administer multiple tasks to measure an ability selected on theoretical grounds. Using multiple indicators, though, has the inconvenient effect of lengthening test batteries and, as a result, the testing time. Therefore, the application of large test batteries has been criticized on the ground that the extended testing can negatively affect performance due to ego-depletion and/or fatigue, especially toward the end of the battery (Green, Strobach, & Schubert, 2014), thereby obscuring potential transfer effects. To directly address this issue, De Simoni, Luethi, Oberauer, & von Bastian (n.d.) reanalyzed the data from four studies (Herkert, 2011; von Bastian & Eschen, 2016; von Bastian & Oberauer, 2013; von Bastian, Souza, & Gade, 2016) in which participants were engaged in various challenging tasks over a period of 2.5 to 4.5 hours. In each of the included studies, half of the participants completed the battery in reverse order, thereby allowing for comparisons of cognitive performance in the beginning of testing to performance toward the end of the battery. Only one out of 58 task comparisons showed evidence in favor of depletion effects, with comparable performance even in the most extreme cases where tasks were completed either first or last in the test battery. Therefore, there was no evidence of negative consequences of the extended testing time.

Designing an Active Control Intervention

As argued by Hicks and Engle in Chapter 1, confounds between nonspecific intervention effects and true transfer effects can be avoided only by using active control groups (see also Dougherty, Hamovitz, & Tidwell, 2016). Thus, from a methodological standpoint, it is surprising that still only about half of the published training studies compare the effects of their intervention to an active control group (estimate based on recent meta-analyses, such as those from Au et al., 2015; Karbach & Verhaeghen, 2014; Lampit, Hallock, & Valenzuela, 2014). Critically, this is a methodological, not an empirical question. In other words, statistically similar meta-analytic effect-size estimates for studies employing active control groups and those using passive control groups do not imply that no confounding nonspecific intervention effects are at work in any single study. Hence, to convincingly argue that an intervention improved the cognitive ability targeted, the ideal control group should differ only in the ability or process that is being trained (see also von Bastian & Oberauer, 2014). More specifically, the optimal control group should complete an alternative intervention that is (1) as believable in regard to its potential to improve cognitive performance (including the variability and/or adaptiveness of difficulty), (2) as enjoyable (including providing similarly encouraging feedback), (3) as intensive, and (4) administered in the same context (e.g., computer-based vs. pen-and-paper, at home vs. in the lab) as the experimental group's intervention.

Obviously, designing such an optimal control intervention is challenging and sometimes requires prioritizing one methodological aspect over another (e.g., keeping stimulus material rather than adaptiveness constant across groups; Zimmermann, von Bastian, Martin, Röcke, & Eschen, 2016). For example, in our own work, we first used perceptual matching as a control intervention (von Bastian & Oberauer, 2013), for which we found large improvements in processing speed—an important component of WM performance (Schmiedek, Oberauer, Wilhelm, Süß, & Wittmann, 2007). To avoid underestimating effects of WM training, we moved toward employing nonspeeded tasks, such as trivia quizzes (von Bastian & Eschen, 2016) and visual search tasks (von Bastian, Langer, et al., 2013). Because trivia quizzes are at risk to be more enjoyable than WM tasks, visual search tasks currently seem to be the best available option. Visual search is known to be uncorrelated to WM capacity across a variety of different tasks (Kane, Poole, Tuholski, & Engle, 2006).

A problem that cannot be completely solved through the use of an active control group is expectancy effects (Oken et al., 2008), since, strictly speaking, cognitive training participants cannot be blinded completely to the condition they are allocated to. Based on their exposure to the training tasks, participants from the experimental and control condition might form different expectations regarding the scope of generalization of the cognitive improvements to other, probably similar, tasks. For example, participants undergoing WM training might expect improvements in other, similar tasks in the test battery, whereas visual search training may not lead to such expectations, thereby potentially yielding confounds between effects caused by the intervention and those caused by differing beliefs. To minimize differing beliefs between groups, we therefore administered test versions of both the experimental tasks (i.e., criterion tasks) and the active control training tasks (von Bastian & Eschen, 2016; von Bastian, Langer, et al., 2013).

Evaluating Evidence in Light of Small Sample Sizes

A notoriously persistent methodological issue in training research is low statistical power due to small sample sizes. For example, in the meta-analysis by Au et al. (2015), the treatment group included in the N-back training studies had on average only 20 (± 8) participants. Similarly, the median treatment group size across the 52 training studies included in the meta-analysis by Lampit, Hallock, and Valenzuela (2014) is only 22. Critically, low statistical power increases the likelihood of false-negative, false-positive, and inflated effects, thereby posing major problems for interpreting meta-analytic results (cf. Bogg & Lasecki, 2015). We discuss each of these issues before suggesting the use of BFs as an alternative approach for evaluating the evidence of the current literature.

First, low statistical power reduces the chance of discovering genuine effects (Hartgerink, Wicherts, & van Assen, 2017). Hence, meta-analyses reporting the absence of evidence for the effectiveness of WM training are possibly distorted by the majority of studies' not being sufficiently powered to detect small but true effects. For example, even for a medium-sized true effect (Cohen's $\delta = 0.50$), statistical power for the typical group size of $n = 22$ in the training literature (Lampit et al., 2014) is only 0.38 (i.e., 38 out of 100 tests can theoretically be expected to yield significant effects). With the ESCI Excel macro, Cumming (2011) provided a great tool for visualizing the relationship between statistical power and the likelihood of statistically significant results through simulating replications of the same experiment. Using ESCI, we ran a series of 25 simulations of the same experiment assuming Cohen's $\delta = 0.50$ and group $n = 22$ (for an example with slightly more statistical power, see Dienes, 2014). Figure 4.1 illustrates what Cumming labeled the "dance of the p values": the p values for each of the 25 experiments range from $p < .001$ to $p = .997$, and only six of them were smaller than 0.05—notable for a medium-sized true effect that is most likely larger than what would be expected for transfer effects.

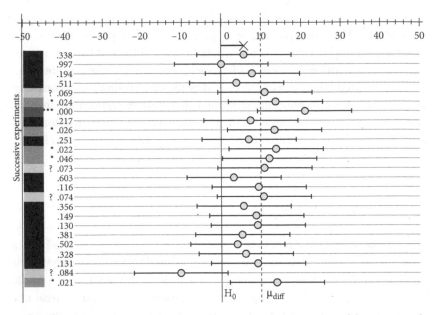

Figure 4.1. "Dance" of the *p* values. Shown are the *p* values and confidence intervals for 25 successive simulations of an experiment for which there is a true underlying effect (Cohen's δ = 0.50, group *n* = 22, and power = 0.38). Simulations and the figure were generated with the ESCI software provided by Cumming (2011).

Second, as laid out in detail by Button et al. (2013), low power also increases the risk of statistically significant results' being false-positives (i.e., the null hypothesis was rejected incorrectly). The probability of a statistically significant result's reflecting a true effect can be expressed in terms of the positive predictive value (PPV; Ioannidis, 2005):

$$PPV = \frac{(1-\beta) \times R}{(1-\beta) \times R + \alpha}$$

where $(1 - \beta)$ is the statistical power, R is the prestudy odds of the effect probed being truly non-null, and α is the type I error. Translated to WM training research, the PPV depends on how likely it is that a given transfer effect is a true effect as well as on the statistical power to detect this effect. Sticking to the above example, consider the case where statistical power is 0.38 (Cohen's δ = 0.50 and group *n* = 22). We cannot determine the prestudy odds *R*, but for the sake of this example, assume that one out of four training regimens can truly produce transfer effects. With $R = 0.25$, the resulting PPV is

$$PPV = \frac{0.38 \times 0.25}{0.38 \times 0.25 + 0.05} = 0.66.$$

In this example, where power is 0.38, this means that out of all significant transfer effects detected, only 66% are true-positives, with the other 34% consequently being false-positives. Now, suppose we increase the group sample size to 62 to gain a power of 0.80. In this case, the PPV would increase to

$$PPV = \frac{0.80 \times 0.25}{0.80 \times 0.25 + 0.05} = 0.80$$

thereby reducing the false-positive rate from 34% to 20%. Notably, in this set of examples, the prestudy odds are relatively optimistic, and more conservative values would yield even higher false-positive rates. Thus, given the low statistical power in WM training research, a considerable proportion of discovered effects can be expected to be false-positives.

A third problem of small sample sizes is their proneness to produce inflated effect sizes (Ioannidis, 2008). This is particularly problematic because authors of studies with small samples but significant findings often argue that statistical power must have been sufficient given the considerable effect size found. In contrast to this fairly common assumption, however, effect-size estimates can vary excessively and in particular when power is low. Halsey, Curran-Everett, Vowler, and Drummond (2015) impressively illustrated the relationship between effect-size estimates and sample size with a series of simulations. Their artificial data set comprised two populations, A and B, which differed with a true effect size of 0.50. From this data set, 1,000 random samples were drawn with varying ns (10, 30, 64, or 100 participants per group) and were statistically tested for group differences. For $n = 30$ (yielding a power of 0.48), 97% of effect sizes were overestimated, with significant effects ranging from 0.44 to 1.23. Even when power was very high (0.94 for $n = 100$), 55% of all effect-size estimates were still larger than the true effect and ranged from 0.28 to 1.07. Therefore, because of the low statistical power in training studies, most effect sizes can be expected to be overestimated, heavily distorting the results of meta-analyses.

In summary, small Ns not only are risky in terms of type I and type II errors, but also often produce overestimated effect sizes. Given the high prevalence of small Ns in the current body of WM training literature, neither p values nor effect sizes are sufficiently informative to estimate the effectiveness of WM training. We therefore suggest using BFs as an alternative approach for evaluating the evidence.

An Alternative Approach for Evaluating Evidence: The BF

The BF quantifies the strength of evidence for a hypothesis relative to another hypothesis on a continuous scale. More specifically, the BF is the ratio of the probability of the data under one (typically the alternative) hypothesis to the probability of the data under another (typically the null) hypothesis (Jeffreys, 1961; Kass & Raftery, 1995). Critically, the hypotheses in the numerator and denominator can be swapped. Hence, in contrast to frequentist statistics where a nonsignificant p value can only indicate the absence of evidence, the BF allows for evaluating the strength of evidence in favor of either the alternative or the null hypothesis (for a detailed discussion, see Dienes, 2014). Furthermore, because the BF is a probability ratio, it varies on a truly continuous scale. In contrast to p values, though, BFs are less variable when power is low (Dienes, 2014) and are therefore better suited to evaluate the evidence from small-N studies. BFs can range from 0 to ∞, with the absolute BF value reflecting how many times more likely the data are under the hypothesis in the numerator relative to the hypothesis in the denominator on a continuous scale. A value of 1 reflects perfect ambiguity, as the probability of the data is exactly the same under one or the other hypothesis. BFs larger than 1 reflect evidence favoring the hypothesis in the numerator, and, conversely, the closer the BF is to 0, the stronger the evidence in favor of the hypothesis in the denominator. For example, when testing the hypothesis that two groups differ against the null hypothesis, a BF of 10 reflects that the data are 10 times more likely under the hypothesis that the difference exists than under the null hypothesis. As guidance, Wetzels et al. (2011) suggested a set of verbal categories for interpreting the size of BFs (see Table 4.1).

The use of verbal labels has been criticized because it encourages interpreting BFs as categorical thresholds rather than as the continuous strength of evidence

Table 4.1. *Suggested Verbal Descriptions for Interpreting Bayes Factors*

Bayes Factor	Verbal Description
1 to 3	Ambiguous
3 to 10	Substantial
10 to 30	Strong
30 to 100	Very strong
>100	Decisive

Note. Adapted from Wetzels et al. (2011).

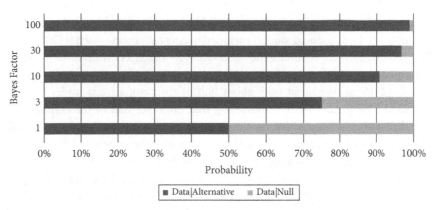

Figure 4.2. Bayes factors (BFs) from Table 4.1 translated into probabilities of the data under the alternative hypothesis compared to the null hypothesis to visualize the continuity of the BF scale. For example, a BF of 100 translates into a probability greater than 99% for the data under the alternative hypothesis relative to the null hypothesis; hence, the evidence clearly favors the alternative hypothesis (e.g., differences between two groups). The smaller the BF, the less clearly the evidence supports either hypothesis. For example, a BF of 3—labeled as substantial evidence—translates into a 75% probability for the favored hypothesis. On the extreme end, a BF of 1 translates into a 50/50 probability of the data under both hypotheses. Thus, tossing a coin would be as informative in terms of evidence for a hypothesized effect as a statistical result that yielded a BF of 1.

(Morey, 2015). Translating the BF into probabilities puts the verbal categories back into their continuous context (see Figure 4.2).

Importantly, what is considered to be a decisive BF generally depends on the standards of evidence in a given field of research and the context of a finding (Kass & Raftery, 1995). Some claims may require stronger evidence than others to be perceived as convincing; what is strong evidence in one case may not be sufficient in another (for a detailed discussion, see Morey, 2015). For example, in the context of training research, stronger evidence may be required for a claim to be convincing when a training regimen was evaluated only in comparison to a passive control group.

How Strong Is the Evidence in the Case of WM Training?

As of yet, only a few training studies have reported BFs (for some exceptions, see De Simoni & von Bastian, 2018; Guye, De Simoni, & von Bastian, 2017; Guye & von Bastian, 2017; Sprenger et al., 2013; von Bastian & Oberauer, 2013).

Recently, Dougherty, Hamovitz, and Tidwell (2016) reevaluated the evidence from the N-back training studies included in Au et al.'s meta-analysis (2015) with Bayesian analyses. Whereas the meta-analysis based on traditional methods resulted in a small but significant training effect on fluid intelligence, Dougherty et al. (2016) showed that only about half of the comparisons under scrutiny yielded a BF greater than 3, indicating that the other half of the comparisons were insufficiently sensitive to clearly support either hypothesis—potentially due to a lack of power. Indeed, another reanalysis of the same set of 20 studies by Bogg and Lasecki (2015) supports this proposition.

As Au et al.'s meta-analysis (2015) focused on transfer effects of N-back training in young adults, it is possible that the evidence is stronger for other cognitive training regimens or in older adults. Therefore, we evaluated the strength of evidence produced by the training studies included in the meta-analysis by Lampit et al. (2014). In their meta-analysis, Lampit et al. (2014) included 52 randomized-controlled trials using various cognitive training approaches, 28 of which measured transfer to WM tasks (e.g., digit span backward, operation span, or updating), and 29 of which measured transfer to executive functions and reasoning tasks (e.g., Stroop, shifting, or matrix reasoning). To compute BFs, we first derived the t-values from the effect-size estimates (Hedge's g) provided by Lampit et al. (2014) with the following formula:

$$t = g \times \frac{\sqrt{n1 \times n2}}{\sqrt{n1 + n2}}$$

where g is Hedge's g, and $n1$ and $n2$ are the sample sizes of the two training groups. BFs were then computed using the BF package (Morey, Rouder, & Jamil, 2015; see also Rouder, Speckman, Sun, Morey, & Iverson, 2009) for R (R Core Team, 2014) with a medium prior scale ($r = \sqrt{2} / 2$). The results are presented in Figure 4.3. Note that BFs favoring the null (i.e., BFs < 1) are expressed as 1/BF to facilitate their interpretation. BFs greater than 3 are shown in darker shades of gray (BF printed on the left end are in favor of the alternative hypothesis and BF printed on the right end are in favor of the null hypothesis). To be consistent with Dougherty et al. (2016), studies are presented separately for those with active and those with passive controls.

As illustrated by the left panel in Figure 4.3, the majority of studies produced only weak evidence for either hypothesis. Notably, 14 out of the 19 studies resulting in BFs smaller than 3 featured group sizes smaller than $n = 22$. In contrast, eight out of the nine studies that produced interpretable evidence featured larger samples, with the two studies with BFs in a range of what Wetzels et al. (2011) refer to as very strong evidence including more than 30 participants per group (Buschkuehl et al., 2008; Shatil, Mikulecká, Bellotti, & Bureš, 2014). Those

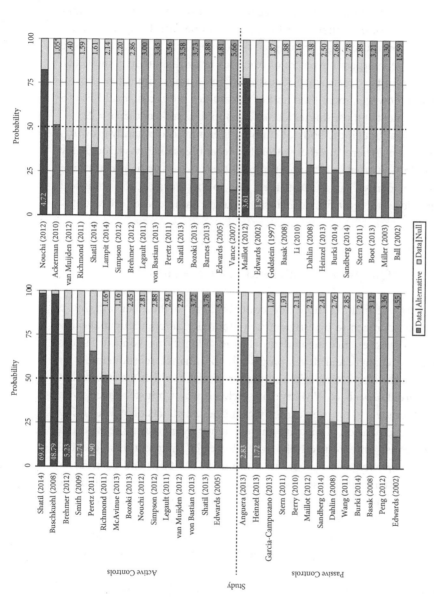

Figure 4.3. Strength of evidence expressed in Bayes factors (BFs) for transfer to working memory (left) and executive functions and reasoning (right) as a function of the control group employed in the studies included in Lampit et al. (2014). Numbers on bars depict BFs in favor of the alternative hypothesis (left) or null hypothesis (right), with darker shades of gray for BFs greater than 3. BFs followed by asterisks denote evidence favoring the alternative hypothesis where the control group outperformed the experimental group. Only a few of the included studies produced evidence that clearly supports either hypothesis (i.e., BFs greater than 3), thereby reducing the overall interpretability of the empirical basis for meta-analytic effect-size estimates.

BFs of an interpretable size appear to favor the existence of transfer effects of training to WM tasks, particularly when focusing on studies including active control groups.

Regarding transfer to measures of executive functions and reasoning (right panel in Figure 4.3), the pattern of BFs is more in line with Dougherty et al.'s findings for N-back training in young adults (Dougherty et al., 2016). More specifically, most interpretable pieces of evidence favor the null hypothesis (11 out 13 studies), in particular when focusing on the studies with active control groups. Again, however, more than half of the studies produced small BFs (16 out of 29), and most of those studies featured group sizes smaller than $n = 22$.

In summary, although the body of training literature is constantly growing, the results presented here, together with other recent reevaluations (Dougherty et al., 2016), indicate that only a subset of published studies actually contribute interpretable evidence. This is most likely due to the prevalent low statistical power. Therefore, to draw firm conclusions, larger sample sizes are strongly needed.

Question 4: Contributions to the Field

Our work aims at making three key contributions to the field: offering a theory-driven approach to investigate the potentially differentially malleable mechanisms of experience-dependent change, identifying context factors influencing the change, and developing tools to enhance reproducibility of training research. First, to extend our earlier work in which we found that training different aspects of WM capacity yields distinct transfer effects (von Bastian & Oberauer, 2013), we currently examine the effects of WM training that targets the expansion of capacity and efficiency by increasing set size and reducing response time, respectively. To evaluate the intervention's potential in expanding capacity, near transfer to untrained WM tasks and far transfer to related cognitive abilities, such as reasoning, is assessed. In addition, we measure performance in three candidate WM mechanisms potentially affected by training: resolving interference arising from previously presented items, efficiency of the removal of irrelevant information from WM, and efficiency in switching the focus of attention between items in WM (De Simoni & von Bastian, 2018).

A second focus of our current work is on context factors potentially influencing experience-dependent change. These can roughly be classified into two categories: intervention-specific features and individual differences (cf. von Bastian & Oberauer, 2014). For example, in a recent study on intervention-specific features, we found that training with individually adaptive task difficulty yielded training effects comparable to those induced by training with randomly varied task difficulty (von Bastian & Eschen, 2016). Besides differences in the design of training interventions, individual differences—either in addition to, or independent of, training-induced changes—are likely to contribute to the differences observed after training. Although the role of individual differences in the context of WM training has received some attention in the field (Bürki, Ludwig, Chicherio, & de Ribaupierre, 2014; Studer-Luethi, Jaeggi, Buschkuehl, & Perrig, 2012; Zinke et al., 2014), studies with sufficiently large sample sizes are still scarce. Therefore, in some of our current projects, we aim to assess how interindividual differences (e.g., in motivation and personality) potentially predict training performance (Guye et al., 2017). By assessing individual differences and understanding their influence on training outcomes, we aim at contributing to the development of more effective and individually tailored training interventions.

Our third contribution is the development of Tatool (see www.tatool-web. com; von Bastian, Locher, & Ruflin, 2013), an open-source software package (licensed under GNU GPL v3.0) for programming and running experiments and training studies online and offline. The goal is to make running computer-based experiments and training studies as easy and accessible as possible and to facilitate sharing of materials and procedures. Since its launch in 2015, the Tatool Web community already counts more than 1,000 active researchers, and its continuously growing library features multiple cognitive tasks for reuse by fellow researchers. Given that training research is particularly resource demanding, we hope that Tatool Web helps to foster sharing of materials, international collaboration, and enhanced financial independence.

Conclusion

Training research still suffers from pervasive methodological issues, such as the use of single tasks for measuring cognitive abilities and inadequate control groups. Furthermore, the majority of previous studies are vastly underpowered with small sample sizes, resulting in only weak Bayesian evidence in either direction. In addition, even the more successful training interventions are often based on only vague theoretical frameworks, rendering it difficult to evaluate whether the interventions have expanded capacity or efficiency limits. Therefore, the current body of literature cannot conclusively support claims that WM training does or does not improve cognitive abilities and stresses the need for theory-driven, methodologically sound studies with larger sample sizes.

References

Au, J., Sheehan, E., Tsai, N., Duncan, G. J., Buschkuehl, M., & Jaeggi, S. M. (2015). Improving fluid intelligence with training on working memory: A meta-analysis. *Psychonomic Bulletin & Review, 22*(2), 366–377. doi:10.3758/s13423-014-0699-x

Barrett, L. F., Tugade, M. M., & Engle, R. W. (2004). Individual difference in working memory capacity and dual-process theories of the mind. *Psychological Bulletin, 130*(4), 553–573. doi:10.1037/0033-2909.130.4.553

Bogg, T., & Lasecki, L. (2015). Reliable gains? Evidence for substantially underpowered designs in studies of working memory training transfer to fluid intelligence. *Frontiers in Psychology, 5*(1589), 1–8. doi:10.3389/fpsyg.2014.01589

Bürki, C. N., Ludwig, C., Chicherio, C., & de Ribaupierre, A. (2014). Individual differences in cognitive plasticity: An investigation of training curves in younger and older adults. *Psychological Research, 78*(6), 821–835. doi:10.1007/s00426-014-0559-3

Buschkuehl, M., Jaeggi, S. M., Hutchison, S., Perrig-Chiello, P., Däpp, C., Müller, M., . . . Perrig, W. J. (2008). Impact of working memory training on old-old adults. *Psychology and Aging, 23*(4), 743–753.

Button, K. S., Ioannidis, J. P. A., Mokrysz, C., Nosek, B. A., Flint, J., Robinson, E. S. J., & Munafò, M. R. (2013). Power failure: Why small sample size undermines the reliability of neuroscience. *Nature Reviews Neuroscience, 14*, 365–376. doi:10.1038/nrn3475

Conway, A. R. A., Kane, M. J., & Engle, R. W. (2003). Working memory capacity and its relation to general intelligence. *Trends in Cognitive Sciences, 7*(12), 547–552. doi:10.1016/j.tics.2003.10.005

Cowan, N. (1995). *Attention and memory: An integrated framework.* New York, NY: Oxford University Press.

Cumming, G. (2011). *Understanding the new statistics: Effect sizes, confidence intervals, and meta-analysis.* Abingdon, U.K.: Routledge.

De Simoni, C., Luethi, M. S., Oberauer, K., & von Bastian, C. C. (n.d.). *No consistent evidence for ego-depletion effects across multiple cognitive tasks and domains.* Manuscript submitted for publication.

De Simoni, C., & von Bastian, C. C. (2018). Working memory updating and binding training: Bayesian evidence supporting the absence of transfer. *Journal of Experimental Psychology: General, 147*(6), 829–858. doi:10.1037/xge0000453

Dienes, Z. (2014). Using Bayes to get the most out of non-significant results. *Frontiers in Psychology, 5*(781), 1–17. doi:10.3389/fpsyg.2014.00781

Dougherty, M. R., Hamovitz, T., & Tidwell, J. W. (2016). Reevaluating the effectiveness of n-back training on transfer through the Bayesian lens: Support for the null. *Psychonomic Bulletin & Review, 23*(1), 306–316. doi:10.3758/s13423-015-0865-9

Ecker, U. K. H., Lewandowsky, S., & Oberauer, K. (2014). Removal of information from working memory: A specific updating process. *Journal of Memory and Language, 74*, 77–90. doi:10.1016/j.jml.2013.09.003

Engle, R. W., Tuholski, S. W., Laughlin, J. E., & Conway, A. R. A. (1999). Working memory, short-term memory, and general fluid intelligence: A latent-variable approach. *Journal of Experimental Psychology: General, 128*(3), 309–331. doi:10.1037/0096-3445.128.3.309

Green, S. C., Strobach, T., & Schubert, T. (2014). On methodological standards in training and transfer experiments. *Psychological Research, 78*(6), 756–772. doi:10.1007/s00426-013-0535-3

Guye, S., De Simoni, C., & von Bastian, C. C. (2017). Do individual differences predict change in cognitive training performance? A latent growth curve modeling approach. *Journal of Cognitive Enhancement*, *1*(4), 374–393. doi:10.1007/s41465-017-0049-9

Guye, S., Röcke, C., Mérillat, S., von Bastian, C. C., & Martin, M. (2016). Plasticity in different age groups: Adult lifespan. In T. Strobach & J. Karbach (Eds.), *Cognitive training: An overview of features and applications* (pp. 45–55). Berlin, Germany: Springer.

Guye, S., & von Bastian, C. C. (2017). Working memory training in older adults: Bayesian evidence supporting the absence of transfer. *Psychology and Aging*, *32*(8), 732–746. doi:10.1037/pag0000206

Halsey, L. G., Curran-Everett, D., Vowler, S. L., & Drummond, G. B. (2015). The fickle P value generates irreproducible results. *Nature Methods*, *12*(3), 179–185. doi:10.1038/nmeth.3288

Hartgerink, C. H. J., Wicherts, J. M., & van Assen, M. A. L. M. (2017). Too good to be false: Nonsignificant results revisited. *Collabra: Psychology*, *3*(1), 1–18. doi:10.1525/collabra.71

Herkert, J. (2011). *Trainings und Transfereffekte des Arbeitsgedächtnisses [Training and transfer effects of working memory]*. Unpublished master's thesis, University of Zurich, Zurich, Switzerland.

Ioannidis, J. P. A. (2005). Why most published research findings are false. *PLOS Medicine*, *2*(8), e124. doi:10.1371/journal.pmed.0020124

Ioannidis, J. P. A. (2008). Why most discovered true associations are inflated. *Epidemiology*, *19*(5), 640–648. doi:10.1097/EDE.0b013e31818131e7

Jaeggi, S. M., Buschkuehl, M., Jonides, J., & Perrig, W. J. (2008). Improving fluid intelligence with training on working memory. *Proceedings of the National Academy of Sciences of the United States of America*, *105*(19), 6829–6833. doi:10.1073/pnas.0801268105

Jeffreys, H. (1961). *Theory of probability* (3rd ed.). Oxford, U.K.: Oxford University Press.

Kane, M. J., Poole, B. J., Tuholski, S. W., & Engle, R. W. (2006). Working memory capacity and the top-down control of visual search: Exploring the boundaries of "executive attention." *Journal of Experimental Psychology: Learning, Memory, and Cognition*, *32*(4), 749–777. doi:10.1037/0278-7393.32.4.749

Karbach, J., & Verhaeghen, P. (2014). Making working memory work: A meta-analysis of executive-control and working memory training in older adults. *Psychological Science*, *25*(11), 2027–2037. doi:10.1177/0956797614548725

Kass, R. E., & Raftery, A. E. (1995). Bayes factors. *Journal of the American Statistical Association*, *90*(430), 773–795.

Klingberg, T., Fernell, E., Olesen, P. J., Johnson, M., Gustafsson, P., Dahlström, K., . . . Westerberg, H. (2005). Computerized training of working memory in children with ADHD—A randomized, controlled trial. *Journal of the American Academy of Child & Adolescent Psychiatry*, *44*(2), 177–186. doi:10.1097/00004583-200502000-00010

Klingberg, T., Forssberg, H., & Westerberg, H. (2002). Training of working memory in children with ADHD. *Journal of Clinical and Experimental Neuropsychology*, *24*(6), 781–791. doi:10.1076/jcen.24.6.781.8395

Kyllonen, P. C., & Christal, R. E. (1990). Reasoning ability (is little more than) working-memory capacity?! *Intelligence*, *14*(4), 389–433. doi:10.1016/S0160-2896(05)80012-1

Lampit, A., Hallock, H., & Valenzuela, M. (2014). Computerized cognitive training in cognitively healthy older adults: A systematic review and meta-analysis of effect modifiers. *PLOS Medicine*, *11*(11), e1001756. doi:10.1371/journal.pmed.1001756

Licini, C. (2014). *Verbesserung der Lernfähigkeit durch gezieltes Arbeitsgedächtnistraining [Improvement of the ability learn through working memory training]*. Unpublished master's thesis, University of Zurich, Zurich, Switzerland.

Moody, D. E. (2009). Can intelligence be increased by training on a task of working memory? *Intelligence, 37*(4), 327–328. doi:10.1016/j.intell.2009.04.005

Morey, R. D. (2015). On verbal categories for the interpretation of Bayes factors. *BayesFactor*. Retrieved from http://bayesfactor.blogspot.com/2015/01/on-verbal-categories-for-interpretation.html.

Morey, R. D., Rouder, J. N., & Jamil, T. (2015). Computation of Bayes factors for common designs. *BayesFactor: An R package for Bayesian data analysis*. Retrieved from http://bayesfactorpcl.r-forge.r-project.org.

Oberauer, K. (2006). Is the focus of attention in working memory expanded through practice? *Journal of Experimental Psychology: Learning, Memory, and Cognition, 32*(2), 197–214. doi:10.1037/0278-7393.32.2.197

Oberauer, K. (2009). Design for a working memory. In B. Ross (Ed.), *The psychology of learning and motivation: Advances in research and theory* (Vol. 51, pp. 45–100). New York, NY: Academic Press.

Oberauer, K., & Hein, L. (2012). Attention to information in working memory. *Current Directions in Psychological Science, 21*(3), 164–169. doi:10.1177/0963721412444727

Oberauer, K., Süß, H.-M., Wilhelm, O., & Wittmann, W. W. (2008). Which working memory functions predict intelligence? *Intelligence, 36*(6), 641–652. doi:10.1016/j.intell.2008.01.007

Oken, B. S., Flegal, K., Zajdel, D., Kishiyama, S., Haas, M., & Peters, D. (2008). Expectancy effect: Impact of pill administration on cognitive performance in healthy seniors. *Journal of Clinical and Experimental Neuropsychology, 30*(1), 7–17. doi:10.1080/13803390701775428

R Core Team. (2014). *R: A language and environment for statistical computing. Vienna, Austria: R Foundation for Statistical Computing*. Retrieved from http://www.R-project.org.

Rouder, J. N., Speckman, P. L., Sun, D., Morey, R. D., & Iverson, G. (2009). Bayesian t tests for accepting and rejecting the null hypothesis. *Psychonomic Bulletin & Review, 16*(2), 225–237. doi:10.3758/PBR.16.2.225

Schmiedek, F., Oberauer, K., Wilhelm, O., Süß, H.-M., & Wittmann, W. W. (2007). Individual differences in components of reaction time distributions and their relations to working memory and intelligence. *Journal of Experimental Psychology: General, 136*(3), 414–429. doi:10.1037/0096-3445.136.3.414

Shatil, E., Mikulecká, J., Bellotti, F., & Bureš, V. (2014). Novel television-based cognitive training improves working memory and executive function. *PLoS One, 9*(7), e101472. doi:10.1371/journal.pone.0101472

Shipstead, Z., Redick, T. S., & Engle, R. W. (2012). Is working memory training effective? *Psychological Bulletin, 138*(4), 628–654. doi:10.1037/a0027473

Sprenger, A. M., Atkins, S. M., Bolger, D. J., Harbison, J. I., Novick, J. M., Chrabaszcz, J. S., . . . Dougherty, M. R. (2013). Training working memory: Limits of transfer. *Intelligence, 41*(5), 638–663. doi:10.1016/j.intell.2013.07.013

Studer-Luethi, B., Jaeggi, S. M., Buschkuehl, M., & Perrig, W. J. (2012). Influence of neuroticism and conscientiousness on working memory training outcome. *Personality and Individual Differences, 53*(1), 44–49. doi:10.1016/j.paid.2012.02.012

Süß, H.-M., Oberauer, K., Wittmann, W. W., Wilhelm, O., & Schulze, R. (2002). Working-memory capacity explains reasoning ability—and a little bit more. *Intelligence, 30,* 261–288. doi:10.1016/S0160-2896(01)00100-3

Verhaeghen, P., Cerella, J., & Basak, C. (2004). A working memory workout: How to expand the focus of serial attention from one to four items in 10 hours or less. *Journal of Experimental Psychology: Learning, Memory, and Cognition, 30,* 1322–1337. doi:10.1037/0278-7393.30.6.1322

von Bastian, C. C., & Eschen, A. (2016). Does working memory training have to be adaptive? *Psychological Research, 80*(2), 181–194. doi:10.1007/s00426-015-0655-z

von Bastian, C. C., Langer, N., Jäncke, L., & Oberauer, K. (2013). Effects of working memory training in young and old adults. *Memory & Cognition, 41*(4), 611–624. doi:10.3758/s13421-012-0280-7

von Bastian, C. C., Locher, A., & Ruflin, M. (2013). Tatool: A Java-based open-source programming framework for psychological studies. *Behavior Research Methods, 45*(1), 108–115. doi:10.3758/s13428-012-0224-y

von Bastian, C. C., & Oberauer, K. (2013). Distinct transfer effects of training different facets of working memory capacity. *Journal of Memory and Language, 69,* 36–58. doi:10.1016/j.jml.2013.02.002

von Bastian, C. C., & Oberauer, K. (2014). Effects and mechanisms of working memory training: A review. *Psychological Research, 78*(6), 803–820. doi:10.1007/s00426-013-0524-6

von Bastian, C. C., Souza, A. S., & Gade, M. (2016). No evidence for bilingual cognitive advantages: A test of four hypotheses. *Journal of Experimental Psychology: General, 145*(2), 246–258. doi:10.1037/xge0000120

Wetzels, R., Matzke, D., Lee, M. D., Rouder, J. N., Iverson, G. J., & Wagenmakers, E.-J. (2011). Statistical evidence in experimental psychology: An empirical comparison using 855 t-tests. *Perspectives on Psychological Science, 6*(3), 291–298. doi:10.1177/1745691611406923

Zimmermann, K., von Bastian, C. C., Martin, M., Röcke, C., & Eschen, A. (2016). Transfer effects after process-based object-location memory training in healthy older adults. *Psychology and Aging, 31*(7), 798–814. doi:10.1037/pag0000123

Zinke, K., Zeintl, M., Rose, N. S., Putzmann, J., Pydde, A., & Kliegel, M. (2014). Working memory training and transfer in older adults: Effects of age, baseline performance, and training gains. *Developmental Psychology, 50*(1), 304–315. doi:10.1037/a0032982

SECTION II
NEUROCOGNITIVE PERSPECTIVE

5

Neuroscience Perspectives
on Cognitive Training .

Stefanie E. Kuchinsky and Henk J. Haarmann

Some recent philosophers seem to have given their moral approval
to these deplorable verdicts that affirm that the intelligence of an individual is a fixed quantity, a quantity that cannot be augmented. We
must protest and react against this brutal pessimism; we will try to
demonstrate that it is founded on nothing.

—Alfred Binet,
Les idées modernes sur les enfants (1909, p. 141)

Introduction

More than a hundred years ago, Alfred Binet and his colleagues sought to quantify intelligence to identify children who could benefit from additional or alternative forms of education. Although Binet opined that intellect is at least
partially malleable (Binet, 1909), the debate regarding the extent to which cognitive skills can be improved through training continues today. Indeed, this issue
has important social implications, particularly in the way education and training
interventions are developed for both clinical and nonclinical populations.

Training-based interventions have been investigated across many of the perceptual, cognitive, and other functions that support human behavior. For example,
training protocols to improve speech recognition, which declines with age and
hearing loss, have targeted a range of skills, from frequency discrimination and
temporal processing (for a review, see Wright & Zhang, 2009) to selective attention
and working memory (WM; Sweetow & Henderson-Sabes, 2006). Recent reviews
suggest that the greatest transfer of benefit to daily life arises from training domain-
general cognitive functions within task-specific materials (e.g., training WM using
auditory stimuli; Henshaw & Ferguson, 2015). Executive functions (e.g., WM, selective attention, cognitive control), which broadly serve to regulate mental function in support of diverse behaviors (for a review, see Jurado & Rosselli, 2007), have
thus been the focus of many training studies (Hsu, Novick, & Jaeggi, 2014).

Evaluating the effectiveness of training that aims to improve one or more types of executive function has theoretical and financial consequences even for healthy adults. The availability of "brain-training" applications has risen dramatically in the past decade, with their global markets forecast to reach $6 billion by 2020 (Sharp Brains, 2013). Scientists are hard pressed to evaluate claims of training-related improvements in cognition and "brain health" at this extraordinary pace. Indeed, many of the top cognitive scientists and neuroscientists from around the world signed a consensus statement cautioning scientists and consumers alike that more thorough research is needed to understand the impact training may have on cognitive functioning in everyday life (Stanford Center on Longevity, 2014). Additional researchers, while noting the promise of studies showing neuroplastic changes with training, have similarly expressed that more quality research is needed, particularly to justify claims made by brain-training companies (see, for example, cognitivetrainingdata.org).

This chapter aims to spark a discussion regarding how cognitive neuroscience research can aid in the evaluation and development of effective cognitive training protocols. In particular, we pose questions relating to whether training-related neural plasticity (i.e., changes in brain function and structure in response to experience) could be used to facilitate the identification and targeting of the neural systems (e.g., those for WM and other executive functions) that both support performance on a desired outcome task (e.g., speech recognition) and are alterable via training.

As is described in the remainder of this chapter, noninvasive assessments of training-related neural plasticity in humans have employed magnetic resonance imaging (MRI) and electroencephalography (EEG), which are differentially sensitive to which neural systems are engaged during a task and when the engagement occurs. In particular, with millimeter spatial resolution, MRI can be used to examine gray- and white-matter volume, density, or other aspects of neural structural integrity, while functional MRI (fMRI) can be used to examine changes in blood flow while participants perform a task.

The blood flow response in fMRI is relatively slow (peaking in ~4–6 seconds) and thus suboptimal for addressing questions about the temporal dynamics of cognitive processing. In contrast, EEG records patterns of the summed electrical firing of thousands of neurons millisecond by millisecond. Although the EEG technique does not readily lend itself to precise spatial localization, when EEG is combined with MRI-constrained source modeling, spatial localization can be much improved (Dale & Sereno, 1993). EEG measures include changes in spectral content (i.e., power, phase, and coherence) in different frequency bands and event-related potentials (ERPs), with characteristic scalp distributions believed to reflect specific cognitive functions. Event-related spectral changes and ERPs have been related to specific cognitive operations (Kok, 1997). Other approaches also provide unique methodological perspectives for understanding the neural systems that support training-related improvements in cognition, such as training-related changes in dopamine receptor binding shown via positron electron tomography (McNab et al., 2009) and spatiotemporal activity in children with dyslexia measured via magnetoencephalography (MEG; Simos et al., 2002).

Multiple challenges have emerged from behavioral studies of cognitive training that neuroscience techniques may help to address, including: establishing the extent to which cognitive training benefits exist for trained tasks and materials, transfer to untrained tasks and materials, persist for extended periods of time, and are effective across a range of individuals. Cognitive neuroscience research has begun not only to tackle these challenges but also to pose new questions. For example, can training benefits be maximized via regulating or stimulating the neural systems that support behavior? How might our current approaches to cognitive training be significantly altered by novel and developing cognitive neuroscience methodologies?

In addition to presenting examples of such research, the chapter highlights the value of adopting a neural perspective and identifying key challenges that neuroscience studies on cognitive training should address in the future. To provide context, the chapter primarily focuses on common, noninvasive neuroscience approaches for evaluating the benefits of cognitive training in healthy adults. Although training-related neural-based changes have been observed for many perceptual and cognitive functions, to provide a representative example of the direction of research in the field, the majority of the cited work pertains to WM, one of the most widely targeted processes in cognitive training. WM is an executive function that is critical for maintaining, updating, and processing information and thus supports many complex cognitive abilities (Baddeley, 2003).

In what follows, we consider the extent to which cognitive neuroscience research facilitates our ability to address the following questions.

Does Training Effectively Target Desired Cognitive Systems?

A basic question in developing effective cognitive training is: What are the particular cognitive and neural systems targeted by a training protocol? For example, if the goal of training is to improve WM, the neural functions that typically support good WM performance might be predicted to change for trained tasks. As described below, cognitive neuroscience research may help elucidate the specific cognitive functions that are affected by training and that predict performance improvements on trained functions.

Previous research has observed that activity in frontal and posterior parietal cortices and in portions of the subcortical basal ganglia is increasingly engaged with greater WM capacity (D'Esposito, Postle, & Rypma, 2000; Gray, Chabris, & Braver, 2003; Klingberg, Forssberg, & Westerberg, 2002a; McNab & Klingberg, 2008; Pessoa, Gutierrez, Bandettini, & Ungerleider, 2002; Wager & Smith, 2003). Similarly, neuropsychological and structural neuroimaging data from children, older adults, and clinical populations have shown that the integrity of white matter underlying the frontoparietal cortices relates to WM capacity (Chua, Wen, Slavin, & Sachdev, 2008; Kanaan et al., 2005; Klingberg, 2006; Moseley, 2002). Based on these findings, researchers might then ask to what extent activity within such cortical and subcortical WM-related regions is modulated by WM training and whether it is predictive of training-related improvements in behavioral indices of WM capacity.

Adopting such an approach may shed light on the long-standing debate about whether WM capacity can be improved (Binet, 1909; Jaeggi, Buschkuehl, Jonides, & Perrig, 2008; Klingberg, 2010; Shipstead, Redick, & Engle, 2012; Thorndike & Woodworth, 1901). In particular, it may be possible to use objective neural markers to determine whether training improves the effectiveness and efficiency or capacity of specific trained cognitive functions themselves or whether it evokes instead a shift in strategy or adaptation to general task demands (e.g., via comparing improvements for trainees made aware of the need to suppress responses to irrelevant items in a WM task versus those without such instruction).

Concerns about the impact of explicit strategies notwithstanding, in the domain of training WM and other cognitive functions, progress has been made in observing changes in the neural markers of the trained functions. For example, extending evidence of training-related neural changes in macaques (Rainer & Miller, 2000) to humans, Olesen, Westerberg, and Klingberg (2004) observed greater activity in middle frontal gyrus and superior and inferior parietal cortices with increases in WM following five weeks of WM training. These changes emerged gradually across five fMRI scans obtained throughout the training. However, studies by Dahlin and colleagues have shown that while WM training yields increases in subcortical activity, frontoparietal activity decreases following training (Dahlin, Bäckman, Neely, & Nyberg, 2009). Similarly, Thompson and colleagues have found reduced sensitivity of frontoparietal networks to WM load following WM training, although greater functional connectivity between subsystems of these regions (i.e., executive control and dorsal attention networks; Thompson, Waskom, & Gabrieli, 2016). (We return to the issue of expected directionality of effects in the section "For Whom Is Cognitive Training Most Effective?")

The impact of WM training has also been observed in terms of structural changes in brain areas that support WM. For example, following two months of daily WM training, changes in white matter integrity within the intraparietal sulcus (bordering the superior and inferior parietal cortices) and in the anterior portion of the corpus callosum (which connects the bilateral dorsolateral prefrontal cortices) have been observed (Takeuchi et al., 2010). Together, these functional and structural results suggest a set of frontal, parietal, and subcortical regions that are engaged in WM practice and are expected to change with training.

EEG has also been employed to understand the specific neural functions engaged in cognitive training. For example, Maclin and colleagues (2011) found that intense video-game practice resulted in more efficient allocation of attention, as indicated by changes in the P300 ERP and alpha spectral power during an oddball task. In a single-session training study, Benikos, Johnstone, and Roodenrys (2013) suggested that training-related changes in the P300 were optimized at moderate levels of training task difficulty and that more rapid perceptual learning was observable in an enhanced P200 ERP over central-parietal cortex. This study illustrates a useful feature of EEG, which is its ability to identify multiple indicators of the operation of real-time cognitive component functions. Inhibitory control training-induced neuroplasticity has also been shown to alter the topography of the P200 ERP, whose sources were estimated to involve frontobasal executive networks (Manuel, Bernasconi, & Spierer, 2013), similar to those observed to support WM capacity with MRI.

Training-related changes in EEG measures of the spectral power and coherence of neural electrical activity have also been observed. Combined cognitive and physical training has been shown to reduce age-related cognitive declines in healthy adults by improving the real-time synchronization of the activity of the two hemispheres (Frantzidis, Ladas, Vivas, Tsolaki, & Bamidis, 2014). Multitasking improvements during video-game training with older adults have also been found to correspond to an increase in midline frontal theta power and a change in frontal-posterior coherence (increased executive functioning) in the same frequency band (Anguera et al., 2013). This pattern suggested that trainees benefited from greater suppression of a neural system typically associated with task disengagement (default-mode network) and from a corresponding increase in frontoparietal executive functioning.

Indeed, EEG studies are also beginning to clarify the information-processing mechanisms through which cognitive training may increase the effective capacity of WM. For example, one study of visual WM practice found evidence for a filtering feedback loop between prefrontal control and posterior brain regions (Liesefeld, Liesefeld, & Zimmer, 2014). In support of previous hypotheses (McNab & Klingberg, 2008), this work suggested that, after distractors are detected via engagement of posterior neural regions, a prefrontal mechanism is activated, which in turn prevents subsequent unnecessary parietal storage of distractor information. Pre-to-post EEG measures revealed learning-induced changes in the efficiency of this prefrontal mechanism. By blocking out distractor information and thus leaving more of the available capacity for the storage and processing of task-relevant representations, such a mechanism is predicted to improve the relative effective storage capacity of WM, even if the underlying storage capacity of all items combined remains unchanged.

As these results illustrate, electrophysiological studies have the potential to reveal which of several executive functions are positively affected by cognitive training. It is also of interest to know which functions are not affected by the training to better understand the specificity of its impact. For example, one EEG study provided young and older adults with extensive training via 12 different cognitive tasks (Grandy et al., 2013). While the training produced considerable performance gains, it did not change the individual alpha frequency, which is a trait and state marker of a person's mental preparedness predicting attention and memory retrieval (Bazanova & Vernon, 2014).

Thus, research suggests that both the function and structure of regions that support WM and other cognitive abilities are affected by training, at least on tasks that are similar to those trained. However, even further evidence that training improves processing capacities per se (i.e., rather than the motivation to allocate them to a task) may be observed when the benefits of training transfer to untrained tasks or materials.

Do the Benefits of Cognitive Training Transfer?

A common test of the effectiveness of cognitive training is whether trainees improve on materials and tasks that they did not explicitly practice. Training-related transfer to increasingly dissimilar measures, often discussed in terms of "near" to "far" transfer (Laker, 1990), implies that training on just one kind of task will improve performance in a number of domains. WM has thus been a common target of cognitive training programs because it has been shown to contribute to performance in a wide array of tasks, ranging from speech perception (Akeroyd, 2008), to planning and problem-solving (Shah & Miyake, 1999), and to spatial reasoning and general fluid intelligence (Kane et al., 2004).

The behavioral evidence for transfer has been mixed and minimally depends on the nature of the training program and of the tasks used to evaluate outcome (Melby-Lervåg & Hulme, 2013; Morrison & Chein, 2011; Shipstead et al., 2012; Sprenger et al., 2013). Cognitive neuroscience research appears to offer a unique perspective on this ongoing debate. For example, the variable transfer effects suggest a possible misalignment in the cognitive systems that were trained versus tested in some studies. Using cognitive neuroscience methods, researchers can test the hypothesis that greater transfer is associated with greater overlap in the neural systems that support training and outcome performance (see, for example, Klingberg, 2010).

Cognitive neuroscience research may be particularly informative in evaluating the benefits of training, because it facilitates researchers' understanding of not just whether, but why, training is effective in improving cognition. Revealing the neural systems that are engaged during training and determining their process-specific contribution is critical to making predictions about whether training benefit will transfer to other tasks. Transfer to novel materials or tasks may be predicted to the extent that both sets engage similar neural substrates and cognitive processes (Hussey & Novick, 2012; Jonides, 2004).

For example, Burgess, Gray, Conway, and Braver (2011) observed that approximately 25% of the shared variance on tests of WM and general fluid intelligence was predicted by individual differences in activity within prefrontal and parietal cortices in response to interfering lure trials (i.e., irrelevant items that are similar to target stimuli). This finding suggests that interference control processing, and the frontoparietal neural systems that underlie it, may serve as a predictor of the training-related transfer across such tasks.

In further support of this overlap hypothesis, Dahlin, Neely, Larsson, Bäckman, and Nyberg (2008a) targeted another cognitive and neural mechanism related to WM: updating. They employed updating training, during which participants performed a running span task, reporting the last four items presented in a sequence of varying length. Training was predicted to transfer only to outcome tasks that engaged similar subcortical striatal regions in support of updating. The authors reported training-related improvements on the WM n-back task, during which participants report whether each item in a sequence matches the item that was presented N-items prior (here, three), thus requiring a continuous updating of the contents of WM. Indeed, overlapping striatal regions were engaged in the updating and n-back tasks at pretesting, and these regions were increasingly engaged following training. However, no transfer was observed for the Stroop task, which requires participants to report the color of the ink a word is printed in while inhibiting reading the text itself (e.g., the word *red* written in blue ink) and did not engage similar striatal regions. Thus, training involved context updating, which running span and n-back (but not Stroop) share, but not competition-resolution processes (e.g., conflict detection, inhibitory control), which may be engaged in the n-back and Stroop outcomes but less so in the trained running span task. This work supports the hypothesis that transfer may be predicted to the extent that the training and outcome tasks of interest engage overlapping neural systems. Moreover, it illustrates the utility of brain data in illuminating which of several specific cognitive processes underlie transfer of training to ostensibly different tasks.

The neural bases of training transfer may involve changes in individual cognitive functions and in their interactions with other functions. The results from several EEG studies illustrate this point by showing that transfer is mediated by changes in the functional connectivity among a network of brain regions (Kundu, Sutterer, Emrich, & Postle, 2013). However, additional research is necessary to identify the specific cognitive processes that interact in this network across different forms of cognitive training. For example, one EEG study reported that transfer of WM training to short-term memory in the visual-spatial domain was mediated by increased functional connectivity in frontoparietal and parieto-occipital networks that occurs with high memory load (Kundu et al., 2013). A study that used a different form of cognitive training (i.e., trainees executed sequences of complex movements) reported transfer to improvements on a test of creativity (Dotan Ben-Soussan, Glicksohn, Goldstein, Berkovich-Ohana, & Donchin, 2013). Following training, only the complex-movement training group showed greater intra- and interhemispheric alpha coherence compared to control groups that followed simple forms of the training. Thus, the nature of training-related changes in neural function likely affects neural connectivity among domain-general and task-specific functions.

Taken together, this research illustrates that understanding the neural systems engaged in training and outcome tasks of interest is critical to predicting the training-related transfer to novel tasks and materials for at least two reasons. First, outcome tasks that do not share underlying neurocognitive mechanisms with training tasks would not be predicted to benefit. Second, changes in activity within and across different brain regions may mediate transfer. Targeting the specific cognitive systems for which improved outcomes are desired may also maximize the efficiency of training programs (Buschkuehl, Jaeggi, & Jonides, 2012), with the potential for larger, more long-lasting benefits in less training time.

Do the Benefits of Cognitive Training Persist?

To the extent that training improves cognitive abilities (rather than transient attentional states, for example), outcome benefits should persist beyond the final training session. However, as is the case for physical exercise, a few sessions of practice are unlikely to yield lifelong improvements. Understanding the stability of training benefits may help to determine how much, and when, retraining is needed to optimize task performance.

A practical limitation on evaluating training persistence with cognitive neuroscience techniques is the high cost of repeating neuroimaging sessions. Indeed, few MRI studies have examined the impact of training beyond pre- and posttraining testing sessions. Looking to domains outside of WM, evidence suggests that training-related changes in brain function can be relatively transient. For example, in a series of studies in which participants were trained in juggling for three months, midtemporal and posterior parietal brain regions that represent visual motion increased in size after training but returned to their original size after practice was stopped for three months (Boyke, Driemeyer, Gaser, Buchel, & May, 2008; Draganski et al., 2004; Driemeyer, Boyke, Gaser, Buchel, & May, 2008). Importantly, a return to baseline was observed in juggling performance as well. Nevertheless, future neuroscience studies are likely to elucidate what types of behavioral-training approaches promote persistence by examining whether and how they affect neural markers of training efficacy. Unlike juggling, the behavioral, and thus perhaps neural, effects of WM training may persist even without continuous practice. In WM-training studies of children, younger adults, and older adults, near-transfer benefits have been shown to persist anywhere from three to 18 months (Dahlin, Nyberg, Bäckman, & Neely, 2008b; Klingberg, Forssberg, & Westerberg, 2002b; Li et al., 2008). Training-related neural changes should persist at least as long (if not longer) than the behavioral benefits of training. Behavioral booster training could then be designed to target the biomarkers of such changes in the hope of achieving accelerated retraining behavioral performance.

While less costly to conduct than MRI studies, EEG studies have just begun to examine neural markers of persistence of cognitive and WM training. As discussed above, Anguera et al. (2013) identified a neural marker of the remediation of an age-related deficit in cognitive control through the practice of a multitasking video game: enhanced midline frontal theta power and frontal-posterior theta coherence. The behavioral performance gains were preserved six months later. Although the expectation might be for this neural marker to be preserved as well, it was not assessed in what was otherwise a landmark study. However, another EEG study did examine a neural marker for the persistence of the effects of attention training, including monitoring, detecting, and resolving conflict control (Rueda, Checa, & Combita, 2012). In this study, 5-year-old children showed improved attention control immediately after 10 computerized training sessions compared to a nonactive control group. This behavioral effect transferred to untrained tasks and was observed two months after the completion of training. The neural marker for the effect was the latency and scalp distribution of an ERP component observed in a flanker task, the N2, which is sensitive to detection of conflict by the anterior cingulate and its signaling to prefrontal cortex. The N2 findings indicated that the training sped up the N2 and shifted it more frontally, accelerating the establishment of children's conflict detection and signaling system. Moreover, this neural marker of the training, like the behavioral results, persisted until at least two months after training. A future study incorporating an active control group would corroborate these findings.

For Whom Is Cognitive Training Most Effective?

Using neural measures to predict individual differences in transfer and persistence may provide a complementary avenue for addressing the previous questions. Although not well-explored in the WM literature, brain-based individual difference estimates of pre-to-post training-related changes (e.g., in frontoparietal activity) could be used to predict a variety of behavioral outcomes. Such approaches have been effective in evaluating the persistence of rehabilitation benefits (Hoeft et al., 2011; Landi, Baguear, & Della-Maggiore, 2011; Parisi et al., 2014) and training-related transfer (Nikolaidis, Voss, Lee, Vo, & Kramer, 2014).

In addition, individual-difference measures of brain structure or function obtained prior to the start of training could be used to predict the likelihood of a positive training outcome. For example, Basak and colleagues assessed individual differences in the volume of two brain regions involved in executive control functions (i.e., the dorsolateral prefrontal cortex and anterior cingulate gyrus) prior to participants' practice of complex video-game strategies (Basak, Voss, Erickson, Boot, & Kramer, 2011). Individual differences in these particular brain structures predicted the acquisition of the game tasks. It may be feasible to use such results to better prepare individuals to succeed in the training by making the pretraining brain activation profile of nonresponders resemble that of training responders, possibly through the application of brain stimulation to relevant brain regions (see the next section, "Can Behavioral Training Be Augmented With Brain Regulation and Stimulation?").

Another source of individual variability that may affect training outcomes is that, due to the complex mapping between brain structure and function, different people may, to varying degrees, engage different regions of cortex to support a given cognitive ability. Indeed, evidence that two tasks engage similar neural regions across a group of participants does not directly reflect the degree of overlap within any given participant. For this reason, individual differences in MRI studies may be evaluated through anatomically or functionally defined regions of interest within each individual, such as with the use of functional localizer tasks (Poldrack, 2007; Saxe, Brett, & Kanwisher, 2006). One recent study observed evidence of overlap at the group and individual levels between WM, Stroop, and language tasks that engage cognitive-control-related regions of the left inferior frontal gyrus (Hsu, Jaeggi, & Novick, 2017). However, it remains to be tested whether the extent of overlap within an individual predicts training-related benefits.

Individual differences may be even more important to consider in populations (e.g., children, older adults, individuals with cognitive impairments) who exhibit greater intersubject variability in both their behavioral performance and in neural systems that support it (MacDonald, Nyberg, & Bäckman, 2006). Many commercial brain-training programs claim they can be used to slow or reverse the declines that are typically observed in healthy aging (e.g., in memory, attention, speed of processing, speech recognition; Deary et al., 2009; Dubno, Lee, Matthews, & Mills, 1997). Yet, some research has found that, although both younger and older adults can benefit from training, younger adults may be more likely to exhibit neural changes and transfer than older adults (Dahlin et al., 2008a; Nyberg et al., 2003). Indeed, older adults may experience greater challenges (Nyberg et al., 2003) or exhibit greater willingness to engage in suboptimal strategies (Hertzog, Touron, & Hines, 2007) than younger adults.

Individual variation poses a number of challenges for ensuring that training programs are effective across a population. For example, variability in training-related engagement and associated outcomes appear to relate to individual variation in intrinsic motivation and holding the belief that intelligence is malleable (Jaeggi, Buschkuehl, Shah, & Jonides, 2014). Additionally, individual differences in the cognitive abilities to be trained may hinder a priori predictions regarding the direction of training-related neural changes; individuals with lower WM capacity may theoretically be either more likely (e.g., because they have more room to improve) or less likely (e.g., because they require more training) to benefit from WM training. The degree of experience or expertise with a cognitive domain likely plays a large role in the direction of training-related outcomes (Brefczynski-Lewis, Lutz, Schaefer, Levinson, & Davidson, 2007). However, the outcomes of cognitive training studies have yet to be optimized by taking neural markers of these differences into account, in particular by honing in on behavioral-training regimes that change the neural profile of nonexperts into that of experts.

Training studies using EEG have also shown that individual differences in pretraining neural electrical activity predict the success of subsequent behavioral training (Mathewson et al., 2012; Wu, Srinivasan, Kaur, & Cramer, 2014). Wu and colleagues assessed baseline EEG coherence between the primary motor cortex and other brain regions (i.e., left parietal cortex and left frontal-premotor cortex). They found that this measure of real-time neural synchronization predicts the outcome of training on a new motor task (i.e., pursuit rotor task) above and beyond baseline behavior and demographics. They suggested that this predictive relationship reflects the effects of visuomotor integration capacity and motor-planning strategies (Wu et al., 2014). Pretraining frontal alpha waves have also been found to predict transfer from trained to untrained cognitive-control tasks (Mathewson et al., 2012). Moreover, this predictive relationship remained strong even after the behavioral performance scores during the same pretraining period were statistically controlled for, indicating that pretraining brain waves can predict training success and transfer above and beyond pretraining behavioral data alone. Frontal alpha brain waves may fulfill this predictive function because they are likely to reflect the degree of engagement of top-down cognitive-control processes (Mathewson et al., 2012). These promising results not only help to determine in advance for whom training will be effective but also suggest ways of better preparing individuals for training, for example, by using EEG neurofeedback to enhance frontal alpha waves.

To address individual differences, particularly in pretraining performance, most cognitive training studies continuously adapt training difficulty to each participant's accuracy level, often in real time (Anguera et al., 2013; Jaeggi et al., 2008; Sprenger et al., 2013). Adaptive WM training has generally resulted in greater benefits than nonadaptive protocols (Brehmer, Westerberg, & Bäckman, 2012; Holmes, Gathercole, & Dunning, 2009; Klingberg et al., 2005). Klingberg (2010) suggested that such adaptive paradigms are needed to ensure that each participant is engaged but not overloaded during training, thus maximizing benefit. The cognitive neuroscience literature seems to support the need for individually adapted tasks, given that peak neural activity often occurs at moderate levels of task difficulty (Poldrack et al., 2001; Wild et al., 2012; Zekveld, Heslenfeld, Festen, & Schoonhoven, 2006).

A caveat in developing neural markers of training effectiveness is evident in a recent review by Buschkuehl and colleagues (2012). The authors examined the evidence for WM-training transfer via metrics of changes in brain function, resting-state blood flow, brain structure, and dopaminergic function. Across more than 20 experiments, no consistent pattern emerged in terms of whether training resulted in increased, decreased, or redistributed patterns of these neural measures. The authors suggest a number of factors that may contribute to such variability, including differences in the nature and extent of training and in individual differences among trainees. Indeed, research in other domains has shown that the starting point of the learner matters. For example, while novices learning to meditate tend to engage attention-related brain regions to a greater degree after training, meditation experts show significantly less activity in these regions than trainees (Brefczynski-Lewis et al., 2007). Thus, initial increases in task-related effort may ultimately lead to greater efficiency. A similar noninverse U-shaped relationship between WM training and neural activity has been observed in frontoparietal regions (Hempel et al., 2004). This nonlinear trajectory is important to consider when developing a priori hypotheses about the effect of training on neural function and structure. In particular, Hsu and colleagues have noted that nonlinear trajectories of the development of executive functions during development, and therefore plasticity across the lifespan, may contribute to variability in function in individuals and with cognitive training (Hsu et al., 2014). They suggest that functional connectivity analyses may help address these issues by examining how interactions across multiple cognitive systems are differentially affected by training over time.

Can Behavioral Training Be Augmented With Brain Regulation and Stimulation?

Noninvasive methods for directly changing the neural-electrical activity of the brain hold great promise as a tool in research and development (R&D) on cognitive training. These methods help elucidate the neural markers of behavioral cognitive-training methods by causing changes in them and examining the effect on cognitive functions. It is then possible to design behavioral cognitive training to better target those cognitive functions and their underlying neural substrates. Additionally, brain-stimulation methods can complement the behavioral-training methods, either by providing an alternative approach to improving cognitive function (Brunoni & Vanderhasselt, 2014) or by augmenting the effect of the behavioral methods beyond what could be achieved through behavioral training alone (Krause & Cohen Kadosh, 2013). EEG neurofeedback (NFB) is a method for self-regulating the power and coherence of brain waves in a frequency-specific manner. NFB using fMRI has also demonstrated improvements in auditory attention with real-time feedback from previously identified attention-modulated regions of auditory cortex (Hinds et al., 2011; Thompson et al., 2009). Additional noninvasive brain-stimulation methods for improving cognitive function include transcranial direct current stimulation (tDCS), repetitive transcranial magnetic stimulation (rTMS), and transcranial electrical stimulation (TES).

EEG NFB trains a person to augment an aspect of their EEG spectral activity by being made aware of its intensity, often by listening to a changing tone (Fell et al., 2002) or watching a visual shape or color patch (Escolano, Olivan, Lopez-del-Hoyo, Garcia-Campayo, & Minguez, 2012) that reflects the real-time fluctuation in this intensity. EEG NFB can be applied alone or in combination with fMRI-based NFB, which modulates regional blood flow (Zotev, Phillips, Yuan, Misaki, & Bodurka, 2014). EEG NFB studies have confirmed the following frequency-specific positive contributions of increased spectral power to cognitive function (Gruzelier, 2014): Increased theta band power (4–7 Hz) over central parietal cortex improves consolidation of procedural motor memory (Reiner, Rozengurt, & Barnea, 2014), while over prefrontal cortex, it improves attention and recognition memory in young and older adults (Wang & Hsieh, 2013). Increased low-beta band power over sensory motor cortex (12–15 Hz) improves sustained attention in adults (Egner & Gruzelier, 2001; 2004; Vernon et al., 2003), semantic WM in adults (Vernon et al., 2003), word recognition in children (Barnea, Rassis, & Zaidel, 2005), psychomotor skills in eye surgeons (Ros et al., 2009), visual-spatial rotation in adults (Doppelmayr & Weber, 2011), and declarative word-pair memory in adults (Hoedlmoser et al., 2008). Increased low-beta band power over right frontal and parietal cortex improves sustained attention and increases gray matter volume and structural connectivity of these areas (Ghaziri et al., 2013). Increased upper-alpha band power (10–12 Hz) improves mental spatial rotation (Hanslmayr, Sauseng, Doppelmayr, Schabus, & Klimesch, 2005; Zoefel, Huster, & Herrmann, 2011), particularly over parieto-occipital cortex, and WM (Escolano et al., 2012; Nan et al., 2012). Increased lower- to mid-beta band power (12–20 Hz) improves familiarity-based episodic recognition (Keizer, Verment, & Hommel, 2010a; Keizer, Verschoor, Verment, & Hommel, 2010b). Increased gamma band power (30–70 Hz) improves visual-feature binding, fluid intelligence, and recollection-based episodic recognition (Keizer, Verment, et al., 2010; Keizer, Verschoor, et al., 2010).

TDCS and rTMS are noninvasive brain-stimulation methods for decreasing and increasing the net excitability of the cortex through weak electrical current delivered by scalp electrodes and magnetic pulses emanating from a coil held near the skull, respectively (Brunoni & Vanderhasselt, 2014; Klimesch, Sauseng, & Gerloff, 2003). Brunoni and Vanderhasselt conducted a careful meta-analysis of TDCS and rTMS studies that aimed to increase the activity of the dorsolateral prefrontal cortex (DLPFC) and thereby performance on n-back. This WM task critically relies on the DLPFC to maintain and update task-relevant goals and rules. Results showed that rTMS delivered immediately prior to n-back decreased reaction time, increased correct responses, and decreased error responses, with a low effect size in healthy individuals and a somewhat larger effect size across the tested clinical populations, including groups with Parkinson's disease, schizophrenia, and depression. TDCS showed the same results, except that they were limited to reaction times. Most studies reported no side effect of the stimulation, such as treatment intolerance or headache. However, the studies did not involve repeated stimulation sessions. Careful participant monitoring would have to be ensured in future research that examines whether repeated stimulation sessions increase the effect size. Another possibility that remains to be examined is whether single-session stimulation combined with behavioral cognitive training results in performance benefits that are larger than either treatment modality alone, or, alternatively, whether the combination could reduce the number of training sessions. The latter effect would be particularly relevant for participants who lack motivation to persist with behavioral-training sessions.

What Neuroscience Approaches on the Horizon Might Change the Cognitive Training Game?

The studies discussed in this chapter clearly illustrate the benefits of applying cognitive neuroscience to cognitive training. This approach has begun to identify biomarkers of training-induced performance gains and their transfer to untrained tasks, persistence over time, and their absence or presence, for example in nonresponders and responders. Consequently, an additional source of information for improving behavioral cognitive training is now rapidly becoming available. Moreover, the methodological limitations that make it difficult in some cases to compare studies and to draw strong conclusions from them are surmountable, provided the required research gets the funding support it needs. For example, as is pointed out in this chapter and as others have emphasized (Brunoni & Vanderhasselt, 2014), neuroscience training studies could be improved by increasing sample size; by controlling for individual differences in strategies, motivation, expertise, and other factors; by employing more similar procedures and materials; and by adopting standardized safeguards to avoid false positives, which is paramount when analyzing large neural data sets.

However, such methodological improvements, while necessary, would not take advantage of the full potential of neuroscience, given major developments in the field that are either very new or on the horizon. These developments could become game-changers in cognitive training, possibility leading to (1) bolstering cognitive training with information about the quality of item-specific knowledge representations in neural systems (KRNS), (2) identifying and using neurophysiological states that may facilitate information processing during cognitive training, and (3) adopting a process-specific approach to training transfer that would identify the individual differences in engagement levels that various tasks have on the neural regions and dynamic networks that support such functions. These possibilities are of special interest because they go beyond using neuroscience to obtain converging evidence that helps to develop and evaluate behavioral-only approaches to cognitive training; they also add value that cannot be obtained from behavioral approaches alone.

The majority of neuroscience studies of cognitive training use the dominant cognition-function approach to research and have yet to leverage the knowledge representations in neural systems approach (i.e., KRNS, after the Intelligence Advanced Research Projects Activity program that sponsored the research). The former uses brain activity during contrasting conditions as an index of the contribution of a cognitive function without necessarily specifying the nature of how information is represented at any given moment during real-time processing.

However, research adopting the KRNS approach suggests that it would be feasible to use such knowledge to improve cognitive training. This research has successfully applied machine-learning approaches to identify the neural patterns of activation that are evoked by specific concepts and categories of concepts, communicated visually or verbally through images or words (Damarla & Just, 2012; Just, Cherkassky, Aryal, & Mitchell, 2010) as they are acquired (Bauer & Just, 2015). In principle, the KRNS approach could therefore generate novel biomarkers of the degree and precision of the activation of item representations during cognitive training (e.g., WM training) alone or in combination with item learning (e.g., WM training using foreign language vocabulary as items). The same information could also be applied in brain–computer interface (BCI) approaches to cognitive training, in which a trainee might make responses by just thinking about intending to do so via BCI systems.

The brain stimulation and neurofeedback approach has identified neurophysiological states of the brain that facilitate cognitive processing, specifically states related to arousal, attention, effort, and task engagement. It may be feasible to optimize cognitive training through a biofeedback approach that encourages the learner to maintain an appropriate neurophysiological state during behavioral training, for example, by presenting materials for behavioral-task practice only, or more clearly, once the appropriate states are achieved. For example, words presented during WM training could be presented with better visual contrast when a person's neurophysiology indicates they are less externally and internally distracted. Some mental strategies may be more effective during neurofeedback training than others (Kober, Witte, Ninaus, Neuper, & Wood, 2013; Witte, Kober, Ninaus, Neuper, & Wood, 2013). Identifying these strategies can help trainees dynamically regulate their neurophysiological state as needed and reduce the continued reliance on neurofeedback. This will require a multimodal methodological approach for theoretical and applied reasons, combining MRI and MEG with methods that can be fielded more easily, in particular pupillometry, EEG, and optical imaging. The latter methods are less expensive, more portable, and easier to use than MRI and MEG. The EEG approach has been applied in peak-performance training for top athletes, who benefit from a better mental preparedness state immediately before, for example, executing planned shots in golf and marksmanship (Stikic et al., 2014). In addition, pupillometry could be used as a measure of cognitive effort to ensure that trainees are optimally engaged, neither too drowsy nor too excited for best task performance (Aston-Jones & Cohen, 2005; Gilzenrat, Nieuwenhuis, Jepma, & Cohen, 2010) and optimal training outcomes (see the section "Does Training Effectively Target Desired Cognitive Systems?"). Finally, more information is needed about which correlated neuro-electrical states across individuals improve performance in collaborative cognition, such as team problem-solving (Stevens et al., 2012), especially since the majority of the current cognitive-training studies approach tasks as if they were mostly performed by individuals alone in everyday life.

Moreover, it may be feasible to improve training transfer through a complementary approach that does not assume that a particular brain region, engaged by a specific cognitive component function, supports only that function. Indeed, it has been suggested that brain regions involved in cognition constitute a highly dynamic network in which any particular brain region may be engaged in not one, but different, specific cognitive component functions, each one evoked by task-dependent interactions with other brain regions (Anderson, 2015). While this possibility remains to be systematically tested, it creates the prospect that a specific cognitive component function can be practiced by identifying training tasks that may not engage that function but nevertheless improve the processing efficiency of the same brain region.

The state of current research suggests that a better neuroscientific understanding of transfer and persistence effects in cognitive training is needed. On first approximation, it would seem that trainees would be interested in permanently improving a cognitive component function that benefits performance in a range of tasks. However, what if the specific function in question benefits performance in one class of tasks but impairs performance in another class of tasks? For example, increased focused attention may benefit explicit but not implicit learning (Stillman, Feldman, Wambach, Howard, & Howard, 2014; Whitmarsh, Udden, Barendregt, & Petersson, 2013). In such cases, a trainee would not benefit from a fixed increase in the efficiency and effectiveness of the cognitive component function. Rather, there would be a number of desired outcomes, including a better regulation of when and how much to engage the function and availability of a higher maximum level of efficiency and effectiveness in its deployment. To illustrate these points in terms of the well-known analogy of the spotlight of attention, the director of the spotlight would benefit from an improved ability to regulate its width and intensity and from a permanent increase in its maximum available intensity. Indeed, the ability to switch between a focused spotlight and a broader, scanning mode of attention is critical to maximizing task utility and behavioral performance and is regulated by the locus coeruleus norepinephrine system (LC-NE), in which function can be indexed via a variety of previously discussed metrics (e.g., P300 ERP, pupil dilation, frontal cortical activity). Loss of LC-NE function has negative cognitive consequences for flexibly shifting attention, as, for example, occurs in individuals with attention deficit disorders or with Parkinson's disease (Aston-Jones & Cohen, 2005).

There is also a need to gain a better understanding of the domain generality of the specific cognitive component functions targeted by the training and of the type, frequency, and intensity of booster training needed to maintain maximum benefits. Accordingly, as long as the task demands of trained and nontrained tasks emphasize reliance on the same cognitive component functions, one may expect transfer, even if the two tasks lack surface similarity. However, since there are individual differences in the strategies used to adjust to task demands, individuals may differ in what particular training task(s) most effectively engage a particular cognitive component function (Jaeggi et al., 2014; Riding & Sadler-Smith, 1997). Tailoring the cognitive training to such individual differences could therefore enhance outcome. These possibilities remain to be examined. The cognitive neuroscience approach offers a direct window into the spatial-temporal patterns of brain activation and the mediating anatomical structures that underlie various aspects of transfer and persistence.

Conclusion

The neuroscience approach to cognitive training has already made valuable contributions to the literature and stands to become a game-changer in this area. This is evident from discussing the neural markers of the effects of behavioral training of WM and its contributing cognitive component functions. Observed effects include changes in the brain mechanisms that mediate performance improvements on trained tasks, transfer to untrained tasks, and persistence of performance benefits after training. Therefore, the neuroscience approach can help validate behavioral training by using converging evidence from data that reflect the precise neural architecture of cognition.

Critically, the neuroscience science approach has an added value beyond what can be derived from behavioral data alone. Information about biomarkers can be used to improve behavioral training by using information about how the neural substrate of cognition changes in response to practice and transfers the resulting benefits in a persistent manner to untrained tasks, with individual differences in level of expertise and motivation taken into account. Executing this research is an exciting and tractable challenge for future researchers interested in this endeavor and its theoretical and practical implications. The neural approach to cognitive training has even greater potential for success if it is combined with existing related methods, including KRNS or neural semantics, neurofeedback, and brain stimulation for inducing desired neurophysiological states, and machine-learning approaches for tailoring training to the neural profile of individual participants.

References

Akeroyd, M. A. (2008). Are individual differences in speech reception related to individual differences in cognitive ability? A survey of twenty experimental studies with normal and hearing-impaired adults. *International Journal of Audiology*, 47(Suppl. 2), S53–71. doi:10.1080/14992020802301142

Anderson, M. L. (2015). *After phrenology: Neural reuse and the interactive brain.* Cambridge, MA: MIT Press.

Anguera, J. A., Boccanfuso, J., Rintoul, J. L., Al-Hashimi, O., Faraji, F., Janowich, J., . . . Gazzaley, A. (2013). Video game training enhances cognitive control in older adults. *Nature, 501*(7465), 97–101. doi:10.1038/nature12486

Aston-Jones, G., & Cohen, J. D. (2005). An integrative theory of locus coeruleus-norepinephrine function: Adaptive gain and optimal performance. *Annual Review of Neuroscience, 28*, 403–450. doi:10.1146/annurev.neuro.28.061604.135709

Baddeley, A. (2003). Working memory: Looking back and looking forward. *Nature Reviews Neuroscience, 4*(10), 829–839. doi:10.1038/nrn1201

Barnea, A., Rassis, A., & Zaidel, E. (2005). Effect of neurofeedback on hemispheric word recognition. *Brain and Cognition, 59*(3), 314–321. doi:10.1016/j.bandc.2004.05.008

Basak, C., Voss, M. W., Erickson, K. I., Boot, W. R., & Kramer, A. F. (2011). Regional differences in brain volume predict the acquisition of skill in a complex real-time strategy videogame. *Brain and Cognition, 76*(3), 407–414. doi:10.1016/j.bandc.2011.03.017

Bauer, A. J., & Just, M. A. (2015). Monitoring the growth of the neural representations of new animal concepts. *Human Brain Mapping, 36*(8), 3213–3226. doi:10.1002/hbm.22842

Bazanova, O. M., & Vernon, D. (2014). Interpreting EEG alpha activity. *Neuroscience and Biobehavioral Reviews, 44*, 94–110. doi:10.1016/j.neubiorev.2013.05.007

Benikos, N., Johnstone, S. J., & Roodenrys, S. J. (2013). Short-term training in the go/nogo task: Behavioural and neural changes depend on task demands. *International Journal of Psychophysiology, 87*(3), 301–312. doi:10.1016/j.ijpsycho.2012.12.001

Binet, A. (1909). *Les idées modernes sur les enfants.* Paris, France: Flammarion.

Boyke, J., Driemeyer, J., Gaser, C., Buchel, C., & May, A. (2008). Training-induced brain structure changes in the elderly. *Journal of Neuroscience, 28*(28), 7031–7035. doi:10.1523/JNEUROSCI.0742-08.2008

Brefczynski-Lewis, J. A., Lutz, A., Schaefer, H. S., Levinson, D. B., & Davidson, R. J. (2007). Neural correlates of attentional expertise in long-term meditation practitioners. *Proceedings of the National Academy of Sciences, 104*(27), 11483–11488. doi:10.1073/pnas.0606552104

Brehmer, Y., Westerberg, H., & Bäckman, L. (2012). Working-memory training in younger and older adults: Training gains, transfer, and maintenance. *Frontiers in Human Neuroscience, 6*, 63. doi:10.3389/fnhum.2012.00063

Brunoni, A. R., & Vanderhasselt, M.-A. (2014). Working memory improvement with non-invasive brain stimulation of the dorsolateral prefrontal cortex: A systematic review and meta-analysis. *Brain and Cognition, 86*(1), 1–9. doi:10.1016/j.bandc.2014.01.008

Burgess, G. C., Gray, J. R., Conway, A. R., & Braver, T. S. (2011). Neural mechanisms of interference control underlie the relationship between fluid intelligence and working memory span. *Journal of Experimental Psychology: General, 140*(4), 674–692. doi:10.1037/a0024695

Buschkuehl, M., Jaeggi, S. M., & Jonides, J. (2012). Neuronal effects following working memory training. *Developmental Cognitive Neuroscience, 2*(Suppl. 1), S167–179. doi:10.1016/j.dcn.2011.10.001

Chua, T. C., Wen, W., Slavin, M. J., & Sachdev, P. S. (2008). Diffusion tensor imaging in mild cognitive impairment and alzheimer's disease: A review. *Current Opinion in Neurology, 21*(1), 83–92. doi:10.1097/WCO.0b013e3282f4594b

Dahlin, E., Bäckman, L., Neely, A. S., & Nyberg, L. (2009). Training of the executive component of working memory: Subcortical areas mediate transfer effects. *Restorative Neurology and Neuroscience, 27*(5), 405–419. doi:10.3233/RNN-2009-0492

Dahlin, E., Neely, A. S., Larsson, A., Bäckman, L., & Nyberg, L. (2008a). Transfer of learning after updating training mediated by the striatum. *Science, 320*(5882), 1510–1512. doi:10.1126/science.1155466

Dahlin, E., Nyberg, L., Bäckman, L., & Neely, A. S. (2008b). Plasticity of executive functioning in young and older adults: Immediate training gains, transfer, and long-term maintenance. *Psychology of Aging, 23*(4), 720–730. doi:10.1037/a0014296

Dale, A. M., & Sereno, M. I. (1993). Improved localization of cortical activity by combining EEG and MEG with MRI cortical surface reconstruction: A linear approach. *Journal of Cognitive Neuroscience, 5*(2), 162–176. doi:10.1162/jocn.1993.5.2.162

Damarla, S. R., & Just, M. A. (2012). Decoding the representation of numerical values from brain activation patterns. *Human Brain Mapping, 34*(10), 2624–2634. doi:10.1002/hbm.22087

Deary, I. J., Corley, J., Gow, A. J., Harris, S. E., Houlihan, L. M., Marioni, R. E., . . . Starr, J. M. (2009). Age-associated cognitive decline. *British Medical Bulletin, 92*(1), 135–152. doi:10.1093/bmb/ldp033

D'Esposito, M., Postle, B. R., & Rypma, B. (2000). Prefrontal cortical contributions to working memory: Evidence from event-related fMRI studies. *Experimental Brain Research, 133*(1), 3–11. doi:10.1007/s002210000395

Doppelmayr, M., & Weber, E. (2011). Effects of SMR and theta/beta neurofeedback on reaction times, spatial abilities, and creativity. *Journal of Neurotherapy, 15*(2), 115–129. doi:10.1080/10874208.2011.570689

Dotan Ben-Soussan, T., Glicksohn, J., Goldstein, A., Berkovich-Ohana, A., & Donchin, O. (2013). Into the square and out of the box: The effects of Quadrato Motor Training on creativity and alpha coherence. *PLoS One, 8*(1), e55023. doi:10.1371/journal.pone.0055023

Draganski, B., Gaser, C., Busch, V., Schuierer, G., Bogdahn, U., & May, A. (2004). Changes in grey matter induced by training. *Nature, 427*(6972), 311–312. doi:10.1038/427311a

Driemeyer, J., Boyke, J., Gaser, C., Buchel, C., & May, A. (2008). Changes in gray matter induced by learning—Revisited. *PLoS One, 3*(7), e2669. doi:10.1371/journal.pone.0002669

Dubno, J. R., Lee, F. S., Matthews, L. J., & Mills, J. H. (1997). Age-related and gender-related changes in monaural speech recognition. *Journal of Speech, Language, and Hearing Research, 40*(2), 444–452. doi:10.1044/jslhr.4002.444

Egner, T., & Gruzelier, J. H. (2001). Learned self-regulation of EEG frequency components affects attention and event-related brain potentials in humans. *Neuroreport, 12*(18), 4155–4159. doi:10.1097/00001756-200112210-00058

Egner, T., & Gruzelier, J. H. (2004). EEG biofeedback of low beta band components: Frequency-specific effects on variables of attention and event-related brain potentials. *Clinical Neurophysiology, 115*(1), 131–139. doi:10.1016/S1388-2457(03)00353-5

Escolano, C., Olivan, B., Lopez-del-Hoyo, Y., Garcia-Campayo, J., & Minguez, J. (2012). Double-blind single-session neurofeedback training in upper-alpha for cognitive enhancement of healthy subjects. *Conference Proceedings of the IEEE Engineering in Medicine and Biology Society, 2012*, 4643–4647. doi:10.1109/EMBC.2012.6347002

Fell, J., Elfadil, H., Klaver, P., Roschke, J., Elger, C. E., & Fernandez, G. (2002). Covariation of spectral and nonlinear EEG measures with alpha biofeedback. *International Journal of Neuroscience, 112*, 1047–1057. doi:10.1080/00207450290026049

Frantzidis, C. A., Ladas, A.-K. I., Vivas, A. B., Tsolaki, M., & Bamidis, P. D. (2014). Cognitive and physical training for the elderly: Evaluating outcome efficacy by means of neurophysiological synchronization. *International Journal of Psychophysiology, 93*(1), 1–11. doi:10.1016/j.ijpsycho.2014.01.007

Ghaziri, J., Tucholka, A., Larue, V., Blanchette-Sylvestre, M., Reyburn, G., Gilbert, G., . . . Beauregard, M. (2013). Neurofeedback training induces changes in white and gray matter. *Clinical EEG and Neuroscience, 44*(4), 265–272. doi:10.1177/1550059413476031

Gilzenrat, M. S., Nieuwenhuis, S., Jepma, M., & Cohen, J. D. (2010). Pupil diameter tracks changes in control state predicted by the adaptive gain theory of locus coeruleus function. *Cognitive and Affective Behavioral Neuroscience, 10*(2), 252–269. doi:10.3758/CABN.10.2.252

Grandy, T. H., Werkle-Bergner, M., Chicherio, C., Schmiedek, F., Lövdén, M., & Lindenberger, U. (2013). Peak individual alpha frequency qualifies as a stable neurophysiological trait marker in healthy younger and older adults. *Psychophysiology, 50*(6), 570–582. doi:10.1111/psyp.12043

Gray, J. R., Chabris, C. F., & Braver, T. S. (2003). Neural mechanisms of general fluid intelligence. *Nature Neuroscience, 6*(3), 316–322. doi:10.1038/nn1014

Gruzelier, J. H. (2014). EEG-neurofeedback for optimising performance. I: A review of cognitive and affective outcome in healthy participants. *Neuroscience and Biobehavioral Reviews, 44*, 124–141. doi:10.1016/j.neubiorev.2013.09.015

Hanslmayr, S., Sauseng, P., Doppelmayr, M., Schabus, M., & Klimesch, W. (2005). Increasing individual upper alpha power by neurofeedback improves cognitive performance in human subjects. *Applied Psychophysiology and Biofeedback, 30*(1), 1–10. doi:10.1007/s10484-005-2169-8

Hempel, A., Giesel, F. L., Garcia Caraballo, N. M., Amann, M., Meyer, H., Wüstenberg, T., . . . Schröder, J. (2004). Plasticity of cortical activation related to working memory during training. *American Journal of Psychiatry, 161*(4), 745–747. doi:10.1176/appi.ajp.161.4.745

Henshaw, H., & Ferguson, M. (2015). How does auditory training work? Joined-up thinking and listening. *Seminars in Hearing, 36*(4), 237–249. doi:10.1055/s-0035-1564456

Hertzog, C., Touron, D. R., & Hines, J. C. (2007). Does a time-monitoring deficit influence older adults' delayed retrieval shift during skill acquisition? *Psychology and Aging, 22*(3), 607–624. doi:10.1037/0882-7974.22.3.607

Hinds, O., Ghosh, S., Thompson, T. W., Yoo, J. J., Whitfield-Gabrieli, S., Triantafyllou, C., & Gabrieli, J. D. (2011). Computing moment-to-moment bold activation for real-time neurofeedback. *Neuroimage, 54*(1), 361–368. doi:10.1016/j.neuroimage.2010.07.060

Hoedlmoser, K., Pecherstorfer, T., Gruber, G., Anderer, P., Doppelmayr, M., Klimesch, W., & Schabus, M. (2008). Instrumental conditioning of human sensorimotor rhythm (12–15 Hz) and its impact on sleep as well as declarative learning. *Sleep, 31*(10), 1401–1408. doi:10.5665/sleep/31.10.1401

Hoeft, F., McCandliss, B. D., Black, J. M., Gantman, A., Zakerani, N., Hulme, C., . . . Gabrieli, J. D. (2011). Neural systems predicting long-term outcome in dyslexia. *Proceedings of the National Academy of Sciences, 108*(1), 361–366. doi:10.1073/pnas.1008950108

Holmes, J., Gathercole, S. E., & Dunning, D. L. (2009). Adaptive training leads to sustained enhancement of poor working memory in children. *Developmental Science*, 12(4), F9–15. doi:10.1111/j.1467-7687.2009.00848.x

Hsu, N. S., Jaeggi, S. M., & Novick, J. M. (2017). A common neural hub resolves syntactic and non-syntactic conflict through cooperation with task-specific networks. *Brain and Language*, 166, 63–77. doi:10.1016/j.bandl.2016.12.006

Hsu, N. S., Novick, J. M., & Jaeggi, S. M. (2014). The development and malleability of executive control abilities. *Frontiers in Behavioral Neuroscience*, 8, 221. doi:10.3389/fnbeh.2014.00221

Hussey, E. K., & Novick, J. M. (2012). The benefits of executive control training and the implications for language processing. *Frontiers in Psychology*, 3, 158. doi:10.3389/fpsyg.2012.00158

Jaeggi, S. M., Buschkuehl, M., Jonides, J., & Perrig, W. J. (2008). Improving fluid intelligence with training on working memory. *Proceedings of the National Academy of Sciences*, 105(19), 6829–6833. doi:10.1073/pnas.0801268105

Jaeggi, S. M., Buschkuehl, M., Shah, P., & Jonides, J. (2014). The role of individual differences in cognitive training and transfer. *Memory and Cognition*, 42(3), 464–480. doi:10.3758/s13421-013-0364-z

Jonides, J. (2004). How does practice make perfect? *Nature Neuroscience*, 7(1), 10–11. doi:10.1038/nn0104-10

Jurado, M. B., & Rosselli, M. (2007). The elusive nature of executive functions: A review of our current understanding. *Neuropsychology Review*, 17(3), 213–233. doi:10.1007/s11065-007-9040-z

Just, M. A., Cherkassky, V. L., Aryal, S., & Mitchell, T. M. (2010). A neurosemantic theory of concrete noun representation based on the underlying brain codes. *PLoS One*, 5(1), e8622. doi:10.1371/journal.pone.0008622

Kanaan, R. A., Kim, J. S., Kaufmann, W. E., Pearlson, G. D., Barker, G. J., & McGuire, P. K. (2005). Diffusion tensor imaging in schizophrenia. *Biological Psychiatry*, 58(12), 921–929. doi:10.1016/j.biopsych.2005.05.015

Kane, M. J., Hambrick, D. Z., Tuholski, S. W., Wilhelm, O., Payne, T. W., & Engle, R. W. (2004). The generality of working memory capacity: A latent-variable approach to verbal and visuospatial memory span and reasoning. *Journal of Experimental Psychology: General*, 133(2), 189–217. doi:10.1037/0096-3445.133.2.189

Keizer, A. W., Verment, R. S., & Hommel, B. (2010). Enhancing cognitive control through neurofeedback: A role of gamma-band activity in managing episodic retrieval. *Neuroimage*, 49(4), 3404–3413. doi:10.1016/j.neuroimage.2009.11.023

Keizer, A. W., Verschoor, M., Verment, R. S., & Hommel, B. (2010). The effect of gamma enhancing neurofeedback on the control of feature bindings and intelligence measures. *International Journal of Psychophysiology*, 75(1), 25–32. doi:10.1016/j.ijpsycho.2009.10.011

Klimesch, W., Sauseng, P., & Gerloff, C. (2003). Enhancing cognitive performance with repetitive transcranial magnetic stimulation at human individual alpha frequency. *European Journal of Neuroscience*, 17(5), 1129–1133. doi:10.1046/j.1460-9568.2003.02517.x

Klingberg, T. (2006). Development of a superior frontal-intraparietal network for visuo-spatial working memory. *Neuropsychologia*, 44(11), 2171–2177. doi:10.1016/j.neuropsychologia.2005.11.019

Klingberg, T. (2010). Training and plasticity of working memory. *Trends in Cognitive Sciences, 14*(7), 317–324. doi:10.1016/j.tics.2010.05.002

Klingberg, T., Fernell, E., Olesen, P. J., Johnson, M., Gustafsson, P., Dahlström, K., . . . Westerberg, H. (2005). Computerized training of working memory in children with ADHD—A randomized, controlled trial. *Journal of the American Academy of Child and Adolescent Psychiatry, 44*(2), 177–186. doi:10.1097/00004583-200502000-00010

Klingberg, T., Forssberg, H., & Westerberg, H. (2002a). Increased brain activity in frontal and parietal cortex underlies the development of visuospatial working memory capacity during childhood. *Journal of Cognitive Neuroscience, 14*(1), 1–10. doi:10.1162/089892902317205276

Klingberg, T., Forssberg, H., & Westerberg, H. (2002b). Training of working memory in children with ADHD. *Journal of Clinical and Experimental Neuropsychology, 24*(6), 781–791. doi:10.1076/jcen.24.6.781.8395

Kober, S. E., Witte, M., Ninaus, M., Neuper, C., & Wood, G. (2013). Learning to modulate one's own brain activity: The effect of spontaneous mental strategies. *Frontiers in Human Neuroscience, 7*, 695. doi:10.3389/fnhum.2013.00695

Kok, A. (1997). Event-related-potential (ERP) reflections of mental resources: A review and synthesis. *Biological Psychiatry, 45*, 19–56. doi:10.1016/S0301-0511(96)05221-0

Krause, B., & Cohen Kadosh, R. (2013). Can transcranial electrical stimulation improve learning difficulties in atypical brain development? A future possibility for cognitive training. *Developmental Cognitive Neuroscience, 6*(100), 176–194. doi:10.1016/j.dcn.2013.04.001

Kundu, B., Sutterer, D. W., Emrich, S. M., & Postle, B. R. (2013). Strengthened effective connectivity underlies transfer of working memory training to tests of short-term memory and attention. *Journal of Neuroscience, 33*(20), 8705–8715. doi:10.1523/JNEUROSCI.5565-12.2013

Laker, D. R. (1990). Dual dimensionality of training transfer. *Human Resource Development Quarterly, 1*(3), 209–223. doi:10.1002/hrdq.3920010303

Landi, S. M., Baguear, F., & Della-Maggiore, V. (2011). One week of motor adaptation induces structural changes in primary motor cortex that predict long-term memory one year later. *Journal of Neuroscience, 31*(33), 11808–11813. doi:10.1523/JNEUROSCI.2253-11.2011

Li, S. C., Schmiedek, F., Huxhold, O., Röcke, C., Smith, J., & Lindenberger, U. (2008). Working memory plasticity in old age: Practice gain, transfer, and maintenance. *Psychology and Aging, 23*(4), 731–742. doi:10.1037/a0014343

Liesefeld, A. M., Liesefeld, H. R., & Zimmer, H. D. (2014). Intercommunication between prefrontal and posterior brain regions for protecting visual working memory from distractor interference. *Psychological Science, 25*(2), 325–333. doi:10.1177/0956797613501170

MacDonald, S. W., Nyberg, L., & Bäckman, L. (2006). Intra-individual variability in behavior: Links to brain structure, neurotransmission and neuronal activity. *Trends in Neurosciences, 29*(8), 474–480. doi:10.1016/j.tins.2006.06.011

Maclin, E. L., Mathewson, K. E., Low, K. A., Boot, W. R., Kramer, A. F., Fabiani, M., & Gratton, G. (2011). Learning to multitask: Effects of video game practice on electrophysiological indices of attention and resource allocation. *Psychophysiology, 48*(9), 1173–1183. doi:10.1111/j.1469-8986.2011.01189.x

Manuel, A. L., Bernasconi, F., & Spierer, L. (2013). Plastic modifications within inhibitory control networks induced by practicing a stop-signal task: An electrical neuroimaging study. *Cortex, 49*(4), 1141–1147. doi:10.1016/j.cortex.2012.12.009

Mathewson, K. E., Basak, C., Maclin, E. L., Low, K. A., Boot, W. R., Kramer, A. F., . . . Gratton, G. (2012). Different slopes for different folks: Alpha and delta EEG power predict subsequent video game learning rate and improvements in cognitive control tasks. *Psychophysiology, 49*(12), 1558–1570. doi:10.1111/j.1469-8986.2012.01474.x

McNab, F., & Klingberg, T. (2008). Prefrontal cortex and basal ganglia control access to working memory. *Nature Neuroscience, 11*(1), 103–107. doi:10.1038/nn2024

McNab, F., Varrone, A., Farde, L., Jucaite, A., Bystritsky, P., Forssberg, H., & Klingberg, T. (2009). Changes in cortical dopamine D1 receptor binding associated with cognitive training. *Science, 323*(5915), 800–802. doi:10.1126/science.1166102

Melby-Lervåg, M., & Hulme, C. (2013). Is working memory training effective? A meta-analytic review. *Developmental Psychology, 49*(2), 270–291. doi:10.1037/a0028228

Morrison, A. B., & Chein, J. M. (2011). Does working memory training work? The promise and challenges of enhancing cognition by training working memory. *Psychonomic Bulletin & Review, 18*(1), 46–60. doi:10.3758/s13423-010-0034-0

Moseley, M. (2002). Diffusion tensor imaging and aging—A review. *NMR in Biomedicine, 15*(7-8), 553–560. doi:10.1002/nbm.785

Nan, W., Rodrigues, J. P., Ma, J., Qu, X., Wan, F., Mak, P.-I., . . . Rosa, A. (2012). Individual alpha neurofeedback training effect on short term memory. *International Journal of Psychophysiology, 86*(1), 83–87. doi:10.1016/j.ijpsycho.2012.07.182

Nikolaidis, A., Voss, M. W., Lee, H., Vo, L. T., & Kramer, A. F. (2014). Parietal plasticity after training with a complex video game is associated with individual differences in improvements in an untrained working memory task. *Frontiers in Human Neuroscience, 8*, 169. doi:10.3389/fnhum.2014.00169

Nyberg, L., Sandblom, J., Jones, S., Neely, A. S., Petersson, K. M., Ingvar, M., & Bäckman, L. (2003). Neural correlates of training-related memory improvement in adulthood and aging. *Proceedings of the National Academy of Sciences, 100*(23), 13728–13733. doi:10.1073/pnas.1735487100

Olesen, P. J., Westerberg, H., & Klingberg, T. (2004). Increased prefrontal and parietal activity after training of working memory. *Nature Neuroscience, 7*(1), 75–79. doi:10.1038/nn1165

Parisi, L., Rocca, M. A., Mattioli, F., Copetti, M., Capra, R., Valsasina, P., . . . Filippi, M. (2014). Changes of brain resting state functional connectivity predict the persistence of cognitive rehabilitation effects in patients with multiple sclerosis. *Multiple Sclerosis, 20*(6), 686–694. doi:10.1177/1352458513505692

Pessoa, L., Gutierrez, E., Bandettini, P., & Ungerleider, L. (2002). Neural correlates of visual working memory: fMRI amplitude predicts task performance. *Neuron, 35*(5), 975–987. doi:10.1016/S0896-6273(02)00817-6

Poldrack, R. A. (2007). Region of interest analysis for fMRI. *Social Cognitive and Affective Neuroscience, 2*(1), 67–70. doi:10.1093/scan/nsm006

Poldrack, R. A., Temple, E., Protopapas, A., Nagarajan, S., Tallal, P., Merzenich, M., & Gabrieli, J. D. E. (2001). Relations between the neural bases of dynamic auditory processing and phonological processing: Evidence from fMRI. *Journal of Cognitive Neuroscience, 13*(5), 687–697. doi:10.1162/089892901750363235

Rainer, G., & Miller, E. K. (2000). Effects of visual experience on the representation of objects in the prefrontal cortex. *Neuron, 27*(1), 179–189. doi:10.1016/S0896-6273(00)00019-2

Reiner, M., Rozengurt, R., & Barnea, A. (2014). Better than sleep: Theta neurofeedback training accelerates memory consolidation. *Biological Psychology, 95,* 45–53. doi:10.1016/j.biopsycho.2013.10.010

Riding, R. J., & Sadler-Smith, E. (1997). Cognitive style and learning strategies: Some implications for training design. *International Journal of Training and Development, 1*(3), 199–208. doi:10.1111/1468-2419.00020

Ros, T., Moseley, M. J., Bloom, P. A., Benjamin, L., Parkinson, L. A., & Gruzelier, J. H. (2009). Optimizing microsurgical skills with EEG neurofeedback. *BMC Neuroscience, 10,* 87. doi:10.1186/1471-2202-10-87

Rueda, M. R., Checa, P., & Combita, L. M. (2012). Enhanced efficiency of the executive attention network after training in preschool children: Immediate changes and effects after two months. *Developmental Cognitive Neuroscience, 2*(Suppl. 1), S192–204. doi:10.1016/j.dcn.2011.09.004

Saxe, R., Brett, M., & Kanwisher, N. (2006). Divide and conquer: A defense of functional localizers. *Neuroimage, 30*(4), 1088–1096. doi:10.1016/j.neuroimage.2005.12.062

Shah, P., & Miyake, A. (1999). Models of working memory: An introduction. In A. Miyake & P. Shah (Eds.), *Models of working memory: Mechanisms of active maintenance and executive control* (pp. 1–27). New York, NY: Cambridge University Press.

Sharp Brains. (2013). Executive summary: Infographic on the digital brain health market 2012–2020. Retrieved from http://sharpbrains.com/executive-summary/

Shipstead, Z., Redick, T. S., & Engle, R. W. (2012). Is working memory training effective? *Psychological Bulletin, 138*(4), 628–654. doi:10.1037/a0027473

Simos, P. G., Fletcher, J. M., Bergman, E., Breier, J. I., Foorman, B. R., Castillo, E. M., . . . Papanicolaou, A. C. (2002). Dyslexia-specific brain activation profile becomes normal following successful remedial training. *Neurology, 58*(8), 1203–1213. doi:10.1212/WNL.58.8.1203

Sprenger, A. M., Atkins, S. M., Bolger, D. J., Harbison, J. I., Novick, J. M., Chrabaszcz, J. S., . . . Dougherty, M. R. (2013). Training working memory: Limits of transfer. *Intelligence, 41*(5), 638–663. doi:10.1016/j.intell.2013.07.013

Stanford Center on Longevity. (2014). A consensus on the brain training industry from the scientific community. Retrieved from http://longevity3.stanford.edu/blog/2014/10/15/the-consensus-on-the-brain-training-industry-from-the-scientific-community/.

Stevens, R., Galloway, T., Wang, P., Berka, C., Tan, V., Wohlgemuth, T., . . . Buckles, R. (2012). Modeling the neurodynamic complexity of submarine navigation teams. *Computational and Mathematical Organization Theory, 19*(3), 346–369. doi:10.1007/s10588-012-9135-9

Stikic, M., Berka, C., Levendowski, D. J., Rubio, R. F., Tan, V., Korszen, S., . . . Wurzer, D. (2014). Modeling temporal sequences of cognitive state changes based on a combination of EEG-engagement, EEG-workload, and heart rate metrics. *Frontiers in Neuroscience, 8,* 342. doi:10.3389/fnins.2014.00342

Stillman, C. M., Feldman, H., Wambach, C. G., Howard, J. H., Jr., & Howard, D. V. (2014). Dispositional mindfulness is associated with reduced implicit learning. *Consciousness and Cognition, 28,* 141–150. doi:10.1016/j.concog.2014.07.002

Sweetow, R. W., & Henderson-Sabes, J. (2006). The need for and development of an adaptive listening and communication enhancement (LACE) program. *Journal of the American Academy of Audiology, 17*(8), 538–558. doi:10.3766/jaaa.17.8.2

Takeuchi, H., Sekiguchi, A., Taki, Y., Yokoyama, S., Yomogida, Y., Komuro, N., . . . Kawashima, R. (2010). Training of working memory impacts structural connectivity. *Journal of Neuroscience, 30*(9), 3297–3303. doi:10.1523/JNEUROSCI.4611-09.2010

Thompson, T. W., Hinds, O., Ghosh, S., Lala, N., Triantafyllou, C., Whitfield-Gabrieli, S., & Gabrieli, J. (2009). Training selective auditory attention with real-time fMRI feedback. *Neuroimage, 47*(S1), S65. doi:10.1016/S1053-8119(09)70339-8

Thompson, T. W., Waskom, M. L., & Gabrieli, J. D. (2016). Intensive working memory training produces functional changes in large-scale frontoparietal networks. *Journal of Cognitive Neuroscience, 28*(4), 575–588. doi:10.1162/jocn_a_00916

Thorndike, E. L., & Woodworth, R. S. (1901). The influence of improvement in one mental function upon the efficiency of other functions (I). *Psychological Review, 8*(3), 247–261. doi:10.1037/h0074898

Vernon, D., Egner, T., Cooper, N., Compton, T., Neilands, C., Sheri, A., & Gruzelier, J. (2003). The effect of training distinct neurofeedback protocols on aspects of cognitive performance. *International Journal of Psychophysiology, 47*(1), 75–85. doi:10.1016/S0167-8760(02)00091-0

Wager, T. D., & Smith, E. E. (2003). Neuroimaging studies of working memory: A meta-analysis. *Cognitive, Affective and Behavioral Neuroscience, 3*(4), 255–274. doi:10.3758/CABN.3.4.255

Wang, J.-R., & Hsieh, S. (2013). Neurofeedback training improves attention and working memory performance. *Clinical Neurophysiology, 124*(12), 2406–2420. doi:10.1016/j.clinph.2013.05.020

Whitmarsh, S., Udden, J., Barendregt, H., & Petersson, K. M. (2013). Mindfulness reduces habitual responding based on implicit knowledge: Evidence from artificial grammar learning. *Consciousness and Cognition, 22*(3), 833–845. doi:10.1016/j.concog.2013.05.007

Wild, C. J., Yusuf, A., Wilson, D. E., Peelle, J. E., Davis, M. H., & Johnsrude, I. S. (2012). Effortful listening: The processing of degraded speech depends critically on attention. *Journal of Neuroscience, 32*(40), 14010–14021. doi:10.1523/JNEUROSCI.1528-12.2012

Witte, M., Kober, S. E., Ninaus, M., Neuper, C., & Wood, G. (2013). Control beliefs can predict the ability to up-regulate sensorimotor rhythm during neurofeedback training. *Frontiers in Human Neuroscience, 7*, 478. doi:10.3389/fnhum.2013.00478

Wright, B. A., & Zhang, Y. (2009). A review of the generalization of auditory learning. *Philosophical Transactions of the Royal Society B: Biological Sciences, 364*(1515), 301–311. doi:10.1098/rstb.2008.0262

Wu, J., Srinivasan, R., Kaur, A., & Cramer, S. C. (2014). Resting-state cortical connectivity predicts motor skill acquisition. *Neuroimage, 91*, 84–90. doi:10.1016/j.neuroimage.2014.01.026

Zekveld, A. A., Heslenfeld, D. J., Festen, J. M., & Schoonhoven, R. (2006). Top-down and bottom-up processes in speech comprehension. *Neuroimage, 32*(4), 1826–1836. doi:10.1016/j.neuroimage.2006.04.199

Zoefel, B., Huster, R. J., & Herrmann, C. S. (2011). Neurofeedback training of the upper alpha frequency band in EEG improves cognitive performance. *Neuroimage, 54*(2), 1427–1431. doi:10.1016/j.neuroimage.2010.08.078

Zotev, V., Phillips, R., Yuan, H., Misaki, M., & Bodurka, J. (2014). Self-regulation of human brain activity using simultaneous real-time fMRI and EEG neurofeedback. *Neuroimage, 85*, 985–995. doi:10.1016/j.neuroimage.2013.04.126

6

Working Memory Training and Transcranial Direct Current Stimulation

Jacky Au, Martin Buschkuehl, and Susanne M. Jaeggi

Introduction

In Chapter 5, Kuchinsky and Haarmann provide an important historical perspective on cognitive training, noting the interest in psychological interventions to improve cognitive abilities in the early 20th century, during the time of Alfred Binet, forefather of modern IQ testing. However, scientific interest in this matter extends farther back than that, with documented attempts at mental enhancement via electrical stimulation methods dating back centuries earlier. Luigi Galvani's discovery of bioelectricity in the late 1700s brought widespread attention to the possibility of harnessing electricity to affect physiology after he successfully induced muscle twitches in dead frogs. This spurred on the perhaps over-exuberant use of electrical stimulation to treat all types of physical and mental maladies, ranging from headaches to neuroses and even blindness (Elliot, 2014). Despite anecdotal reports of both mental and physical improvements, the absence of strong scientific guidance and the unfavorable risk/reward ratio due to the high levels of current used at the time spurred skepticism and dampened enthusiasm, and the technology fell out of favor throughout most of the 1900s.

Although a few animal experiments were conducted in the latter half of the 20th century (e.g., Albert, 1966; Bindman, Lippold, & Redfearn, 1964; Creutzfeldt, Fromm, & Kapp, 1962), followed by a few human trials in depression that went largely unnoticed (Lippold & Redfearn, 1964; Lolas, 1977), it was not until the turn of the millennium that widespread interest in electrical brain stimulation resurged, beginning with a group of researchers from Germany who demonstrated the ability to reliably, selectively, and noninvasively modulate the excitability of the human motor cortex via weak direct currents introduced across the scalp (Nitsche & Paulus, 2000). The researchers' approach was to measure motor evoked potentials (MEPs) recorded after eliciting a small muscle contraction near the right pinky finger using a well-established protocol involving transcranial magnetic stimulation (Rothwell, 1991; Rothwell et al., 1987). When the researchers preceded this protocol with a weak electric current delivered to the motor cortical region corresponding to the representation of the right pinky finger, they found that the electrical activity recorded from the hand selectively increased or decreased in a manner dependent on the polarity of the delivered current. The same research group followed up their initial study by demonstrating that the excitability changes in the motor cortex were detectable by magnetic resonance imaging (MRI; Baudewig, Nitsche, Paulus, & Frahm, 2001), were replicable in the visual cortex (Antal, Kincses, Nitsche, Bartfai, & Paulus, 2004; Antal, Nitsche, & Paulus, 2001), and could also affect motor and visuomotor learning and plasticity paradigms (Antal, Nitsche, et al., 2004; Nitsche, Schauenburg, et al., 2003; Rosenkranz, Nitsche, Tergau, & Paulus, 2000).

This gradual evolution from the study of neurophysiological to behavioral effects led to the hypothesis that brain stimulation could also affect higher cognitive functions. Moving on from the motor and visual cortices, researchers next demonstrated that stimulation of the dorsolateral prefrontal cortex (DLPFC), a region well known to be involved with many higher cognitive functions, enhanced accuracy on a working memory (WM) task (Fregni et al., 2005). However, stimulation of the motor cortex—as done in previous studies—had no effect on WM, supporting the idea that stimulation effects are specific to the function of the targeted brain region. These seminal studies paved the way for a renaissance of brain stimulation research, with many new researchers in the field seeking to replicate and extend the initial findings, causing an exponential increase in the number of studies published on the topic since the turn of the millennium (Voytek & Gazzaley, 2013). Thus, an ancient fascination re-emerged under the auspices of modern-day scientific rigor.

The time frame of the resurgence propitiously matched up with the rise of interest in computerized cognitive training that Kuchinsky and Haarmann noted. Therefore, an enticing avenue of research is to evaluate the synergistic combination of brain stimulation and cognitive training in service of a common underlying goal—cognitive enhancement. Kuchinsky and Haarmann (Chapter 5) state that their main aim is to "spark a discussion about how cognitive neuroscience research can aid in the evaluation and development of effective cognitive training protocols" (p. 98). Accordingly, this chapter aims to contribute to the discussion with respect to the cognitive neuroscience of brain stimulation. In doing so, we emphasize work from our own laboratory that focuses both on WM training and transcranial direct current stimulation (tDCS).

What Is Transcranial Direct Current Stimulation?

tDCS is one of the most commonly used and extensively researched methods of transcranial electrical stimulation (tES). Typically, it involves the application of two sponge electrodes,[1] an anode and a cathode, to the scalp. tDCS is thought to modify cortical excitability by altering the relative ionic distribution across neural membranes. Importantly, the stimulation is not enough to elicit neural firing, but it can lead to polarity-specific increases or decreases in the resting membrane potential underneath the anode or cathode, respectively (Stagg & Nitsche, 2011). Much evidence supports this anodal-excitatory/cathodal-inhibitory (AeCi) model of membrane polarization, drawn mostly from early in vivo animal studies (Bindman et al., 1964; Creutzfeldt et al., 1962). More recently, pharmacological studies in humans have demonstrated sodium-channel blockade of neural membranes is sufficient to abolish anodal but not cathodal tDCS effects by restricting neural firing (Liebetanz, Nitsche, Tergau, & Paulus, 2002; Nitsche, Fricke, et al., 2003). However, as it turns out, the AeCi model is an oversimplification. Although it extrapolated well to the simple motor studies of the early 2000s that involved stimulation of isolated and relatively self-contained brain regions, studies in the cognitive domain, which involve intricate functional networks, present a more complex, often nonlinear relationship between behavior and stimulation polarity (Jacobson, Koslowsky, & Lavidor, 2012). Therefore, actual mechanisms of action are likely more complicated, involving not only changes in resting membrane potential, but also changes in the synaptic micro-environment, such as alterations in GABAergic and glutamatergic activity, as well as long-term-potentiation-like plasticity (Stagg & Nitsche, 2011).

This chapter focuses on the implementation of tDCS to enhance and inform research on WM training; thus, a more thorough mechanistic account is not within its scope. However, interested readers are referred to other sources for current, albeit incomplete, knowledge on the underlying mechanisms of tDCS (Fertonani & Miniussi, 2016; Stagg & Nitsche, 2011). While respecting the intricacies and unknowns of the inner workings of electrical stimulation on the brain, we move forward under the premise that tDCS is able to directly affect the electrophysiological profile of the brain, and we provide evidence that this in turn can influence behavior, given the right parameters.

Cognitive Effects of Transcranial Direct Current Stimulation

Given early demonstrations of the efficacy of tDCS in affecting cortical excitability across a variety of domains, including motor (Baudewig et al., 2001; Nitsche & Paulus, 2000; Rosenkranz et al., 2000), visual (Accornero, Li Voti, La Riccia, & Gregori, 2007; Antal, Kincses, et al., 2004; Antal, Nitsche, et al., 2004),

---

[1] Although this is the most common setup, other montages exist. See the chapter's conclusion for discussion.

and somatosensory (Matsunaga, Nitsche, Tsuji, & Rothwell, 2004), the prospect of harnessing these excitability changes for cognitive enhancement became the logical next step for many researchers. This idea was particularly appealing due to the subthreshold nature of the neuromodulatory effects. Since the weak currents of tDCS do not directly cause action potentials but merely increase the probabilistic firing rate of target neurons, tDCS is a natural tool for potentiating ongoing cognitive activity. Accordingly, it has become commonplace to administer tDCS concomitantly with a cognitive task to specifically enhance the neural networks activated by that task.

One of the first such studies was done by Fregni et al. (2005), who paired 10 minutes of 1-mA tDCS[2] with a 3-back task. The 3-back task is a specific variant of the family of N-back tasks, which are popular paradigms in the WM field and require monitoring of a continuous stream of stimuli to indicate whether the current stimulus matches the one N positions back in the stream. Fregni et al. (2005) showed that anodal stimulation to the left DLPFC resulted in increased 3-back accuracy compared to a sham condition in which participants were connected to a tDCS device but current was turned off after 15–30 seconds without their knowledge. Since then, research on tDCS for cognitive enhancement has boomed, with many successful studies (reviewed in Bennabi et al., 2014; Coffman, Clark, & Parasuraman, 2014), although others were less so (cf. Antal, Keeser, Priori, Padberg, & Nitsche, 2015; Chhatbar & Feng, 2015; Horvath, Forte, & Carter, 2015b). Nevertheless, the mixed effects aggregate into reliable signals across studies, especially for more vulnerable populations, such as neuropsychiatric patients and older adults, where meta-analyses show moderate to large effect sizes (Dedoncker, Brunoni, Baeken, & Vanderhasselt, 2016; Hsu, Ku, Zanto, & Gazzaley, 2015; Summers, Kang, & Cauraugh, 2015). Effect sizes are smaller and more variable in healthy young adults (Dedoncker et al., 2016; Hill, Fitzgerald, & Hoy, 2016; Mancuso, Ilieva, Hamilton, & Farah, 2016).

Determining the cause of the inconsistent results is a challenge, mainly because tDCS protocols have many degrees of freedom, much like other common cognitive neuroscience techniques, such as functional magnetic resonance imaging (fMRI) and electroencephalography (EEG; Carp, 2012; Churchill et al., 2012; Nuwer, 1988). There are a multitude of parameters that remain to be optimized, resulting in much methodological heterogeneity between studies, such as differences in current intensity, stimulated region, electrode size, electrode montage, length of stimulation, number of sessions, choice of cognitive

---

[2] The current applied to the scalp is measured in milliamperes (mA), and in tDCS applications with humans, it typically ranges between 1 and 2 mA. In this particular study, the current was set to 1 mA.

task, and others. It is possible, and even plausible, that a more thorough mechanistic understanding can help optimize some of the parameters to increase the intervention's effect size in future studies.

For example, despite the small effect sizes found among healthy young adults, some evidence suggests that the studies that applied higher levels and densities of current for longer amounts of time achieved greater enhancements (Dedoncker et al., 2016; Hill et al., 2016). More relevant to the purposes of this chapter, though, is evidence indicating that the strongest effects in healthy young adults manifested from the pairing of WM training with tDCS (Elmasry, Loo, & Martin, 2015; Mancuso et al., 2016). Since this line of work is also the focus in our own laboratory, we review the supporting evidence, drawing both from the literature and our own work to demonstrate how tDCS can be leveraged to enhance and inform WM training (see also Au, Karsten, Buschkuehl, & Jaeggi, 2017).

Pairing WM Training With Transcranial Direct Current Stimulation

The conjunctive use of WM training and tDCS typically involves a pretest → posttest intervention design (as do the two fields upon which it draws). The intervention usually involves stimulating participants while they are training on a WM task and ranges in duration anywhere from one to 10 sessions in the current literature. The tDCS + training group is usually contrasted with a sham + training group, a no-training group, or both. Sham stimulation is generally achieved by ramping the current up for the first 30 seconds of the training, so that participants feel the initial sensations, but participants are then unaware that the current has been turned off after the 30 seconds. Since the sensations associated with tDCS are often subtle and tend to attenuate after the initial ramp-on period, the protocol lends itself well to participant blinding in sham procedures (Russo, Wallace, Fitzgerald, & Cooper, 2013).

The pretests and posttests consist of testing on transfer tasks in the absence of stimulation to assess the generalizability of training benefits. The transfer tasks involve measures of WM or a theoretically related domain and may or may not overlap with the trained WM task. Therefore, when evaluating the effects of such designs, at least two issues are important to consider with respect to the timing of tDCS administration. One is the effect on WM performance during stimulation (online effects) and another is the effect on performance after stimulation (offline effects). These parameters can influence WM training and transfer results, which have implications for the applicability and utility of the field at large.

Online and Offline Effects

Online stimulation effects are thought to be driven primarily (but not exclusively) by excitability shifts in neural membrane potential, which allow greater neural recruitment in a target region of interest during WM-task performance. The majority of single-session WM-tDCS studies rely on these effects, based on the premise that tDCS works best on an active brain by potentiating ongoing cognitive activity. However, these designs leave untapped an important element of the tDCS effect on cognition: offline effects. Offline effects are measured by task performance after a period of stimulation and seem to at least partly overlap with the mechanisms of online effects, at least in the minutes after stimulation. For example, pharmacological studies show that sodium-channel blockers, which act to desensitize neural membranes and restrict depolarization, are able to abolish offline effects minutes after stimulation in the same way they abolish the effects of online stimulation (Liebetanz et al., 2002; Nitsche, Fricke, et al., 2003). This suggests tDCS induces residual membrane polarization that takes some time to return to baseline even after the electrodes are removed, which explains why performance enhancement on behavioral tasks can persist for up to an hour (Hsu, Zanto, Anguera, Lin, & Gazzaley, 2015; Nitsche, Nitsche, et al., 2003; Nitsche & Paulus, 2001; Ohn et al., 2008).

There are also distinct mechanisms to offline effects, as evidenced by use of NMDA antagonists, which are capable of blocking offline, but not online, effects (Stagg & Nitsche, 2011). Since NMDA receptors are critical for learning and memory formation, work with NMDA antagonists suggests the potential for consolidative processes to have a role in the manifestation of offline effects. In fact, a critical study demonstrated that stimulating during a WM task on Day 1 manifested task-related enhancements the next day (Martin, Liu, Alonzo, Green, & Loo, 2014). Since no stimulation was administered on Day 2, the enhancements can be considered offline effects. Importantly, the next-day enhancements were demonstrated only when participants received online, but not offline, stimulation on Day 1. The authors argued that online stimulation concurrent with WM engagement on Day 1 was necessary to promote offline consolidation of task-relevant WM networks that boosted performance on Day 2. Therefore, WM training studies, which regularly make use of measuring both online effects (i.e., the training task) and offline effects (i.e., transfer tests), may be optimally suited to more consistently manifest tDCS effects compared to single-session studies.

tDCS Effects on Training

When evaluating the use of tDCS for WM training, the session-by-session performance on the trained task is the first area in which to expect enhancements. If tDCS is to be of any added value to the outcomes of WM training, expecting it to enhance the quality of training is reasonable, much like an athlete would expect a nutritional supplement to enhance the quality of workouts. Using tDCS to boost training effects in this manner not only allows researchers to better probe the extent to which performance on a particular training regimen is malleable, but also potentially increases the effect size of transfer to untrained tasks, assuming that more gain in training leads to more pronounced transfer effects (cf. Jaeggi, Buschkuehl, Jonides, & Shah, 2011; Novick, Hussey, Teubner-Rhodes, Harbison, & Bunting, 2014).

Work from our laboratory confirms that tDCS can be used to improve training performance in such a manner (Au, Katz, et al., 2016). We randomly assigned healthy young, college-aged adults to receive 25 minutes of 2-mA stimulation either to the left or right DLPFC while training on the N-back task (see Au et al., 2015). We compared the tDCS + training groups to a sham + training group. Over seven training days, we found that both tDCS + training groups outperformed the sham + training group, although they performed similarly relative to each other. We further demonstrated that the effects could not be explained by self-reported motivational factors or alertness, and that participants could not reliably distinguish their condition, thus minimizing the potential impact of placebo effects.

We are not alone in demonstrating this effect in healthy young adults. Richmond, Wolk, Chein, and Olson (2014) also found similar training effects during a 10-day WM training regimen using left DLPFC stimulation. However, they found significant effects only on the verbal portion of the training, not the visual. The relevance of this verbal/visual distinction is discussed later on, but, for now, we note that evidence for training enhancements through tDCS is sometimes mixed, much as in the single-session studies. For example, Martin and colleagues (2013) also trained participants on an N-back task while stimulating the left DLPFC, but the group differences they observed were not significant after controlling for differences in baseline performance. However, important methodological differences exist between these studies and ours that should be considered. For example, Richmond and colleagues trained participants mostly offline (i.e., after stimulation) while ours trained online. Also, Martin and colleagues used a much more difficult dual N-back task that renders improvement more difficult to discern. These differences notwithstanding, both studies still showed strong trends and respectable effect sizes in the training data in favor of the active stimulation groups, suggesting at least part of the mixed replication evidence may be attributable to sampling and power issues.

The enhancements observed in training performance with the use of tDCS, beyond merely being suggestive of greater transfer results (as discussed below), have further ramifications for the study of consolidation and plasticity with respect to the training of WM. Such consolidation mechanisms are known to play an integral role in determining the efficacy of WM training, irrespective of tDCS. For example, a period of sleep after training has been shown to enhance WM performance the next day (Kuriyama, Mishima, Suzuki, Aritake, & Uchiyama, 2008; Scullin, Trotti, Wilson, Greer, & Bliwise, 2012), and levels of dopamine, particularly in the prefrontal cortex, have been shown to correlate positively with training performance (Bäckman & Nyberg, 2013; Bellander et al., 2011; Soderqvist et al., 2012). Both sleep and dopamine levels play critical roles in the consolidation of learning and memory (Walker & Stickgold, 2004; Wise, 2004), and it is reasonable to assume that these and similar processes impose biological constraints on the degree of plasticity observed after WM training.

The putative offline effects of tDCS on enhancing consolidation may present a viable route for mitigating these constraints. These effects have already been demonstrated several times in longitudinal studies of motor-skill learning, in which training gains seem to accumulate over multiple sessions (Boggio et al., 2006; Reis et al., 2009). Martin and colleagues (2014) were the first to begin exploring this in the cognitive domain, demonstrating WM improvements 24 hours after stimulation. We took this one step further in our work by demonstrating cumulative gains over seven days of N-back training, with a steeper rate of improvement in the stimulated groups than in the sham groups. This suggests that offline learning effects occurred between sessions as a result of stimulation. In fact, we posit that the offline effects may explain some of the inconsistency in the literature among single-session studies. It may be that a significant part of the stimulation effect is not captured because studies did not follow up on participants the next day. In support of this idea, our data do not begin to show significant differences between active and sham groups until several days into the training regimen, after putative tDCS-enhanced consolidation processes have had a chance to take effect.

One practical application of tDCS consolidation effects is in studying the role of spacing to inform optimal training schedules. A broad and enduring literature extending back more than a century (Ebbinghaus, 1913) promulgates the benefits of spacing out learning over time, as opposed to massed learning all at once, to take advantage of consolidation mechanisms in between learning blocks. Despite these well-known effects, the optimal training schedule has still been a relatively understudied factor in WM training studies. Wang, Zhou, and Shah (2014) were among the first to explore this in a systematic manner, varying the spacing of 20 sessions across two, five, 10, or 20 days. Although they succeeded in demonstrating that maximally spaced sessions promoted the greatest transfer benefit, they were unable to demonstrate any differences in training performance as a function of training schedule. Outside the domain of WM, Stafford and Dewar (2014) found that spaced video-game training benefited performance in a massive online database of over 800,000 participants. However, due to the quasi-experimental nature of this online study, they were unable to control the parameters of spacing, and participants self-selected their own training schedules, thereby limiting the amount of inference that can be drawn from the study.

Despite the difficulty in convincingly demonstrating this well-established effect in the context of standard WM training protocols, we reported that the addition of tDCS can manifest spacing effects after a weekend break. Since our seven-day training was naturally interrupted by a weekend, during which participants did not come into the lab, we analyzed the gain in performance between Friday and Monday, as compared to consecutive training days, controlling for the total number of previous training sessions. We found that participants performed significantly better after the weekend break if they had been receiving active tDCS. We did not find these effects with sham stimulation, corroborating previous results (Wang et al., 2014) that WM training alone is insufficient to manifest convincing spacing effects on training performance. Nevertheless, in light of the pervasive literature on spacing effects (e.g., Cepeda, Pashler, Vul, Wixted, & Rohrer, 2006; Reynolds & Glaser, 1964) and the limited evidence to date with training studies (Stafford & Dewar, 2014; Wang et al., 2014), it is still reasonable to assume that spacing effects in WM training exist, even if they are difficult to detect using standard protocols. The use of tDCS, therefore, may facilitate our ability to detect these effects and may inform the design of future training studies, particularly because spacing effects may have direct ramifications for manifesting transfer effects (Wang et al., 2014).

tDCS Effects on Transfer

Although a demonstration of enhanced training performance is a critical first step in determining the feasibility and utility of tDCS for WM training, the ultimate goal of many training studies is to demonstrate transfer of training to untrained tasks. Such transfer can be understood along a spectrum ranging from near to far transfer (Barnett & Ceci, 2002). The former represents transfer of training gain onto similar, but untrained, tasks of WM. Such effects are generally accepted to occur (Karbach & Verhaeghen, 2014; Schwaighofer, Fischer, & Bühner, 2015; Weicker, Villringer, & Thone-Otto, 2016), thus demonstrating that training can render improvements in WM performance at the construct level. Far transfer, on the other end of the spectrum, represents transfer of training gain to different cognitive domains, such as attention or fluid intelligence, and is more controversial. Although far transfer can sometimes be difficult to detect in primary studies (e.g., Chooi & Thompson, 2012; Harrison et al., 2013;

Redick et al., 2013), it is commonly reported in meta-analyses (e.g., Au et al., 2015; Karbach & Verhaeghen, 2014; Weicker et al., 2016). This pattern of results suggests that far-transfer effects exist but are too small to be reliably detected by individual studies in the field, which typically employ small sample sizes that underpower the analyses (Bogg & Lasecki, 2014). Accordingly, a few large-scale training studies with sample sizes in the hundreds or thousands (Corbett et al., 2015; Hardy et al., 2015; Schmiedek, Lövdén, & Lindenberger, 2010; Willis et al., 2006) have been successful in demonstrating far transfer into domains like reasoning and, for older adults, activities of daily living (but see also Owen et al., 2010; Roberts et al., 2016). The logistics of running a training study of this magnitude are daunting, requiring substantial investments from the participants as well as the researchers. Not only is funding an important limiting factor, but attrition rates are high, both in terms of the participants, who are asked to come into the laboratory to engage in demanding cognitive tasks multiple times per week over an extended period of time, and for the research staff, many of whom are young students or recent graduates who eventually move on in their careers during the years that such studies generally take to complete. Therefore, since practical constraints often limit both the length of training interventions and the sample sizes recruited, the field would benefit from a catalyst that can both expedite and intensify transfer effects. There is evidence that argues for the use of tDCS as one such catalyst.

The first studies pairing tDCS with WM training in healthy young adults had some limited success (Martin et al., 2013; Richmond et al., 2014). Despite some partial success in demonstrating enhanced training effects, neither study was able to demonstrate any differential transfer effects when comparing the tDCS + training group to the sham + training group. However, both studies found transfer to untrained WM measures in the contrast between tDCS + training and no-training controls. These effects were not found in the contrast between sham + training and no-training, suggesting that standard training paradigms were insufficient to reveal transfer effects, but that the addition of tDCS facilitated the detection of effects that may have otherwise been obscured by noise.

Although these initial studies suggested limits to the enhancements that tDCS can confer, since the effects were unable to manifest over and above the effects of regular training, other studies since then have been more successful. For example, our study demonstrated that such transfer can occur under the right conditions (Au, Katz, et al., 2016). We compared both our tDCS groups (right DLPFC and left DLPFC) to the sham group, all of whom trained on a visuospatial WM task, and found selective transfer to two untrained visual WM tasks at posttest, but only in the tDCS group who received right, but not left, DLPFC stimulation. Interestingly, we also observed a modest increase on the forward digit-span task, but only in the group receiving left DLPFC stimulation. Though this effect did not reach significance in the contrast against the sham group, it was significant in the contrast against the right DLPFC group. We interpret these combined findings amid previous evidence that supports a functional lateralization of the DLPFC such that the right hemisphere is more associated with visual WM performance and the left with verbal WM performance (Smith, Jonides, & Koeppe, 1996; Wager & Smith, 2003). Although both tDCS groups improved comparably on the trained task, we posit that only the right-DLPFC group experienced offline consolidation of a visuospatial WM network that enhanced performance at posttest on visual WM tests. Our findings also complement those of Richmond and colleagues (2014), who found that left DLPFC stimulation was able to enhance verbal, but not spatial, WM performance. Together, the combined evidence highlights the specificity of the tDCS effect and suggests that strategic targeting can enhance the strength of the effect (Bikson & Rahman, 2013).

The promising evidence that tDCS facilitates transfer in WM training studies—as demonstrated with healthy young adults—is even more compelling in older adults. This is consistent with the single-session literature, where meta-analyses also report stronger effects of tDCS in older adults than in younger adults (Hsu, Ku, et al., 2015; Summers et al., 2015). For example, Park, Seo, Kim, and Ko (2014) reported that 10 days of WM training improved transfer to verbal WM performance in healthy older adults over 65 years of age who received anodal stimulation bilaterally to the right and left DLPFC (relative to sham + training). Moreover, the improvements persisted for at least one month posttraining, lending further credence to the notion that tDCS may operate offline to enhance consolidation and retention. Similarly, Jones, Stephens, Alam, Bikson, and Berryhill (2015) applied frontoparietal anodal stimulation to healthy older adults over the age of 55 during WM training. All participants, regardless of whether they received sham or active stimulation, benefited from the training at posttest and demonstrated general transfer to trained and untrained tasks of WM and inhibitory control. Thus, the effects of tDCS did not improve upon the effects of standard training, which was already quite effective in this study. Importantly, however, the same test battery was repeated one month after the intervention period, and while the sham + training group began to show signs of regressing toward their baseline performance, the tDCS + training group still demonstrated transfer effects, which were significantly different than findings in the sham + training group at that time point. Therefore, while there appear to be natural limits to the longevity of standard training, as would be expected from any training program whether cognitive, physical, or otherwise, tDCS may be able to prolong the effects to increase their durability.

To support and extend this idea, the same group ran a follow-up study, also on healthy older adults, and found far-transfer effects one month after intervention with 2-mA stimulation to the right DLPFC (Stephens & Berryhill, 2016). The far-transfer effects occurred for laboratory-based measures (processing speed, arithmetic, and cognitive flexibility) as well as ecologically valid tasks of everyday living, such as calendar planning and driving awareness. No such effects were found in the sham + training group, nor even in a weaker tDCS + training group who received only 1-mA stimulation. Since this study did not measure performance immediately after training ended, but one month later instead, how the sham + training group would have fared immediately after training is unclear. Therefore, it is uncertain whether the effects present in the 2-mA tDCS group represent transfer over and above the sham group or whether tDCS merely enhances the durability of transfer benefits over a month-long period. Regardless, either scenario would render tDCS a useful adjunct to WM training, and the consistent demonstration of long-term effects across multiple studies supports the practicality of the combined intervention for cognitive enhancement in both young and older adults.

How Does Transcranial Direct Current Stimulation Inform Cognitive Training Research?

Kuchinsky and Haarmann (Chapter 5) argue, much as we do, that cognitive neuroscience techniques can do much to assist in the design and evaluation of cognitive training research. They distilled the cognitive training field into a few of the most pertinent questions that researchers try to answer and discussed them in light of the data cognitive neuroscience has to offer. We now further this discussion from the perspective of the field of tDCS.

Impact of WM Training and tDCS on Cognitive Processes

The first question is whether cognitive training can effectively target desired cognitive systems. This is a contentious issue in behavioral studies, because it is virtually impossible to design a cognitive training task that isolates only one cognitive process. Multiple processes are always involved, and, as Kuchinsky and Haarmann point out, mismatches in neural overlap between training and transfer tasks can obscure results. Moreover, improvement on a cognitive task can arise from any of a number of different sources besides a change in the desired cognitive system. Alternative explanations include, but are not limited to, expectancy effects (Boot, Simons, Stothart, & Stutts, 2013), a shift in strategy use (Dunning & Holmes, 2014; Kelly, Foxe, & Garavan, 2006; Morrison & Chein, 2011), improvements in unrelated perceptual/motor skills, or even test–retest effects (Colom et al., 2010). However, tDCS confers a degree of specificity to its effects through its ability to physically target a desired brain region with known functions (Bikson et al., 2013). For example, changes in MEP, one of the earliest and best-studied effects of tDCS, are induced by stimulation over the primary motor cortex, but not other regions (Horvath, Forte, & Carter, 2015a). Conversely, WM improvements are observed with stimulation over the DLPFC or parietal cortex (Dedoncker et al., 2016; Jones & Berryhill, 2012; Jones et al., 2015; Tremblay et al., 2014) but not the motor cortex (Fregni et al., 2005). Moreover, even stimulation to the same region (e.g., the DLPFC) can target distinct behavioral outcomes depending on the initial state of the brain (Tremblay et al., 2014). The brain state can be manipulated, for example, by introducing different cognitive tasks during stimulation, each of which can put the DLPFC into a different prepotent state depending on the particular neural networks primed by the task. It should also be noted, though, that the state-dependency of tDCS is not well understood and can also lead to unpredictable outcomes that reduce reliability between studies (Tremblay et al., 2014). Nevertheless, evidence suggests that both the anatomical and state-selectivity of tDCS can affect cognitive-training outcomes, such as using right-DLPFC stimulation during a visuospatial WM-training task to selectively boost visual, but not verbal, WM outcomes, or left-DLPFC stimulation to boost verbal, but not visuospatial, WM-training performance. Additionally, given that tDCS alone, uncoupled from a cognitive task, does not usually result in any measureable cognitive changes in healthy participants (Andrews, Hoy, Enticott, Daskalakis, & Fitzgerald, 2011; Bikson et al., 2013; Martin et al., 2013), the selective enhancements to verbal and visual WM performance suggest that the training tasks in question are able to target the desired cognitive systems, and that the effects are boosted with the strategic placement of tDCS.

Impact of WM Training and tDCS on Daily Life

If it can be assumed that the training task effectively targets the relevant cognitive processes, then the next questions to ask address practicality. Does the training generalize to untrained tasks, and do these effects persist over time? The mixed nature of transfer effects, and the possible underlying causes, are mentioned several times in this chapter and by Kuchinsky and Haarmann (Chapter 5). The effects are generally robust at the near-transfer level (Karbach & Verhaeghen, 2014; Melby-Lervåg & Hulme, 2013; Schwaighofer et al., 2015; Weicker et al., 2016) but are hotly debated at the far-transfer level (Au, Buschkuehl, Duncan, & Jaeggi, 2016; Au et al., 2015; Melby-Lervåg & Hulme, 2015). The question, therefore, is not whether WM training is effective, but the degree to which it is effective and the translatability or practicality of these effects.

From this perspective, a catalyst that can facilitate training efficacy can increase the signal-to-noise ratio in the field and allow more probing investigations into the limits of training and transfer. We argue throughout this chapter that tDCS can serve as one such catalyst and we provide evidence of its ability to manifest transfer effects that would otherwise be obscured in standard training regimens relative to no-training groups (Martin et al., 2013; Richmond et al., 2014). Moreover, we demonstrate that, under the right conditions, the tDCS effect can even manifest over and above regular training effects when tDCS + training groups are compared to sham + training groups (Au, Katz, et al., 2016; Park et al., 2014). Furthermore, not only might tDCS strengthen the effect of training, but it also carries with it putative offline effects on consolidation that can enhance the durability of training benefits. Training enhancements have remained present for up to a year (Au, Katz, et al., 2016; Katz et al., 2017), and transfer effects have maintained for up to one month (Jones et al., 2015; Martin et al., 2013; Stephens & Berryhill, 2016), including far transfer to real-world measures (Stephens & Berryhill, 2016), which have been traditionally very difficult to demonstrate with standard training studies (but see Rebok et al. 2014, Weicker et al., 2016; Willis et al., 2006).

Individual Differences of WM Training and tDCS

Amid the mixed WM-training literature, a question arises regarding whether training is more effective for some individuals than for others (Jaeggi, Buschkuehl, Shah, & Jonides, 2014). Individual differences in motivation and personality have been found to play a role, as have age and various neural markers, such as genotype, resting-state EEG coherence, and frontal alpha-wave power (discussed in Chapter 5). These individual differences, which separate low from high performers, present a rewarding avenue of research for understanding the underlying mechanisms of what makes training successful. Uncovering the mechanisms may allow us to adapt training protocols to be more accessible to otherwise low-performing populations, as has already been demonstrated twice (Katz et al., 2017; Looi et al., 2016). For example, Kuchinsky and Haarmann (Chapter 5) discuss the prospect of manipulating the brain states of low performers to more closely resemble those of high performers. Although much remains to be discovered concerning the precise neural profile of a "high performer" (cf. Jaeggi et al., 2007), tDCS, with its established neuromodulatory effects, presents a viable option to explore this issue.

Aside from the behavioral effects already discussed, measured electrophysiological changes have also been reported after stimulation of WM-relevant areas. For example, Accornero and colleagues (2014) stimulated the DLPFC at rest and demonstrated polarity-dependent changes in the mean frequency of brain oscillations, while Romero-Lauro and colleagues (2014) used anodal stimulation on the posterior parietal cortex and observed increased evoked potentials elicited by transcranial magnetic stimulation in a procedure similar to the MEP studies described previously. Furthermore, Peña-Gómez and colleagues (2012) reported decreased coherence within the default-mode network and increased coherence within the anti-correlated network after anodal stimulation over the DLPFC. In the healthy brain, these two resting-state networks tend to correlate inversely with each other and play complex roles in many cognitive functions, such as learning and memory. The direction of the described tDCS effects on their functional connectivity implies a brain state better primed for cognitive demand (Keeser et al., 2011; Peña-Gómez et al., 2012), consistent with other studies showing increases in beta-frequency power during resting DLPFC stimulation (Mangia, Pirini, & Cappello, 2014; Song, Shin, & Yun, 2014). In fact, the application of tDCS during cognitive demand (i.e., online stimulation) also has similar neuromodulatory effects that increase the readiness of the WM network, particularly increasing event-related alpha- and theta-power concomitant with behavioral WM improvements (Heimrath, Sandmann, Becke, Müller, & Zaehle, 2012; Zaehle, Sandmann, Thorne, Jäncke, & Herrmann, 2011). Power in these frequency bands has been associated with WM performance (Klimesch, 1999) and has been observed to increase as a function of training (Gevins, Smith, McEvoy, & Yu, 1997). Therefore, stimulation to WM-relevant areas of the brain does seem to alter the electrophysiological profile of the brain in ways consistent with supporting greater WM performance.

Despite the promising electrophysiological effects of tDCS, as measured across multiple studies, efforts to manipulate the brains of low performers to resemble those of high performers are stymied by a lack of mechanistic understanding of the neuromodulatory effects of tDCS as well as the neural profile that distinguishes successful from unsuccessful cognitive training. However, at least with respect to the latter issue, one individual difference factor, age, has reasonably well-established neural correlates. Age-associated WM decline is uncontroversial, and numerous studies indicate that older adults do not gain as much from training WM as do younger adults (Borella et al., 2014; Burki, Ludwig, Chicherio, & de Ribaupierre, 2014; Dahlin, Neely, Larsson, Bäckman, & Nyberg, 2008). The deficits in WM performance and plasticity are typically associated with hyperactivity in the DLPFC bilaterally, which is thought to reflect compensatory mechanisms to cope with demanding WM tasks (Berlingeri, Danelli, Bottini, Sberna, & Paulesu, 2013; Cabeza, 2002; Reuter-Lorenz & Cappell, 2008). Additionally, functional connectivity in resting-state networks known to subserve cognition, such as the default mode network, is reduced (Damoiseaux et al., 2008; Ng, Lo, Lim, Chee, & Zhou, 2016).

Remarkably, an exciting study led by Meinzer and colleagues showed that a single session of anodal tDCS over frontal cortex not only improved cognitive performance on a word-generation task but also temporarily reversed many of the age-related neural changes as measured by fMRI (Meinzer, Lindenberg, Antonenko, Flaisch, & Flöel, 2013). In fact, after application of tDCS, the behavioral performance, prefrontal activity, and resting-state functional connectivity in the group of older adults studied were all statistically indistinguishable from those of a group of young adults. Although not all studies have been as successful in manifesting such promising results in the aging population (Nilsson, Lebedev, & Lövdén, 2015), converging evidence suggests tDCS can be especially effective for older adults and other low-performing populations (Dedoncker et al., 2016; Hill et al., 2016; Hsu, Ku, et al., 2015; Looi et al., 2016; Summers et al., 2015). Moreover, studies pairing training with tDCS have so far been quite successful in the older adult population, with durable gains and far-transfer benefits seen in both laboratory tasks as well as real-life activities of daily living (Jones et al., 2015; Park et al., 2014; Stephens & Berryhill, 2016). Therefore, the use of tDCS may allow WM-training researchers to broaden their reach to include populations typically underserved by traditional methods of computerized training.

Final Remarks

Throughout this chapter, we provide evidence for the efficacy of WM training and tDCS for cognitive enhancement and we argue that their combined use produces synergistic effects that can both enhance our understanding of cognitive plasticity as well as promote greater training and transfer gains. Yet, although tDCS is promising, tDCS in its modern manifestation is still a relatively new technology, and much more work needs to be carried out to understand it and subsequently optimize cognitive interventions.

Another important avenue for future research is to replicate tDCS-enhanced training effects using standard protocols to evaluate whether a relationship can be delineated that quantifies the added boost of tDCS. For example, one simple question to explore is how many training sessions without tDCS are required to replicate the effects of a given number of training sessions with tDCS. Clarifying such relationships would go a long way toward being able to adapt the research findings from the tDCS field into standard cognitive training, which ultimately is more practical and accessible to daily life.

Also, although most of the studies described herein use standard bipolar/ bifocal tDCS montages with two electrodes of opposite polarity, it is worth noting the various other iterations that exist. For example, high-definition tDCS

(HD-tDCS) involves a small, gel-based electrode surrounded by a ring of return electrodes. In essence, this allows unifocal stimulation, with polarity determined by the center electrode. This setup provides greater focality and can target deeper brain structures than traditional tDCS setups can, all without the need to worry about the possible confounding effects of a return electrode of opposite polarity. The trade-off is a decreased margin for error in electrode placement and fairly precise localization of the desired brain region. In addition, other devices also offer multifocal stimulation that can target multiple brain areas simultaneously. Such technology can be especially advantageous for high-level cognitive research, such as WM training, which requires simultaneous activation of broad networks rather than isolated brain regions.

Moreover, although tDCS has been the most extensively researched method of tES and has produced promising effects across multiple studies, it is not the only form of brain stimulation currently being explored. Other techniques, such as transcranial alternating current stimulation (tACS) and transcranial random noise stimulation (tRNS), are also suited to induce neuroplastic effects but remain less studied than tDCS (cf. Paulus, 2011). Concerning WM training, we are not aware of any studies that have combined WM training with other tES methods (e.g., tACS or tRNS), but given the current pace of the field, it is conceivable that such studies will surface sooner rather than later, providing indications of how promising these methods will be as additional tools to enhance neuroplasticity and further cognitive training research.

Funding and Conflicts of Interest

This work was supported by a National Science Foundation Graduate Research Fellowship (Grant No. DGE-1321846) to J. A. Additionally, M. B. and J. A. are employed at the MIND Research Institute, whose interest is related to this work, and S. M. J. has an indirect financial interest in the MIND Research Institute.

References

Accornero, N., Capozza, M., Pieroni, L., Pro, S., Davi, L., & Mecarelli, O. (2014). EEG mean frequency changes in healthy subjects during prefrontal transcranial direct current stimulation. *Journal of Neurophysiology, 112*(6), 1367–1375. doi:10.1152/jn.00088.2014

Accornero, N., Li Voti, P., La Riccia, M., & Gregori, B. (2007). Visual evoked potentials modulation during direct current cortical polarization. *Experimental Brain Research, 178*(2), 261–266. doi:10.1007/s00221-006-0733-y

Albert, D. J. (1966). The effects of polarizing currents on consolidation of learning. *Neuropsychologia, 4*(1), 65–77. doi:10.1016/0028-3932(66)90021-2

Andrews, S. C., Hoy, K. E., Enticott, P. G., Daskalakis, Z. J., & Fitzgerald, P. B. (2011). Improving working memory: The effect of combining cognitive activity and anodal transcranial direct current stimulation to the left dorsolateral prefrontal cortex. *Brain Stimulation*, 4(2), 84–89. doi:10.1016/j.brs.2010.06.004

Antal, A., Keeser, A., Priori, A., Padberg, F., & Nitsche, M. A. (2015). Conceptual and procedural shortcomings of the systematic review "Evidence that transcranial direct current stimulation (tDCS) generates little-to-no reliable neurophysiologic effect beyond MEP amplitude modulation in healthy human subjects: A systematic review" by Horvath and co-workers. *Brain Stimulation*, 8(4), 846–849. doi:10.1016/j.brs.2015.05.010

Antal, A., Kincses, T. Z., Nitsche, M. A., Bartfai, O., & Paulus, W. (2004). Excitability changes induced in the human primary visual cortex by transcranial direct current stimulation: Direct electrophysiological evidence. *Investigative Ophthalmology and Visual Science*, 45(2), 702–707. doi:10.1167/iovs.03-0688

Antal, A., Nitsche, M. A., Kruse, W., Kincses, T. Z., Hoffmann, K. P., & Paulus, W. (2004). Direct current stimulation over V5 enhances visuomotor coordination by improving motion perception in humans. *Journal of Cognitive Neuroscience*, 16(4), 521–527. doi:10.1162/089892904323057263

Antal, A., Nitsche, M. A., & Paulus, W. (2001). External modulation of visual perception in humans. *Neuroreport*, 12(16), 3553–3555. doi:10.1097/00001756-200111160-00036

Au, J., Buschkuehl, M., Duncan, G. J., & Jaeggi, S. M. (2016). There is no convincing evidence that working memory training is NOT effective: A reply to Melby-Lervåg and Hulme (2015). *Psychonomic Bulletin and Review*, 23(1), 331–337. doi:10.3758/s13423-015-0967-4

Au, J., Karsten, C., Buschkuehl, M., & Jaeggi, S. M. (2017). Optimizing transcranial direct current stimulation protocols to promote long-term learning. *Journal of Cognitive Enhancement*, 1(1), 1–8. doi:10.1007/s41465-017-0007-6

Au, J., Katz, B., Buschkuehl, M., Bunarjo, K., Senger, T., Zabel, C., . . . Jonides, J. (2016). Enhancing working memory training with transcranial direct current stimulation. *Journal of Cognitive Neuroscience*, 28(9), 1–14. doi:10.1162/jocn_a_00979

Au, J., Sheehan, E., Tsai, N., Duncan, G. J., Buschkuehl, M., & Jaeggi, S. M. (2015). Improving fluid intelligence with training on working memory: A meta-analysis. *Psychonomic Bulletin and Review*, 22(2), 366–377. doi:10.3758/s13423-014-0699-x

Bäckman, L., & Nyberg, L. (2013). Dopamine and training-related working-memory improvement. *Neuroscience and Biobehavioral Reviews*, 37(9), 2209–2219. doi:10.1016/j.neubiorev.2013.01.014

Barnett, S. M., & Ceci, S. J. (2002). When and where do we apply what we learn? A taxonomy for far transfer. *Psychological Bulletin*, 128(4), 612–637. doi:10.1037//0033-2909.128.4.612

Baudewig, J., Nitsche, M. A., Paulus, W., & Frahm, J. (2001). Regional modulation of BOLD MRI responses to human sensorimotor activation by transcranial direct current stimulation. *Magnetic Resonance in Medicine*, 45(2), 196–201. doi:10.1002/1522-2594(200102)45:23.0.CO;2-1

Bellander, M., Brehmer, Y., Westerberg, H., Karlsson, S., Furth, D., Bergman, O., . . . Bäckman, L. (2011). Preliminary evidence that allelic variation in the *LMX1A* gene influences training-related working memory improvement. *Neuropsychologia*, 49(7), 1938–1942. doi:10.1016/j.neuropsychologia.2011.03.021

Bennabi, D., Pedron, S., Haffen, E., Monnin, J., Peterschmitt, Y., & Van Waes, V. (2014). Transcranial direct current stimulation for memory enhancement: From clinical research to animal models. *Frontiers in Systems Neuroscience, 8,* 159. doi:10.3389/fnsys.2014.00159

Berlingeri, M., Danelli, L., Bottini, G., Sberna, M., & Paulesu, E. (2013). Reassessing the HAROLD model: Is the hemispheric asymmetry reduction in older adults a special case of compensatory-related utilisation of neural circuits? *Experimental Brain Research, 224*(3), 393–410. doi:10.1007/s00221-012-3319-x

Bikson, M., & Rahman, A. (2013). Origins of specificity during tDCS: Anatomical, activity-selective, and input-bias mechanisms. *Frontiers in Human Neuroscience, 7,* 688. doi:10.3389/fnhum.2013.00688

Bindman, L. J., Lippold, O. C., & Redfearn, J. W. (1964). The action of brief polarizing currents on the cerebral cortex of the rat (1) during current flow and (2) in the production of long-lasting after-effects. *Journal of Physiology, 172*(3), 369–382. doi:10.1113/jphysiol.1964.sp007425

Bogg, T., & Lasecki, L. (2014). Reliable gains? Evidence for substantially underpowered designs in studies of working memory training transfer to fluid intelligence. *Frontiers in Psychology, 5,* 1589. doi:10.3389/fpsyg.2014.01589

Boggio, P. S., Ferrucci, R., Rigonatti, S. P., Covre, P., Nitsche, M., Pascual-Leone, A., & Fregni, F. (2006). Effects of transcranial direct current stimulation on working memory in patients with Parkinson's disease. *Journal of the Neurological Sciences, 249*(1), 31–38. doi:10.1016/j.jns.2006.05.062

Boot, W. R., Simons, D. J., Stothart, C., & Stutts, C. (2013). The pervasive problem with placebos in psychology: Why active control groups are not sufficient to rule out placebo effects. *Perspectives on Psychological Science, 8*(4), 445–454. doi:10.1177/1745691613491271

Borella, E., Carretti, B., Cantarella, A., Riboldi, F., Zavagnin, M., & De Beni, R. (2014). Benefits of training visuospatial working memory in young-old and old-old. *Developmental Psychology, 50*(3), 714–727. doi:10.1037/a0034293

Burki, C. N., Ludwig, C., Chicherio, C., & de Ribaupierre, A. (2014). Individual differences in cognitive plasticity: An investigation of training curves in younger and older adults. *Psychological Research, 78*(6), 821–835. doi:10.1007/s00426-014-0559-3

Cabeza, R. (2002). Hemispheric asymmetry reduction in older adults: The HAROLD model. *Psychology and Aging, 17*(1), 85–100. doi:10.1037//0882-7974.17.1.85

Carp, J. (2012). On the plurality of (methodological) worlds: Estimating the analytic flexibility of fMRI experiments. *Frontiers in Neuroscience, 6,* 149. doi:10.3389/fnins.2012.00149

Cepeda, N. J., Pashler, H., Vul, E., Wixted, J. T., & Rohrer, D. (2006). Distributed practice in verbal recall tasks: A review and quantitative synthesis. *Psychological Bulletin, 132*(3), 354–380. doi:10.1037/0033-2909.132.3.354

Chhatbar, P. Y., & Feng, W. (2015). Data synthesis in meta-analysis may conclude differently on cognitive effect from transcranial direct current stimulation. *Brain Stimulation, 8*(5), 974–976. doi:10.1016/j.brs.2015.06.001

Chooi, W. T., & Thompson, L. A. (2012). Working memory training does not improve intelligence in healthy young adults. *Intelligence, 40*(6), 531–542. doi:10.1016/j.intell.2012.07.004

Churchill, N. W., Yourganov, G., Oder, A., Tam, F., Graham, S. J., & Strother, S. C. (2012). Optimizing preprocessing and analysis pipelines for single-subject fMRI: 2.

Interactions with ICA, PCA, task contrast and inter-subject heterogeneity. *PLoS One*, 7(2), e31147. doi:10.1371/journal.pone.0031147

Coffman, B. A., Clark, V. P., & Parasuraman, R. (2014). Battery powered thought: Enhancement of attention, learning, and memory in healthy adults using transcranial direct current stimulation. *Neuroimage*, 85(Pt. 3), 895–908. doi:10.1016/j.neuroimage.2013.07.083

Colom, R., Quiroga, M. A., Shih, P. C., Martinez, K., Burgaleta, M., Martinez-Molina, A., . . . Ramirez, I. (2010). Improvement in working memory is not related to increased intelligence scores. *Intelligence*, 38(5), 497–505. doi:10.1016/j.intell.2010.06.008

Corbett, A., Owen, A., Hampshire, A., Grahn, J., Stenton, R., Dajani, S., . . . Ballard, C. (2015). The effect of an online cognitive training package in healthy older adults: An online randomized controlled trial. *Journal of the American Medical Directors Association*, 16(11), 990–997. doi:10.1016/j.jamda.2015.06.014

Creutzfeldt, O. D., Fromm, G. H., & Kapp, H. (1962). Influence of transcortical DC-currents on cortical neuronal activity. *Experimental Neurology*, 5(6), 436–452. doi:10.1016/0014-4886(62)90056-0

Dahlin, E., Neely, A. S., Larsson, A., Bäckman, L., & Nyberg, L. (2008). Transfer of learning after updating training mediated by the striatum. *Science*, 320(5882), 1510–1512. doi:10.1126/science.1155466

Damoiseaux, J. S., Beckmann, C. F., Arigita, E. J. S., Barkhof, F., Scheltens, P., Stam, C. J., . . . Rombouts, S. A. R. B. (2008). Reduced resting-state brain activity in the "default network" in normal aging. *Cerebral Cortex*, 18(8), 1856–1864. doi:10.1093/cercor/bhm207

Dedoncker, J., Brunoni, A. R., Baeken, C., & Vanderhasselt, M. A. (2016). A systematic review and meta-analysis of the effects of transcranial direct current stimulation (tDCS) over the dorsolateral prefrontal cortex in healthy and neuropsychiatric samples: Influence of stimulation parameters. *Brain Stimulation*, 9(4), 501–517. doi:10.1016/j.brs.2016.04.006

Dunning, D. L., & Holmes, J. (2014). Does working memory training promote the use of strategies on untrained working memory tasks? *Memory and Cognition*, 42(6), 854–862. doi:10.3758/s13421-014-0410-5

Ebbinghaus, H. (1913). *Memory: A contribution to experimental psychology* (H. A. Ruger & C. E. Bussenius, Trans.). New York, NY: Teacher's College, Columbia University. (Original work published 1885)

Elliot, P. (2014). Electricity and the brain: An historical evaluation. In R. C. Kadosh (Ed.), *The stimulated brain: Cognitive enhancement using non-invasive brain stimulation* (pp. 3–33). London, UK: Elsevier.

Elmasry, J., Loo, C., & Martin, D. (2015). A systematic review of transcranial electrical stimulation combined with cognitive training. *Restorative Neurology and Neuroscience*, 33(3), 263–278. doi:10.3233/RNN-140473

Fertonani, A., & Miniussi, C. (2016). Transcranial electrical stimulation: What we know and do not know about mechanisms. *Neuroscientist*, 23(2), 109–123. doi:10.1177/1073858416631966

Fregni, F., Boggio, P. S., Nitsche, M., Bermpohl, F., Antal, A., Feredoes, E., . . . Pascual-Leone, A. (2005). Anodal transcranial direct current stimulation of prefrontal cortex enhances working memory. *Experimental Brain Research*, 166(1), 23–30. doi:10.1007/s00221-005-2334-6

Gevins, A., Smith, M. E., McEvoy, L., & Yu, D. (1997). High-resolution EEG mapping of cortical activation related to working memory: Effects of task difficulty, type of processing, and practice. *Cerebral Cortex*, 7(4), 374–385. doi:10.1093/cercor/7.4.374

Hardy, J. L., Nelson, R. A., Thomason, M. E., Sternberg, D. A., Katovich, K., Farzin, F., & Scanlon, M. (2015). Enhancing cognitive abilities with comprehensive training: A large, online, randomized, active-controlled trial. *PLoS One, 10*(9), e0134467. doi:10.1371/journal.pone.0134467

Harrison, T. L., Shipstead, Z., Hicks, K. L., Hambrick, D. Z., Redick, T. S., & Engle, R. W. (2013). Working memory training may increase working memory capacity but not fluid intelligence. *Psychological Science, 24*(12), 2409–2419. doi:10.1177/0956797613492984

Heimrath, K., Sandmann, P., Becke, A., Müller, N. G., & Zaehle, T. (2012). Behavioral and electrophysiological effects of transcranial direct current stimulation of the parietal cortex in a visuo-spatial working memory task. *Frontiers in Psychiatry, 3*, 56. doi:10.3389/fpsyt.2012.00056

Hill, A., Fitzgerald, P. B., & Hoy, K. E. (2016). Effects of anodal transcranial direct current stimulation on working memory: A systematic review and meta-analysis of findings from healthy and neuropsychiatric populations. *Brain Stimulation, 9*(2), 197–208. doi:10.1016/j.brs.2015.10.006

Horvath, J. C., Forte, J. D., & Carter, O. (2015a). Evidence that transcranial direct current stimulation (tDCS) generates little-to-no reliable neurophysiologic effect beyond MEP amplitude modulation in healthy human subjects: A systematic review. *Neuropsychologia, 66*, 213–236. doi:10.1016/j.neuropsychologia.2014.11.021

Horvath, J. C., Forte, J. D., & Carter, O. (2015b). Quantitative review finds no evidence of cognitive effects in healthy populations from single-session transcranial direct current stimulation (tDCS). *Brain Stimulation, 8*(3), 535–550. doi:10.1016/j.brs.2015.01.400

Hsu, W. Y., Ku, Y., Zanto, T. P., & Gazzaley, A. (2015). Effects of noninvasive brain stimulation on cognitive function in healthy aging and Alzheimer's disease: A systematic review and meta-analysis. *Neurobiology of Aging, 36*(8), 2348–2359. doi:10.1016/j.neurobiolaging.2015.04.016

Hsu, W. Y., Zanto, T. P., Anguera, J. A., Lin, Y. Y., & Gazzaley, A. (2015). Delayed enhancement of multitasking performance: Effects of anodal transcranial direct current stimulation on the prefrontal cortex. *Cortex, 69*, 175–185. doi:10.1016/j.cortex.2015.05.014

Jacobson, L., Koslowsky, M., & Lavidor, M. (2012). tDCS polarity effects in motor and cognitive domains: A meta-analytical review. *Experimental Brain Reseach, 216*(1), 1–10. doi:10.1007/s00221-011-2891-9

Jaeggi, S. M., Buschkuehl, M., Etienne, A., Ozdoba, C., Perrig, W. J., & Nirkko, A. C. (2007). On how high performers keep cool brains in situations of cognitive overload. *Cognitive, Affective, and Behavioral Neuroscience, 7*(2), 75–89. doi:10.3758/CABN.7.2.75

Jaeggi, S. M., Buschkuehl, M., Jonides, J., & Shah, P. (2011). Short- and long-term benefits of cognitive training. *Proceedings of the National Academy of Sciences, 108*(25), 10081–10086. doi:10.1073/pnas.1103228108

Jaeggi, S. M., Buschkuehl, M., Shah, P., & Jonides, J. (2014). The role of individual differences in cognitive training and transfer. *Memory and Cognition, 42*(3), 464–480. doi:10.3758/s13421-013-0364-z

Jones, K. T., & Berryhill, M. E. (2012). Parietal contributions to visual working memory depend on task difficulty. *Frontiers in Psychiatry, 3*, 81. doi:10.3389/fpsyt.2012.00081

Jones, K. T., Stephens, J. A., Alam, M., Bikson, M., & Berryhill, M. E. (2015). Longitudinal neurostimulation in older adults improves working memory. *PLoS One, 10*(5), e0129751. doi:10.1371/journal.pone.0121904

Karbach, J., & Verhaeghen, P. (2014). Making working memory work: A meta-analysis of executive-control and working memory training in older adults. *Psychological Science, 25*(11), 2027–2037. doi:10.1177/0956797614548725

Katz, B., Au, J., Buschkuehl, M., Abagis, T., Zabel, C., Jaeggi, S. M., & Jonides, J. (2017). Individual differences and long-term consequences of tDCS-augmented cognitive training. *Journal of Cognitive Neuroscience, 29*(9), 1498–1509. doi:10.1162/jocn_a_01115

Keeser, D., Meindl, T., Bor, J., Palm, U., Pogarell, O., Mulert, C., . . . Padberg, F. (2011). Prefrontal transcranial direct current stimulation changes connectivity of resting-state networks during fMRI. *Journal of Neuroscience, 31*(43), 15284–15293. doi:10.1523/JNEUROSCI.0542-11.2011

Kelly, C., Foxe, J. J., & Garavan, H. (2006). Patterns of normal human brain plasticity after practice and their implications for neurorehabilitation. *Archives of Physical Medicine and Rehabilitation, 87*(12), S20–S29. doi:10.1016/j.apmr.2006.08.333

Klimesch, W. (1999). EEG alpha and theta oscillations reflect cognitive and memory performance: A review and analysis. *Brain Research Reviews, 29*(2–3), 169–195. doi:10.1016/S0165–0173(98)00056–3

Kuriyama, K., Mishima, K., Suzuki, H., Aritake, S., & Uchiyama, M. (2008). Sleep accelerates the improvement in working memory performance. *Journal of Neuroscience, 28*(40), 10145–10150. doi:10.1523/JNEUROSCI.2039-08.2008

Liebetanz, D., Nitsche, M. A., Tergau, F., & Paulus, W. (2002). Pharmacological approach to the mechanisms of transcranial DC-stimulation-induced after-effects of human motor cortex excitability. *Brain, 125*(Pt. 10), 2238–2247. doi:10.1093/brain/awf238

Lippold, O. C., & Redfearn, J. W. (1964). Mental changes resulting from the passage of small direct currents through the human brain. *British Journal of Psychiatry, 110*, 768–772. doi:10.1192/bjp.110.469.768

Lolas, F. (1977). Brain polarization: Behavioral and therapeutic effects. *Biological Psychiatry, 12*(1), 37–47.

Looi, C. Y., Duta, M., Brem, A. K., Huber, S., Nuerk, H. C., & Kadosh, R. C. (2016). Combining brain stimulation and video game to promote long-term transfer of learning and cognitive enhancement. *Scientific Reports, 6*, Art no. 22003. doi:10.1038/srep22003

Mancuso, L. E., Ilieva, I. P., Hamilton, R. H., & Farah, M. J. (2016). Does transcranial direct current stimulation improve healthy working memory? A meta-analytic review. *Journal of Cognitive Neuroscience, 28*(8), 1063–1089. doi:10.1162/jocn_a_00956

Mangia, A. L., Pirini, M., & Cappello, A. (2014). Transcranial direct current stimulation and power spectral parameters: A tDCS/EEG co-registration study. *Frontiers in Human Neuroscience, 8*, 601. doi:10.3389/fnhum.2014.00601

Martin, D. M., Liu, R., Alonzo, A., Green, M., & Loo, C. K. (2014). Use of transcranial direct current stimulation (tDCS) to enhance cognitive training: Effect of timing of stimulation. *Experimental Brain Research, 232*(10), 3345–3351. doi:10.1007/s00221-014-4022-x

Martin, D. M., Liu, R., Alonzo, A., Green, M., Player, M. J., Sachdev, P., & Loo, C. K. (2013). Can transcranial direct current stimulation enhance outcomes from cognitive training? A randomized controlled trial in healthy participants. *International Journal of Neuropsychopharmacology, 16*(9), 1927–1936. doi:10.1017/S1461145713000539

Matsunaga, K., Nitsche, M. A., Tsuji, S., & Rothwell, J. C. (2004). Effect of transcranial DC sensorimotor cortex stimulation on somatosensory evoked potentials in humans. *Clinical Neurophysiology, 115*(2), 456–460. doi:10.1016/S1388–2457(03)00362–6

Meinzer, M., Lindenberg, R., Antonenko, D., Flaisch, T., & Flöel, A. (2013). Anodal transcranial direct current stimulation temporarily reverses age-associated cognitive decline and functional brain activity changes. *Journal of Neuroscience, 33*(30), 12470–12478. doi:10.1523/JNEUROSCI.5743-12.2013

Melby-Lervåg, M., & Hulme, C. (2013). Is working memory training effective? A meta-analytic review. *Developmental Psychology, 49*(2), 270–291. doi:10.1037/a0028228

Melby-Lervåg, M., & Hulme, C. (2015). There is no convincing evidence that working memory training is effective: A reply to Au et al. (2014) and Karbach and Verhaeghen (2014). *Psychonomic Bulletin and Review, 23*(1), 324–330. doi:10.3758/s13423-015-0862-z

Morrison, A. B., & Chein, J. M. (2011). Does working memory training work? The promise and challenges of enhancing cognition by training working memory. *Psychonomic Bulletin and Review, 18*(1), 46–60. doi:10.3758/s13423-010-0034-0

Ng, K. K., Lo, J. C., Lim, J. K., Chee, M. W., & Zhou, J. (2016). Reduced functional segregation between the default mode network and the executive control network in healthy older adults: A longitudinal study. *Neuroimage, 133,* 321–330. doi:10.1016/j.neuroimage.2016.03.029

Nilsson, J., Lebedev, A. V., & Lövdén, M. (2015). No significant effect of prefrontal tDCS on working memory performance in older adults. *Frontiers in Aging Neuroscience, 7,* 230. doi:10.3389/fnagi.2015.00230

Nitsche, M. A., Fricke, K., Henschke, U., Schlitterlau, A., Liebetanz, D., Lang, N., . . . Paulus, W. (2003). Pharmacological modulation of cortical excitability shifts induced by transcranial direct current stimulation in humans. *Journal of Physiology, 553*(Pt. 1), 293–301. doi:10.1113/jphysiol.2003.049916

Nitsche, M. A., Nitsche, M. S., Klein, C. C., Tergau, F., Rothwell, J. C., & Paulus, W. (2003). Level of action of cathodal DC polarisation induced inhibition of the human motor cortex. *Clinical Neurophysiology, 114*(4), 600–604. doi:10.1016/S1388-2457(02)00412-1

Nitsche, M. A., & Paulus, W. (2000). Excitability changes induced in the human motor cortex by weak transcranial direct current stimulation. *Journal of Physiology, 527*(Pt. 3), 633–639. doi:10.1111/j.1469-7793.2000.t01-1-00633.x

Nitsche, M. A., & Paulus, W. (2001). Sustained excitability elevations induced by transcranial DC motor cortex stimulation in humans. *Neurology, 57*(10), 1899–1901. doi:10.1212/WNL.57.10.1899

Nitsche, M. A., Schauenburg, A., Lang, N., Liebetanz, D., Exner, C., Paulus, W., & Tergau, F. (2003). Facilitation of implicit motor learning by weak transcranial direct current stimulation of the primary motor cortex in the human. *Journal of Cognitive Neuroscience, 15*(4), 619–626. doi:10.1162/089892903321662994

Novick, J. M., Hussey, E., Teubner-Rhodes, S., Harbison, J. I., & Bunting, M. F. (2014). Clearing the garden-path: Improving sentence processing through cognitive control training. *Language and Cognitive Processes, 29*(2):186–217. doi:10.1080/01690965.2012.758297

Nuwer, M. R. (1988). Quantitative EEG: I. Techniques and problems of frequency analysis and topographic mapping. *Journal of Clinical Neurophysiology, 5*(1), 1–43. doi:10.1097/00004691-198801000-00001

Ohn, S. H., Park, C. I., Yoo, W. K., Ko, M. H., Choi, K. P., Kim, G. M., . . . Kim, Y. H. (2008). Time-dependent effect of transcranial direct current stimulation on the enhancement of working memory. *Neuroreport, 19*(1), 43–47. doi:10.1097/WNR.0b013e3282f2adfd

Owen, A. M., Hampshire, A., Grahn, J. A., Stenton, R., Dajani, S., Burns, A. S., . . . Ballard, C. G. (2010). Putting brain training to the test. *Nature, 465*(7299), 775–778. doi:10.1038/nature09042

Park, S. H., Seo, J. H., Kim, Y. H., & Ko, M. H. (2014). Long-term effects of transcranial direct current stimulation combined with computer-assisted cognitive training in healthy older adults. *Neuroreport, 25*(2), 122–126. doi:10.1097/WNR.0000000000000080

Paulus, W. (2011). Transcranial electrical stimulation (tES - tDCS; tRNS, tACS) methods. *Neuropsychological Rehabilitation, 21*(5), 602–617. doi:10.1080/09602011.2011.557292

Peña-Gómez, C., Sala-Lonch, R., Junqué, C., Clemente, I. C., Vidal, D., Bargalló, N., . . . Bartrés-Faz, D. (2012). Modulation of large-scale brain networks by transcranial direct current stimulation evidenced by resting-state functional MRI. *Brain Stimulation, 5*(3), 252–263. doi:10.1016/j.brs.2011.08.006

Rebok, G. W., Ball, K., Guey, L. T., Jones, R. N., Kim, H.-Y., King, J. W., . . . Willis, S. L. (2014). Ten-year effects of the advanced cognitive training for independent and vital elderly cognitive training trial on cognition and everyday functioning in older adults. *Journal of the American Geriatrics Society, 62*(1), 16–24. doi:10.1111/jgs.12607

Redick, T. S., Shipstead, Z., Harrison, T. L., Hicks, K. L., Fried, D. E., Hambrick, D. Z., . . . Engle, R. W. (2013). No evidence of intelligence improvement after working memory training: A randomized, placebo-controlled study. *Journal of Experimental Psychology: General, 142*(2), 359–379. doi:10.1037/a0029082

Reis, J., Schambra, H. M., Cohen, L. G., Buch, E. R., Fritsch, B., Zarahn, E., . . . Krakauer, J. W. (2009). Noninvasive cortical stimulation enhances motor skill acquisition over multiple days through an effect on consolidation. *Proceedings of the National Academy of Sciences, 106*(5), 1590–1595. doi:10.1073/pnas.0805413106

Reuter-Lorenz, P. A., & Cappell, K. A. (2008). Neurocognitive aging and the compensation hypothesis. *Current Directions in Psychological Science, 17*(3), 177–182. doi:10.1111/j.1467-8721.2008.00570.x

Reynolds, J. H., & Glaser, R. (1964). Effects of repetition and spaced review upon retention of a complex learning-task. *Journal of Educational Psychology, 55*(5), 297–308. doi:10.1037/h0040734

Richmond, L. L., Wolk, D., Chein, J., & Olson, I. R. (2014). Transcranial direct current stimulation enhances verbal working memory training performance over time and near transfer outcomes. *Journal of Cognitive Neuroscience, 26*(11), 2443–2454. doi:10.1162/jocn_a_00657

Roberts, G., Quach, J., Spencer-Smith, M., Anderson, P. J., Gathercole, S., Gold, L., . . . Wake, M. (2016). Academic outcomes 2 years after working memory training for children with low working memory: A randomized clinical trial. *JAMA Pediatrics, 170*(5), e154568. doi:10.1001/jamapediatrics.2015.4568

Romero Lauro, L. J., Rosanova, M., Mattavelli, G., Convento, S., Pisoni, A., Opitz, A., . . . Vallar, G. (2014). TDCS increases cortical excitability: Direct evidence from TMS-EEG. *Cortex, 58*, 99–111. doi:10.1016/j.cortex.2014.10.022

Rosenkranz, K., Nitsche, M. A., Tergau, F., & Paulus, W. (2000). Diminution of training-induced transient motor cortex plasticity by weak transcranial direct current stimulation in the human. *Neuroscience Letters, 296*(1), 61–63. doi:10.1016/S0304-3940(00)01621-9

Rothwell, J. C. (1991). Physiological studies of electric and magnetic stimulation of the human brain. *Electroencephalography and Clinical Neurophysiology, Supplement, 43*, 29–35.

Rothwell, J. C., Thompson, P. D., Day, B. L., Dick, J. P., Kachi, T., Cowan, J. M., & Marsden, C. D. (1987). Motor cortex stimulation in intact man. 1. General characteristics of EMG responses in different muscles. *Brain, 110*(Pt. 5), 1173–1190. doi:10.1093/brain/110.5.1173

Russo, R., Wallace, D., Fitzgerald, P. B., & Cooper, N. R. (2013). Perception of comfort during active and sham transcranial direct current stimulation: A double blind study. *Brain Stimulation, 6*(6), 946–951. doi:10.1016/j.brs.2013.05.009

Schmiedek, F., Lövdén, M., & Lindenberger, U. (2010). Hundred days of cognitive training enhance broad cognitive abilities in adulthood: Findings from the COGITO Study. *Frontiers in Aging Neuroscience, 2*, 27. doi:10.3389/fnagi.2010.00027

Schwaighofer, M., Fischer, F., & Bühner, M. (2015). Does working memory training transfer? A meta-analysis including training conditions as moderators. *Educational Psychologist, 50*(2), 138–166. doi:10.1080/00461520.2015.1036274

Scullin, M. K., Trotti, L. M., Wilson, A. G., Greer, S. A., & Bliwise, D. L. (2012). Nocturnal sleep enhances working memory training in Parkinson's disease but not Lewy body dementia. *Brain, 135*(9), 2789–2797. doi:10.1093/brain/aws192

Smith, E. E., Jonides, J., & Koeppe, R. A. (1996). Dissociating verbal and spatial working memory using PET. *Cerebral Cortex, 6*(1), 11–20. doi:10.1093/cercor/6.1.11

Soderqvist, S., Nutley, S. B., Peyrard-Janvid, M., Matsson, H., Humphreys, K., Kere, J., & Klingberg, T. (2012). Dopamine, working memory, and training induced plasticity: Implications for developmental research. *Developmental Psychology, 48*(3), 836–843. doi:10.1037/a0026179

Song, M., Shin, Y., & Yun, K. (2014). Beta-frequency EEG activity increased during transcranial direct current stimulation. *Neuroreport, 25*(18), 1433–1436. doi:10.1097/WNR.0000000000000283

Stafford, T., & Dewar, M. (2014). Tracing the trajectory of skill learning with a very large sample of online game players. *Psychological Science, 25*(2), 511–518. doi:10.1177/0956797613511466

Stagg, C. J., & Nitsche, M. A. (2011). Physiological basis of transcranial direct current stimulation. *Neuroscientist, 17*(1), 37–53. doi:10.1177/1073858410386614

Stephens, J. A., & Berryhill, M. E. (2016). Older adults improve on everyday tasks after working memory training and neurostimulation. *Brain Stimulation, 9*(4), 553–559. doi:10.1016/j.brs.2016.04.001

Summers, J. J., Kang, N., & Cauraugh, J. H. (2015). Does transcranial direct current stimulation enhance cognitive and motor functions in the ageing brain? A systematic review and meta-analysis. *Ageing Research Reviews, 25*, 42–54. doi:10.1016/j.arr.2015.11.004

Tremblay, S., Lepage, J. F., Latulipe-Loiselle, A., Fregni, F., Pascual-Leone, A., & Theoret, H. (2014). The uncertain outcome of prefrontal tDCS. *Brain Stimulation, 7*(6), 773–783. doi:10.1016/j.brs.2014.10.003

Voytek, B., & Gazzaley, A. (2013). Stimulating the aging brain. *Annals of Neurology, 73*(1), 1–3. doi:10.1002/ana.23790

Wager, T. D., & Smith, E. E. (2003). Neuroimaging studies of working memory: A meta-analysis. *Cognitive, Affective, and Behavioral Neuroscience, 3*(4), 255–274. doi:10.3758/CABN.3.4.255

Walker, M. P., & Stickgold, R. (2004). Sleep-dependent learning and memory consolidation. *Neuron, 44*(1), 121–133. doi:10.1016/j.neuron.2004.08.031

Wang, Z., Zhou, R., & Shah, P. (2014). Spaced cognitive training promotes training transfer. *Frontiers in Human Neuroscience, 8*, 217. doi:10.3389/fnhum.2014.00217

Weicker, J., Villringer, A., & Thone-Otto, A. (2016). Can impaired working memory functioning be improved by training? A meta-analysis with a special focus on brain injured patients. *Neuropsychology, 30*(2), 190–212. doi:10.1037/neu0000227

Willis, S. L., Tennstedt, S. L., Marsiske, M., Ball, K., Elias, J., Koepke, K. M., . . . ACTIVE Study Group. (2006). Long-term effects of cognitive training on everyday functional outcomes in older adults. *JAMA, 296*(23), 2805–2814. doi:10.1001/jama.296.23.2805

Wise, R. A. (2004). Dopamine, learning and motivation. *Nature Reviews Neuroscience, 5*(6), 483–494. doi:10.1038/nrn1406

Zaehle, T., Sandmann, P., Thorne, J. D., Jäncke, L., & Herrmann, C. S. (2011). Transcranial direct current stimulation of the prefrontal cortex modulates working memory performance: Combined behavioural and electrophysiological evidence. *BMC Neuroscience, 12*, 2. doi:10.1186/1471-2202-12-2

7

Cognitive Training

Component Processes and Criteria for Change

Kristine B. Walhovd, Anders M. Fjell, and Lars Nyberg

Introduction

In Chapter 5, Kuchinsky and Haarmann consider a number of important questions related to cognitive and working memory (WM) training, including whether there are benefits of cognitive training and, if so, whether the benefits transfer, persist, and can be further augmented by brain regulation and stimulation. We can draw positive conclusions with respect to these questions, at least in part. Kuchinsky and Haarmann also discuss neuroscience approaches on the horizon that might change the cognitive training game. While the questions and points they raise are important, attention should also be paid to additional basic, yet difficult, questions in cognitive training to move the field forward.

This chapter first emphasizes the need to acknowledge multiple component processes of WM. Next, the chapter considers how the field needs a better account of what a training effect actually means (Chein & Morrison, 2010). The discussion then turns to how to design training studies to allow for measurement of specificity and temporal aspects of effects. Also, there is a need to delineate standard criteria for change, because the criteria can be calculated in multiple ways. Finally, in addition to being a vehicle for studying change, training effects can yield unique information about the state of the individual nervous system.

Will the Real Component Process Please Stand Up?

A common conceptual model is to equate WM with executive function and executive function with intelligence. Such a conceptualization is reflected in accounts of WM as part of the G factor (Chuderski & Necka, 2012; Davies et al., 2015), but such an emphasis may prove a challenge for furthering the field of cognitive training. Few would disagree that higher-order mental operations involve WM, but these operations also tap multiple other processes, such as perception, processing speed, and long-term memory (Healey, Crutchley, & Kahana, 2014; Robitaille et al., 2013; Tadin, 2015).

On what grounds can primacy of WM be assumed? According to a "component processes" view of WM, no processes and, correspondingly, no brain structures are unique or specific to WM (Eriksson, Vogel, Lansmer, Bergström, & Nyberg, 2015). Rather, WM emerges from different combinations of processes that, in other constellations, can be functionally described in other terms (e.g., perceptual and long-term memory representations). This will vary widely across tasks and will be modality- or at least domain-specific to a certain extent (Michalka, Kong, Rosen, Shinn-Cunningham, & Somers, 2015). WM is a temporarily enhanced accessibility state of a representation, regardless of the kind of representation (Eriksson et al., 2015). Hence, in principle, WM training, regardless of task, could involve training any or all of the component processes, and it is thus hard to claim supremacy of WM training over any other cognitive training program. This goes also for training programs directed at other processes. For example, episodic memory training (e.g., Engvig et al., 2014) will inevitably involve WM (see also Nyberg et al., 2003).

Indeed, as pointed out by Eriksson and colleagues (Eriksson et al., 2015), the relation between WM and long-term memory is dynamic. For example, the degree of medial temporal lobe (MTL) involvement in WM tasks can depend on task features, such as requirements for binding, relational processing and load (Jeneson & Squire 2012; Olson, Page, Moore, Chatterjee, & Verfaellie, 2006). Thus, we cannot claim that WM is the key to improved general cognitive performance based on results from studies on WM training. This leads to the fundamental question of which, if any, is the most essential part: will the real component process please stand up? That is, we need to identify which processes are the most important, and also their relative roles, or we might as well conclude like Eminem: *Let´s all stand up.*

Thus, a further emphasis on component processes is needed. While perhaps all or most forms of cognitive training are likely to engage WM to some extent, cognitive training effects cannot readily be reduced to "pure" WM in isolation, and WM should not be treated as a single construct or as equivalent to intelligence. A process-based approach is likely to further refine our understanding of the neural basis of cognition and WM at cellular and systems levels (Eriksson et al., 2015). For example, as also noted by Kuchinsky and Haarmann in Chapter 5, we have previously found (Dahlin, Nyberg, Bäckman, & Neely, 2008) that among different WM tasks expected to engage executive control processes and activate frontoparietal circuits, transfer was highly restrictive and based on a striatal-updating network only. That is, transfer from letter-memory training was restricted to an N-back task and was not seen for a Stroop inhibition task, which did not involve updating and striatal activation. These and related findings may not only circle in on process specificity as a basis for transfer of training effects, they may also lead to the question of whether it is really the executive aspect of WM that WM studies are training.

The fact that separate component processes need to be understood does not mean that a large-scale training intervention can only aim at extremely select processes. Attempts at this may yield low ecological validity and high drop-out rates. Rather, to help identify essential neural and cognitive components in performance improvement, training studies should use a combination of features with close follow-up, temporal sequencing of on- and off-training, and multimodal imaging to measure structural and functional changes, along with a broad battery of cognitive tests to track possible differential effects and transfer.

What Does a Training Effect Really Mean?

As noted by Kuchinsky and Haarmann in Chapter 5, the effect of WM training has been observed as structural changes in brain areas that support it. However, while correspondence between anatomical localization of training effects and brain regions found to be active during certain cognitive tasks has sometimes been observed, specificity of such changes can hardly be claimed, given that no brain regions are unique or specific to WM (Eriksson et al., 2015). Indeed, as all of human cognition is bound to engage multiple brain areas and processes (see Karadottir and Walhovd, 2014), any cognitive training regime is bound to do the same.

How May Training Studies Be Designed to Allow for Measurement of Specificity and Temporal Aspects of Effects?

In our view, an "add-one" or "leave-one-out" design comparing multiple training regimes with different features can start narrowing in on specific cognitive processes. This can be carried out in studies that apply, in addition to a training and a passive control condition, an active control condition, where features of the training program not hypothesized to constitute the critical component process are all incorporated in an alternative training program. This will allow for testing regarding any differential effects seen. To heighten sensitivity to training-specific changes, given the enormous baseline individual variation in brain and cognition (Nyberg, Lövdén, Riklund, Lindenberger, & Bäckman, 2012; Walhovd et al., 2011), such conditions can be incorporated or combined in a time-series crossover design, where persons alternate between A and B or C, serving as their own controls. Given the continuous age-related changes in the central nervous system across the lifespan (Fjell et al., 2015; Storsve et al., 2014), it is important that time-series crossover designs are properly balanced with regard to sequencing of A-B-A-B vs. B-A-B-A and so forth. With time-series designs of sufficient length and pacing, the temporal characteristics of training effects can also be better studied.

Criteria for Change

A pressing issue in training studies is how to best measure outcome. In the search for neural changes, we believe that measures of behavioral improvement deserve further attention, as the relation between measures depends on the reliability and validity of the mesurescomponents. In principle, outcome can be measured in several different ways:

- Performance after training (i.e., an absolute score)
- The difference in performance before and after training (i.e., an absolute difference score)
- The increase in performance after training relative to baseline, such as a ratio (*time point 2 / time point 1*) or the proportion of increase over baseline [(*time point 2 – time point 1*) / *time point 1*, which is a ratio or percentage score]
- The variance in outcome independent of baseline (e.g., when removing baseline variance in a training group and using standardized residuals; i.e., a residual score)

The choice of training effect measure may vary, but it should be dependent on the aim of analyses, as well as awareness of the possible relationships between the measures. For example, absolute scores can be used to ensure that participants reached some criterion indicating that they have performed the training tasks correctly and indeed acquired a new skill. An example would be when persons originally naïve to juggling can sustain juggling a three-ball cascade for at least 60 seconds (Draganski et al., 2004).

While such scores can be used as a basis for inclusion/exclusion in group comparisons of brain changes relating to presence or absence of training, they are less suited to relate to magnitude or localization of brain characteristics or changes in relation to training gains, which may be preferable when one wants to investigate functional relationships directly. Absolute difference scores may be suitable as a way of measuring change (Engvig et al., 2010). However, they do not take into account possible vast differences in relative improvement across persons if significant baseline differences are present in the sample and a wide range of scores is possible, so a person who increases memory 100% (e.g., from 10 to 20) can, in principle, get the same score as the person who increases by 25% (e.g., from 40 to 50).

Relative, percentage, or ratio scores may be attractive to express proportional gain. They may be used to compare training effects in groups with different baseline functioning, such as patients relative to healthy control groups (Engvig et al., 2014). Ratio scores do not eliminate the influence of baseline variance in analyses, though. To the extent that pretraining (x) and posttraining (y) scores were not related in a sample, the pretraining scores will still be (spuriously) correlated with posttraining ratio (y/x) scores, as there is an intrinsic relationship between the ratio y/x and its denominator x (Curran-Everett, 2013; Van Petten, 2004). Only residual scores will reflect training effects independent of baseline scores. The residual variance in posttraining memory scores will, by definition, be uncorrelated with pretraining memory scores. Such an approach is well-suited if a broad sample with different levels of functioning is available (e.g., a highly heterogeneous patient group) and one wants to know whether cognitive performance increments as a result of a training intervention relate to brain structure or activity regardless of the pre-existing differences (Engvig, Fjell, Westlye, Skaane, et al., 2012).

The goal here is not to define what type of criteria or scores for change should invariably be used, but to emphasize that measures of change must be tailored to the performance distributions pre- and posttraining in the specific sample, as well as to their interrelationship.

Behavioral Training Effects—A Brain Barometer?

The vast individual differences in normal brain and cognitive function at all points in life render it difficult to make comparisons across individuals, whether within the framework of intervention/control, different types of interventions, or even comparisons for diagnostic purposes, such as the presence or absence of neurodegenerative disease. Devoid of baseline information, as we often are in real-life settings (e.g., clinical assessments), we are left with only cross-group comparisons, means, and their considerable standard deviations, which results in much uncertainty. Interpretation of cross-sectional group differences can be further critically hampered by selection effects (Nyberg et al., 2010). Knowledge that even short-term training interventions can produce changes in brain and cognition (Zatorre, Fields, & Johansen-Berg, 2012) yields a unique tool, enabling "speeded longitudinal data" with a relatively high degree of control.

An intriguing idea is that capacity for change is a highly sensitive biomarker for the state of the nervous system. Individual variability may yield valuable information far beyond what an absolute level of performance at any given time can. This is illustrated, for example, in a memory training study where patients with subjective memory impairment showed, on average, training gains relative to controls, but with much variance; some did not show any benefit at all, while others had substantial improvement (Engvig, Fjell, Westlye, Skaane, et al., 2012; Engvig et al., 2014). It turned out that hippocampal volume at baseline was a unique predictor of training gains, and, combined with baseline depressive symptoms, baseline CA2/3-volume explained 42% of the variation in recall improvement (Engvig, Fjell, Westlye, Mobergeet, et al., 2012). It is known that a subset of patients with subjective memory impairment eventually go on to develop neurodegenerative conditions, and the sensitivity of cognitive plasticity as a marker of central nervous system integrity should be further investigated. Also, cognitive plasticity could be a valuable future focus of training studies more broadly, as it may relate to neural mechanisms or changes otherwise not readily observed.

Conclusions and Future Directions

Regardless of how it is targeted, any cognitive training will involve a vast number of component processes, and their identification should be a goal of further studies. Sensitivity of measurement specificity and temporal aspects of effects may be facilitated by experimental time-series crossover designs. How training effects are calculated, and whether the outcome measures are made dependent or independent of baseline differences in functioning, should receive further attention. Beyond being a vehicle for studying change, training effects can yield unique information about the state of the individual nervous system, hence serving as a valuable biomarker.

An interesting challenge for future studies concerns ways of boosting the magnitude of the effect of various interventions, as well as their generalization (transfer) beyond the specifics of the intervention program. This challenge can likely be met in several different ways (e.g., identifying very basic, fundamental component processes that are taxed by a broad range of tasks and situations and trying to strengthen such processes by training). One example could be programs involving the ability to control attention, arguably a process of broad use and applicability; some studies have explained findings of strong and broad effects of cognitive training by referring to the possibility that their training program was able to support attentional processes (Jaeggi, Buschkuehl, Jonides, & Perrig, 2008). Thus, fluid intelligence and WM can be related through attentional control processes (Gray, Chabris, & Braver, 2003), for example, and training attentional control through one task (e.g., WM) could be beneficial for the second task—in this case fluid intelligence (Jaeggi et al., 2008).

Another way to strengthen the effectiveness of cognitive interventions could be to broaden the scope of training. For example, rather than developing interventions that foster a single critical WM processing component (e.g., updating; Dahlin et al., 2008), it is possible to have programs that challenge multiple components (e.g., updating, shifting, inhibition). Some evidence suggests that, for younger adults, training multiple components provides additional benefits beyond single-domain training (Sandberg, Rönnlund, Nyberg, & Stigsdotter Neely, 2014). However, the same study showed that five weeks of training multiple executive processes was insufficient to foster transfer beyond the very near in older adults. To foster that, it may be necessary to train over longer time and/or combine the cognitive intervention with other kinds of intervention. In the context of aging and cognitive decline, a particularly interesting combination could be that of cognitive and physical interventions. One recent example is a study that investigated multidomain intervention (i.e., diet, exercise, cognitive training, vascular risk monitoring) over two years and found a significant effect on the test scores of older adults on a comprehensive neuropsychological test battery (Ngandu et al., 2015). These findings are promising and indicate that cognitive training and stimulation could, on its own or in combination with other forms of intervention, be an important ingredient in future efforts toward brain maintenance.

References

Chein, J. M., & Morrison, A. B. (2010). Expanding the mind's workspace: Training and transfer effects with a complex working memory span task. *Psychonomic Bulletin and Review, 17*(2), 193–199. doi:10.3758/PBR.17.2.193

Chuderski, A., & Necka, E. (2012). The contribution of working memory to fluid reasoning: Capacity, control, or both? *Journal of Experimental Psychology, Learning, Memory, and Cognition, 38*(6), 1689–1710. doi:10.1037/a0028465

Curran-Everett, D. (2013). Explorations in statistics: The analysis of ratios and normalized data. *Advances in Physiology Education, 37*(3), 213–219. doi:10.1152/advan.00053.2013

Dahlin, E., Nyberg, L., Bäckman, L., & Neely, A. S. (2008). Plasticity of executive functioning in young and older adults: Immediate training gains, transfer, and long-term maintenance. *Psychology and Aging, 23*(4), 720–730. doi:10.1037/a0014296

Davies, G., Armstrong, N., Bis, J. C., Bressler, J., Chouraki, V., Giddaluru, S., . . . Deary, I. J. (2015). Genetic contributions to variation in general cognitive function: A meta-analysis of genome-wide association studies in the CHARGE consortium (*N* = 53,949). *Molecular Psychiatry, 20*(2), 183–192. doi:10.1038/mp.2014.188

Draganski, B., Gaser, C., Busch, V., Schuierer, G., Bogdahn, U., & May, A. (2004). Neuroplasticity: Changes in grey matter induced by training. *Nature, 427*(6972), 311–312. doi:10.1038/427311a

Engvig, A., Fjell, A. M., Westlye, L. T., Moberget, T., Sundseth, Ø., Larsen, V. A., & Walhovd, K. B. (2010). Effects of memory training on cortical thickness in the elderly. *Neuroimage, 52*(4), 1667–1676. doi:10.1016/j.neuroimage.2010.05.041

Engvig, A., Fjell, A. M., Westlye, L. T., Moberget, T., Sundseth, Ø., Larsen, V. A., & Walhovd, K. B. (2012). Memory training impacts short-term changes in aging white matter: A longitudinal diffusion tensor imaging study. *Human Brain Mapping, 33*(10), 2390–2406. doi:10.1002/hbm.21370

Engvig, A., Fjell, A. M., Westlye, L. T., Skaane, N. V., Dale, A. M., Holland, D., . . . Walhovd, K. B. (2014). Effects of cognitive training on gray matter volumes in memory clinic patients with subjective memory impairment. *Journal of Alzheimer's Disease, 41*(3), 779–791. doi:10.3233/JAD-131889

Engvig, A., Fjell, A. M., Westlye, L. T., Skaane, N. V., Sundseth, Ø., & Walhovd, K. B. (2012). Hippocampal subfield volumes correlate with memory training benefit in subjective memory impairment. *Neuroimage, 61*(1), 188–194. doi:10.1016/j.neuroimage.2012.02.072

Eriksson, J., Vogel, E. K, Lansner, A., Bergström, F., & Nyberg, L. (2015). Neurocognitive architecture of working memory. *Neuron, 88*(1), 33–46. doi:10.1016/j.neuron.2015.09.020

Fjell, A. M., Grydeland, H., Krogsrud, S. K., Amlien, I., Rohani, D. A., Mork, L, . . . Walhovd, K. B. (2015). Development and aging of cortical thickness correspond to genetic organization patterns. *Proceedings of the National Academy of Sciences, 112*(50), 14562–14567. doi:10.1073/pnas.1508831112

Gray, J. R., Chabris, C. F., & Braver, T. S. (2003). Neural mechanisms of general fluid intelligence. *Nature Neuroscience, 6*(3), 316–322. doi:10.1038/nn1014

Healey, M. K., Crutchley, P., & Kahana, M. J. (2014). Individual differences in memory search and their relation to intelligence. *Journal of Experimental Psychology: General, 143*(4), 1553–1569. doi:10.1037/a0036306

Jaeggi, S. M., Buschkuehl, M., Jonides, J., & Perrig, W. J. (2008). Improving fluid intelligence with training on working memory. *Proceedings of the National Academy of Sciences, 105*(19), 6829–6833. doi:10.1073/pnas.0801268105

Jeneson, A., & Squire, L. R. (2012). Working memory, long-term memory, and medial temporal lobe function. *Learning and Memory, 19*(1), 15–25. doi:10.1101/lm.024018.111

Karadottir, R. T., & Walhovd, K. B. (2014). The CNS white matter. *Neuroscience, 276*, 1. doi:10.1016/j.neuroscience.2014.03.010

Michalka, S. W., Kong, L., Rosen, M. L., Shinn-Cunningham, B. G., & Somers, D. C. (2015). Short-term memory for space and time flexibly recruit complementary sensory-biased frontal lobe attention networks. *Neuron, 87*(4), 882–892. doi:10.1016/j.neuron.2015.07.028

Ngandu, T., Lehtisalo, J., Solomon, A., Levälahti, E., Ahtiluoto, S., Antikainen, R., . . . Kivipelto, M. (2015). A 2 year multidomain intervention of diet, exercise, cognitive training, and vascular risk monitoring versus control to prevent cognitive decline in at-risk elderly people (FINGER): A randomised controlled trial. *The Lancet, 385*(9984), 2255–2263. doi:10.1016/S0140-6736(15)60461–5

Nyberg, L., Lövdén, M., Riklund, K., Lindenberger, U., & Bäckman, L. (2012). Memory aging and brain maintenance. *Trends in Cognitive Science, 16*(5), 292–305. doi:10.1016/j.tics.2012.04.005

Nyberg, L., Salami, A., Andersson, M., Eriksson, J., Kalpouzos, G., Kauppi, K., ... Nilsson, L. G. (2010). Longitudinal evidence for diminished frontal cortex function in aging. *Proceedings of the National Academy of Sciences, 107*(52), 22682–22686.

Nyberg, L., Sandblom, J., Jones, S., Neely, A. S., Petersson, K. M., Ingvar, M., & Backman, L. (2003). Neural correlates of training-related memory improvement in adulthood and aging. *Proceedings of the National Academy of Sciences, 100*(23), 13728–13733. doi:10.1073/pnas.1735487100

Olson, I. R., Page, K., Moore, K. S., Chatterjee, A., & Verfaellie, M. (2006). Working memory for conjunctions relies on the medial temporal lobe. *Journal of Neuroscience, 26*(17), 4596–4601. doi:10.1523/JNEUROSCI.1923-05.2006

Robitaille, A., Piccinin, A. M., Muniz-Terrera, G., Hoffman, L., Johansson, B., Deeg, D. J., ... Hofer, S. M. (2013). Longitudinal mediation of processing speed on age-related change in memory and fluid intelligence. *Psychology and Aging, 28*(4), 887–901. doi:10.1037/a0033316

Sandberg, P., Rönnlund, M., Nyberg, L., & Stigsdotter Neely, A. (2014). Executive process training in young and old adults. *Aging, Neuropsychology, and Cognition, 21*(5), 577–605. doi:10.1080/13825585.2013.839777

Storsve, A. B., Fjell, A. M., Tamnes, C. K., Westlye, L. T., Overbye, K., Aasland, H. W., & Walhovd, K. B. (2014). Differential longitudinal changes in cortical thickness, surface area and volume across the adult life span: Regions of accelerating and decelerating change. *Journal of Neuroscience, 34*(25), 8488–8498. doi:10.1523/JNEUROSCI.0391-14.2014

Tadin, D. (2015). Suppressive mechanisms in visual motion processing: From perception to intelligence. *Vision Research, 115*(Pt. A), 58–70. doi:10.1016/j.visres.2015.08.005

Van Petten, C. (2004). Relationship between hippocampal volume and memory ability in healthy individuals across the lifespan: Review and meta-analysis. *Neuropsychologia, 42*(10), 1394–1413. doi:10.1016/j.neuropsychologia.2004.04.006

Walhovd, K. B., Westlye, L. T., Amlien, I., Espeseth, T., Reinvang, I., Raz, N., ... Fjell, A. M. (2011). Consistent neuroanatomical age-related volume differences across multiple. *Neurobiology of Aging, 32*(5), 916–932. doi:10.1016/j.neurobiolaging.2009.05.013

Zatorre, R. J., Fields, R. D., & Johansen-Berg, H. (2012). Plasticity in gray and white: Neuroimaging changes in brain structure during learning. *Nature Neuroscience, 15*(4), 528–536. doi:10.1038/nn.3045

SECTION III
DEVELOPMENTAL PERSPECTIVE

8

Review of the Evidence on, and Fundamental Questions About, Efforts to Improve Executive Functions, Including Working Memory

Adele Diamond and Daphne S. Ling

Table of Contents

Introduction

Efforts to improve executive functions (EFs)—which include selective attention, self-control, working memory (WM), cognitive flexibility, and reasoning—to remediate deficits, improve academic performance, improve productivity, increase the likelihood of healthy choices and quality of life, and head off, slow, or reverse cognitive decline during aging. This systematic review is the most extensive review to date of interventions, programs, and approaches that have tried to improve EFs. Previous reviews have focused on one type of intervention, for example, the large literature cognitive training approaches to improving EFs or on physical-activity approaches to improving EFs. These reviews have also often concentrated only on children or only on adults. The review here looks at all the different methods that have been tried for improving EFs and at all ages.

In total, 179 studies (reported across 193 papers) from all over the world (North and South America, Europe, South and East Asia, the Middle East, and Oceania) are included. If a study a) evaluated a method to improve EFs, b) was published in English in a peer-reviewed journal by or before 2015, c) had at least one objective EF outcome measure, d) had least eight people per group, e) included a control group and compared EF improvement and/or posttest performance in the experimental and control groups, f) was not simply correlational, and g) involved more exposure to the approach or program than a single session, it is reviewed here. Since our primary focus is normal development and aging, we excluded all studies of participants with brain damage or dementia. We included studies with persons with attention deficit hyperactivity disorder (ADHD), since ADHD is primarily a problem with EFs, and a small random sampling of studies of individuals with other clinical conditions, such as depression or autism, or individuals who had a learning disorder. Tabulations were done both excluding results for clinical populations and including them.

The findings reveal some surprises. Perhaps the biggest surprise is that a relatively understudied approach—**mindfulness practices involving movement** (Chinese mind-body practices, taekwondo, t'ai chi, and Quadrato Motor Training)—yielded the strongest results for improving EFs.[1] Mindfulness practices involving movement produced the best results for improving EFs across all four different metrics we used for judging strength of EF benefits. When results were taken as reported, even including potentially spurious ones, mindful movement practices still produced the best results on two of the four metrics (see Table 8.1). Table 8.2 omits studies where positive results might not have survived the needed corrections for multiple comparisons or data analyses reflecting the level at which they randomized. These results are far better than those for any other approach to improving EFs. Often, initial findings look strong but then do not hold up in subsequent studies, so there is a chance that this category looks strongest because of the relatively small number of studies that have investigated it thus far. However, right now, all eight studies of mindful movement practices (100%) have found at least suggestive evidence of EF improvement. No other approach to improving EFs can claim that.

Tables 8.1 and 8.2 report results across our four metrics for 13 of the types of interventions we investigated. This review also looks at neurofeedback, combinations of aerobic exercise with other things, and programs using drama, music, photography, quilting, or Experience Corps*, but there were too few studies of each of those to include them in Tables 8.1 or 8.2.

In Table 8.1, **promising school programs** comprise the only approach to come in first or second on all four metrics. In Table 8.2, promising school programs comes in second every time, behind mindful movement practices. *Both approaches show results superior to those for all cognitive training interventions targeting EFs.* School programs have produced much better results for improving inhibitory control than any other approach. That is important because inhibitory control seems to be the EF most predictive of long-term outcomes.

Public school programs targeting EF skills are able to reach more children, more economically, and more fairly (in that ability to pay is irrelevant) than any other approach to improving EFs. When EF training is embedded in activities throughout the school day, children are challenged on diverse EFs under

---

[1] Yoga forms its own category in our review because there were a sufficient number of studies of yoga to make that possible. EF benefits from yoga have generally been disappointing, although a few studies found outstanding results. It is unclear why there is such a discrepancy across studies, but it might have to do with how yoga was taught (as a mindfulness practice or just as a physical activity) and/or characteristics of the instructor.

Table 8.1. Summary of Results for All EFs Assessed (Including Reasoning/Fluid Intelligence) Across All Program and Intervention Types

	Percent of Studies Finding Even Suggestive[1] Evidence of EF Benefits (# of Studies)	Percent of Studies Finding Clear[2] Evidence of EF Benefits (# of Studies)	Percent of EF Outcome Measures on which Experimental Group Improved More Than Control Group (# of Measures)	Percent Of EF Measures on which Experimental Group Performed Better at Posttest Than Control Group (# of Measures)
Cogmed Training	60% (15)	23% (13)	42% (138)	28% (104)
N-back Training	46% (13)	31% (13)	24% (93)	20% (91)
Computerized Complex-Span Training	25% (4)[3]	0% (4)	27% (30)	24% (29)
Task-Switching Training	20% (5)	0% (5)	47% (51)	24% (42)
Other Computerized Cognitive Training (including commercial products)[4]	44% (27)	13% (24)	29% (223)	13% (196)
Noncomputerized Cognitive Training	67% (12)	20% (10)	45% (74)	30% (60)
Plain Aerobic Exercise	31% (16) ⎱ 43%(35)	6% (16) ⎱ 7%(30)	17% (70) ⎱ 27%(151)	11% (64) ⎱ 13%(111)
Aerobic Exercise with Cognitive and/or Motor Skill Demand[5]	53% (19) ⎰	7% (14) ⎰	36% (81) ⎰	15% (47) ⎰
Resistance Training	22% (9)	0% (8)	25% (36)	7% (30)
Yoga	43% (7) ⎱ 73%(15)	14% (7) ⎱ 51%(55)	38% (32)[6] ⎱ 51%(55)	23% (35) ⎱ 31%(51)
Mindfulness Practices Involving Movement (other than yoga)	100% (8) ⎰	29% (7) ⎰	70% (23) ⎰	50% (16) ⎰

(continued)

Table 8.1. Continued

	Percent of Studies Finding Even Suggestive[1] Evidence of EF Benefits (# of Studies)	Percent of Studies Finding Clear[2] Evidence of EF Benefits (# of Studies)	Percent of EF Outcome Measures on which Experimental Group Improved More Than Control Group (# of Measures)	Percent Of EF Measures on which Experimental Group Performed Better at Posttest Than Control Group (# of Measures)
More Sedentary Mindfulness Practices	61% (23)	17% (23)	36% (91)	30% (96)
Promising School Programs[7]	75% (8)	57% (7)	61% (28)[8]	53% (38)

Note. There were too few studies in any of the following categories to include them here, although they appear in Tables 8.3 and 8.4 and are discussed in the chapter: interventions that combined aerobic exercise with other interventions, neurofeedback, theater, piano, photography, quilting, and Experience Corps.

[1] Suggestive = more EF improvement or better EF posttest performance than control group on ≥ 50% of measures.

[2] Clear = more EF improvement and better EF posttest performance than control group on ≥ 67% of measures. Whenever a study reported ≥ 67% of measures showing positive results for improvement or posttest and did not provide any data on the other, that study is not included in calculations here because it is possible the results of the study might have met our criteria for "clear" had the results not reported been included.

[3] Six complex-span training studies are included in the review. Two were noncomputerized and are included under "noncomputerized training" in Table 8.1 rather than under computerized complex-span training.

[4] Other Computerized Cognitive Training includes both interventions classified as miscellaneous computerized cognitive training and commercial computerized cognitive training products, including the noncommercial BrainGame Brian.

[5] If the FITKids studies are counted as three separate, independent studies, then for enriched aerobic exercise, the results would be 52% (21) for suggestive evidence, 6% (16) for clear evidence, 35% (91) for improvement, and 14% (57) for posttest.

[6] One yoga study did not do pretesting.

[7] Included in the Promising School Programs category are the following school programs: Attention Academy, Chicago School Readiness Program (CSRP), MindUP, Montessori, PATHS, and Tools of the Mind.

[8] Two studies of School Programs did not do pretesting.

Table 8.2. Summary of Results for EFs Assessed (Including Reasoning/Fluid Intelligence) Across All Program and Intervention Types, Omitting Studies Whose Positive Results Might Not Have Held up Had They Corrected for Multiple Comparisons or Conducted Data Analyses Reflecting the Level at Which They Randomized

	Percent of Studies Finding Even Suggestive[1] Evidence of EF Benefits (# of Studies)	Percent of Studies Finding Clear[2] Evidence of EF Benefits (# of Studies)	Percent of EF Outcome Measures on which Experimental Group Improved More Than Control Group (# of Measures)	Percent of EF Measures on which Experimental Group Performed Better at Posttest Than Control Group (# Of Measures)
Cogmed Training	54% (13)	27% (11)	36% (103)	28% (69)
N-back Training	30% (10)	30% (10)	18% (72)	18% (72)
Computerized Complex-Span Training	33% (3)	0% (3)	30% (10)	22% (9)
Task-switching Training	20% (5)	0% (5)	47% (51)	24% (42)
Other Computerized Cognitive Training (including commercial products)[3]	45% (22)	10% (20)	33% (145)	14% (125)
Noncomputerized Cognitive Training	67% (12)	20% (10)	45% (74)	30% (61)
Plain Aerobic Exercise	31% (16) } 41%(34)	6% (16) } 7%(29)	17% (70) } 26%(145)	11% (64) } 13%(105)
Aerobic Exercise with Cognitive and/or Motor Skill Demand	50% (18)	8% (13)	33% (75)	17% (41)
Resistance Training	22% (9)	0% (8)	25% (36)	7% (30)
Yoga	20% (5) } 60%(10)	20% (5) } 33%(9)	16% (19) } 40%(30)	14% (22) } 26%(27)
Mindfulness Practices Involving Movement (other than yoga)	100% (5)	50% (4)	82% (11)	80% (5)

(continued)

Table 8.2. Continued

	Percent of Studies Finding Even Suggestive[1] Evidence of EF Benefits (# of Studies)	Percent of Studies Finding Clear[2] Evidence of EF Benefits (# of Studies)	Percent of EF Outcome Measures on which Experimental Group Improved More Than Control Group (# of Measures)	Percent of EF Measures on which Experimental Group Performed Better at Posttest Than Control Group (# Of Measures)
More Sedentary Mindfulness Practices	59% (22)	18% (22)	38% (86)	28% (85)
Promising School Programs[4]	67% (6)	40% (5)	53% (19)	52% (25)

Note. There were too few studies in any of the following categories to include them here, although they appear in Tables 8.3 and 8.4 and are discussed in the chapter: interventions that combined aerobic exercise and other things, neurofeedback, theater, piano, photography, quilting, and Experience Corps.

[1] Suggestive = more EF improvement or better EF posttest performance than control group on ≥ 50% of measures.

[2] Clear = more EF improvement and better EF posttest performance than control group on ≥ 67% of measures. Whenever a study reported ≥ 67% of measures showing positive results for improvement or posttest and did not provide any data on the other, that study is not included in calculations here because it is possible the results of that study might have met our criteria for "clear" had the results not reported been included.

[3] Other Computerized Cognitive Training includes both interventions we classified as miscellaneous computerized cognitive training and commercial computerized cognitive training products, including the noncommercial BrainGame Brian.

[4] Included in the Promising School Programs category are the following school programs: Attention Academy, Chicago School Readiness Program (CSRP), MindUP, Montessori, PATHS, and Tools of the Mind.

very diverse circumstances. That is important for improvement on multiple EFs and for being able to generalize skills to novel situations. School programs are also able to provide greater doses, frequency, and duration than most other approaches to improving EFs. The data suggest that this combination of a great deal of training and practice under diverse circumstances pays off.

Despite much hype in the popular press and even some influential reviews in high-profile journals, there is a glaring lack of evidence that interventions tried thus far of **resistance training** or **aerobic exercise** consistently improve EFs. Across all the different methods investigated thus far for improving EFs, only resistance training and "plain" aerobic exercise (e.g., running or brisk walking) fall in the bottom half on *all* four measures we used to assess intervention efficacy in both Tables 8.1 and 8.2. (Results are slightly better for aerobic exercise with more cognitive or motor-skill challenges. It shows better results than plain aerobic exercise on three of the four metrics, with comparable results on the fourth. However, it still falls in the bottom half of interventions on three of the four metrics.) *No study of resistance training* and only two studies each of plain aerobic exercise and aerobic exercise with more cognitive or motor-skill challenges found clear evidence of EF benefits. Across all EF outcome measures, participants in resistance training or plain aerobic exercise improved more than control participants on only 17% to 25% of the measures. Compare that to mindfulness movement practices, task switching, or promising school programs, where across all EF outcome measures the experimental group improved more than the control group on 82%, 48%, and 53% of the measures, respectively (see Table 8.2). These results probably reflect how these types of physical-activity interventions have been structured rather than that aerobic activity does not benefit EFs. Persons who are more physically fit and people who spend more time doing physical activity consistently show better EFs. Engaging in physical activity might be driving EF benefits in ways that most intervention studies have not been capturing. (Hypotheses about that are offered in this chapter.)

Another approach that has received less media attention, **noncomputerized cognitive training**, looks potentially promising. Of the 13 approaches listed in Tables 8.1 and 8.2, it ranked third. It fell in the top 50% of programs on all four metrics in both Table 8.1 and Table 8.2. *Noncomputerized cognitive training has produced better EF results than any type of computerized cognitive training.* Across all studies of noncomputerized cognitive training, 67% report at least suggestive evidence of EF benefits, but only a few of those studies used blinded assessment. Note that all three approaches producing the best EF results involve more in-person interaction than computerized cognitive training. Perhaps some of the success of noncomputerized training has to do with the greater degree of instructor–trainee interaction when training is not computerized. On the other hand, perhaps there is just more room for unintentional biases of the trainers to affect the results when the training is not computerized.

Despite much fuss about possible benefits of **N-back training** for improving fluid intelligence, *only one N-back training study* with an active control group (out of six) found more improvement or better posttest performance on any measure of fluid intelligence in participants compared with control subjects. Compared to no-treatment control groups results look better, but still less than half of N-back studies found evidence of any benefit to fluid intelligence.

The computerized training approach most successful at improving EFs is Cogmed*. It ranked in the top 50% of programs on all four metrics in both Table 8.1 and Table 8.2, the only computerized method to do so. It is the only method to consistently show *sustained* near-transfer benefits. Benefits to WM from Cogmed have been shown to last for 3 to 6 months and even for a year. Benefits from Cogmed are narrow, though, extending only to the aspects of WM trained and perhaps some aspects of attention. Cogmed is marketed as being beneficial to children with ADHD, yet its generalization to ADHD symptomatology has not been confirmed by blinded observers or objective measures.

Results from three different studies suggest that the mentoring component of Cogmed may play a greater role in Cogmed's benefits than people have appreciated. The control version of Cogmed (where difficulty does not increase) also includes interaction with mentors, but it usually produces less benefit than the standard, adaptive version of Cogmed. Is mentoring then irrelevant to the benefits or might the mentors not expect similar benefits from the control condition? Interacting with an adult who believes in the efficacy of the training and expects you to improve is probably critical.

In all age groups, cognitive training, both computerized and noncomputerized, improves the cognitive skills on which one trains. There does not appear to be an age too young or too old. There is very limited evidence of transfer to untrained skills, however.

If someone has a specific deficit in WM (as can be common with aging), Cogmed or N-back training might be quite beneficial. There has been very little study of Cogmed with older adults, but WM deteriorates earlier and more severely during aging than most other cognitive skills. The few studies of Cogmed and N-back training with older adults suggest that such targeted cognitive training might be especially beneficial for that subset of the population.

It is clear that generally, sessions of 30 to 40 minutes (min) yield better EF outcomes than sessions shorter than 30 min, and that is true both for cognitive training and physical activity (although Quadrato Motor Training provides a notable exception). It is not clear, however, that even longer sessions yield better results. For aerobic exercise, the evidence suggests that sessions longer than an hour yield fewer benefits than sessions of 45 to 60 min (of which about 30–40 min is aerobic).

We predict that many activities not yet studied will likely improve EFs. We also predict that the way an activity is done and the human qualities of the mentors or trainers (such as how enjoyable they make the activity, their supportiveness, and their ability to communicate their unwavering faith in the participants and the program), as well as whether the activity is personally meaningful and relevant, inspiring a deep commitment and emotional investment from participants to the activity and to one another, will likely prove more decisive than what the activity is. We are impressed with the potential benefits of real-world activities, such as sports, theater, and Experience Corps*, that engender deep commitments, bring joy, build self-confidence and pride, challenge EFs, and build community. We would like to see more studies of these and other real-world activities, including more that are done outdoors in nature.

EFs certainly can be improved—at every age from infancy through old age. We are only at the beginning, however, of understanding what characterizes the approaches that are most successful and how success differs by type of approach, EF domain, and/or subject characteristics. We have hardly begun to explore how to make benefits generalize further and last longer. Much has been revealed about what works to improve EFs and what does not, but this is only the tip of the iceberg.

Executive Functions (EFs)

Before discussing the general principles that can be gleaned from the vast literature relevant to improving EFs, it is important to define EFs and to explain why it is important to try to improve them.

EFs (also called executive control or cognitive control) refer to a family of interrelated, top-down processes needed to concentrate and pay attention, when "going on autopilot" or relying on instinct or intuition would be ill-advised, insufficient, or impossible (Diamond, 2006, 2013; Espy, 2004; Hughes, 2005; Jacques & Marcovitch, 2010). There is general agreement that there are three core EFs (inhibitory control, WM, and cognitive flexibility; Diamond, 2013; Miyake et al., 2000; Lehto, Juujärvi, Kooistra, & Pulkkinen, 2003; Logue & Gould, 2013; see Figure 8.1). Using EFs is effortful. It is easier to continue doing what one has been doing than to change or to put thought into what to do next. It is easier to give into temptation than to resist it.

One core EF is inhibition (also called inhibitory control), under which are usually categorized both self-control (behavioral inhibition or response inhibition) and interference control (including selective attention [also called

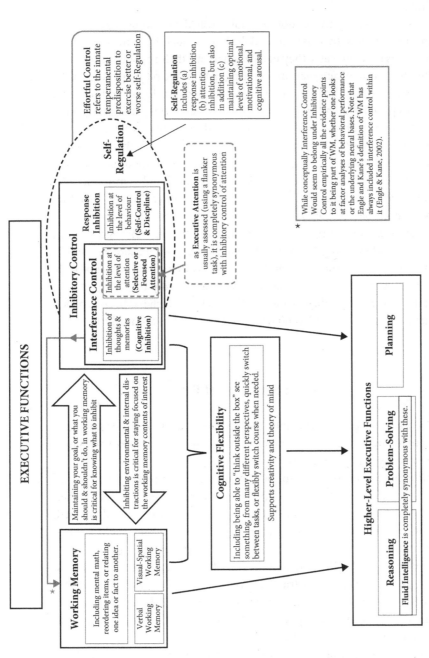

Figure 8.1. The components that together comprise executive functions (EFs) and the relation of EFs to other concepts. The two primary EFs are WM and inhibitory control, which together make cognitive flexibility possible. From the three core EFs, the higher order EFs of reasoning, problem-solving, and planning are built. Inhibitory control is usually thought to consist of response (or behavioral) inhibition and interference control, but there is increasing empirical evidence that interference control is more closely aligned with WM than with inhibitory control.

Reprinted from Diamond, A. (2016). Why assessing and improving executive functions early in life is critical. In P. McCardle, L. Freund, & J. A. Griffin (Eds.), *Executive Function in Preschool-age Children: Integrating Measurement, Neurodevelopment, and Translational Research* (pp. 11–43). Washington, DC: American Psychological Association.

executive attention or focused attention] and cognitive inhibition). Self-control involves control over one's behavior and control over one's emotions in the service of controlling one's behavior. Self-control is about resisting temptations and not acting impulsively. It is about suppressing a dominant response, or one's first inclination, and giving a more appropriate response instead. The strong inclination might be, for example, to reflexively strike back at someone who has hurt your feelings, blurt out an inappropriate remark, cut in line, or for a visitor from North America reflexively looking left when crossing the street in the United Kingdom (where looking right is the correct response). Importantly, self-control involves waiting before speaking or acting so that what comes out is a considered response rather than an impulsive one.

One aspect of interference control is inhibitory control of attention (interference control at the level of perception). It enables us to focus on what we choose and ignore distractions (Posner & DiGirolamo, 1998; Theeuwes, 1991). We need such selective attention at a cocktail party when we want to screen out all but one voice. Another aspect of interference control is cognitive inhibition—suppressing extraneous or unwanted thoughts or memories, resisting proactive interference from information acquired earlier, and resisting retroactive interference from items presented later (Anderson & Levy, 2009; Postle, Brush, & Nick, 2004).

WM, a second core EF, involves holding information in mind and mentally working with that information (Baddeley, 1992; Baddeley & Hitch, 1994, Cohen, Pearlstein et al., 1997; D'Esposito et al., 1995, 1998; Owen, Evans, & Petrides, 1996; Petrides, 1994; 1995; Smith & Jonides, 1999; Smith, Jonides, Marshuetz, & Koeppe, 1998). Translating instructions into action plans requires WM, as does updating one's thinking or planning, mentally re-ordering a to-do list, calculating a route, considering alternatives, or relating one piece of information to another. Two types of WM are distinguished by content—verbal WM and non-verbal (visuospatial) WM (Alloway, Gathercole, & Pickering, 2006; Baddeley, 1992).[2]

---

[2] The term *working memory* is not always used the same way. In this book, there is a chapter by Engle who uses WM to mean maintaining information in mind, sometimes also including manipulating information (e.g., Unsworth & Engle, 2007), although not always (e.g., Engle & Kane, 2004), but always including an element absent from the definition above—interference control (blocking or inhibiting other information from entering that active state; Engle & Kane, 2004). That is, Engle defines WM as holding information in mind (and perhaps also manipulating it) in the presence of interference that must be inhibited (Unsworth & Engle, 2007). As the note in Figure 8.1 indicates, empirical evidence is increasingly in support of the perspective of Engle and colleagues.

When WM is defined in this way, it also applies to holding information in mind while performing mental operations on other information (D'Esposito & Postle, 2015; Unsworth & Engle, 2007). An example of WM conceived of in this way would be holding in mind a question or comment you want to raise while you are trying to follow what others are saying.

Short-term memory (STM) involves just holding information in mind. WM also involves holding information in mind but while also performing mental operations. WM and STM cluster on separate factors in factor analyses of children, adolescents, and adults (Alloway, Gathercole, Willis, & Adams, 2004; Gathercole, Pickering, Knight, & Stegmann, 2004). They are linked to different neural subsystems. For example, WM relies more on dorsolateral prefrontal cortex (DLPFC), while maintaining information in mind but not manipulating it (as long as the number of items is not suprathreshold) does not require DLPFC (D'Esposito, Postle, Ballard, & Lease, 1999; Eldreth et al., 2006; Smith & Jonides, 1999). WM and STM also show different developmental progressions. STM develops earlier and faster (Diamond, 1995; Davidson, Amso, Anderson, & Diamond, 2006).

There are often misunderstandings about which tasks tap WM or STM. Forward Span tasks (which require recalling items in the order in which they were presented) assess STM because no manipulation of the information is needed. Re-ordering span tasks (mentally ordering the items according to some criterion, such as size) assess WM because they require manipulating the information presented. They are relatively pure tests of WM in the sense that they require little else besides WM.

N-back tasks, especially 2- or 3-back versions (e.g., "Press when you see an A two items after seeing an X" or "Press when any letter is repeated two items after that letter was shown"), require maintaining the ordinal position of items while continually entering incoming stimuli into WM, comparing them to new stimuli, and continually updating ordinal position and deleting old items so that one's WM capacity is not over-run. N-back tasks are more difficult WM tasks and require maintenance + manipulation + interference control. In addition, they require response inhibition when lures appear (e.g., an A one item after seeing an X), so they are not pure measures of WM. Further evidence that they do not require WM alone is that training on N-back tests may improve inhibitory control (attentional control and response inhibition) more than it improves WM (Jaeggi, Buschkuehl, Jonides, & Shah, 2012).

Complex-span tasks (e.g., counting span, reading span, and operations span) require holding in mind a piece of information from each previous trial in chronological order while performing the required mental operation on each trial. For example, the counting span task requires, on each trial, counting out loud all the blue dots (ignoring the yellow ones) and then announcing in chronological

One example of "the ability to keep a representation active, particularly in the face of interference and distraction" (Engle, Tuholski, Laughlin, & Conway, 1999, p. 309), i.e., maintenance plus interference control, would be spatial span tasks with a masking stimulus. For example, several boxes in a grid might be simultaneously illuminated momentarily (a masking stimulus) on each trial between presentation of a sequence of boxes in the grid being illuminated and when the subject can respond by touching the boxes in the order in which they were illuminated.

order the totals for blue dots on all previous trials. These are difficult tasks and clearly require maintenance plus manipulation. The problem is that they can also require inhibition (e.g., on the counting span, selectively attending to the blue dots and ignoring the yellow ones) and always require task switching (e.g., switching from the task of counting or reading to the task of reciting ordered information held in mind). Diamond has argued that complex-span tasks might more properly be considered measures of EFs in general, rather than WM in particular (Diamond, 2013).

The third core EF is cognitive flexibility (also called mental flexibility, set shifting, or task switching). One aspect of cognitive flexibility is being able to change perspectives, being able to see something from different perspectives. For example, from one perspective, the A and C in the string AC15 go together because they are both letters. From another perspective, the C and 1 go together because they are both in italics. From yet another perspective, the A and the 1 go together because they are both first in their respective series. The ability to see different perspectives could also come into play in a situation in which two people seemingly share little in common, perhaps because they have different religious beliefs or political affiliations, but from other perspectives they might share commonalities in their taste in music or commitment to social justice.

Another aspect of cognitive flexibility involves changing how you think about something ("thinking outside the box"). If there's a problem no one has been able to solve, one might think outside the box to conceive of the problem, frame the problem in a new way, or come up with a new way of attacking it. Cognitive flexibility enables us to flexibly adjust to changing demands or priorities, take advantage of a sudden, unexpected opportunity (serendipity), and overcome sudden, unexpected problems.

From these three core EFs, higher-order EFs, such as reasoning, problem-solving, and planning, are built (Collins & Koechlin, 2012; Lunt et al., 2012; see Figure 8.1). The family of EFs depend on prefrontal cortex and other neural regions (especially the anterior cingulate cortex and parietal cortex) with which prefrontal cortex is interconnected (Aron, Behrens, Mith, Frank, & Poldrack, 2007; Cole & Schneider, 2007; Eisenberg & Berman, 2010; Leh, Petrides, & Strafella, 2010; Niendam et al., 2012; Zanto, Rubens, Thangavel, & Gazzaley, 2011).

Why It Is Important to Improve EFs

EFs are necessary in our ever-evolving world. Self-control is vital for a civil society where people abide by rules and norms, resisting temptations not to do so. Without inhibitory control, we would be at the mercy of impulse, old habits of thought or action (conditioned responses), and environmental stimuli that pull us this way or that. Thus, inhibitory control makes it possible for us to change and to choose how to react and behave, rather than being "unthinking" creatures of habit. While it is by no means easy, it makes it possible for us to avoid saying or doing things we would later regret.

WM is critical for making sense of anything that unfolds over time, for that always requires holding in mind what happened earlier and relating it to what comes later. Note that anything that involves language (oral or written) unfolds over time. WM is also critical for reasoning and problem-solving because those require holding items in mind to see their interrelations and to explore novel recombinations. Cognitive flexibility is the core of critical and creative problem-solving.

All of life's aspects require the presence of mind to wait before speaking or acting, being able to resist impulsively reacting, staying focused despite distraction, seeing tasks through to completion although tempted not to, holding alternatives in mind so one can look at a situation from multiple perspectives (creatively coming up with new ways to attack problems), flexibly changing course when needed, and seizing opportunities when they unexpectedly arise. As societies become increasingly complex and fast-paced, having good WM and cognitive flexibility allows us to keep up with information processing, contribute to meaningful conversations, reason, solve problems, read critically, and see things from another person's perspective. This is especially important in an increasingly virtual, borderless world, where people from different backgrounds, cultures, political ideologies, languages, and beliefs interact with one another on a daily basis through the Internet.

Research confirms that EFs are critical for *school readiness* (they are even more critical than IQ or entry-level reading or math; Blair, 2002; Blair & Razza, 2007; Carlson & Moses, 2001; Hughes & Ensor, 2008; Kochanska, Murray, & Coy, 1997; Morrison, Ponitz, & McClelland, 2010), *success in school from the earliest grades through university* (Alloway & Alloway, 2010; Borella, Carretti, & Pelgrina, 2010; Duncan et al., 2007; Fiebach, Ricker, Friederici, & Jacobs, 2007; Hamre & Pianta, 2001; Loosli, Buschkuehl, Perrig, & Jaeggi, 2012; McClelland et al., 2007; Nicholson, 2007; Savage, Cornish, Manly, & Hollis, 2006; St Clair-Thompson & Gathercole, 2006), *career success* (Bailey, 2007), *making and keeping friends* (Hughes & Dunn, 1998), *marital harmony* (Eakin et al., 2004), and *good health* (Crescioni et al., 2011; Cserjési, Luminet, Poncelet, & Schafer, 2009; Hall, Crossley, & D'Arcy, 2010; Miller, Barnes, & Beaver, 2011; Moffitt et al., 2011; Perry et al., 2011; Riggs, Spruijt-Metz, Sakuma, Chour, & Pentz, 2010).

An influential meta-analysis questioned whether improving EFs improves academic achievement (Jacob & Parkinson, 2015) but that analysis included no intervention studies (just longitudinal and cross-sectional ones) and the authors did two things that would greatly reduce the size of any effect found. One, in their analyses of causal relationships, they excluded studies that did not control for IQ. However, fluid intelligence and EFs correlate about 0.8 to 1.0 (Conway, Kane, & Engle, 2003; Stauffer, Ree, & Caretta, 1996). Indeed, in one study of patients with neurological and neuropsychiatric conditions impairing EFs, when the effects of fluid intelligence were partialled out, no clinical deficit remained on EF measures (Roca et al., 2014). Most IQ tests now assess both crystallized and fluid intelligence; to the extent that controlling for IQ resulted in controlling for the fluid intelligence aspect of IQ, studies were essentially guaranteeing no effect of EFs. Second, in a school context, rarely does a program only target EFs and not academic skills. Jacobs and Parkinson set too high a bar in specifying that if a school program worked on both, then there's no evidence that training EFs per se helps. What one needs to ask is: If the program is during the regular school day, does targeting EFs *in addition to academic subjects* yield greater benefits than a comparable program that only targets academic subjects?

Intervention studies show that improving EFs does indeed improve academic performance. For example, when the Chicago School Readiness Project was delivered in Head Start preschools, it was able to improve the EFs of disadvantaged children by the end of that preschool year (Raver et al., 2008, 2011). Those children continued to perform better in math and reading than their peers who had attended regular Head Start for the next 3 years, and those academic gains were mediated almost entirely through improved EFs (Li-Grining, Raver, & Pess, 2011). Similarly, the Tools of the Mind curriculumdelivered in kindergartens in Massachusetts not only improved EFs by the end of kindergarten, but also improved reading, vocabulary, and math more than did regular kindergarten, and that difference in academic progress was still evident in first grade (indeed, for reading, it was only evident in first grade; Blair & Raver, 2014).

Graduating from high school has become increasingly important, especially with a competitive job market and high rates of unemployment. It is becoming more and more difficult for someone without a high school degree to find work (Friedman et al., 2007; Winstok, 2009). Children with poor EFs are far less likely to graduate from high school (Friedman et al., 2007; Friedman, Miyake, Robinson, & Hewitt, 2011; Moffitt et al., 2011), which ultimately has a major economic impact on them and the entire nation. The estimated cost savings to Canada if high school graduation increased by only 1% is more than $7.7 billion (Hankivsky, 2008).

With a large sample of over 14,000 youths, Miller et al. (2011) found that those who had poorer inhibitory control were "exponentially more likely" to suffer from nine of the ten adverse health conditions they examined, including asthma, high cholesterol, high blood pressure, and cancer. People with better EFs generally enjoy a better quality of life (Archontaki, Lewis, & Bates, 2013; Brown & Landgraf, 2010; Davis, Marra, Najafzadeh, & Lui-Ambrose, 2010; Moffitt, 2012; Tangney, Baumeister, & Boone, 2004) and live longer (Hall et al., 2010).

In a study of 1,000 children born in the same city in the same year, thos with worse inhibitory control as children (those who were less persistent, more impulsive, and had poorer attention regulation)later as adolescents were more likely to smoke, have unplanned pregnancies, and drop out of school. As adults 30 years later, they were likely to earn less,have worse health three times more likely to be addicted to drugs),be a single parent (twice as likely), and commit more crimes (four times more likely to have a criminal record) than those with better inhibitory control as children, controlling for IQ, gender, social class, home lives, and family circumstances growing up (Moffitt et al., 2011). They were also less likely to be happy as adults (Moffitt, 2012).

Evidence shows that EF deficits often do not disappear on their own and can grow larger (Nigg et al., 2006; O'Shaughnessy, Lane, Gresham, & Beebe-Frankenberger, 2003; Riggs, Blair, & Greenberg, 2003). Early EF deficits can lead to a negative self-image and maladaptive behaviors that can be extremely difficult to correct (Broidy et al., 2003; Nagin & Tremblay, 1999). EFs during early childhoo often predict adult outcomes better than does IQ or SES (Moffitt et al., 2011). Research suggests that early EF gains can reduce the later incidence of school failure, substance abuse, addiction, aggression, crime, and other antisocial or inappropriate behaviors (Hall et al., 2010; Nagin & Tremblay, 1999; Olson, Smeroff, Kerr, Lopez, & Wellman, 2005; Vitaro, Barker, Brendgen, & Tremblay, 2012).

Being able to enhance EF development early in a child's life is critical because it affects the trajectory (the negative or positive feedback loop) on which a child is launched. Indeed, Moffitt et al. (2011) predicted "that interventions that achieve even small improvements in [the inhibitory control component of EFs] for individuals could shift the entire distribution of outcomes in a salutary direction and yield large improvements in health, wealth, and crime rate for a nation" (p. 2694).

Also, any approach that slows the decline of EFs with aging, or reverses it, would improve the quality of life for millions of people. EFs are the last abilities to fully mature during development and the first to start to deteriorate in adulthood (as early as one's forties; de Luca & Leventer, 2008). Poor WM and attention are among the most common complaints of older adults and among the cognitive problems that most negatively impact their lives (e.g., Park & Payer, 2000; Reuter-Lorenz et al., 2001). A survey of older adults in the United States found that their number-one health-related concern was not being "mentally sharp" (National Council on Aging, 2015). Heading off those problems, slowing their advance, or reversing the decline, even if only a little, could have major health and economic benefits.

Principles of Experimental Design and Principles for Interpreting Results Often Violated in Training or Intervention Studies

a. Suppose study participants were given a passage to read at Time 1 and thengiven the same passage to read at Time 2, and they read it faster at Time 2. If we had them wash dishes between Times 1 and 2, would we be justified in concluding that washing dishes improves reading speed? Of course not. As obvious as this seems, more than a few training studies display a similar logic.

To conclude that what individuals did between Times 1 and 2 produced the improvement at Time 2, evidence of differential improvement is needed (e.g., those who washed dishes improved more in reading speed than those who did not wash dishes, or those who spent more time washing dishes improved more in reading speed than those who spent less time washing dishes). Without differential improvement (i.e., a group X change interaction; Chang, Tsai, Chen, & Hung, 2013) we cannot know if what people did between Times 1 and 2 caused the improvement, or if the improvement was simply due to practice effects from having taken the tests before, or, in the case of children, from developmental changes.

Thus, a control group is essential. Studies that show improvements from Time 1 to Time 2 but have assigned all participants to the training group are fundamentally unconvincing. Finding that participants did better after an intervention than before is insufficient to determine that the intervention helped. Comparable improvements might have been seen without training.

b. If two groups in a study both show completely comparable improvement and performance, we cannot know why that was found. Similarly, if the two groups in a study both received the same training, but one had longer sessions or more sessions, yet both improved comparably, one cannot conclude that the training helped and that length of sessions or their duration did not matter. That may be the case, but it could also be that the training didn't help at all, regardless of whether participants got more or less of it; there is no way of knowing. To conclude that the training helped, a group that does not show comparable improvements is needed.

c. Another all-too-common problem with treatment studies involving young adults or children is that a greater benefit to the experimental group than to the control group is based primarily on the control group performing worse at post-test than they performed on the same measures at pretest, rather than the experimental groupshowing much improvement. For studies with older adults, where a decline might be expected, it might well be that the intervention reduced the rate of decline or stopped it. However, for children and young adults, it is very problematic if one group performs worse the second time they are tested. Such results should be viewed with caution, indeed skepticism.

d. If the control group started out performing better on outcome measures and at posttest the experimental group had caught up to the control group, it is difficult to conclude with any surety that the intervention helped, rather than what whe are seeing is regression to the mean or normal differences in developmental timetables. For example, children in the control group might have experienced a spurt in EF development before the study started (hence they were better at pretest) and the experimental group might have experienced a spurt in EF development during the time of the study, independent of the intervention. For example, if children in one group but not in another were walking at 9 months, and then by 15 months children in both groups were comparable in walking, that could easily be due to normal developmental processes rather than any program that those not walking at 9 months received. When Iit is impressive that an experimental group caught up to a control group is when the experimental group is a clinical population and the control group consists of a nonclinical population (such as typically developing children) because one · would not expect such catch-up in the normal course of events without the aid of some program.

An absolutely superb review of computer-based cognitive training approaches (Simons et al., 2016) that came out after our review had been submitted to the publisher makes a similar point: "The control group should be comparable to the experimental group before treatment or the results of the intervention will be effectively uninterpretable—that is, differential gains after training could just reflect those different starting points (Redick & Webster, 2014)."

A successful intervention should ideally produce both of theseresults: (1) significantly more improvement in the experimental group than the control group from before to after the intervention (i.e., better change scores), and (2) significantly better postintervention performance by the experimental group than the control group (i.e., better posttest scores). If the degree of change is significantly better, but posttest performance is not significantly better than the control group, that often (although not always) means that the intervention group started off worse and caught up or the control group inexplicably, mysteriously got worse. If the degree of change is no different, but the experimental group showed better posttest performance than the control group, that could be because the experimental group simply mantained the advantage they started with. Thus, what is needed are significant group differences both for final test results and for degree of improvement on the tests.

e. Another all-too-frequent problem is for studies to have a no-treatment, or business-as-usual, group as their only control group. People often get a boost from a change, any change (e.g., the Hawthorne effect; McCarney et al., 2007) and from receiving attention (as when they are trained on something new). Therefore, people in the control group should also receive a new program and similar amounts of attention. (See Boot, Simons, Stothart, & Stutts, 2013, for a similar argument.)

To compare a program or intervention to no treatment sets a very low bar for determining whether an intervention really worked. Anything might be better than nothing, and anything new might produce better results than business as usual. It is still unclear what the optimal control group would be (see the section "Limitations" in the section "Discussion"), but it is almost universally agreed that a control group should be actively engaged in something new and different (i.e., it should be an active control group), and ideally control subjects should get the same amount of attention and have the same expectations for benefits. See Simons et al. (2016) for a similar, and even more strongly worded, exposition on this point.

f. A related point is that people often do better simply because they believe what they are doing will yield improvements—a placebo effect (Boot et al., 2013), or expectancies become self-fulfilling (Jenner, 1990; Rosenthal & Jacobsen, 1968). Accordingly, the expectation of success on the dependent measures should be as high among members of the control group(s) and those interacting with them as among the experimental group.

g. Ideally, neither the subjects nor program providers (the trainers) should know which group is the experimental one and which is the control group (i.e., they should be blind to whether they are part of the group expected to show the most benefits or not).

h. Testers and raters should also be blind to who is in the group expected to do better. In studies involving children, when the outcome measures are adult ratings of, or reports on, children, if those adults are not blind to the children's group assignments, the raters might be inclined to think they saw greater improvement in children in the intervention group because of the expectation that they would improve more. Alas, too often training studies that use adult ratings as an outcome measure use adults who are fully knowledgeable about which groups the children were in.

Just as the raters of participants' performance should be blinded, people administering the outcome measures should also be blinded to participants' group assignments or the study's hypotheses, or both, because their expectations that participants in one group will perform better can cause those participants to perform better when otherwise they would not (Kit, Mateer, Tuokko, & Spencer-Rodgers, 2014; Pfungst, 1911; Rosenthal & Jacobsen, 1968; Rydell, Van Loo, & Boucher, 2014).

Textbooks often cite the need for double blinding (neither the participants in the study nor the people testing or evaluating them afterward should know who was in the experimental group), but when an intervention is delivered by people, a third group should be blind to this as well, and that is the people providing the intervention and the control conditions.

i. Spurious benefits can be seen if low performers drop out of the intervention group and higher performers remain. Be concerned if a study has a high attrition rate, especially since it usually indicates that something about the intervention was insufficient to keep people engaged (perhaps it was too boring or too demanding).

j. Although attrition is often reported, compliance rarely is (only 35% of studies reviewed here reported compliance). How often individuals attended the condition to which they were assigned is likely to affect how much they benefitted from that condition. *Studies should report data on compliance and should look at the relation between attendance and degree of benefit.*

k. Fidelity to the program by the people administering it is also important and is rarely monitored or reported. Two different people can administer the 'same' program in different ways and produce very different outcomes. Whether the people administering the program believe in the program and in the ability of participants in the program to succeed can also have important effects on results (Frank, 1961; Rosenthal & Jacobsen, 1968).

l. Too often, intervention studies have been underpowered, with too few participants per group.

m. Occasionally, interventions with a great many participants per group report only significance levels and not effect sizes. A study with a great many participants per group can find even the most trivial, minimal difference to be significant at $p \leq .05$. That is too low a bar.

n. Many intervention studies reviewed here that have used cluster-randomized designs (e.g., randomizing schools or classes to condition) have ignored the design when analyzing their data; they analyzed their data as if they had randomized individuals to condition. When this error in data analysis is made, many results reported as significant would not have reached significance had the appropriate data analysis been conducted. (A footnote in Column 1 of Tables 8.3 and 8.4 indicates when a study has done this.)

o. Several studies reviewed here that conducted multiple comparisons in their data analyses did not correct for that in their significance testing. When many comparisons are made, one might find a significant difference on one or more just by chance. Therefore, one needs to include a correction, such as dividing the 0.05 significance level by the number of comparisons made. (A footnote in Column 1 of Tables 8.3 and 8.4 alerts the reader when such a correction was not made for any study with ≥ 5 between-group comparisons on EF outcome measures. This was a relatively low bar to pass, as many would set the bar at simply 5 or more comparisons, not limiting it to just the outcome variables of interest, which in this case were EF measures.)

p. All too often, nontraining studies (correlational studies that look at whether, for example, people who exercise more have better EFs or whether students who play in the orchestra have better EFs than students who do not) are discussed as if they show causality (that more exercise or playing in the orchestra helps one to have better EFs). *Causal inferences from such studies are unjustified.* It could be that students with better EFs are more likely to choose to be in the orchestra or are disciplined enough to exercise more. For example, among adults who had been doing t'ai chi, aerobic exercise, or meditation for 5 years or more, at least three times a week, for at least 30 min per session, those who meditated or did t'ai chi, but not those who did aerobic exercise, showed significantly better cognitive flexibility (smaller switching costs) than sedentary controls (Hawkes, Manselle, & Woollacott, 2013). Alas, we cannot know if some characteristic that caused participants to select their particular activity was responsible for the difference in cognitive flexibility, rather than the activity itself.

q. If participants in a training study are free to choose which group they would like to be in, then that training study is vulnerable to the same criticism—persons who chose the training condition might have differed at the outset in a way important to the outcome. Thus, for example, Verghese et al. (2003) recruited participants who were ≥ 75 years old and were dementia-free, and who regularly did one or another leisure activity, such as social dancing or reading and crossword puzzles. Five years later, the researchers found that participants who had been doing social dance showed a significantly greater reduction in the risk for dementia than participants who did any other activity. Even though the participants were followed longitudinally, we cannot conclude that dancing caused the reduced risk of dementia, because participants who chose social dance may have differed from other participants in other ways that were responsible for the difference in their risk of dementia.

r. If participants have no say in their group assignment, however, and if the characteristics of participants in the different groups are carefully matched on variables thought to be potentially relevant to the outcome, there is no particular benefit to random assignment unless the number of participants is very large. Indeed, random assignment when dealing with small numbers (such as 20 or fewer per group) can lead to problematic differences across groups. Random assignment is only likely to ensure comparable groups when numbers are large. Stratifying randomization by participant variables potentially important to the results can help yield more comparable groups while still preserving the advantages of randomization. Remember, though, that even groups well matched on variables thought to be relevant might not be matched on a critical variable because the experimenter had not realized it might be important.

s. The other problem with assigning participants to condition is that some participants might not want to do the activity to which they are assigned and might do it reluctantly or actively resist doing it, thus reducing the size of the effect. One way to address this is to include only participants who express interest in the intervention and then randomly assign half to the intervention and half to an attractive alternative. It is sometimes argued that teachers do not typically have a say in which curriculum they teach, so random assignment accurately reflects teachers' lack of choice. However, teachers who might be opposed to a new experimental curricular change and implement it only half-heartedly in a study might be much more willing to adopt it later after evidence of its effectiveness has been demonstrated. Randomly assigning teachers to an unproven program for a research study may not accurately reflect how they would implement the same program after studies have found promising results using it.

t. Studies of the effects of physical-activity interventions too rarely gather data on participants' activities outside the intervention. For example, children might be involved in sports, take dance lessons, bike to school, or bike a paper delivery route, etc. Participants in physical-activity studies should be asked about their activities and, when possible, should be asked to wear a simple device that monitors their amount and intensity of physical activity.

Studies Included in this Systematic Review

To locate studies for review, we searched PubMed and PsycNET for all publications that had any keyword or word in the title or abstract from both of the following sets: Set 1—*evaluate, evaluation, intervention, program, randomized control trial, train*, or *training*; Set 2—*attention* (apart from ADHD), *cognitive control, cognitive flexibility, executive function, inhibition, inhibitory control, fluid intelligence, mental flexibility, reasoning, self-control, self-regulation, set shifting, task switching*, or *working memory*. For all relevant articles found, we searched the reference lists for additional relevant publications. The inclusion criteria for assigning a study to the review were:

1. The study had to include at least one EF outcome measure.
2. At least one of the EF measures had to be an objective behavioral measure. Studies that included *only* self-, parental, and/or teacher ratings or questionnaires were not included.

3. The study had to be done on humans.
4. The study had to look at benefits other than improvements on the task(s) practiced during the intervention. There is much evidence that if someone practices a task or procedure, that person gets better at that task. We were interested in whether there was improvement in a basic cognitive ability that generalized at least to similar tasks.
5. A report of the study had to be published in English in a peer-reviewed journal by, or before, 2015.
6. The study could not be purely correlational (e.g., looking at the EFs of people who happen to be doing something, such as meditating or exercising), because causality cannot be determined from correlations.
7. The study must have included a control group. Without a control group, it is not possible to determine if improvements would have been found even if participants had not received the intervention.
8. The study had to provide data that enabled us to compare the improvement and/or posttest performance of the experimental and control groups. We did not include studies, for example, that reported whether each of the groups individually showed EF improvements, but never compared the relative sizes of those improvements. The point of a control group is to be able to determine whether an effect might have occurred anyway, even without the specific intervention of interest. To see that both groups improved, but not to know whether the intervention group improved significantly more, does not enable us to answer that question.
9. The study must not have examined *only* acute effects (i.e., immediate benefits after doing something only once). Such immediate benefits are usually transitory, lasting only minutes.
10. The study must have had at least eight participants per condition.
11. The study must not have had a design problem so severe that it is impossible to draw any conclusion about the experimental condition (e.g., administering twice as many sessions to the experimental group as the control group).

We included all studies with healthy subjects or with subjects with an EF disorder, such as ADHD, that met the above criteria. Since we are primarily interested in normal development, we randomly selected 10% of the studies conducted with other patient populations (thus many studies of older adults with a diagnosed ailment are not included here). We excluded studies of persons with brain damage, stroke, severe cognitive decline, or dementia. In the Appendix, which appears online (URL: http://www.devcogneuro.com/tables/supplemental.html), we list for each broad category (e.g., cognitive training or physical activity) the many studies we considered, but excluded, along with the reasons why each of the studies was excluded (i.e., which inclusion criteria each of the excluded studies had not met).

From the 403 articles that met our search criteria, we found, 179 studies that met criteria for inclusion in this systematic review. Their results were reported in a total of 193 peer-reviewed, published research papers. Tables 8.3 and 8.4 summarize the study characteristics and cognitive outcomes found in the studies. Given their level of detail (Tables 8.3 and 8.4 are quite long [over 100 pages]), these tables are provided online only.

These tables provide rich detail so that readers can check our conclusions against the data and explore hypotheses of their own. These tables summarize all cognitive and academic outcomes (not just EF outcomes), but they do not include any other outcomes (e.g., social, emotional, behavioral, motor skills, physical fitness, or personality outcomes). We also do not report on improvements on the trained task (although if the authors mention a lack of improvement there, we mention that in the text). We only looked at improvements on untrained tasks, since we are interested in improvements in a cognitive ability rather than facilitation in the performance of a specific task. In the text, we primarily discuss EF outcomes.

Table 8.3 (appears in online: URL: http://www.devcogneuro.com/tables/supplemental.html) includes only studies that had an active control group. Table 8.3a provides details on the participants and methods. Table 8.3b provides the results for near transfer. Table 8.3c provides the results for far transfer. Results for EF outcome measures are always provided before results for other measures.

Table 8.4 (appears in online: URL: http://www.devcogneuro.com/tables/supplemental.html) includes studies that compared the experimental condition to only a no-treatment or business-as-usual control group, a lower bar for an intervention to pass to say an effect was significant. In general, studies in Table 8.4 included only two groups (the experimental condition and no treatment). The studies by Ball et al. (2002), Brown, Liu-Ambrose, Tate, & Lord (2009), Klusmann et al. (2010), and Mortimer et al. (2012) also appear in Table 8.4, although they included other groups, because their analyses compared intervention groups only to the no-treatment group, not to one another. Table 8.4a provides details on the subjects and methods. Table 8.4b provides the results for both near and far transfer. Results for EF outcome measures are always provided before results for other measures.

Principles That Govern EF Training, Whatever the Form

Principle 1

There is good news: EFs can be improved. Moreover, it appears that improving EFs is possible across the lifespan and by different methods. The many reviews all more or less agree on that point (see Table 8.5).

Table 8.5. Review Papers on Different Methods for Improving EFs

Topic Reviewed	Authors and Year of Review
Cognitive Training	Au, Buschkuehl, Duncan, & Jaeggi, 2016; Au et al., 2015; Baltes & Lindenberger, 1988; Buitenweg, Murre, & Ridderinkhof, 2012; Cortese et al., 2015; Karbach & Verhaeghen, 2014; Kelly et al., 2014; Klingberg, 2010; Kueider, Parisi, Gross, & Rebok, 2012; Melby-Lervåg & Hulme, 2012; Melby-Lervåg, Redick, & Hulme, 2016; Morrison & Chein, 2011; Noack, Lövdén, Schmiedek, & Lindenberger, 2009; Rapport, Orban, Kofler, & Friedman, 2013; Redick & Lindsey, 2013; Reijnders, van Heugten, & van Boxtel, 2013; Schubert, Strobach, & Karbach, 2014; Schwaighofer, Fischer, & Bühner, 2015; Shipstead, Hicks, & Engle, 2012; Shipstead, Redick, & Engle, 2012; Simons et al., 2016; Spencer-Smith & Klingberg, 2015; Spierer, Chavan, & Manuel, 2013; Stine-Morrow & Basak, 2011; Tardif & Simard, 2011; von Bastian & Oberauer, 2013
Physical Exercise Training	Angevaren, Aufdemkampe, Verhaar, Aleman, & Vanhees, 2008; Best, 2010; Bustamante, Williams, & Davis, 2016; Chaddock, Pontifex, Hillman, & Kramer, 2011; Chang, Pan, Chen, Tsai, & Huang, 2012; Colcombe & Kramer, 2003; Donnelly et al., 2016; Fedewa & Ahn, 2011; Gates, Singh, Sachdev, & Valenzuela, 2013; Hillman, Erickson, & Kramer, 2008; Karbach & Verhaeghen, 2014; Kramer & Erickson, 2007; Penedo & Dahn, 2005; Scherder et al., 2014; Smith et al., 2010; Snowden et al., 2011; Streiner, 2009; Tomporowski, Lambourne, & Okumura, 2011; Tomporowski, McCullick, Pendleton, & Pesce, 2015; Tseng, Gau, & Lou, 2011; van Uffelen, Chinapaw, Hopman-Rock, & van Mechelen, 2008; Verburgh, Königs, Scherder, & Oosterlaan, 2014; Voss, Nagamatsu, Liu-Ambrose, & Kramer, 2011, Young, Angevaren, Rusted, & Tabet, 2015
Mindfulness	Mak, Whittingham, Cunnington, & Boyd, 2017; Ng, Ho, Chan, Yong, & Yeo, 2017; Zenner, Herrnleben-Kurz, & Walach, 2014; Zoogman, Goldberg, Hoyt, & Miller, 2015
Multiple Methods or Other Methods	Barenberg, Berse, & Dutke, 2011; Bryck & Fisher, 2012; Burke, 2010; Cortese et al., 2016; Diamond & Lee, 2011; Etnier, Nowell, Landers, & Sibley, 2006; Etnier et al., 1997; Gothe & McAuley, 2015; Green & Bavalier, 2008; Greenberg & Harris, 2012; Hindin & Zelinski, 2012; Howard-Jones, 2014; Law, Barnett, Yau, & Gray, 2014; Lustig, Shah, Seidler, & Reuter-Lorenz, 2009; Moreau & Conway, 2013; Muraven, 2010; Rabipour & Raz, 2012; Riccio & Gomes, 2013; Sedlmeier et al., 2012; Sonuga-Barke et al., 2013; Stine-Morrow & Basak, 2011

Principle 2

The bad news is that transfer of training is narrow. Individuals improve on what they practice and that generalizes to untrained tasks where the same skills they practiced are required. Benefits are fairly specific, rather than general, and rarely transfer to untrained skills. Across all intervention approaches, regardless of the skills targeted, participants improve on the skills they practice, and that usually transfers to other contexts where those same skills are needed (narrow transfer), but people rarely improve on what they have not practiced. This has been the nearly universal conclusion of reviews for the past 25 years (see Table 8.6). Simon et al. (2016) went farther and concluded that transfer is only to tasks nearly identical to the training tasks. Indeed, as we have pointed out previously, "It is not even clear that training nonverbal WM transfers to verbal WM or that training nonverbal analogical-reasoning transfers to nonverbal gestalt reasoning on Raven's Matrices (e.g., Bergman Nutley, Söderqvist, Ottersen, Grill, & Klingberg, 2012)" (Diamond & Ling, 2016, p. 36).

At one extreme, some worry that WM training may not improve WM at all, but simply train task-specific strategies or response patterns useful for similar tasks but nothing else (Harrison et al., 2013; Simon et al., 2016). At the other extreme, some read the evidence as indicating that WM training can improve intelligence (that is, fluid intelligence, as assessed by tests of reasoning; Jaeggi, Buschkuehl, Jonides, & Perrig, 2008; for a meta-analysis, see Au et al., 2015). For example, an excellent meta-analysis of N-back training studies (Au et al., 2015) and a meta-analysis of diverse "process-based" EF and WM training programs (Karbach & Verhaeghen, 2014)[3] both found small but significant improvements in fluid intelligence (i.e., reasoning). However, other excellent meta-analyses (Melby-Lervåg & Hulme, 2012; Melby-Lervåg et al., 2016) found no convincing evidence of generalization of WM training to other cognitive skills and concluded that training produces "short-term, specific training effects that do not generalize" (Melby-Lervåg & Hulme, 2012, p. 270).

As an example, of the Cogmed computerized training studies that met criteria for inclusion in the present review, fully 79% of the 15 studies that looked at near transfer found at least suggestive evidence of it, but only 33% of the 12 studies that looked at far transfer found even suggestive evidence of it (see Figure 8.2). Of the 10 N-back training studies that looked at near transfer, 40% found at least suggestive evidence of it. Of the 11 N-back studies that looked at far transfer, only 30% found clear or suggestive evidence of it.

---

[3] Karbach and Verhaeghen's (2014) meta-analysis is not discussed further in this paper because it included many studies that did not meet criteria for inclusion in the systematic review. For example, they included studies did not include a control group (e.g., Buchler, Hoyer, & Cerella, 2008), they only looked at performance on the training tasks (e.g., Dorbath, Hasselhorn, & Titz, 2011), they did not include any EF outcome measure (e.g., Mahncke et al., 2006), or they only looked at acute effects from a single training session (Karbach, Mang, & Kray, 2010).

Table 8.6. Conclusions About the Presence or Absence of Far Transfer Drawn by Reviews

Authors (Year of Review)	Conclusion
Baltes & Lindenberger (1988)	"Effects of cognitive training do not spread beyond the boundaries of the task spaces trained."
Park, Gutchess, Meade, & Stine-Morrow (2007)	"Improving speed through training has not resulted in a global improvement in a range of cognitive processes—improvements have thus far been limited to only the trained ability."
Green & Bavalier (2008)	"Learning tends to be quite specific to the trained regimen and does not transfer to even qualitatively similar tasks."
Noack et al. (2009)	"Cognitive intervention studies . . . have failed to observe generalizable performance improvements. . . . Evidence for far positive transfer is almost entirely absent."
Diamond & Lee (2011)	"EF training appears to transfer (i.e., produce benefits to performance of tasks other than the task used in training), but transfer . . . thus far has been narrow."
Melby-Lervåg & Hulme (2012)	"Meta-analyses indicated that the programs produced reliable short-term improvements in WM skills. . . . There was no convincing evidence of the generalization of WM training to other skills."
Melby-Lervåg et al. (2016)	"Working memory training programs appear to produce short-term, specific training effects that do not generalize to measures of 'real-world' cognitive skills. . . . Attempts to increase working memory capacity by repetitively practicing simple memory tasks on a computer are unlikely to lead to generalized cognitive benefits."
Simons et al. (2016)	"Practice generally improves performance, but only for the practiced task or nearly identical ones; practice generally does not enhance other skills, even related ones."

Only one of the seven N-back studies *with an active control group* that looked at far transfer found more improvement or better posttest performance among the N-back trainees than active controls on any far-transfer measure. The lone exception (Stephenson & Halpern, 2013) found a positive result on only one of their four far-transfer measures, although all four were measures of fluid intelligence. Indeed, the N-back training study with the greatest dose, duration,

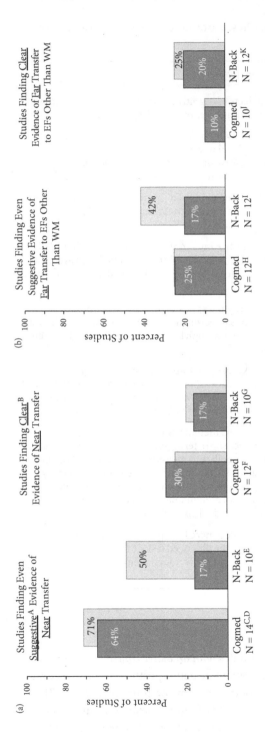

Figure 8.2. Success rates for improving EFs of the two most heavily studied computerized cognitive training approaches. Figure 8.2a presents the success rates for near-transfer measures. Figure 8.2b presents the results for EF far-transfer measures (including reasoning). The darker bars in the foreground present the results without the studies with possibly spurious positive results. The lighter bars in the background present the results for all studies. Studies omitted for having positive results that might not have held up were those that had not corrected for multiple comparisons or that had not conducted data analyses reflecting the level at which they randomized. [A] Suggestive evidence = more EF improvement *or* better EF posttest performance than control group on ≥50% of measures. [B] Strong, or clear, evidence = more EF improvement *and* better EF posttest performance than control group on ≥ 67% of measures. Whenever a study reported > 67% of measures showing positive results for improvement or posttest and did not provide any data on the other, that study is not included in calculations of strong evidence because it is possible the results of that study might have met our criteria had the results not reported been included. The

number of studies per group were: For Cogmed, suggestive evidence of near transfer: omitting some studies = 11; all studies = 14. For N-back, suggestive evidence of near transfer, omitting some studies = 6; all studies = 10. For Cogmed, strong evidence of near transfer: omitting some studies = 6; all studies = 10. For N-back, strong evidence of near transfer: omitting some studies = 12; all studies = 12. For N-back, suggestive evidence of far transfer: omitting some studies = 10; all studies = 10. For Cogmed, strong evidence far transfer: omitting some studies = 10; all studies = 10. For N-back, suggestive evidence of far transfer: omitting some studies = 12. For near transfer: For _Cogmed_, the studies with the needed statistical analyses that found suggestive evidence of near transfer were: Green et al. (2012), Holmes et al. (2009), Hovik et al. (2013), Klingberg et al. (2005), and Thorell et al. (2009). Across all Cogmed studies, those reporting suggestive evidence of near transfer were: Bergman-Nutley and Klingberg (2014), Bergman-Nutley et al. (2011), Bigorra, Garolera, Guijarro, and Hervás (2015), Brehmer, Westerberg, and Bäckman (2012), Dunning, Holmes, and Gathercole (2013), Green et al. (2012), Gropper, Gotlieb, Kronitz, and Tannock (2014), Holmes, Gathercole, and Dunning (2009), Hovik, Saunes, Aarlien, and Egeland (2013), Klingberg et al. (2005), and Thorell et al. (2009). For N-back, the studies with the needed statistical analyses that found suggestive evidence of near transfer were Li et al. (2008) and Pugin et al. (2014). Across all N-back studies, those reporting suggestive evidence of near transfer were: Jaeggi et al. (2008), Li et al. (2008), Pugin et al. (2014), Redick et al. (2013), and Stepankova et al. (2014). For _Cogmed_, the studies with the needed statistical analyses that found strong evidence of near transfer were: Green et al. (2012), Holmes et al. (2009), Hovik et al. (2013), Klingberg et al. (2005), and Thorell et al. (2009). Across all Cogmed studies, those reporting strong evidence of near transfer were: Green et al. (2012), Holmes et al. (2009), Hovik et al. (2013), Klingberg et al. (2005), and Thorell et al. (2009). For N-back, the study with the needed statistical analyses that found strong evidence of near transfer was Li et al. (2008). Across all N-back studies, those reporting strong evidence of near transfer were: Jaeggi et al. (2008), Li et al. (2008), and Stepankova et al. (2014). For far transfer to EFs other than WM, including reasoning/fluid intelligence: For Cogmed, the studies with the needed statistical analyses that found suggestive evidence of far transfer were Green et al. (2012) and Klingberg et al. (2005). Across all Cogmed studies, those reporting suggestive evidence of far transfer were: Dunning et al. (2013), Green et al. (2012), Gropper et al. (2014), and Klingberg et al. (2005). For N-back, the studies with the needed statistical analyses that found suggestive evidence of far transfer were Jaeggi et al. (2010) and Rudebeck et al. (2012). Across all N-back studies, those reporting suggestive evidence of far transfer were: Jaeggi et al. (2008), Jaeggi, Buschkuehl, Perrig, and Meier (2010), Rudebeck, Bo, Ormond, O'Reilly, and Lee (2012), Stepankova et al. (2014), and Stephenson and Halpern (2013). For Cogmed, the study with the needed statistical analyses that found strong evidence of far transfer was Green et al. (2012). Across all Cogmed studies, those reporting strong evidence of far transfer were Green et al. (2012) and Klingberg et al. (2005). For N-back, the studies with the needed statistical analyses that found strong evidence of far transfer were Jaeggi et al. (2010) and Rudebeck et al. (2012). Across all N-back studies, those reporting strong evidence of far transfer were: Jaeggi et al. (2008, 2010), Rudebeck et al. (2012), and Stepankova et al. (2014).

and frequency (Redick et al., 2013) found neither more improvement nor better posttest performance on any of several fluid-intelligence/reasoning measures compared to controls, although they found the strongest near-transfer results of any N-back study with an active control group.

Similarly, task-switching training improves task switching, but not WM or reasoning. Reasoning training improves reasoning but not memory or speed of processing.

As the reader will see in our discussion below of each training method, there is only extremely spotty evidence for far transfer to untrained cognitive skills for any training method. The rare claims of far transfer have not held up to scrutiny. For example, the much-cited study by Jaeggi et al. (2008) has failed twice to be replicated by the same group (Buschkuehl, 2011) and had methodological problems (Moody, 2009; Redick et al., 2013; Shipstead et al., 2012; Tidwell, Dougherty, Chrabaszcz, Thomas, & Mendoza, 2014). No study has looked at transfer after WM training to a test of reasoning controlling for the WM components of the reasoning test.

Moreau and Conway (2014) suggest a very plausible explanation for why many cognitive training studies find such narrow transfer. Their suggestion deserves to be tested empirically. WM is typically trained using minor variations of laboratory tasks that challenge WM. These tests are repetitive and predictable; the timing of displays, type of stimuli, and response requirements stay the same. "Intense practice [on these] exacerbates the importance of domain-specific processes, because these tasks, when administered repeatedly, allow honing strategies or skills rather than tapping domain-general processes" (Moreau & Conway, 2014, p. 334). Following this logic, interspersing very different types of WM and EF challenges and reducing the predictability of what is presented or what is required to generate a correct response should improve both the generalizability and longevity of the effects of cognitive training, which echoes findings from the older learning literature for both cognitive skills (Bransford, Franks, Morris, & Stein, 1977) and motor skills (Kerr & Booth, 1978).

Another possibility is that training skills for arbitrary, decontextualized use (as in computerized WM training) may be the problem. *Engaging in a real-world activity* may be critical. We are less motivated to learn skills for which we have no direct use. We evolved to be able to learn to help us act in the world, to do what we need to do. We are more motivated and learn something at a deeper level when we *need* that deeper level of understanding for a real-world use we care about (Cordova & Lepper, 1996; Freeman et al., 2014; Olson, 1964). It may be that what is needed is to engage participants in activities they care deeply about, where improving EFs is needed for what they want to do.

Consistent with that, Gathercole, Dunning, and Holmes (2012) suggested that "a fruitful approach may be to provide the trainee with practice in transferring their newly learned strategies to other situations that more directly simulate the everyday classroom demands on working memory.... We are therefore working on developing a complex task environment designed to bridge the gap between highly structured and relatively artificial activities that load WM and its flexible use in educational settings" (p. 202).

To see widespread benefits, diverse skills must be practiced. Where broader benefits have been found, it is usually true that there was narrow transfer of each of the practiced skills. For example, broad EF benefits have been found for a variant of taekwondo (Lakes & Hoyt, 2004) that trained diverse EF skills. In their meta-analysis of 15 randomized control trials (RCTs), Cortese et al. (2015) found that WM training had little or no effect on ADHD symptoms or other cognitive skills, even ones thought to rely on WM (such as reading and math). Their meta-analysis found, however, that interventions that targeted multiple cognitive skills had large effects on ADHD symptoms. Indeed, Cortese and colleagues call for the development and evaluation of multicomponent training models as a critical priority.

In an older meta-analysis of 25 studies (not all of which were RCTs and five of which had no control group), Rapport et al. (2013) also concluded that "claims regarding the academic, behavioral, and cognitive benefits associated with extant cognitive training programs are unsupported in ADHD" (p. 1237). Contrary to Cortese et al., however, Rapport et al. included studies in their review of programs that targeted EFs but did not produce improvements on any cognitive processes they targeted and so naturally failed to show far transfer.

Principle 3

To see benefits, continued challenge (i.e., adaptive practice) is critical. If participants are not challenged to keep improving, but simply continue doing what is easy, minimal benefit is seen. This is not only true for EF training and has been known for some time. It applies to all skills and ages. From studying experts in many different fields, Ericsson has repeatedly found that key to their prowess was continually pushing themselves to keep working at the outer limit of their competence and comfort zones (Ericsson, Nandagopal, & Roring, 2009; this is what Vygotsky, 1978, termed the "zone of proximal development"). The trick, of course, is not to discourage or frustrate someone by presenting too great a challenge (as can happen when aiming for the zone of proximal development but overreaching it). Yet, undershooting risks not helping participants progress as fast and far as they might or boring them and causing them to lose interest. Often participants are allowed to determine how fast and far they progress; the key then is to motivate them to keep pushing themselves to improve.

Principle 4

Studies have demonstrated that EF benefits can last months or even years (Ball et al., 2002; Bigorra et al., 2015; Borella et al., 2010; Brehmer et al., 2012; Carretti, Borella, Zavagnin, & de Beni, 2013; Dovis, van der Oord, Wiers, & Prins, 2015; Dunning et al., 2013; Gropper et al., 2014; Holmes et al., 2009; Klingberg et al., 2005; Li et al., 2008; Li-Grining et al., 2011; Plemons, Willis, & Baltes,1978; Pugin et al., 2014; Roberts et al., 2016; Van der Donk, Hiemstra-Beernink, Tjeenk-Kalff, van der Leij, & Lindauer, 2015; van der Oord, Ponsioen, Geurts, Ten Brink, & Prins, 2014; Willis et al., 2006), but naturally they grow smaller as the time since training increases. It would be unrealistic to expect benefits to last for a long time after practice ends. If you had gone to the gym regularly and improved your physical fitness, you would not expect the fitness benefits to last indefinitely if you stopped working out. On the other hand, if someone keeps using and challenging EF skills, then, presumably, benefits could last indefinitely. Use it or lose it. Also, academic benefits from improving EFs have occasionally increased over time or only shown up later (Bigorra et al., 2015; Blair & Raver, 2014; Holmes et al., 2009; Li-Grining et al., 2011).

Principle 5

Those more behind on EFs benefit the most from any intervention—within limits. Thus, persons with poorer WM or smaller WM spans (Holmes et al., 2009; Sibley & Beilock, 2007; Zinke, Zeintl, Eschen, Herzog, & Kliegel, 2012), worse attention (Flook et al., 2010), worse inhibitory control (Drollette et al., 2014), worse EFs in general (Flook, Goldberg, Pinger, & Davidson, 2015; Lakes & Hoyt, 2004), ADHD (Holmes et al., 2010; Klingberg et al., 2005), or the beginnings of cognitive decline with aging (Colcombe & Kramer, 2003; Kramer & Erickson, 2007) benefit more than others without the same challenges. Indeed, Fedewa and Ahn's (2011) meta-analysis of studies looking at the effects of physical activity on children's achievement and cognitive outcomes found that the mean effect size was twice as large for "cognitively impaired" children as for typically developing ones.

Children who are socioeconomically disadvantaged tend to have worse EFs (Blair & Raver, 2014; Farah et al., 2006; Noble, McCandliss, & Farah, 2007). These children benefit more from EF interventions than children with more socioeconomic advantages (Blair & Raver, 2014; Lillard & Else-Quest, 2006;

Mackey, Hill, Stone, & Bunge, 2011; Raver et al., 2008, 2011). This is seen in particularly stark relief in the Blair and Raver (2014) study, where effect sizes for children from diverse socioeconomic backgrounds rarely exceeded 0.1 standard deviation, but effect sizes for lower-income children were as high as 0.8! (See Figures 8.3a and 8.3b.) In all the aforementioned studies, similar subjects in the control group did not show similar gains, so disproportionate benefits to those farthest behind is not due to simple regression to the mean. Since those farthest behind on EFs tend to benefit the most from any EF intervention, and since less socioeconomically advantaged children tend to have poorer EFs, EF training might be a good way to reduce societal disparities in academic achievement and/or health associated with social disparities in EFs, especially if the EF training can be done early, before self-images of not being capable become established.

Extreme groups, such as children with a very low IQ or adults with severe cognitive decline, have benefitted less from cognitive training than have those

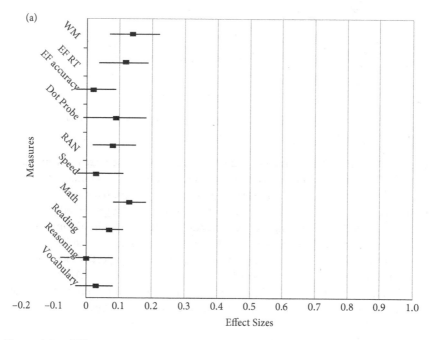

Figure 8.3a. Effect-size estimates including all schools in the study for main effects of Tools of the Mind versus the control group at the end of kindergarten in Blair and Raver (2014). Note that no effect size exceeds 0.2. Error bars represent ± 1 standard error. RT = reaction time; RAN = rapid automatic naming.

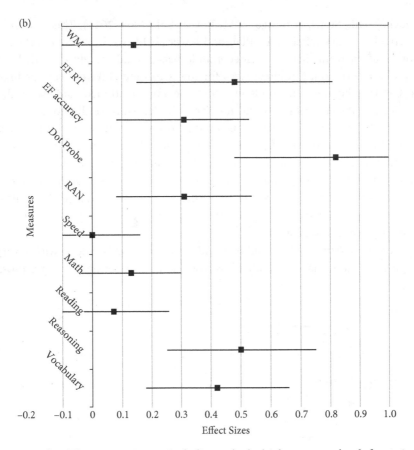

Figure 8.3b. Effect-size estimates including only the high poverty schools for main effects of Tools of the Mind versus the control group at the end of kindergarten in Blair and Raver (2014). Note that effect sizes now are as large as 0.8. Comparing Figures 8.3a and 8.3b, it is clear that Tools of the Mind had a far greater effect on EFs and academic skills among economically disadvantaged children than among more economically advantaged children. Errors bars represent ± 1 standard error. RT = reaction time; RAN = rapid automatic naming. Psychological Association. Figure 8.3 is reprinted, with permission, from Diamond, A., & Ling, D. S. (2016). Conclusions about interventions, programs, and approaches for improving EFs that appear justified and those that, despite much hype, do not. *Developmental Cognitive Neuroscience, 18*, 34–48.

with more intact cognitive skills or just the beginnings of dementia (Colcombe & Kramer, 2003; Söderqvist et al., 2012). That may be because the training was too demanding for them.

Principle 6

Duration matters. Within the range of durations studied, generally more weeks of computerized cognitive training has produced better results than fewer weeks. For example, training using Rise of Nations* (90-min sessions; three sessions per week) for 5 weeks produced better EF benefits than doing the same for only 2½ weeks (Basak, Boot, Voss, & Kramer, 2008). Perhaps one reason that Cogmed training has generally been more successful at improving EFs than N-back training is that the duration of Cogmed training is usually longer (5–8 weeks vs. 2–5 weeks).

Among the three mindfulness retreats for which EF benefits were investigated, the one that lasted the longest produced the best EF outcomes (MacLean et al., 2010), but the retreats also differed in type of mindfulness practiced, how long a mindfulness session lasted, and outcome measures. One year of resistance training has been found to improve inhibition (Stroop task performance), whereas only a half year into the training, that benefit was not evident (Liu-Ambrose et al., 2010). Masley, Roetzheim, and Gualtieri (2009), who studied extremely short durations of 5 to 7 days versus 3 to 4 days of a comprehensive program that included exercise, found that more days produced better EF results than fewer days.

There is evidence that at least four school curricular programs (MindUP™, Tools of the Mind™, PATHS™, and CSRP) do an excellent job at improving inhibitory control in young children. These programs include scaffolding, training, practicing, and challenging inhibitory control throughout a good part of the school day over the entire school year.

In short, Ericsson's conclusion about the critical importance of the amount of time spent practicing (with difficulty progressively increasing) for becoming really good at anything (Ericsson, 2006; Ericsson et al., 2009; Ericsson & Towne, 2010) appears to apply to EF skills just as it does to all the skills Ericsson studied.

One exception is EF benefits from aerobic exercise interventions. There is no evidence of greater EF benefits from longer aerobic exercise programs (see Table 8.7). This is contrary to the conclusion by Colcombe and Kramer (2003) that cognitive benefits to older adults were greater from longer aerobic exercise interventions, as many subsequent studies since their review have shown.

Principle 7

Dose (the length of individual sessions) appears to matter as well; 30 min or more seems better than less than 30 min. (The results for Quadrato Motor Training, with only 7-min sessions, are a marked exception however [Ben-Soussan,

Table 8.7. Duration, Dose, and Frequency of Aerobic Exercise Interventions Broken Down by Plain Aerobics (e.g., Brisk Walking) Versus Enriched Aerobics and by Whether EF Benefits Were Found

Studies where benefits were found on at least half the EF measures

EF Benefits?	Study	Compared to AC or NT?	Duration (weeks)	Dose in Minutes [AE portion in brackets]	Frequency per week	N	Age Range (years)	Mean Age (years)[1]	Was a demanding EF measure used?
Suggestive	Albinet et al., 2010	AC	12	60 [40]	3	12	65-78	71	no
Suggestive	Dustman et al., 1984[2]	AC	16	60 [n/a]	3	14	55-70	60	YES
Clear	Kramer et al., 1999	AC	24	? [n/a]	?	62	60-75	67	YES
Clear	Moul et al., 1995[2]	AC	16	30-40 [30-40]	5	10	65-72	69	no
Suggestive	Tuckman & Hinkle, 1986	NT	12	30 [30]	3	77	8-12	10	no

Studies where benefits were not found at all or were found on less than half of EF measures

EF Benefits?	Study	Compared to AC or NT?	Duration (weeks)	Dose in Minutes [AE portion in brackets]	Frequency per week	N	Age Range (years)	Mean Age (years)[1]	Was a demanding EF measure used?
0	Blumenthal et al., 1989	NT	16	60 [45]	3	34	60-83	67	no
0	Erickson et al., 2011, Leckie et al., 2014, McAuley et al., 2011	AC	52	40 [10-40]	3	65	55-80	67	YES
0	Fabre et al., 2002	AC	8	60 [45]	2	8	60-76	66	no
<50%	Fisher et al., 2011[2]	NT	10	60 [60]	2	32	5-7	6	YES
0	Legault et al., 2011	AC	17	60 [40]	2	18	70-85	76	YES

(continued)

Studies where benefits were NOT found at all or were found on less than half of EF measures

EF Benefits?	Study	Compared to AC or NT?	Duration (weeks)	Dose in Minutes [AE portion in week brackets]	Frequency per week	N	Age Range (years)	Mean Age (years)[1]	Was a demanding EF measure used?
0	Mortimer et al., 2012[2]	NT	40	50 [30]	3	30	60-79	68	no
0	Oken et al., 2006	NT	26	90 [60]	1	45	65-85	72	no
0	Schmidt et al., 2015	AC	6	45 [45]	2	60	10-12	11	YES
0	Smiley-Oyen et al., 2008	AC	40	45-50 [25-30]	3	28	65-79	70	no
0	Voelcker-Rehage et al., 2011	AC	52	60 [35-50]	3	15	63-79	70	no
Means			27	57 [42]	2	34		57 [70]	

Studies where benefits were found on at least half the EF measures

Means			16	46 [35]	4	35		55 [67]	

Aerobic Exercise Enriched with Cognitive and/or Motor Skill Demands

Studies where benefits were found on at least half the EF measures

EF Benefits?	Study	Compared to AC or NT?	Duration (weeks)	Dose (min) [AE portion in brackets]	Frequency per week	N	Age Range (years)	Mean Age (years)[1]	Was a demanding EF measure used?
Clear	Chang et al., 2014	NT	8	90 [40]	2	15	5-10	8.5	no

Studies where benefits were NOT found at all or were found on less than half of EF measures

EF Benefits?	Study	Compared to AC or NT?	Duration (weeks)	Dose in Minutes [AE portion in week brackets]	Frequency per week	N	Age Range (years)	Mean Age (years)[1]	Was a demanding EF measure used?
<50%	Chaddock-Heyman et al., 2013, Hillman et al., 2014, Kamijo et al., 2011	NT	36	120 [77]	5	14	8-9	9	no

Table 8.7. Continued

Studies where benefits were found on at least half the EF measures

EF Benefits?	Study	Compared to AC or NT?	Duration (weeks)	Dose in Minutes [AE portion in brackets]	Frequency per week	N	Age Range (years)	Mean Age (years)[1]	Was a demanding EF measure used?
Suggestive	Chuang et al., 2015	AC	13	30[30]	3	8	65–75	68	no
Suggestive	Gallotta et al., 2015[2,3]	AC	20	60[30]	2	52	8–11	9.5	no
Suggestive	Kim et al., 2011[2]	NT	26	60[45]	2	26	60–78	68	YES
Suggestive	Maillot et al., 2012	NT	12	60[60]	2	16	65–78	74	YES
Suggestive	Moreau et al., 2015[2]	AC	8	60[40]	3	22	18–52	30	no
Suggestive	Predovan et al., 2012	NT	13	60[15–40]	3	25	57–80	68	YES
Suggestive	Staiano et al., 2012	NT	10	30[30]	1	18	15–19	16.5	index or latent

Studies where benefits were not found at all or were found on less than half of EF measures

EF Benefits?	Study	Compared to AC or NT?	Duration (weeks)	Dose in Minutes [AE portion in brackets]	Frequency per week	N	Age Range (years)	Mean Age (years)[1]	Was a demanding EF measure used?
0	Dalziell et al., 2015	NT	16	60[n/a]	2	23	9–10	10	no
<50%	Davis et al., 2007, 2011	NT	13	40[35]	5	44	7–11	9	YES
<50%	Klusman et al., 2010	NT	24	90[30]	3	91	70–93	74	no
<50%	Krafft, Pierce, et al., 2014; Krafft, Schaeffer, et al., 2014	AC	32	40[40]	7	22	8–11	9.8	YES
0	Legault et al., 2011	AC	17	60[40]	2	18	70–85	76	YES
0	Marmeleira et al., 2009	NT	12	60[60]	3	16	60–82	68	no

Clear	Study									
Clear	Williams & Lord, 1997	NT	42	50–55 [35]	2	94	≥60	72	YES	
	Pesce et al., 2013	NT	26	60 [60]	1	83	5-10	7	YES	<50%
	Schmidt et al., 2015	AC	6	45 [45]	2	57	10-12	11	YES	<50%
Means[4]			17	56 [38]	2	31		46 [70]		
Means[5]			20	64 [48]	3	41		30 [73]		

Note. AC = active control group; NT = no treatment group; AE = aerobic exercise. A demanding measure = a measure like the Wisconsin Card Sort Test or Tower of London, on which group differences are often more easily found than on easier EF tasks.

Clear = more EF improvement and better EF posttest performance than control group on ≥ 67% of measures.

Suggestive = more EF improvement or better EF posttest performance than control group on ≥ 50% of measures.

Index or Latent = creating a composite index from multiple EF measures or looking at the latent variable underlying performance on multiple EF measures is noted because those are likely to be more reliable and more sensitive than individual EF measures.

Reed et al. (2010) did not report near transfer EF measures, so is not included in this table.

[1] The number in brackets includes only studies where the mean age was ≥ 60 years.

[2] The authors of this study did not include a correction for multiple comparisons. It is unclear which of their results would remain significant had they done that.

[3] Gallotta et al. (2015) randomized by school but appear to have analyzed the data as if they randomized by individual children.

[4] If the FITKids studies are counted as three separate, independent studies, then for studies of enriched aerobic exercise where benefits were found on at least half the measures, the mean duration would be 18 weeks, mean dose would be 62 min of intervention with 41 min of aerobics, mean frequency would be three times per week, mean number of subjects would be 30, and the mean age of participants would be 42 years.

[5] If the FITKids studies are counted as three separate, independent studies, then for enriched aerobic exercise, the mean duration would be 22 weeks, mean dose would be 70 min of intervention with 47 min of aerobics, mean frequency would be four times per week, mean number of subjects would be 48, and the mean age of participants would be 28 years.

Glicksohn, & Berkovich-Ohana, 2015].) Colcombe and Kramer (2003), in their review of aerobic exercise interventions, reached a similar conclusion: "Short bouts of exercise (< 30 min) had very little impact on cognitive function; the effect at this training duration was not significantly different from zero" (p. 128). Davis et al. (2007, 2011) found better EF outcomes from longer sessions of aerobic games than from shorter ones (40 min per session vs. 20 min; see also McNaughten & Gabbard, 1993, who found greater cognitive benefits from physical exercise for 30 or 40 min vs. 20 min). Note, however, that for aerobic exercise at least, for doses of 30 min or more, there is no evidence that dose matters or that longer sessions produce better results (see Table 8.7).

Cogmed results have been better for children of 7 to 14 years than for children of 4 to 5 years (the former practiced for 30–45 min at a time, the latter for only 15 min). More Cogmed studies than N-back studies have found at least suggestive evidence of EF benefits (79% vs. 40%), and Cogmed sessions generally lasted 30 to 45 min, while N-back sessions generally lasted 15 to 30 min.

There are two exception to training for ≥ 30 min being better than < 30 min. One is the review of N-back studies by Au et al. (2015), which found a trend for shorter N-back training sessions to yield greater benefits than longer ones (within the range of 18.5–60 min, but only two N-back studies had sessions > 30 min). The other is a study by Mawjee and colleagues (Mawjee et al., 2014; Mawjee, Woltering, & Tannock, 2015), who compared 45 min of Cogmed to 15 min, both 5 days a week for 5 weeks, and found no difference. Neither duration produced any improvement, even on near-transfer measures. This is one of only three Cogmed studies *with adults* that looked at EF outcomes (the participants were 18–35 years old). Perhaps Cogmed works better with children than with young adults. Of concern, 23% of experimental group participants dropped out (very high attrition in that group alone). Either age or attrition, alone or in combination, might account for why they failed to find a WM benefit from Cogmed when most studies have.

The developers of the Tools of the Mind preschool and kindergarten curriculum initially tried their program as an add-on to existing curricula so that children did activities designed to improve EFs for roughly an hour a day. Benefits were narrow and specific to the context in which the skills had been practiced. Only when training and practicing of EFs was embedded in all school activities were benefits seen (Bodrova & Leong, 2007). Clements, Sarama, and Layzer (2012) replicated the limited benefits from Tools of the Mind as an add-on. Diamond, Barnett, Thomas, and Munro (2007) and Blair and Raver (2014) replicated marked EF benefits with Tools of the Mind as an all-day curriculum.

Principle 8

Spacing matters. Spaced, or distributed, practice produces better long-term outcomes than massed practice. That is, relatively shorter training or practice sessions spaced out over time produce better outcomes than compressing the training into a shorter time period with longer sessions (this is often referred to as the spacing effect). Skills and content are learned more slowly with spaced practice, but the learning is deeper and lasts longer. This has been shown for many cognitive skills (Cepeda et al., 2006; Rea & Modigliani, 1985) and motor skills (Lee & Magill, 1983; Shea & Morgan, 1979). Penner et al. (2012) provides an example of this with WM training. Their group that trained 2 days a week over 8 weeks (spaced practice) performed better on three near-transfer measures than the group that trained 4 days a week over 4 weeks (massed practice) and than no-treatment controls. Improvement of the massed-practice and no-treatment groups did not differ. Similarly, of three mindfulness retreats, the one that had the most spaced practice— 2 hours a day for 13 weeks (MacLean et al., 2010)—produced better EF results than two other retreats that had more hours of meditation per day over a shorter time—11 hours a day for 1.5 weeks (Chambers, Lo, & Allen, 2008) and 9.8 hours a day for 4 weeks (Zanesco, King, MacLean, & Saron, 2013).

Principle 9

Often, the benefits of an intervention are only seen, or are seen most clearly, on outcome measures that push the limits of participants' EFs (Albinet, Boucard, Bouquet, & Audiffren 2010; Alesi et al., 2016; Chan, Sze, Siu, Lau, & Cheung, 2013; Davis et al., 2007, 2011; Diamond et al., 2007; Hillman et al., 2014; Manjunath & Telles, 2001; Predovan, Fraser, Renaud, & Bherer, 2012; Schmidt, Jäger, Egger, Roebers, & Conzelmann, 2015; Tucha et al., 2011; Westendorp et al., 2014). Complex, multicomponent measures (such as the Tower of London or Wisconsin Card Sort Test, which require multiple EF skills) are often excellent for distinguishing between groups, although because they require multiple EF skills they are not good for isolating which particular EF skill improved. They are good candidates for detecting outcome differences between the intervention and control groups precisely because they tax more than one EF skill.

How Different Approaches to Improving EFs Measure Up

Computerized Cognitive Training

Cogmed

Cogmed and N-back training have been researched more than any other cognitive training method. Of the 14 studies of Cogmed that met criteria for inclusion here and that reported WM outcomes, 10 found improved WM (71%) on at least 50% of their WM outcome measures. That is better than any other computerized training program (although noncomputerized cognitive training shows slightly better results: 75% of 12 studies found training benefits on at least 50% of their near-transfer measures).

An important caveat is that most Cogmed studies used outcome measures very similar to the Cogmed games on which participants trained (e.g., they might differ only in the precise stimuli, with task demands and structure being the same as in Cogmed, such as training participants to recall digits in reverse order and then testing participants on Backward Digit Span). Thus, improvements on Backward Digit Span, Spatial Span, or Corsi Blocks are not that impressive after Cogmed training because training games closely resemble those tasks. On the other hand, transfer to performance on complex-span tasks, which differ more from Cogmed training, was also reported by all 10 Cogmed studies that included one or more complex-span outcome measures.

Another important caveat is that although 70% to 71% of Cogmed studies found more improvement on ≥ 50% of their near-transfer EF measures in Cogmed trainees than in control subjects, only 25% to 33% of Cogmed studies found strong evidence of this (see Table 8.8 and Figure 8.2).

Twelve Cogmed studies administered at least one far-transfer task on which they compared groups. Of the 12, only 10 looked at both improvement and post-test. Of the 10 studies, only one (Green et al., 2012) found clear evidence of far transfer. Compared to a nonincrementing version of Cogmed, Green et al. found reduced inattentiveness in youths with ADHD in a naturalistic setting.

Note, however, that seven other studies of persons with ADHD (*six studies with children*: Bigorra et al., 2015; Chacko et al., 2014; Egeland et al., 2013; Gray et al., 2012; Klingberg et al., 2005; van der Donk et al., 2015; *one study with adults*: Gropper et al., 2014) looked for far transfer but found mixed results. Chacko et al., Gray et al., and van der Donk et al. found no far transfer. Bigorra et al. found some far transfer among children with ADHD—better improvement on a continuous performance test (CPT) and on teacher ratings on the BRIEF of metacognition and of children's ability to initiate an activity and independently problem-solve. Yet Bigorra et al. failed to find far transfer more often than they found it, and it is not clear that all their positive findings

Table 8.8. Summary of Results for Only Near-Transfer EF Outcomes for All Cognitive Training Approaches

	Percent of Studies Finding even Suggestive[1] Evidence of EF Benefits (# of studies)	Percent of Studies Finding Clear[2] Evidence of EF Benefits (# of studies)	Percent of EF Outcome Measures on Which Experimental Group Improved more than Control Group (# of measures)	Percent of EF Measures on Which Experimental Group Performed Better at Posttest than Control Group (# of measures)
A. All Studies Are Included Here				
Cogmed Computerized Training	71% (14)[3]	25% (12)	69% (61)	33% (51)
N-back Computerized Training	50% (10)[4]	20% (10)	34% (35)	31% (35)
Computerized Complex-Span Training	50% (3)[5]	50% (2)	39% (18)	39% (18)
Task-switching Computerized Training	40% (5)	0% (5)	45% (47)	26% (38)
Other Computerized Cognitive Training (including Commercial Products)[6]	50% (26)	13% (23)	33% (168)	15% (150)
Noncomputerized Cognitive Training	75% (12)	30% (10)	47% (49)	38% (45)
B. Studies With Possibly Spurious Results Are Omitted Here[7]				
Cogmed Computerized Training	64% (11)	30% (10)	61% (36)	41% (32)
N-back Computerized Training	17% (6)	17% (6)	28% (29)	28% (29)

(continued)

Table 8.8. Continued

	Percent of Studies Finding even Suggestive[1] Evidence of EF Benefits (# of studies)	Percent of Studies Finding Clear[2] Evidence of EF Benefits (# of studies)	Percent of EF Outcome Measures on Which Experimental Group Improved more than Control Group (# of measures)	Percent of EF Measures on Which Experimental Group Performed Better at Posttest than Control Group (# of measures)
Computerized Complex-Span Training	—[8]	—[8]	—[8]	—[8]
Task-switching Computerized Training	67% (3)	0% (3)	74% (23)	43% (14)
Other Computerized Cognitive Training (including Commercial Products)	43% (23)	10% (21)	29% (143)	12% (130)
Noncomputerized Cognitive Training	73% (11)	33% (9)	46% (41)	39% (36)

[1] Suggestive = more EF improvement or better EF posttest performance than control group.

[2] Clear = more EF improvement and better EF posttest performance than control group on ≥ 67% of measures. Whenever a study reported ≥ 67% of measures showing positive results for improvement or posttest and did not provide any data on the other, that study is not included in calculations here because it is possible the results of that study might have met our criteria for "clear" had the results not reported been included.

[3] Fifteen Cogmed studies are included in our review. One study did not include near-transfer measures and so does not appear in Table 8.8A.

[4] Thirteen N-back training studies are included in our review. Three did not include near-transfer measures.

[5] Six complex-span training studies are included in our review. One study did not include near-transfer measures and so does not appear in Table 8.8A. Two were non-computerized and are included under Noncomputerized Training in Table 8.8A rather than under complex-span training.

[6] Other Computerized Cognitive Training includes both interventions we classified as miscellaneous computerized cognitive training and commercial computerized cognitive training approaches, as well as the noncommercial BrainGame Brian.

[7] Studies whose positive results might not have held up had they corrected for multiple comparisons or conducted data analyses reflecting the level at which they random-ized are omitted here. Studies that made either error but found few, if any, positive results are not omitted, since the errors only increase the likelihood of false positives, not false negatives.

[8] There were too few computerized complex-span studies (only one) to be included here.

would have survived correction for multiple comparisons. They did not find far transfer on the Tower of London, Wisconsin Card Sort, any item on the parent version of the BRIEF, or most scales of the teacher version of the BRIEF. Klingberg et al. found more improvement on the Stroop task, Raven's Colored Matrices, and parental ratings of inattentiveness and hyperactivity/impulsivity, and better posttest scores for Raven's Colored Matrices. Posttest scores between groups for parental ratings of hyperactivity/impulsivity, however, were not significant. There was neither more improvement nor better posttest scores for teacher ratings of inattentiveness and hyperactivity/impulsivity. The far-transfer benefits were no longer present 3 months later in the Klingberg et al. study but were still robustly present 6 months later in the study by Bigorra et al. Indeed, more benefits on the BRIEF showed up later in the Bigorra et al. study than were present initially.

Egeland et al. did not find far transfer on the Stroop or Trail Making tests, any index on Conners' CPT or on the parent or teacher version of the BRIEF, or math. However, they found far transfer to reading, and the benefit was even greater 8 months later. Gropper et al. found EF benefits on self-report measures (Cognitive Failures Questionnaire and Adult ADHD Self-Report Scale [reduced occurrences of off-task behavior]) but no benefit on their one behavioral far-transfer measure (the Ruff 2 and 7 selective attention task).

In typically developing adults, Brehmer et al. (2012) found far transfer on the Cognitive Failures Questionnaire (more improvement and better posttest scores) yet no transfer benefit at all on the Stroop or Raven's Matrices tests. In typically developing schoolchildren, Dunning et al. (2013) reported more improvement and better posttest scores on far transfer to a CPT task after Cogmed versus no treatment, which seems to be due to the no-treatment group getting worse at posttest. No differences were found in change or posttest scores compared to the nonincrementing Cogmed group.[4]

Overall, the results for Cogmed were slightly better for typically developing schoolchildren than for schoolchildren with ADHD: When nonincrementing Cogmed and Cogmed were compared, suggestive WM benefits for Cogmed were found by 100% of studies of children without ADHD but only 75% of studies of children with ADHD. When Cogmed was compared to no treatment, suggestive WM benefits for Cogmed were found by 100% of studies, whether the children participating were typically developing or had ADHD. Strong evidence of EF benefits was similar for typically developing children and those with ADHD: Fifty percent of studies found strong evidence of

---

[4] Brehmer et al. (2012) did not correct for multiple comparisons. Dunning et al. (2013) did not either, nor did they conduct multilevel data analyses reflecting their cluster RCT design. It is unclear if their positive findings would have reached significance had these things been done.

WM benefits with Cogmed compared to nonincrementing Cogmed whether the children had ADHD or not. No studies found strong evidence of WM benefits when Cogmed was compared to no treatment whether the children had ADHD or not.

For some individuals with ADHD, auditory selective attention and auditory WM can be more of a challenge than visual or visual-auditory processing or WM (Fabio, Castriciano, & Rondanini, 2015; Gomes et al., 2012). Thus, training that includes challenging auditory selective attention and WM might be more beneficial for those with ADHD than protocols that are exclusively visual and visual-auditory.

Cogmed seems quite effective at improving the WM skills it trains people on, and those benefits last 3 to 12 months. Benefits have been found to last 3 months (Brehmer et al., 2012; Klingberg et al., 2005), 4 months (Beck, Hanson, Puffenberger, Benninger, & Benninger, 2010), 6 months (Bigorra et al., 2015; Holmes et al., 2009; van der Donk et al., 2015), and even 1 year (Dunning et al., 2013; Roberts et al., 2016). There is no evidence that the benefits last much beyond that, however, and little evidence that benefits extend to other cognitive skills. Three meta-analyses have all concluded that Cogmed does not reduce ADHD symptoms like inattentiveness or hyperactivity/ impulsivity when the raters are blind to who got Cogmed (Cortese et al., 2015; Rapport et al., 2013; Sonuga-Barke et al., 2013). Spencer-Smith and Klingberg (2015) reached a different conclusion but included unblinded parental raters.

A recent RCT of Cogmed with 452 first-graders found that while the WM improvement was still robust after 1 year (which is better than most methods can claim), it was no longer present 2 years later, and those who had trained on Cogmed performed *worse* in math 2 years later than the control group who received regular classroom instruction while their peers were training on Cogmed (Roberts et al., 2016, which was too recent to be included in Table 8.4 or our tabulations).

There is some evidence that Cogmed WM training might reduce inattentiveness in daily living for ADHD patients (Green et al., 2012; Gropper et al., 2014; Spencer-Smith & Klingberg, 2015) but there is a lack of evidence that any WM training improves performance on tests of attention. Positive evidence of reduced inattentiveness in daily living comes primarily from unblinded raters, with Green et al. (2012) an important exception. There is a great need for objective measures of attentiveness in daily life, such as the Restricted Academic Setting Task (RAST) used by Green et al. (2012). The few Cogmed studies that have looked at *objective laboratory measures* of attention have found either no benefit (Chacko et al., 2014; Egeland, Aarlien,

& Saunes, 2013; Gray et al., 2012; Gropper et al., 2014; van der Donk et al., 2015) or more improvement but not better posttest scores (the WM training group caught up: Klingberg et al., 2005; Thorell et al., 2009). No study has found WM training to yield both more improvement and a better end result on any objective laboratory tests of attention than they found in control subjects, except the study by Dunning et al. (2013), where it was not that the WM training particularly improved attention but that the control group mysteriously got worse.

Results on Near-Transfer EF Measures for School-Age Children Trained on Cogmed versus a Nonincrementing *Version of Cogmed*
The best results for Cogmed have been found with school-age children (7–15 years old) using a nonincrementing (a nonadaptive) version of Cogmed as the control condition (where difficulty does not keep increasing). On a total of eight WM measures in two studies (Dunning et al., 2013; Holmes et al., 2009), typically developing children who trained on Cogmed improved more than active controls on 88% of the measures, performed better at posttest than active controls on 38%, and showed better change and posttest scores on 38% (see Table 8.9).

Table 8.10, (appears in online: URL: http://www.devcogneuro.com/tables/supplemental.html) which presents the percentage across all EF measures (except reasoning/fluid intelligence) on which persons trained on Cogmed showed more improvement and/or better posttest results than comparison groups across all studies and ages, broken down by study, appears online.

On 100% of their three WM indices—a composite score for WM of shape or orientation (what they call visual-spatial WM), a composite score for WM for spatial location (what they call visual-spatial STM), and the Counting Span task (what they call verbal WM)—Holmes et al. (2009) found more improvement and better posttest performance for those who trained on Cogmed than among those who played a nonadaptive version of Cogmed, among 10-year-olds with initially poor WM. The Cogmed group also showed significant improvement in mathematical reasoning but that did not show up immediately after training. It was first evident at follow-up testing 6 months later.

For school-age children (7–15 years old) with ADHD, comparing Cogmed to a nonincrementing version of Cogmed on 13 WM measures across four studies (Bigorra et al., 2015; Chacko et al., 2014; Green et al., 2012; Klingberg et al., 2005) those trained on Cogmed improved more than active controls on 62% of the measures. Two studies did not compare whether posttest performance was better for Cogmed versus nonincrementing Cogmed, but for both studies that did (Klingberg et al., 2005; Green et al., 2012), those trained

Table 8.9. Percentage of WM Measures on Which Persons Who Trained on Cogmed Showed More Improvement and/or Better Posttest Results Across All Studies and Ages, Broken Down by Study

Study #	Study Name	Condition of Interest	Comparison Condition	Significantly Better Improvement			Significantly Better Posttest			Significantly Better Posttest Only Including Measures Where This Was Looked at			Both Significantly Better Change and Posttest		
				# Sign.	# Measures	% Sign.	# Sign.	# of Measures	% Sign.	# Sign.	# Measures	% Sign.	# Sign.	# of Measures	% Sign.
			YOUNG CHILDREN (3–6 YEARS OLD)												
1	Bergman Nutley et al., 2011	Cogmed	Nonincrementing Cogmed	2	2	100%	1	2	50%	1	2	50%	1	2	50%
1	Bergman Nutley et al., 2011	Nonverbal reasoning (NVR)	Nonincrementing Cogmed	1	2	50%	1	2	50%	1	2	50%	1	2	50%
1	Bergman Nutley et al., 2011	Cogmed & NVR	Nonincrementing Cogmed	2	2	100%	1	2	50%	1	2	50%	1	2	50%
2	Thorell et al., 2009[A]	Cogmed	No treatment + commercially available computer games, where minimal need for WM or inhibition	2	2	100%	0	2	0%	0	2	0%	0	2	0%
2	Thorell et al., 2009[A]	Training on inhibitory control computer games	No treatment + commercially available computer games, where minimal need for WM or inhibition	0	2	0%	0	2	0%	0	2	0%	0	2	0%

	Study	Treatment	Comparison										
	Totals and Percents for young children who trained specifically on WM using Cogmed compared with nonincrementing Cogmed			4	4	100%	1	4	25%	1	4	25%	25%
	Grand Totals and Percents for young children who trained on Cogmed, NVR, and inhibitory control computer games compared with nonincrementing Cogmed or no treatment + commercially available computer games			7	10	70%	3	10	30%	3	10	30%	30%

SCHOOL-AGE CHILDREN (7–17 YEARS OLD) WITH NO CLINICAL DIAGNOSIS

	Study	Treatment	Comparison										
3	Dunning et al., 2013[B,C]	Cogmed	Nonincrementing Cogmed	4	5	80%	0	5	0%	0	5	0%	0%
4	Holmes et al., 2009[D]	Cogmed	Nonincrementing Cogmed	3	3	100%	3	3	100%	3	3	100%	100%
	Totals and Percents for school-age children with no clinical diagnosis who trained on Cogmed compared with a nonincrementing version of Cogmed			7	8	88%	3	8	38%	3	8	38%	38%
3	Dunning et al., 2013[B]	Cogmed	No treatment	4	5	80%	3	5	60%	3	5	60%	60%

(continued)

Table 8.9. Continued

Study #	Study Name	Condition of Interest	Comparison Condition	Significantly Better Improvement			Significantly Better Posttest			Significantly Better Posttest Only Including Measures Where This Was Looked at			Both Significantly Better Change and Posttest		
				# Sign.	# Measures	% Sign.	# Sign.	# of Measures	% Sign.	# Sign.	# Measures	% Sign.	# Sign.	# of Measures	% Sign.
	Grand Totals and Percents for school-age children with no clinical diagnosis who trained on Cogmed compared with a nonincrementing version of Cogmed or no treatment			11	13	85%	6	13	46%	6	13	46%	6	13	46%
SCHOOL-AGE CHILDREN (7–17 YEARS OLD) WITH ADHD															
5	Gray et al., 2012[C,E]	Cogmed	Academy of Math: Math Training	2	5	40%	1	5	20%	1	5	20%	1	5	20%
	Roberts et al. 2016[F]	Cogmed	No treatment	2	3	67%	2	3	67%	2	3	67%	2	3	67%
6	Bigorra et al., 2015[C,G,H]	Cogmed	Nonincrementing Cogmed	4	6	67%									
7	Chacko et al., 2014[H]	Cogmed	Nonincrementing Cogmed	1	4	25%									
8	Green et al., 2012	Cogmed	Nonincrementing Cogmed	1	1	100%	1	1	100%	1	1	100%	1	1	100%
9	Klingberg et al., 2005	Cogmed	Nonincrementing Cogmed	2	2	100%	2	2	100%	2	2	100%	2	2	100%
	Totals and Percents for school-age children with ADHD who trained on Cogmed compared with a nonincrementing version of Cogmed			8	13	62%	3	3	100%	3	3	100%	3	3	100%

#	Study	Training	Comparison	n	N	%	n	N	%	n	N	%	n	N	%
10	Bergman Nutley & Klingberg, 2014[l]	Cogmed	No treatment (typically developing)	2	2	100%	1	2	50%	1	2	50%	1	2	50%
11	Egeland et al., 2013[j]	Cogmed	No treatment												
12	Hovik et al., 2013	Cogmed	No treatment	3	3	100%	0	3	0%	0	3	0%	0	3	0%
	Grand Totals and Percents for school-age children 7–17 years old with ADHD who trained on Cogmed compared with a nonincrementing version of Cogmed or no treatment			13	18	72%	4	8	50%	4	8	50%	4	8	50%
13	van der Donk et al., 2015	Cogmed	Paying Attention in Class, which combines WM & psychoeducational training	1	5	20%	1	5	20%	1	5	20%	1	5	20%
	Grand Totals and Percents for school-age children with or without a clinical diagnosis who trained on Cogmed compared with only nonincrementing Cogmed and no-treatment controls			24	31	77%	10	21	48%	10	21	48%	10	21	48%
	Grand Totals and Percents for school-age children with or without a clinical diagnosis who trained on Cogmed compared with nonincrementing Cogmed, no treatment, and the two studies with promising active-control conditions			27	41	66%	12	31	39%	12	31	39%	12	31	39%
	ADULTS (18–55 YEARS OLD) WITHOUT CLINICAL DIAGNOSES														
14	Brehmer et al., 2012[C]	Cogmed	Nonincrementing Cogmed	4	4	100%	2	4	50%	2	4	25%	2	4	50%
	ADULTS (18–55 YEARS OLD) WITH ADHD														
15	Gropper et al., 2014[C]	Cogmed	No treatment	2	6	33%	2	6	33%	2	6	33%	2	6	33%
	Grand Totals and Percents for all ages and populations who trained on Cogmed compared with no treatment			11	16	69%	6	16	38%	6	16	38%	6	16	38%

(continued)

Table 8.9. Continued

Study #	Study Name	Condition of Interest	Comparison Condition	Significantly Better Improvement			Significantly Better Posttest			Significantly Better Posttest Only Including Measures Where This Was Looked at			Both Significantly Better Change and Posttest		
				# Sign.	# Measures	% Sign.	# Sign.	# of Measures	% Sign.	# Sign.	# Measures	% Sign.	# Sign.	# of Measures	% Sign.
			OLDER ADULTS (OLDER THAN 55 YEARS) WITHOUT CLINICAL DIAGNOSES												
14	Brehmer et al., 2012	Cogmed	Nonincrementing Cogmed	3	4	75%	0	4	0%	0	4	0%	0	4	0%
	Grand Totals and Percents for all ages and populations who trained on Cogmed compared with a nonincrementing version of Cogmed (N = 10 studies)			26	33	79%	10	23	43%	10	23	43%	10	23	43%
	Grand Totals and Percents for all ages and populations who trained on Cogmed compared with no treatment (N = 7 studies)			13	18	72%	6	18	33%	6	18	33%	6	18	33%
	Grand Totals and Percents across all Cogmed studies (including only nonincrementing Cogmed and no-treatment control conditions)			39	51	76%	16	41	39%	16	41	39%	16	41	39%
	Grand Totals and Percents across all Cogmed studies (including the promising active-control conditions of Gray et al. & van der Donk et al.)			42	61	69%	18	51	35%	18	51	35%	18	51	35%

Note. Results for outcomes other than WM are not included here.

[A] Thorell et al. (2009) had separate no-treatment and active-control groups but they combined the two groups in their analyses. The results for Thorell et al.'s inhibitory control training are listed here, but they were not included in any calculations because it was not WM training.

[B] The authors of this study did not conduct the needed multilevel data analysis. It is unclear how many of their results would remain significant had they done that.

[C] The authors of this study did not correct for multiple comparisons. It is unclear which results, if any, would remain significant had that been done.

[D] All participants had poor WM at study outset.

[E] One might plausibly expect EF benefits from math training, so a failure to find a difference here might be due to both interventions' being beneficial, rather than Cogmed's being ineffectual, thus we have not included the null findings here when calculating totals or percentages, except where otherwise noted.

[F] The study by Roberts et al. (2016) was published after the 2015 cutoff date. We include it here because we think it is important, but we did not include it in calculations of totals or percentages.

[G] One might plausibly expect EF benefits from the "Pay Attention in Class" intervention, so a failure to find a difference here might be due to both interventions' being beneficial, rather than Cogmed's being ineffectual, thus we have not included the null findings here when calculating totals or percentages, except where otherwise noted.

[H] These studies did report the difference between posttest scores.

[I] All participants in the experimental group for Bergman-Nutley & Klingberg (2014) had WM deficits at the outset of the study and most had ADHD.

[J] Egeland et al. (2013) did not report any near-transfer EF results.

on Cogmed performed better at posttest on all (100%) of their three WM measures.

For example, on the composite WM score from the WISC-IV, Green et al. (2012) found more improvement and superior posttest performance among those who trained on Cogmed versus those who played a nonadaptive version of Cogmed among 10-year-olds with ADHD. Also, on an objective measure of inattentive behavior in a naturalistic paradigm designed to simulate attentional demands in the classroom, those trained on Cogmed improved more and performed better at posttest than active controls. Such positive results were obtained even though Green et al. had participants train for a relatively short time (4 weeks), whereas other Cogmed studies have had participants train for 5 to 8 weeks.

Bigorra et al. (2015) created a composite WM score consisting of the two WM subscales from the WISC-IV (Backward Digit Span and Letter-Number Sequencing) plus Backward Spatial Span from the Wechsler Memory Scale-III. They found that 7- to 12-year-old children with ADHD who trained on Cogmed improved significantly more on the WM composite and performed better on Conners' CPT than their peers who played a nonadaptive version of Cogmed. Their relatively better improvement on the WM composite, however, seems to be because the control group mysteriously got worse. Parental ratings on the BRIEF did not differ significantly between the two groups right after training, but 6 months later, the parents of those who had done Cogmed reported more improvements than the parents of controls. Teachers saw some benefits right away in Cogmed-trained children, but more and larger benefits 6 months laterthan compared with their ratings of children in the control group. Between-group comparisons of posttest performance were not reported.

Across all Cogmed studies (regardless of the age of participants or whether they had ADHD or not), those trained on Cogmed, when compared with those who played a nonadaptive version of Cogmed, (a) improved more on 79% of WM measures, (b) performed better at posttest on 43% of WM measures, and (c) showed both better change and better posttest scores on 43% of WM measures (i.e., near-transfer measures). The numbers improve considerably if studies that did not do the needed statistical analyses are excluded. Then, the percentage of measures both for better performance at posttest and for the combination of better posttest performance and more improvement was 89%.

Results on Near-Transfer EF Measures for Cogmed Training of School-Age
Children versus No Treatment

Dunning et al. (2013) conducted the only study to compare Cogmed training to no treatment among typically developing children. Schoolchildren 7 to 9 years old who trained on Cogmed improved more and performed better at posttest than the no-treatment group on four of the five WM measures used (80%). Those benefits, and the benefits found in comparison to nonincrementing Cogmed, were still evident 12 months later (but see Footnote 4).

Three studies compared Cogmed training to no treatment for children with ADHD (Bergman-Nutley & Klingberg, 2014; Egeland et al., 2013; Hovik et al., 2013). Bergman-Nutley and Klingberg found that, after 5 weeks of Cogmed training, 11-year-old children with ADHD not only improved more than, but also outperformed, typically developing peers on a visuospatial WM measure. That's impressive, although on a second visuospatial WM measure, results were less positive.

Egeland et al. (2013) included no near-transfer EF measures (i.e., no WM measures) and found no benefit compared to controls on any of their six far-transfer EF measures.

Hovik et al. (2013) found that, after 5 weeks of Cogmed training, 10-year-olds with ADHD had improved more on all three WM indices they looked at (both verbal and visuospatial measures) than other ADHD children in the no-treatment group. That, too, is impressive.

Across all Cogmed studies with no-treatment or nonadaptive-Cogmed control conditions (regardless of the age of participants or whether they had ADHD), those trained on Cogmed improved more on 69% of the WM measures compared with control subjects, but performed better at posttest than controls on only 33% (and for studies reporting only posttest comparisons: 33%) and showed better change and posttest scores on 37% of WM measures (see Table 8.9).

Results on Near-Transfer EF Measures for Cogmed Training of School-Age
Children versus Another Intervention

The two studies that compared Cogmed training to another intervention both found few differences between the benefits of Cogmed and the other intervention. Gray et al. (2012) compared Cogmed training to special math training for schoolchildren of 7 to 15 years diagnosed with both ADHD and a learning disorder. They found more improvement after Cogmed than math training on two measures very similar to Cogmed games (e.g., Backward Digit Span and Spatial Span), but not on three other WM measures, and not on five tests of other EF skills. Except for Backward Digit Span, on no measure was posttest performance significantly better in the Cogmed group than in the math-training group. Van der Donk et al. (2015) found similar benefits from Cogmed training and an intervention called "Pay Attention in Class" when these were delivered to 8- to

12-year-old schoolchildren with ADHD. Those who received Cogmed training improved more and performed better at posttest on a Forward Spatial Span task, but not on four other WM measures nor on five tests of other EF skills. Either the comparison condition in each of these two studies produced similar benefits to those from Cogmed, or—since there was no business-as-usual or no-treatment group in either study—it could be that neither program produced more benefit than would normally have occurred from practice effects plus 5 weeks of school.

If the programs indeed produced comparable benefits, given how different the three programs were, perhaps it was the increased attention from adults who expected benefits from the program or the excitement about a new program that produced the benefits, rather the content of the training programs per se. That would be consistent with de Jong's finding (de Jong, 2014) that the mentoring component of Cogmed might be more central to its benefits than the computerized training. Of course, one might plausibly expect EF benefits from math training or from the "Pay Attention in Class" intervention, so a failure to find a difference might be due to both interventions' also being beneficial due to their content, rather than the interaction with adults.

Results for Cogmed Training of Adults
Only one study has looked at possible EF benefits from Cogmed among older adults (Brehmer et al., 2012). Brehmer and colleagues studied both younger adults (20–30 years old) and older adults (60–70 years old). They found more improvement on all four (100%) of their near-transfer EF measures (backward digit span, Backward Span Board, Forward Span Board, and Paced Auditory Serial Addition Test—PASAT[5]) among both age groups for the Cogmed group versus the nonincrementing Cogmed control group, with larger differences for younger than older participants. The Cogmed trainees also showed more improvement on a more distal measure: the Cognitive Failures Questionnaire (CFQ). All the improvements were still evident 3 months later. (Brehmer et al., however, did not correct for multiple comparisons.)

However, even on near-transfer measures quite similar to the training tasks (e.g., Backward Digit Span), Brehmer et al. found only one significant posttest score difference between Cogmed trainees and active controls (young adults on Forward Span Board; the other three comparisons did not show a Cogmed benefit), although all four comparisons showed more improvement from Cogmed than from the nonadaptive-training active-control condition. The same was true for the CFQ (more improvement, but not better posttest scores than active controls). In general, the two Cogmed groups started off a bit worse than the two nonincrementing Cogmed groups (except for younger adults on Forward Span

---

[5] Brehmer et al. (2012) considered PASAT a sustained attention test and thus a far-transfer measure, but on this test participants hear single digits every 3 seconds and are to add each new digit to the one immediately prior to it. We consider this a WM measure.

Board and older adults on Backward Digit Span and CFQ), which helps explain why change scores were significantly different while performance after the intervention generally was not.

The only other study of Cogmed in adults meeting our inclusion criteria (besides Mawjee et al., 2014, 2015, discussed above, which only compared more versus less Cogmed) looked at Cogmed training versus no treatment for adults 19 to 52 years old who had ADHD (Gropper et al., 2014). They found that those who did Cogmed reported fewer cognitive failures in everyday life and fewer instances of off-task behavior (and a greater reduction in those) than did control subjects, and that was still true 2 months later. Cogmed trainees also improved more and performed better at posttest on very-near-transfer measures of spatial span and digit span than did no-treatment controls, although they showed no benefits on the PASAT (an auditory serial addition WM task), a CANTAB spatial WM measure, or a measure of selective attention.

Results for Cogmed Training of Children 4 to 6 Years Old

Among children 5 years old, Thorell et al. (2009) found lots of evidence of differential improvement. The experimental group improved more on Forward + backward word span (combined), Forward + backward visuospatial span (combined), and sustained attention than those who did a nonadaptive version of Cogmed or received no treatment. On no measure, however, did Cogmed trainees perform significantly better at posttest than those who trained on a non-adaptive version of Cogmed or those in the no-treatment group.

The other study that looked at Cogmed benefits in very young children was by Bergman Nutley et al. (2011). They studied 4-year-olds and found a benefit from Cogmed on a visuospatial WM task (similar to Corsi Blocks and similar to Cogmed games) compared with active controls. On STM measures, which do not assess WM—Block Design (visuospatial STM) and Forward Word Span (verbal STM)—and on a far-transfer reasoning measure (Odd One Out), they found no benefit from Cogmed WM training. Those trained in reasoning improved more on reasoning than did active controls. Benefits were narrow, not even generalizing from visuospatial to verbal (or even to all visuospatial memory measures), although perhaps if their measures had tapped WM, rather than STM, a benefit might have been found.

Discussion of Results from Cogmed Training Studies

Cogmed works well for improving WM, especially on measures similar to training games in Cogmed itself and on complex-span measures. The one study that looked at improving inhibitory control (Thorell et al., 2009) found disappointing results: Perhaps not enough pretesting, thought, and/or effort had gone into constructing the inhibitory-control training, or the children (ages 4–5 years) were too young, the training sessions were too short (15 min rather than the 30–45 min used for older children), or computer training may not be the best way to improve inhibition. Rueda and colleagues (Rueda, Checa, & Cómbita, 2012; Rueda, Rothbart, McCandliss, Saccomanno, & Posner, 2005) also found disappointing results in their attempt to improve inhibitory control in 4- and 5-year-olds using different computerized training. Blakey and Carroll (2015) administered both computerized WM and inhibitory-control training to 4-year-olds and found that WM improved but inhibitory control did not.

If someone has a circumscribed deficit specific to WM, we recommend Cogmed. The superb recent review by Simons et al. (2016) also concluded that "the randomized controlled trials provide strong evidence that Cogmed training improves performance on other working-memory tasks with similar processing demands" (pp. 147–148), although Simons et al. would be more circumspect in emphasizing benefits only on quite similar tasks, not benefits to WM (even visuospatial WM) generally. For benefits to last, we recommend engaging in continuing, ongoing challenges to WM, or else the WM benefits from Cogmed or anything else will likely disappear in months or a year or two. For benefits to WM more broadly, we recommend training and challenges also to other aspects of WM less emphasized in Cogmed games. If someone has deficits in inhibitory control or diverse EF components, however, we do not recommend Cogmed.

Surprisingly, only one study has looked at Cogmed with older adults. That is a topic crying out for research. WM is one of the first cognitive skills to deteriorate with aging and it often shows the greatest decline (e.g., Hedden & Park, 2001; Wang et al., 2011). Hence Cogmed might be an excellent option for older adults with WM decline, provided they enjoy the Cogmed games enough to keep working at them and have good mentors. In the one study that looked (Brehmer et al., 2012), older adults who trained on Cogmed showed more improvement than controls on all four (100%) of the near-transfer EF measures and on the more distal CFQ, and all improvements were still evident 3 months later. More studies with older adults are needed.

Sometimes the reason something works can be quite different from what anyone expected. Although most studies of Cogmed do not mention the mentoring component, to be certified to administer Cogmed, adults must be trained in, and commit to, mentoring those doing Cogmed. De Jong (2014) found that the mentoring may actually account for the benefits of Cogmed even more than the computerized games. That merits follow-up and further study. It also illustrates that the reason why a program is beneficial should be investigated, rather than assumed. (See the discussion above about the two studies that compared Cogmed to other interventions—the attention from adults might turn out to be more crucial than most people have thought.)

The nonincrementing version of Cogmed includes interaction with adults similar to what occurs with the standard, adaptive version of Cogmed, yet the standard version consistently produces better results than the nonincrementing version. Does this argue against the mentoring component potentially being critical for the benefits? Probably not, because it is unlikely that the mentors expect similar benefits from the control condition. It is probably critical that the mentor believes in the efficacy of the training and expects it to benefit the trainee.

N-Back Training

N-back tasks are explained in the section "EFs Explained" above. Unlike studies of Cogmed, where all but two of the 15 studies (87%) were with children, most studies of N-back training have been with adults (all but two out of 13, or 85%). Whereas 80% of Cogmed studies included an active control group, only 54% of N-back training studies did (albeit most Cogmed studies have used a control group not challenged as much nor presumably expecting as much improvement as those training on Cogmed). Most N-back studies included training for fewer weeks than Cogmed (all but one Cogmed study had 5–8 weeks of training; all but two N-back studies trained participants only 2–5 weeks) and had shorter training sessions—15 to 30 min in 10 out of 13 N-back studies (77%) versus 30 to 45 min for Cogmed sessions in 13 of the 16 studies (81%).[6]

Across all Cogmed studies, more improvement than in comparison conditions was reported on 69% of the near-transfer EF measures and better posttest results than in comparison conditions on 33% of near-transfer EF tasks for which a comparison of posttest results was reported (see Table 8.8a). (Looking only at the studies that included the requisite statistical analyses, the corresponding percents are 61% for more improvement than comparison conditions on WM measures and 38% for better posttest performance on WM measures than comparison conditions for Cogmed. See Table 8.8b.) Seventy-one percent of Cogmed studies found at least suggestive evidence of WM benefits.

Results for N-back training are more disappointing. For N-back training, more improvement than comparison subjects was reported on only 34% of near-transfer EF measures and better posttest performance than comparison subjects on 31% of near-transfer EF measures. (See Table 8.8a. Looking only at the studies that did the requisite statistical analyses, the corresponding percents are 28% for more improvement than comparison conditions on WM measures and 28% for better posttest performance on WM measures than comparison conditions for N-back training. See Table 8.8b.) Fifty percent of all N-back studies (only 17% of N-back studies with the requisite statistical analyses) found at least suggestive evidence of WM benefits (see Figure 8.2). Perhaps if N-back training sessions were longer or continued for more weeks (more like Cogmed), better results would be found. However, N-back studies with longer sessions (30–60 min) have not found EF benefits (Chooi & Thompson, 2012; Kundu, Sutterer, Emrich, &

---

[6] Although note that Mawjee et al. (2014, 2015) compared Cogmed training sessions of 15 and 45 minutes, holding everything else constant, and found no difference in benefits.

Postle, 2013; Lilienthal, Tamez, Shelton, Myerson, & Hale, 2013; Pugin et al., 2014; Thompson et al., 2013). On the other hand, Basak et al.'s (2008) findings with Rise of Nations (that 2–3 weeks was too short to see benefits, although benefits were seen after 4–5 weeks) suggest that continuing training for more weeks might well make a difference.

Results were somewhat better when N-back was compared to no treatment than when it was compared to active-control conditions. *Compared to active-control conditions*, N-back training produced more improvement on only 18% of all near-transfer measures across studies and better posttest performance on only 18% as well. *Compared to no-treatment controls*, N-back training produced more improvement on 42% of all near-transfer measures across studies and better posttest performance on 38%.

Perhaps it is not that surprising that no differences were found in benefits from N-back training versus playing Tetris (Kundu et al., 2013), since both would be expected to challenge EFs and hence improve them. That there were no differences in WM benefits from N-back training versus visual search training (Redick et al., 2013) or versus training multiple-object tracking (Thompson et al., 2013) is more surprising (see Tables 8.11 and 8.12). Both studies included no treatment controls; results were no better comparing N-back to no treatment than to the active-control condition in each of these studies. This suggests that benefits in these two studies seem to have been simply due to practice in taking the outcome measures (which were completed both before and after the weeks of training).

Two N-back training studies have been done with older adults (Li et al., 2008; Stepankova et al., 2014). For the study by Stepankova et al., where the mean age of participants was 68 years (range = 65–74), those trained on N-back showed more improvement and performed better at posttest than no-treatment controls on both (100%) of the EF near-transfer measures (Forward + backward combined digit span task and a challenging WM task [Letter-Number Sequencing, where a series of numbers and letters are presented orally in random order and then the participant is to repeat back the numbers in numerical order followed by the letters in alphabetical order]) regardless of whether participants were assigned to N-back practice more or less often (four vs. two times per week). Those trained on N-back also improved more and performed better at posttest than no-treatment controls on both visuospatial far-transfer measures (matrix reasoning and block design) that Stepankova and colleagues administered. Participants in the Li et al. study were 70 to 80 years old (mean age = 74). Those trained on N-back performed better and improved more on two very-near-transfer nontrained N-back tasks but not on either complex-span task tested compared to no-treatment controls (but, remember, the correlations between N-back and complex-span tests are low, so the lack of transfer may not be indicative of lack of benefits).

Table 8.11. Percentage of WM Measures on Which Persons Who Received N-Back Training Showed More Improvement and/or Better Posttest Results Across All Studies and Ages, Broken Down by Study

Study #	Study Name	Condition of Interest	Comparison Condition	Significantly Better Improvement			Significantly Better Posttest			Significantly Better Posttest Only Including Measures Where This Was Looked at			Both Significantly Better Change and Posttest		
				# Sign.	# of Measures	% Sign.	# Sign.	# of Measures	% Sign.	# Sign.	# of Measures	% Sign.	# Sign.	# of Measures	% Sign.
1	Jaeggi et al., 2011[A]	Single N-back task	Computerized knowledge & vocabulary task												
	SCHOOL-AGE CHILDREN (7–17 YEARS OLD)														
2	Pugin et al., 2014[B]	Single N-back task	No treatment	1	2	50%	1	2	50%	1	2	50%	1	2	50%
	ADULTS (19–55 YEARS OLD)														
3	Chooi & Thompson 2012	Dual N-back task (8 or 20 sessions)	Nonincrementing version of training games	0	1	0%	0	1	0%	0	1	0%	0	1	0%
4	Lilienthal et al., 2013[C]	Dual N-back task	Nonincrementing dual N-back task task	1	4	25%	1	4	25%	1	4	25%	1	4	25%
5	Redick et al., 2013[B,D]	Dual N-back task	Visual search	1	2	50%	1	2	50%	1	2	50%	1	2	50%
6	Stephenson & Halpern, 2013[A,B]	Dual N-back task	Spatial matrix span												
7	Thompson et al., 2013	Dual N-back task	Multiple object tracking	0	3	0%	0	3	0%	0	3	0%	0	3	0%
	Totals and Percents for adults who received N-back training compared with an active-control condition			2	10	20%	2	10	20%	2	10	20%	2	10	20%

(continued)

Table 8.11. Continued

	Study			Significantly Better Improvement			Significantly Better Posttest			Significantly Better Posttest Only Including Measures Where This Was Looked at			Both Significantly Better Change and Posttest		
Study #	Study Name	Condition of Interest	Comparison Condition	# Sign.	# of Measures	% Sign.	# Sign.	# of Measures	% Sign.	# Sign.	# of Measures	% Sign.	# Sign.	# of Measures	% Sign.
8	Kundu et al., 2013[E]	Dual N-back task	Tetris	0	1	0%	0	1	0%	0	1	0%	0	1	0%
	Grand Totals and Percents for adults and children who received N-back training compared with an active-control condition (excluding Kundu et al., 2013)			2	10	20%	2	10	20%	2	10	20%	2	10	20%
	Grand Totals and Percents for adults and children who received N-back training compared with any active-control condition (including Kundu et al., 2013)			2	11	18%	2	11	18%	2	11	18%	2	11	18%
	Grand Totals and Percents for adults and children who received N-back training compared with an active-control condition or no treatment (excluding Kundu et al., 2013)			3	12	25%	3	12	25%	3	12	25%	3	12	25%
	Grand Totals and Percents for adults and children who received N-back training compared with any active control condition or no treatment (including Kundu et al., 2013)			3	13	23%	3	13	23%	3	13	23%	3	13	23%
ADULTS (19–55 YEARS OLD)															
3	Chooi & Thompson, 2012	Dual N-back task (8 or 20 sessions)	No treatment	0	1	0%	0	1	0%	0	1	0%	0	1	0%
9	Jaeggi et al., 2008[B]	Dual N-back task (8, 12, 17, or 19 sessions)	No treatment	1	2	50%	1	2	50%	1	2	50%	1	2	50%

10	Jaeggi et al., 2010	Dual N-back task	No treatment	0	1	0%	0	1	0%	0	1	0%	0	1	0%
10	Jaeggi et al., 2010	Single N-back task	No treatment	0	1	0%	0	1	0%	0	1	0%	0	1	0%
11	Li et al., 2008	Single N-back ask	No treatment	2	3	67%	2	3	67%	2	3	67%	2	3	67%
4	Lilienthal et al., 2013	Dual N-back task	No treatment	1	4	25%	1	4	25%	1	4	25%	1	4	25%
5	Redick et al., 2013[B]	Dual N-back task	No treatment	1	2	50%	0	2	0%	0	2	0%	0	2	0%
12	Rudebeck et al., 2012[A]	Dual N-back task	No treatment												
5	Stephenson & Halpern, 2013[A,B]	Dual N-back task	No treatment												
7	Thompson et al., 2013	Dual N-back task	No treatment	0	3	0%	0	3	0%	0	3	0%	0	3	0%
	Totals and Percents for adults 19–55 years old who received N-back training compared with no treatment			5	17	29%	4	17	24%	4	17	24%	4	17	24%
	Grand Totals and Percents for adults and children who received N-back training compared with no treatment			6	19	32%	5	19	26%	5	19	26%	5	19	26%
	Grand Totals and Percents for Adults (19–55 years old)			7	28	25%	6	28	21%	6	28	21%	6	28	21%
OLDER ADULTS (OLDER THAN 55 YEARS)															
11	Li et al., 2008	Single N-back task	No treatment	2	3	67%	2	3	67%	2	3	67%	2	3	67%

(continued)

Table 8.11. Continued

Study #	Study Name	Condition of Interest	Comparison Condition	Significantly Better Improvement			Significantly Better Posttest			Significantly Better Posttest Only Including Measures Where This Was Looked at			Both Significantly Better Change and Posttest		
				# Sign.	# of Measures	% Sign.	# Sign.	# of Measures	% Sign.	# Sign.	# of Measures	% Sign.	# Sign.	# of Measures	% Sign.
13	Stepankova et al., 2014[B]	Single N-back task (10 or 20 sessions)	No treatment	2	2	100%	2	2	100%	2	2	100%	2	2	100%
	Totals and Percents for older adults who received N-back training compared with no treatment			4	5	80%	4	5	80%	4	5	80%	4	5	80%
	Grand Totals and Percents across all ages who received N-back training compared with no treatment			10	24	42%	9	24	38%	9	24	38%	9	24	38%
	Grand Totals and Percents across all N-back studies (excluding Kundu et al., 2013)			12	34	35%	11	34	32%	11	34	32%	11	34	32%
	Grand Totals and Percents across all N-back studies (including Kundu et al., 2013)			12	35	34%	11	35	31%	11	35	31%	11	35	31%

Note. Results for outcomes other than WM are not included here.

A This study did not include any near-transfer measures. They only looked at reasoning/fluid intelligence (R/FL) measures.

B The authors of this study did not include a correction for multiple comparisons. It is unclear which results, if any, would remain significant had that been done.

C Studies that varied the number of training sessions found no difference by number of sessions on anything relevant to this table, so results across those different conditions are collapsed here.

D The one significant difference here was because the control group mysteriously got worse at posttest on the running letter span task, while those who trained on N-back with difficulty increasing (as well as the no-treatment group) improved.

E One might plausibly expect EF benefits from playing Tetris, so a failure to find a difference here might be due to both interventions' being beneficial, rather than N-back training's being ineffectual, thus we have not included the null findings here when calculating totals or percentages, except for the last line.

Table 8.12. Percentage of EF Measures (Except Reasoning/Fluid Intelligence) on Which Persons Who Received N-Back Training Showed More Improvement and/or Better Posttest Results Than Comparison Groups Across All Studies and Ages, Broken Down by Study

Study #	Study Name	Condition of Interest	Comparison Condition	Significantly Better Improvement			Significantly Better Posttest			Significantly Better Posttest Only Including Measures Where This Was Looked at			Both Significantly Better Change and Posttest		
				# Sign.	# of Measures	% Sign.	# Sign.	# of Measures	% Sign.	# Sign.	# of Measures	% Sign.	# Sign.	# of Measures	% Sign.
				SCHOOL-AGE CHILDREN (7–17 YEARS OLD)											
1	Jaeggi et al., 2011[A]	Single N-back task	Computerized knowledge & vocabulary task												
2	Pugin et al., 2014[B]	Single N-back task	No treatment	1	5	20%	1	5	20%	1	5	20%	1	5	20%
				ADULTS (19–55 YEARS OLD)											
3	Chooi & Thompson, 2012	Dual N-back task (8 or 20 sessions)	Nonincrementing version of training games	0	2	0%	0	2	0%	0	2	0%	0	2	0%
4	Lilienthal et al., 2013[C]	Dual N-back task	Nonincrementing dual N-back task	1	5	20%	1	5	20%	1	5	20%	1	5	20%
5	Redick et al., 2013[B,D]	Dual N-back task	Visual search	1	3	33%	1	3	33%	1	3	33%	1	3	33%
6	Stephenson & Halpern, 2013[A,B]	Dual N-back task	Spatial matrix span												
7	Thompson et al., 2013	Dual N-back task	Multiple object tracking	0	3	0%	0	3	0%	0	3	0%	0	3	0%
	Totals and Percents for adults who received N-back training compared with an active-control condition			2	10	20%	2	10	20%	2	10	20%	2	10	20%

(continued)

Table 8.12. Continued

Study				Significantly Better Improvement			Significantly Better Posttest			Significantly Better Posttest Only Including Measures Where This Was Looked at			Both Significantly Better Change and Posttest		
Study #	Study Name	Condition of Interest	Comparison Condition	# Sign.	# of Measures	% Sign.	# Sign.	# of Measures	% Sign.	# Sign.	# of Measures	% Sign.	# Sign.	# of Measures	% Sign.
8	Kundu et al., 2013[E]	Dual N-back task	Tetris	0	2	0%	0	2	0%	0	2	0%	0	2	0%
	Grand Totals and Percents for adults and children who received N-back training compared with an active-control condition (excluding Kundu et al., 2013)			2	13	15%	2	13	15%	2	13	15%	2	13	15%
	Grand Totals and Percents for adults and children who received N-back training compared with any active-control condition (including Kundu et al., 2013)			2	15	13%	2	15	13%	2	15	13%	2	15	13%
	Grand Totals and Percents for adults and children who received N-back training compared with an active-control condition or no treatment (excluding Kundu et al., 2013)			3	18	17%	3	18	17%	3	18	17%	3	18	17%
	Grand Totals and Percents for adults and children who received N-back training compared with any active-control condition or no treatment (including Kundu et al., 2013)			3	20	15%	3	20	15%	3	20	15%	3	20	15%
ADULTS (19–55 YEARS OLD)															
3	Chooi & Thompson 2012	Dual N-back task (8 or 20 sessions)	No treatment	0	1	0%	0	1	0%	0	1	0%	0	1	0%
9	Jaeggi et al., 2008[B]	Dual N-back task (8, 12, 17, or 19 sessions)	No treatment	1	2	50%	1	2	50%	1	2	50%	1	2	50%

	Study	Task	Comparison												
10	Jaeggi et al., 2010	Dual N-back task	No treatment	0	1	0%	0	1	0%	0	1	0%	0	1	0%
10	Jaeggi et al., 2010	Single N-back task	No treatment	0	1	0%	0	1	0%	0	1	0%	0	1	0%
11	Li et al., 2008	Single N-back task	No treatment	2	3	67%	2	3	67%	2	3	67%	2	3	67%
4	Lilienthal et al., 2013	Dual N-back task	No treatment	1	5	20%	1	5	20%	1	5	20%	1	5	20%
5	Redick et al., 2013[B]	Dual N-back task	No treatment	1	3	33%	0	3	0%	0	3	0%	0	3	0%
12	Rudebeck et al., 2012[A]	Dual N-back task	No treatment												
6	Stephenson & Halpern, 2013[A,B]	Dual N-back task	No treatment												
7	Thompson et al., 2013	Dual N-back task	No treatment	0	3	0%	0	3	0%	0	3	0%	0	3	0%
	Totals and Percents for adults 19–55 years old who received N-back training compared with no treatment			5	19	26%	4	19	21%	4	19	21%	4	19	21%
	Grand Totals and Percents for adults and children who received N-back training compared with no treatment			6	24	25%	5	24	21%	5	24	21%	5	24	21%
	Grand Totals and Percents for Adults (19–55 years old)			7	34	21%	6	34	18%	6	34	18%	6	34	18%
OLDER ADULTS (OLDER THAN 55 YEARS)															
11	Li et al., 2008	Single N-back task	No treatment	2	3	67%	2	3	67%	2	3	67%	2	3	67%

(continued)

Table 8.12. Continued

Study #	Study Name	Condition of Interest	Comparison Condition	Significantly Better Improvement			Significantly Better Posttest			Significantly Better Posttest Only Including Measures Where This Was Looked at			Both Significantly Better Change and Posttest		
				# Sign.	# of Measures	% Sign.	# Sign.	# of Measures	% Sign.	# Sign.	# of Measures	% Sign.	# Sign.	# of Measures	% Sign.
13	Stepankova et al., 2014[B]	Single N-back task (10 or 20 sessions)	No treatment	2	2	100%	2	2	100%	2	2	100%	2	2	100%
	Totals and Percents for older adults who received N-back training compared with no treatment			4	5	80%	4	5	80%	4	5	80%	4	5	80%
	Grand Totals and Percents across all ages who received N-back training compared with no treatment			10	29	34%	9	29	31%	9	29	31%	9	29	31%
	Grand Totals and Percents across all N-back studies (excluding Kundu et al., 2013)			12	42	29%	11	42	26%	11	42	26%	11	42	26%
	Grand Totals and Percents across all N-back studies (including Kundu et al., 2013)			12	44	27%	11	44	25%	11	44	25%	11	44	25%

Note. Results for reasoning/fluid intelligence (R/FL) are not included in Table 8.12 (although they are mentioned in the text) but results for all other EF measures are included here.

[A] This study did not include any near-transfer measures. They only looked at R/FL measures.

[B] The authors of this study did not include a correction for multiple comparisons. It is unclear which results would remain significant had they done that.

[C] Studies that varied the number of training sessions found no difference by number of sessions on anything relevant to this table, so results across those different conditions are collapsed here.

[D] The one significant difference here was because the control group mysteriously got worse at posttest on the running letter span task, while those who trained on N-back with difficulty increasing (as well as the no-treatment group) improved.

[E] One might plausibly expect EF benefits from playing Tetris, so a failure to find a difference here might be due to both interventions' being beneficial, rather than N-back training's being ineffectual, thus we have not included the null findings here when calculating totals or percentages, except for the last line.

Far-Transfer Results for N-Back Training

Every N-back study but one looked for evidence of far transfer. Of those 12 studies (of 13), three (25%) found clear evidence of far transfer (Jaeggi et al., 2010, on Raven's and less so on BOMAT; Rudebeck et al., 2012, on BOMAT; Stepankova et al., 2014, on Matrix Design from WAIS-III). Two found only the slightest hint (Jaeggi et al., 2008; Stephenson & Halpern, 2013). Note that neither Rudebeck et al., Stepankova et al., Jaeggi et al., nor Stephenson and Halpern corrected for multiple comparisons; it is unclear which of their findings, if any, would remain significant had they done that.

Table 8.12, which presents the percentage across all EF measures (except reasoning/fluid intelligence) on which persons trained on N-back showed more improvement and/or better posttest results than comparison groups across all studies and ages, broken down by study, appears online.

Although N-back performance has been found to be fairly highly correlated with performance on fluid-intelligence or reasoning tasks (Gray, Chabris, & Braver, 2003; Jaeggi et al., 2010; Kane, Conway, Miura, & Colflesh, 2007; Schmiedek, Hildebrandt, Lövdén, Lindenberger, & Wilhelm, 2009), of the 11 N-back training studies that looked for far transfer to tests of fluid intelligence or reasoning, most (55%) found none (Chooi & Thompson, 2012; Jaeggi, Buschkuehl, Jonides, & Shah, 2011; Kundu et al., 2013; Pugin et al., 2014; Redick et al., 2013; Thompson et al., 2013). Three of the five that found benefits looked at more than one fluid-intelligence/reasoning measure: One found clear benefits on both measures (Jaeggi et al., 2010). One found some benefit on one measure but none on the other (Jaeggi et al., 2008). The third found more improvement on two of four measures and no better posttest scores on any (Stephenson & Halpern, 2013).

Only one study (Stephenson & Halpern 2013) with an active control group (out of six) found more improvement or better posttest performance on any measure of fluid intelligence or reasoning after N-back training compared to controls. Of the five N-back studies with only a no-treatment control group that looked at fluid-intelligence or reasoning outcomes, four (80%) found suggestive evidence of such far transfer. Redick et al. (2013) found neither more improvement nor better posttest performance from N-back training on any of several fluid-intelligence/reasoning measures (including Raven's Matrices and a test of inferences and analogies) compared to controls trained on a visual search task. This is despite their N-back training continuing for longer and having lengthier and more frequent sessions than any other N-back training study, and even though they and Lilienthal et al. (2013) found the strongest near-transfer results of any N-back study with an active control group. They also found no benefits on measures of multitasking, although training had been with dual-task N-back.

Chooi and Thompson (2012) found neither more improvement nor better posttest performance on Raven's Matrices after N-back training than after sham N-back training or no treatment. However, Chooi and Thompson had a very high attrition rate (35%) and allowed participants to switch groups. Twenty-five percent of those assigned to N-back training or the active-control condition (nonadaptive N-back training) opted to transfer into the no-treatment group. Hence, one should view their results with caution.

Jaeggi et al. (2011) found no benefits on Raven's Matrices from N-back training versus computerized training on knowledge and vocabulary items. They did find, however, that those who improved most during the N-back training also improved most on Raven's and that subset of subjects improved more on Raven's than did subjects in the active-control condition. Pugin et al. (2014) found no benefits on the TONI test of nonverbal fluid intelligence compared to no-treatment controls.

Turning to the five studies with only a no-treatment control group that found some benefits to fluid intelligence/reasoning, Jaeggi et al. (2010) found both more improvement and better posttest performance on Raven's Matrices whether training was with the regular N-back task or a dual N-back paradigm (both groups were considered experimental groups by the authors) compared to no-treatment controls. On the Bochum Matrices Test (BOMAT), however, while those trained with the regular N-back task showed a benefit, those trained on the dual-task version did not.

Jaeggi et al. (2008) found the opposite: No benefit on Raven's (which was administered to those trained on the regular N-back task) but significant improvement on the BOMAT (which was administered to those trained with the dual N-back paradigm) compared to no-treatment controls.[7]

Rudebeck et al. (2012) administered only the BOMAT and found both more improvement and better posttest performance compared to no-treatment controls.[6] That was true for those who showed much improvement during N-back training and also for those who showed little.

Stephenson and Halpern (2013) administered Raven's, Beta-III Matrix Reasoning, Wechsler Abbreviated Scale of Intelligence (WASI) Matrix Reasoning, and Cattell. They found that those who trained on dual N-back improved more on Beta-III than those who trained on a single auditory N-back task (but not a single visual N-back), those who trained on a complex-span task, and no-treatment controls. Those who trained on the dual N-back, the visual single N-back, or the complex-span task improved more on Raven's than did no-treatment controls.[6] All other results for fluid-intelligence measures were

---

[7] Corrections for multiple comparisons were not made; not all their positive results might still be significant had those corrections been made.

negative: There were no group differences on WASI or Cattell, no group difference on posttest scores for any measures, and no other differences between groups on Beta-III or Raven's.

Stepankova et al. (2014) found both more improvement and better posttest performance on Matrix Design (from WAIS-III) than no-treatment controls whether the verbal N-back training was conducted twice a week (for a total of 10 sessions) or four times a week (for a total of 20 sessions).[6]

Results for Near Transfer to WM Tests After N-Back Training
Two N-back training studies with young adults found a benefit on one complex-span task but not on others (compared to no-treatment controls or active controls who trained on a nonadaptive version of the N-back task (Lilienthal et al., 2013) or on a visual-search task (Redick et al., 2013). Redick et al. included only two complex-span tasks, but Lilienthal et al. included four and found a benefit on only one. The one N-back training study with older adults to look at transfer to complex-span tasks found none. It has been found repeatedly that performance on N-back and complex-span tasks is only very weakly correlated (see Jaeggi et al., 2010; Kane et al., 2007; Oberauer, 2005; Roberts & Gibson, 2002; for a meta-analysis, see Redick & Lindsey, 2013). Hence, it is not too surprising that only two out of the eight studies that looked at whether N-back training improved complex-span performance found that it did.

Of the 10 N-back studies that looked at transfer to other WM measures, 60% found benefits. Stepankova et al. (2014) found older adults showed greater improvement and better posttest performance on both a Forward + backward combined digit span task and Letter-Number Sequencing.[7] Jaeggi et al. (2008) found both more improvement and better posttest scores from N-back training (whether regular or dual N-back) on backwards digit span compared to no-treatment controls regardless of whether the number of N-back training sessions was anywhere from 8 to 19.

In contrast, Pugin et al. (2014) found that adolescents (10–16 years old) showed no benefit on Letter-Number Sequencing from N-back training, although on a measure of very narrow transfer (training on a visuospatial N-back and testing on an auditory N-back) they showed more benefit than no-treatment controls. Likewise, Li et al. (2008) found very narrow transfer from N-back training. Li's group trained younger (age 21–30) and older adults (age 70–80) on N-back. Compared to no-treatment controls, those trained on N-back in both age groups improved more and performed better at posttest on a spatial N-back and numerical N-back task. The only memory measure Rudebeck et al. (2012) included assessed episodic memory (non-EF); they found no group difference on that when comparing N-back training to no treatment. They did find, however, that those who improved most during N-back training also improved the most from pre- to posttest on episodic memory. Jaeggi et al. (2008) found no benefit from N-back training (whether regular or dual N-back) on backward digit span compared to no-treatment controls regardless of the number of N-back training sessions (within the range of 8–19 sessions).

In general, results are not encouraging for N-back training's improving performance on complex-span tasks or other WM measures and decidedly mixed for whether it improves reasoning/fluid intelligence. It does improve performance on other N-back tasks. These conclusions apply both to training on a regular, single N-back task and to training on a dual-task variant.

Complex-Span Tasks

Complex-span tasks are explained in the section "EFs Explained" above. WM benefits after computerized complex-span training have generally been better than those after N-back training but not as good as those after Cogmed training (see Table 8.8a). Looking at all EF outcomes (including reasoning/fluid intelligence), computerized complex-span results look more on a par with those for N-back training but fall even farther short of Cogmed than when only WM outcomes are considered (see Tables 8.1 and 8.2).

The complex-span training administered by Loosli et al. (2012) was extremely brief and consisted of ten 12-min sessions (once per day, 5 days a week, for only 2 weeks). Participants were 10 years old. Loosli et al. tested no near-transfer measures and found no benefits on a measure of reasoning/fluid intelligence (TONI). Indeed, although all six complex-span studies looked at far transfer to reasoning/fluid intelligence, Borella et al. (2010) alone reported finding it. The study by Borella et al. is one of two papers reporting results from noncomputerized complex-span training. The other study is Carretti et al. (2013). (The two studies are discussed in the section below on noncomputerized cognitive training.) Except for these two studies from Borella and Carretti's group, no two studies trained on the same complex-span tasks. In sum, complex-span training improves performance on complex-span tasks. It does not improve reasoning/fluid intelligence.

Training on complex-span tasks was not much longer in other studies (15–20 sessions total—comparable to most N-back studies). Two had active-control conditions: Harrison et al. (2013) had two active control groups (simple span and visual search, both of which kept increasing in difficulty). Richmond et al. (2011) had one active control group (trivia learning). The Harrison et al. study had a high attrition rate (37%). None of the other complex-span studies provided data on attrition.

Harrison et al. found that young adults (mean age = 20 years) showed more improvement and performed better at posttest on two other complex-span tasks (reading span and rotation span) when compared to either active control group.

In addition, they performed better than the visual-search control group on the running letter, running spatial, and Keep Track complex-span tasks (although on the Keep Track Task this seems to be due to the visual-search group mysteriously getting worse). Harrison et al. found no transfer on the word or arrow span tasks. Thus, they found benefits on four of six untrained complex-span tasks (67%). They did not correct for multiple comparisons, however. It is unclear which results would have remained significant had they done that. See Tables 8.13 and 8.14.

Richmond et al. (2011) found more improvement and better posttest scores on a reading span task and on repetitions in the California Verbal Learning Task (CVLT) among older adults (mean age = 66 years) trained on complex-span tasks compared to active controls, but no benefit on CVLT accuracy or intrusion errors nor on Backward Digit Span or the Test of Everyday Attention.

Chein and Morrison (2010) found more improvement, but not better posttest scores, on the Stroop task in young adults (mean age = 20 years) trained on complex-span tasks than no-treatment controls.

Table 8.14, which presents the percentage across all EF measures (except reasoning/fluid intelligence) on which persons trained on N-back showed more improvement and/or better posttest results than comparison groups across all studies and ages, broken down by study, appears online.

Task Switching

Task switching, also called set-shifting, involves going back and forth between doing one task and doing another. Typically, although not always, all stimuli in a task-switching task are relevant to each task so participants must switch how they think about the stimuli or what aspect of a stimulus they focus on when switching between tasks.

For near transfer for task switching, we included all three core EFs (inhibitory control, WM, and cognitive flexibility) because all are required for task switching. Note that for Cogmed, N-back, and complex span we only included WM under near transfer. Even so, when studies with possibly spurious positive results are omitted, task switching shows better results for both the number of measures on which more improvement was found and the number of measures on which better posttest performance was found after training compared to the control group, and task switching comes in second, just behind Cogmed, in the percentage of studies finding at least suggestive evidence of near-transfer benefits (see Table 8.8a).

Three task-switching training studies were from Kray's lab (see Table 8.15). By far the best results were found in the first study: Karbach and Kray (2009) found task-switching training (vs. training on a single task with no switching) showed very-near-transfer improvements—there was improved task switching on other

Table 8.13. Percentage of WM Measures on Which Persons Who Received Complex-Span Training Showed More Improvement and/or Better Posttest Results Across All Studies and Ages, Broken Down by Study

Study #	Study Name	Condition of Interest	Comparison Condition	Significantly Better Improvement			Significantly Better Posttest			Significantly Better Posttest Only Including Measures Where This Was Looked at			Both Significantly Better Change and Posttest		
				# Sign.	# of Measures	% Sign.	# Sign.	# of Measures	% Sign.	# Sign.	# of Measures	% Sign.	# Sign.	# of Measures	% Sign.
SCHOOL-AGE CHILDREN (7–15 YEARS OLD)															
1	Loosli et al., 2012[A]	Complex Span	No treatment												
ADULTS (18–55 YEARS OLD)															
2	Harrison et al., 2013[B]	Complex Span	Adaptive simple span training	1	6	17%	1	6	17%	1	6	17%	1	6	17%
2	Harrison et al., 2013[B]	Complex Span	Adaptive visual-search training	4	6	67%	4	6	67%	4	6	67%	4	6	67%
3	Chein & Morrison, 2010[C]	Computerized adaptive verbal and spatial complex WM span task	No treatment												
	Grand Totals and Percents for adults who received complex-span training compared with an active-control condition			5	12	42%	5	12	42%	5	12	42%	5	12	42%
	Grand Totals and Percents for adults who received complex-span training compared with an active-control condition or no treatment			5	12	42%	5	12	42%	5	12	42%	5	12	42%

4	Borella et al., 2010	Complex Span (noncomputerized)	Fill-in paper-and-pencil questionnaires	3	3	100%	3	3	100%	3	3	100%	3	3	100%
5	Carretti et al., 2013	Complex Span (noncomputerized)	Fill-in paper-and-pencil questionnaires	1	1	100%	1	1	100%	1	1	100%	1	1	100%
6	Richmond et al., 2011	Complex Span	Trivia Learning	2	6	33%	2	6	33%	2	6	33%	2	6	33%
	Grand Totals and Percents for older adults who received complex-span training compared with an active control condition			6	10	60%	6	10	60%	6	10	60%	6	10	60%
	Grand Totals and Percents for adults who received complex-span training compared with an active control condition			11	22	50%	11	22	50%	11	22	50%	11	22	50%
	Grand Totals and Percents for adults who received computerized complex-span training compared with an active-control condition or no treatment			7	18	39%	7	18	39%	7	18	39%	7	18	39%
	Grand Totals and Percents for adults who received complex-span training (computerized or noncomputerized) compared with an active-control condition or no treatment			11	22	50%	11	22	50%	11	22	50%	11	22	50%

Note. Results for far-transfer measures, such as reasoning/fluid intelligence are not included here.

[A] Loosli et al. (2012) did not include any near-transfer measures.

[B] The authors of this study did not include a correction for multiple comparisons. It is unclear which results would remain significant had they done that.

[C] Chein and Morrison (2010) did not test the difference between posttest scores, and from their figure we are not able to tell if posttest scores were significantly different.

Table 8.14. Percentage of EF Measures (Except Reasoning Fluid Intelligence) on Which Persons Who Received Complex-Span Training Showed More Improvement and/or Better Posttest Results Than Comparison Groups Across All Studies and Ages, Broken Down by Study

Study #	Study Name	Condition of Interest	Comparison Condition	Significantly Better Improvement			Significantly Better Posttest			Significantly Better Posttest Only Including Measures Where This Was Looked at			Both Significantly Better Change and Posttest		
				# Sign.	# of Measures	% Sign.	# Sign.	# of Measures	% Sign.	# Sign.	# of Measures	% Sign.	# Sign.	# of Measures	% Sign.
SCHOOL-AGE CHILDREN (7–15 YEARS OLD)															
1	Loosli et al., 2012[A]	Complex Span	No treatment												
2	Harrison et al., 2013[B]	Complex Span	Adaptive simple span training	1	7	14%	1	7	14%	1	7	14%	1	7	14%
2	Harrison et al., 2013[B]	Complex Span	Adaptive visual-search training	4	7	57%	4	7	57%	4	7	57%	4	7	57%
ADULTS (18–55) YEARS OLD															
3	Chein & Morrison, 2010[C]	Computerized adaptive verbal and spatial complex WM span task	No treatment	1	1	100%									
	Grand Totals and Percents for adults who received complex-span training compared with an active-control condition			5	14	36%	5	14	36%	5	14	36%	5	14	36%
	Grand Totals and Percents for adults who received complex-span training compared with an active-control condition or no treatment			6	15	40%	5	14	36%	5	14	36%	5	14	36%

OLDER ADULTS (OLDER THAN 55)

	Study	Measure	Measure													
4	Borella et al., 2010	Complex Span (noncomputerized)	Fill-in paper-and-pencil questionnaires	100%	3	3	100%	3	3	100%	3	3	100%	3	3	3
5	Carretti et al., 2013	Complex Span (noncomputerized)	Fill-in paper-and-pencil questionnaires	100%	1	1	100%	1	1	100%	1	1	100%	1	1	1
6	Richmond et al., 2011	Complex Span	Trivia Learning	33%	2	6	33%	2	6	33%	2	6	33%	2	6	
	Grand Totals and Percents for older adults who received complex-span training compared with an active-control condition			60%	6	10	60%	6	10	60%	6	10	60%	6	10	
	Grand Totals and Percents for adults who received complex-span training compared with an active-control condition			46%	11	24	46%	11	24	46%	11	24	46%	11	24	
	Grand Totals and Percents for adults who received computerized complex-span training compared with an active-control condition or no treatment			38%	8	21	35%	7	20	35%	7	20	35%	7	20	
	Grand Totals and Percents for adults who received complex-span training (computerized or noncomputerized) compared with an active-control condition or no treatment			48%	12	25	46%	11	24	46%	11	24	46%	11	24	

Note. Results for reasoning/fluid intelligence (R/FL) are not included in Table 8.14 (although they are mentioned in the text) but results for all other EF measures are included here.

[A] Loosli et al. (2012) only included measures of R/FL.

[B] The authors of this study did not include a correction for multiple comparisons. It is unclear which results would remain significant had they done that.

[C] Chein & Morrison (2010) did not test the difference between posttest scores, and from their figures we are not able to tell if posttest scores were significantly different.

Table 8.15. Percentage of Near-Transfer Measures on Which Persons Who Received Task-Switching Training Showed More Improvement and/or Better Posttest Results Than Comparison Groups Across All Studies and Ages, Broken Down by Study

Study #	Study Name	Condition of Interest	Comparison Condition	Significantly Better Improvement			Significantly Better Posttest			Significantly Better Posttest only including measures where this was looked at			Both Significantly Better Change and Posttest		
				# Sign.	# of Measures	% Sign.	# Sign.	# of Measures	% Sign.	# Sign.	# of Measures	% Sign.	# Sign.	# of Measures	% Sign.
			SCHOOL-AGE CHILDREN (7–15 YEARS OLD) WITH NO CLINICAL DIAGNOSIS												
1	Dörrenbächer et al., 2014	Task switching in a low- or high-motivational setting	Single-task training in a low- or high-motivational setting	0	4	0%	0	4	0%	0	4	0%	0	4	0%
2	Karbach & Kray, 2009	Task switching	Single-task training	5	5	100%	1	5	20%	1	1	100%	1	5	20%
3	Zinke et al., 2012[A]	Task switching & task switching + exercise group	Exercise on a stationary bike only & no treatment	1	9	11%	1	9	0%	1	9	0%	1	9	0%
	Totals and Percents for typically developing children who trained on task switching compared with single-task training			5	9	56%	1	9	11%	1	5	20%	1	9	11%
	Totals and Percents for typically developing children who trained on task-switching compared with any active-control condition			6	18	33%	2	18	11%	2	14	14%	2	18	11%
3	Zinke et al., 2012[A]	Task switching & task switching + exercise group	No treatment	1	9	11%	1	9	0%	1	9	0%	1	9	0%
	Totals and Percents for typically developing children who trained on task switching compared with any active-control condition or no treatment			7	27	26%	3	27	11%	3	23	13%	3	27	11%

SCHOOL-AGE CHILDREN (7–15 YEARS OLD) WITH ADHD

#	Study / Group	Condition	n	N	%	n	N	%	n	N	%	n	N	%
4	Kray et al., 2012	Task switching / Single-task training	2	4	50%	2	4	50%	2	4	50%	2	4	50%
	Grand Totals and Percents for school-age children (with or without a clinical diagnosis) who trained on task switching compared with single-task training		7	13	54%	3	13	23%	3	9	33%	3	13	23%
	Grand Totals and Percents for school-age children (with or without a clinical diagnosis) who trained on Task-switching compared with any active control condition		8	22	40%	4	22	20%	4	18	24%	4	22	20%
	Grand Totals and Percents for school age children (with or without a clinical diagnosis) who trained on task switching compared with any active-ontrol condition or no treatment		9	31	32%	5	31	18%	5	27	20%	5	31	18%

ADULTS (18–55 YEARS OLD)

#	Study / Group	Condition	n	N	%	n	N	%	n	N	%	n	N	%
2	Karbach & Kray, 2009	Task switching / Single-task training	5	5	100%	2	5	40%	2	3	67%	2	5	40%
5	Pereg et al., 2013[A,B]	No treatment / Karbach & Kray's (2009) Task-switching training + verbal self-instruction + training variability	2	6	33%	2	6	33%	2	6	33%	2	6	33%
	Totals and Percents for younger adults who trained on task switching compared with single-task training or no treatment		7	11	64%	4	11	36%	4	9	44%	4	11	36%

OLDER ADULTS (OLDER THAN 55 YEARS)

#	Study / Group	Condition	n	N	%	n	N	%	n	N	%	n	N	%
2	Karbach & Kray, 2009	Task switching / Single-task training	5	5	100%	1	5	20%	1	2	50%	1	5	20%

(continued)

Table 8.15. Continued

Study				Significantly Better Improvement			Significantly Better Posttest			Significantly Better Posttest only including measures where this was looked at			Both Significantly Better Change and Posttest		
Study #	Study Name	Condition of Interest	Comparison Condition	# Sign.	# of Measures	% Sign.	# Sign.	# of Measures	% Sign.	# Sign.	# of Measures	% Sign.	# Sign.	# of Measures	% Sign.
	Grand Total and Percents across all task-switching studies that used single-task training as an active-ontrol condition			17	23	74%	6	23	26%	6	14	43%	6	23	26%
	Grand Total and Percents across all task-switching studies that used an active-ontrol condition			18	32	59%	7	32	22%	7	23	29%	7	32	22%
	Grand Total and Percents across all task-switching studies compared to no treatment			3	15	20%	3	15	20%	3	15	20%	3	15	20%
	Grand Total and Percents across all task-switching studies (including no-treatment controls)			21	47	45%	10	47	21%	10	38	26%	10	47	21%

Note. The only far-transfer EF measures in studies of task switching were of reasoning/fluid intelligence (R/FL); therefore, a second table that included near and far transfer excluding R/FL would be identical to this table and so is not included.

[A] The authors of this study did not include a correction for multiple comparisons. It is unclear which results would remain significant had they done that.

[B] The positive improvement and posttest result for the Manual Stroop task in Pereg et al. (2013) is due to the control group's mysteriously getting worse, not because of much improvement by the experimental group.

tasks (both reduced mixing costs and reduced switching costs) in 8– to 10-year-old children and younger and older adults. The training also produced greater improvements on the Stroop task and on Raven's Matrices for children, and on both verbal and spatial memory for children and young adults compared with controls (although it is unclear if posttest performance differences were significant). Therefore, children seemed to show the most benefits and older adults the least.

Study 2: Kray, Karbach, Haenig, and Freitag (2012) found task-switching training improved performance more than single-task training on task switching on other tasks (very near transfer), including switching on the Stroop task, among 8 to 12-year-old boys with ADHD, but benefits did not transfer to WM (re-ordering digits) or to matrix reasoning/fluid intelligence (Raven's Matrices; unlike what Karbach & Kray had found). Of concern is that there were only 10 subjects per group and attrition was high (33%).

Study 3: Dörrenbächer, Müller, Tröger, and Kray (2014) found that task-switching training improved switching costs on other tasks (very near transfer) more than did single-task training (though not mixing costs) for 8– to 11-year-olds trained in a highly motivating context but not in a low motivational context. Benefits did not transfer to WM (as assessed by an AX-CPT, N-back task, Backward Digit Span, and counting span) or to inhibitory control as indexed by the Stroop test (vs. Kray et al., 2012, who found a benefit on Stroop).

Two studies are from other labs. Pereg, Shahar, and Meiran (2013) found that benefits for young adults (mean age = 24 years) from task-switching training (vs. no treatment) on alternative-runs task switching did not transfer to cued task switching (very near transfer) nor to the Stroop or N-back tasks. (On the verbal Stroop task, there was clearly no benefit; on a manual Stroop task, the appearance of a benefit was due to the no-treatment group's mysteriously getting worse.)

Zinke, Einert, Pfenning, and Kliegel (2012) looked at task-switching training (modeled after Karbach and Kray) alone or preceded by cycling on a stationary bike versus just that aerobic exercise alone or no treatment in early adolescents (mean = 12 years; age range = 10–14 years). Results for the two task-switching conditions were similar. They found that task-switching training improved task switching on a nontrained task (mixing costs were more improved and better at posttest than for those who only exercised or had no treatment). However, task-switching training did not improve inhibitory control or WM as indexed by the N-back, Flanker, and Stroop tests.

Computerized Cognitive Training Using Commercial Brain-Training Products (other than Cogmed) and One Noncommercial Product ("BrainGame Brian")

We found two peer-reviewed studies each of Brainware Safari* and three of Lumosity*, and one each of Rise of Nations*, Wii Big Brain Academy*,

Neuroracer, and Posit Science*. The two studies of Brainware Safari, the study of Rise of Nations, and one of the studies of Lumosity included only no-treatment controls. The studies of Brainware Safari were done in children, the two studies of Lumosity were done with young to middle-aged adults, and the studies of Neuroracer, Rise of Nations, Posit Science, and Wii Big Brain Academy were done with older adults.

Unfortunately, most of these studies looked at benefits only immediately after training, training was quite brief, and half had no active control group. Given those caveats, after only 10 to 12 weeks of only 15 to 30 min of practice 3 to 5 times a week, studies of Lumosity and Avtzon's (2012) study of Brainware Safari report some benefits worthy of note, as does Rise of Nations after a similar number of hours of training but over only 4 to 5 weeks. In more rigorous studies than all the others, after only 5 weeks of BrainGame Brian (done 35 to 50 min a day), several benefits were noted, and after only 4 weeks of Neuroracer (done 60 min three times a week), mixed results were found.

Basak et al. (2008) found that playing Rise of Nations (a real-time strategy videogame) for 24 hours (roughly 16 90-min sessions over 4–5 weeks), but not for 11 hours (roughly seven 90-min sessions over 2–3 weeks), improved cognitive flexibility (task switching), WM + inhibitory control (N-back task), and fluid intelligence/reasoning (Raven's Matrices), although not inhibition as assessed by the Stop-Signal task or WM + interference control as assessed by the Operation Span task, compared with no-treatment controls (mean age of 69 years in both groups).

Wii Big Brain Academy is a video game that presents puzzles meant to challenge logic, reasoning, math computations, and memory. Ackerman, Kanfer, and Calderwood (2010) found that 8 weeks of Wii Big Brain Academy (60 min, five times a week) produced no greater benefits to fluid intelligence/reasoning, crystallized intelligence, or speed of processing than reading in adults 50 to 71 years old.

Kesler et al. (2013) found that immediately after 12 weeks of Lumosity training, women with a mean age of 56 years who were breast cancer survivors (18 months post-chemotherapy) performed better and had improved more on the Wisconsin Card Sort and Verbal Fluency (both of which require multiple EF skills) than those in the no-treatment group. In another study with almost 5,000 adults (mean age of 39 years) all of whom wanted Lumosity but some of whom were assigned to do online crossword puzzles, on almost every outcome measure those who trained on Lumosity for 10 weeks improved significantly more than controls—which is hardly surprising with so many subjects (Hardy et al.02, 2015). The effect sizes for Backward Digit Span (WM) and Raven's Matrices (visuospatial reasoning) were tiny, and controls improved more on grammatical reasoning. Slighty larger effect sizes were found for go/no-go, which assesses inhibitory control (0.16, still minimal) and for arithmetic reasoning and a composite of all the outcome measures (~ 0.25). Whether the groups differed in posttest performance on any measure was not reported.

Just as we thought this was going to press (too late to be included in our tables or tabulations), a study of Lumosity with 128 young adults was accepted for publication (Kable et al., 2017). For 10 weeks, 30 min per day, 5 days a week, participants either trained on Lumosity or played video games. On no outcome measure did posttest performance differ significantly among the groups. However, those who trained on Lumosity showed more improvement than those who played video games or no-treatment controls on the N-back task (which requires WM plus inhibitory control) and reduced false-positives on a CPT (which requires sustained attention), although there was no difference in improvement in task switching, decision-making, delay discounting, or on the Stroop or Stop-Signal tests compared to business-as-usual or video-game participants.

Unlike Rise of Nations, Lumosity, and Wii Big Brain Academy, Brainware Safari is intended for children. Avtzon (2012) reports that 9-year-olds with learning disabilities who played Brainware Safari video games for 12 weeks (30 min per day, 5 days a week) improved more and achieved better posttest scores in verbal and visuospatial WM and on a composite index of EFs than did no-treatment controls. Helms and Sawtelle (2007), who did not report pre- or posttest scores (so we cannot tell if improvements simply reflect catching up to controls), found that 11-year-olds who played Brainware Safari showed more improvement on two measures of WM (Backward Digit Span and auditory WM), planning, and concept formation, although no better math or reading fluency or comprehension than their peers who did not play Brainware Safari.

Clearly, what is needed is a higher bar than no-treatment controls, assessment on more diverse skills, and assessments months and years after training (rather than only immediately after) to determine the potential benefits, and limits to benefits, of any of these commercial products. Also, Basak et al. (2008; Rise of Nations) and Hardy et al. (2015; Lumosity) neglected to correct for multiple comparisons in their data analyses.

By contrast, the study by Dovis et al. (2015) of the noncommercial BrainGame Brian (BGB) is one of the most rigorous studies reviewed here. BGB has WM, inhibitory control, and cognitive flexibility training components. One group of 8- to 12-year-old children with ADHD played BGB with the three components; Group 2 got a version with the latter two components, but the WM component did not increase in difficulty; and Group 3 received nonincrementing versions of all three components. Groups 1 and 2 improved more on the Stop-Signal task (a measure of inhibition) than Group 3 and maintained that benefit 3 months later. Group 1 improved more on Corsi Blocks (a measure of visuospatial WM) than Group 3, with that benefit slightly reduced three months later. Neither Group 1 nor Group 2 improved more than Group 3 on the Stroop task (an inhibitory control measure), Digit Recall (a composite WM-STM measure), Trail-Making (cognitive flexibility), Raven's Matrices (reasoning/fluid intelligence), or parent or teacher ratings of ADHD behaviors. No significant differences in benefits were found for Group 1 versus Group 2.

That followed a pilot study by van der Oord et al. (2014), also with medicated ADHD children, but comparing them only to wait-list controls. Parents, who were not blind to group assignment, felt the children who did BGB improved more on EFs in general. This study is not included in our tabulations because the researchers only used questionnaires and/or self-reports.

In the only test of BGB with children on the autism spectrum, one group was trained on just the WM component, another on just the cognitive flexibility component, and a third group on a mock training control condition where difficulty did not increase (de Vries et al., 2014). No greater gains were found for either BGB condition compared with the control condition.

Anguera et al. (2013) evaluated EF benefits for older adults (mean age = 67, range = 60–85 years) of multitask training using their Neuroracer video game. The two tasks in the video game were to drive a car on a winding road and to respond to a sign only when a green circle was also visible. The active control group did each of the tasks one at a time, dividing their time between the two. Thus, since both groups spent the same amount of time, the control group spent half as much time doing either task. Anguera et al. pointed out that "difficulty was maintained using an adaptive staircase algorithm to independently adjust the difficulty of the 'sign' and 'driving' tasks following each 3-min run based on task performance" and that task engagement on both the multi- and single-task versions "was motivated by rewards given only when both component tasks improved beyond 80% on a given run" (p. 98).

Anguera et al. (2013) found that after 4 weeks, those who played the multi-tasking version improved more in (a) RT variability, but not in RT, on the TOVA test (a measure of sustained attention) and (b) delayed recognition with distraction but not delayed recognition attending to distractors (both measures of WM plus selective attention) than those who played the single-task version or no-treatment controls. The group that trained on multitasking did not improve more in visual WM (Change Detection task), visuospatial WM (Filter task), selective attention (Useful Field of View), or on either of their measures of speed of processing. Posttest scores were not provided, so do not know if there was any group difference on those. Testing at 6 months after training is mentioned, but nothing about performance is provided for that time point.

Being able to drive on a winding road while also attending to traffic lights and road signs is an important real-world skill. Anguera et al. (2013) claim impressive gains in the video version of this from playing Neuroracer. An excellent study would be to put a dash cam on the car of each trainee and see if training on the video game transfers to real-world benefits when behind the wheel.

Barnes et al. (2013) trained older adults (mean age = 74 years) on visual and auditory processing tasks emphasizing both speed and accuracy from the Posit Science task battery. (The Posit Science Corporation was started by neuroscientists Merzenich and Mahncke [Mahncke, Bronstone, & Merzenich, 2006].) Barnes et al.'s active control group watched DVDs of educational lectures. Both the experimental and active-control conditions lasted 12 weeks, with three 60-min sessions per week. Half of each group was also assigned to an aerobic exercise program or stretching and toning, but they collapsed across the exercise conditions in reporting EF outcomes. On Verbal Fluency (whether letter or category), Trails B, or the Flanker task, there were no group differences. Only on the Useful Field of View test (on indices of both selective and divided attention) was there more improvement in the Posit Science group than in the control group.

Other Types of Cognitive Training, Both Computerized and Noncomputerized

Twenty-nine (29) studies fall into this "other" category. Over 50% of the studies (17 studies) found at least suggestive evidence of EF benefits (see Table 8.16). The most impressive results were found for noncomputerized complex-span training (Borella et al., 2010; Carretti et al., 2013). Also noteworthy are three other studies: Röthlisberger, Neuenschwander, Cimeli, Michel, and Roebers (2012) trained children on noncomputerized laboratory EF tests and found more improvement and better posttest results on three of their five untrained EF tasks compared to no-treatment controls. The Advanced Cognitive Training for Independent and Vital Elderly (ACTIVE) study trained older adults on reasoning using noncomputerized cognitive training (Ball et al., 2002; Rebok et al., 2014; Willis et al., 2006). The study found that the benefits to reasoning on nontrained measures was still evident even 5 years later, and the effect size 1 year after training was an impressive 0.40. Johnstone et al. (2012) administered computerized training on self-ordered pointing and go/no-go to children. Although children without a clinical diagnosis improved on only two out of five of the EF outcome measures compared to no-treatment controls, and children with ADHD improved more, and performed better at posttest, on four of the five EF outcome measures (80%) compared to no-treatment controls.

Both studies that had people train on things directly related to their real-world activities found at least suggestive evidence of EF benefits (the ACTIVE study and Wang, Chang, & Su, 2011).

Table 8.16. Percentage of Near-Transfer EF Measures on Which Persons Who Received Other Cognitive Training Showed More Improvement and/or Better Posttest Results Across All Studies And Ages, Broken Down by Study

	Study			Significantly Better Improvement			Significantly Better Posttest			Significantly Better Posttest Only Including Measures Where This Was Looked at			Both Significantly Better Change and Posttest		
Study #	Study Name	Condition of Interest	Comparison Condition	# Sign.	# of Measures	% Sign.	# Sign.	# of Measures	% Sign.	# Sign.	# of Measures	% Sign.	# Sign.	# of Measures	% Sign.
				YOUNG CHILDREN (<7 YEARS OLD) WITH NO CLINICAL DIAGNOSIS											
1	Blakey & Carroll, 2015	Training in WM and inhibitory control	Training in perceptual judgments	1	4	25%	1	4	25%	1	4	25%	1	4	25%
2	Kroesbergen et al., 2014	*Domain-general WM training*	Domain-specific WM training	0	2	0%	0	2	0%	0	1	0%	0	2	0%
3	Rueda et al., 2005	Inhibition training: younger children	Watched children's videos	0	2	0%	0	2	0%	0	2	0%	0	2	0%
3	Rueda et al., 2005	Inhibition training: older children	Watched children's videos	0	2	0%	0	2	0%	0	2	0%	0	2	0%
4	Rueda et al., 2012	Inhibition training	Watched children's videos	0	3	0%	0	3	0%	0	3	0%	0	3	0%
5	Wass et al., 2011	Visual-attention training of infants	Viewed TV clips and images	3	4	75%	3	4	75%	3	4	75%	3	4	75%
2	Kroesbergen et al., 2014	*Domain-general WM training*	No treatment	1	2	50%	0	2	0%				0	2	0%

2	Kroesbergen et al., 2014	*Domain-specific WM training*	No treatment	1	2	50%	0	2	0%	0	2	0%
6	Kyttälä et al., 2015[A]	*WM and counting training*	No treatment	0	3	0%	0	3	0%	0	3	0%
7	Röthlisberger et al., 2012: xxx5-year-olds	*Training on EF laboratory tasks (including Stroop, Card Sort, Trail-Making, and Grass-Snow)*	No treatment	3	5	60%	3	5	60%	3	5	60%
7	Röthlisberger et al., 2012: xxx6-year-olds	*Training on EF laboratory tasks (including Stroop, Card Sort, Trail-Making, and Grass-Snow)*	No treatment	1	5	20%	1	5	20%	1	5	20%
SCHOOL-AGE CHILDREN (7–15 YEARS OLD) WITH NO CLINICAL DIAGNOSIS												
8	Mackey et al., 2011	Computerized & noncomputerized reasoning training	Computerized & noncomputerized speed training	1	1	100%	0	1	0%	0	1	0%
9	Johnstone et al., 2012[B]	Computerized training on self-ordered pointing & go/no-go tasks	No treatment	2	5	40%	2	5	40%	2	5	40%

(continued)

Table 8.16. Continued

Study #	Study Name	Condition of Interest	Comparison Condition	Significantly Better Improvement			Significantly Better Posttest			Significantly Better Posttest Only Including Measures Where This Was Looked at			Both Significantly Better Change and Posttest		
				# Sign.	# of Measures	% Sign.	# Sign.	# of Measures	% Sign.	# Sign.	# of Measures	% Sign.	# Sign.	# of Measures	% Sign.
10	Wong et al., 2014[B]	Computerized WM training (Visuospatial & Auditory)	No treatment	3	4	75%	1	4	25%	1	4	25%	1	4	25%
SCHOOL-AGE CHILDREN (7–15 YEARS OLD) WITH ADHD OR A LEARNING DISABILITY															
11	Alloway et al., 2013	High-frequency Jungle Memory™ WM training	Low-frequency Jungle Memory™ WM Training	2	2	100%	0	2	0%	0	2	0%	0	2	0%
12	Tucha et al., 2011[B]	AixTent computerized attention training	Noncomputerized visual-perception training	1	8	13%	1	8	13%	1	8	13%	1	8	13%
11	Alloway et al., 2013	High-frequency Jungle Memory™ WM training	No treatment	2	2	100%	0	2	0%	0	2	0%	0	2	0%
11	Alloway et al., 2013	Low-frequency Jungle Memory™ WM Training	No treatment	0	2	0%	0	2	0%	0	2	0%	0	2	0%
9	Johnstone et al., 2012[B]	Computerized training on self-ordered pointing & go/no-go tasks	No treatment	4	5	80%	4	5	80%	4	5	80%	4	5	80%

13	Semrud-Clikeman et al., 1999	Training on visual and auditory attention tasks	No treatment (ADHD control group)	2	2	100%	2	2	100%	2	2	100%	2	2	100%
13	Semrud-Clikeman et al., 1999	Training on visual and auditory attention tasks	No treatment (typically developing)	2	2	100%	0	2	0%	0	2	0%	0	2	0%
14	Tamm et al., 2013[B]	Pay Attention! noncomputerized intervention	No treatment	4	8	50%	3	12	25%	3	9	33%	0	12	0%

ADULTS (18–55 YEARS OLD)

15	Owen et al., 2010	Web-based reasoning, planning, and problem-solving training	Web-based adaptive training in memory, attention, visuospatial processing, and math calculations	0	2	0%	0	2	0%	0	2	0%	0	2	0%
15	Owen et al., 2010	Web-based reasoning, planning, and problem-solving training	Answering obscure knowledge questions	0	1	0%	0	1	0%	0	1	0%	0	1	0%
15	Owen et al., 2010	Web-based training in memory, attention, visuospatial processing, and math calculations	Answering obscure knowledge questions	0	1	0%	0	1	0%	0	1	0%	0	1	0%

(continued)

Table 8.16. Continued

Study #	Study Name	Condition of Interest	Comparison Condition	Significantly Better Improvement			Significantly Better Posttest			Significantly Better Posttest Only Including Measures Where This Was Looked at			Both Significantly Better Change and Posttest		
				# Sign.	# of Measures	% Sign.	# Sign.	# of Measures	% Sign.	# Sign.	# of Measures	% Sign.	# Sign.	# of Measures	% Sign.
16	Penner et al., 2012[C]	Distributed-intensity BrainStim	High-intensity BrainStim WM training	3	7	43%	0	7	0%	0	7	0%	0	7	0%
17	Dahlin et al., 2008[B]	Letter memory, updating (verbal & nonverbal), and a complex-span task (Keep Track Task)	No treatment	1	4	25%	1	4	25%	1	4	25%	1	4	25%
16	Penner et al., 2012[B,C]	High-intensity BrainStim WM training	No treatment	0	7	0%	0	7	0%	0	7	0%	0	7	0%
16	Penner et al., 2012[B]	Distributed-intensity BrainStim WM training	No treatment	3	7	43%	0	7	0%	0	7	0%	0	7	0%
18	Schmiedek et al., 2010, 2014[B,D]	Nonadaptive tasks of perceptual speed, episodic memory, WM updating, and complex-span tasks	No treatment	0	5	0%									

	Study	Intervention	Control											
19	Borella et al., 2010	*Noncomputerized complex-span task*	Fill-in paper-and-pencil questionnaires	1	1	100%	3	3	100%	3	3	100%	3	3	100%
20	Buschkuehl et al., 2008	WM computerized training	Nonaerobic muscle training on recumbent bicycle	1	2	50%	1	2	50%	1	2	50%	1	2	50%
21	Carretti et al., 2013	*Noncomputerized complex-span task (categorization WM span)*	Fill-in paper-and-pencil questionnaires	1	1	100%	1	1	100%	1	1	100%	1	1	100%
22	Corbett et al., 2015	Web-based Reasoning and Problem-Solving Cognitive Training (ReaCT)	Web-based game where statements needed to be put in correct order	1	1	100%	0	1	0%	0	1	0%	0	1	0%
22	Corbett et al., 2015	Web-based training in memory, attention, visuospatial processing, and math calculations	Web-based game where statements needed to be put in correct order	0	1	0%	0	1	0%	0	1	0%	0	1	0%
23	Ball et al., 2002; Rebok et al., 2014, and Willis et al., 2006[D]	Reasoning and problem-solving noncomputerized training including real-world tasks	No treatment	1	1	100%									

(continued)

Table 8.16. Continued

Study #	Study Name	Condition of Interest	Comparison Condition	Significantly Better Improvement			Significantly Better Posttest			Significantly Better Posttest Only Including Measures Where This Was Looked at			Both Significantly Better Change and Posttest		
				# Sign.	# of Measures	% Sign.	# Sign.	# of Measures	% Sign.	# Sign.	# of Measures	% Sign.	# Sign.	# of Measures	% Sign.
24	Blieszner et al., 1981[E]	*Inductive reasoning training (noncomputerized)*	No treatment				1	1	100%	1	1	100%			
25	Cheng et al., 2012[F]	*Multidomain cognitive training (general EF training)*	No treatment	1	4	25%	1	4	25%	1	4	25%	1	4	25%
25	Cheng et al., 2012[F]	*Single-domain cognitive training (reasoning)*	No treatment	1	4	25%	1	4	25%	1	4	25%	1	4	25%
17	Dahlin, Nyberg, et al., 2008, Dahlin, Stigsdotter-Neely, et al., 2008[B]	Letter memory, updating (verbal & nonverbal), and a complex-span task (Keep Track Task)	No treatment	0	4	0%	0	4	0%	0	4	0%	0	4	0%
26	Plemons et al., 1978	*Visuospatial reasoning training (noncomputerized)*	No treatment	2	3	67%	1	3	33%	1	3	33%	1	3	33%

#	Study	Control condition	n	Total	%	n	Total	%	n	Total	%	n	Total	%
18	Schmiedek et al., 2010, 2014[B,D]	Nonadaptive tasks of perceptual speed, episodic memory, WM updating, and complex-span tasks	No treatment	2	5	40%								
27	Wang et al., 2011	Cooking task computerized training	Participated in other lab studies (no treatment)	1	2	50%	1	2	50%	1	2	50%	1	2	50%
28	Wilkinson & Yang, 2012	Stroop task computerized training	No treatment	0	2	0%	0	2	0%	0	2	0%	0	2	0%
29	Zinke et al., 2014	WM and EF training	No treatment	0	2	0%	0	2	0%	0	2	0%	0	2	0%
	Grand Total and Percents across all studies examining noncomputerized training compared with any active control condition (excluding Blakey & Carrol, 2015, and Mackey et al., 2011) (N = 3)			4	6	67%	4	6	67%	4	5	80%	4	6	67%
	Grand Total and Percents across all studies examining noncomputerized training compared with any active control condition (including Blakey & Carrol, 2015, and Mackey et al., 2011) (N = 5)			6	11	55%	5	11	45%	5	10	50%	5	11	45%
	Grand Total and Percents across all studies examining noncomputerized training compared with no treatment (N = 10)			19	43	44%	13	47	28%	13	40	33%	9	46	20%

(continued)

Table 8.16. Continued

Study #	Study Name	Condition of Interest	Comparison Condition	Significantly Better Improvement			Significantly Better Posttest			Significantly Better Posttest Only Including Measures Where This Was Looked at			Both Significantly Better Change and Posttest		
				# Sign.	# of Measures	% Sign.	# Sign.	# of Measures	% Sign.	# Sign.	# of Measures	% Sign.	# Sign.	# of Measures	% Sign.
		Grand Total and Percents across all studies examining noncomputerized training (excluding Blakey & Carrol, 2015, and Mackey et al., 2011) compared with any active-control condition or no treatment ($N = 13$)		23	49	47%	17	53	32%	17	45	38%	13	52	25%
		Grand Total and Percents across all studies examining noncomputerized training (including Blakey & Carrol, 2015, and Mackey et al., 2011) compared with any active-control condition or no treatment ($N = 15$)		25	54	46%	18	58	31%	18	50	36%	14	57	25%
		Grand Total and Percents across all studies examining computerized training compared with any active-control condition ($N = 12$)		14	39	36%	7	43	16%	7	41	17%	7	43	16%
		Grand Total and Percents across all studies examining computerized training compared with no treatment ($N = 7$)		17	52	33%	8	42	19%	8	42	19%	8	42	19%
		Grand Total and Percents across all studies examining computerized training compared with any active-control condition or no treatment ($N = 19$)		31	91	34%	15	85	18%	15	83	18%	15	85	18%
		Grand Total and Percents across all studies for which participants trained on activities related to real-world activities compared with no treatment ($N = 2$)		2	3	67%	1	2	50%	1	2	50%	1	2	50%

Grand Total and Percents across all studies examining reasoning training compared with any active-control condition (N=3)	2	5	40%	0	5	0%	0	4	0%	0	5	0%
Grand Total and Percents across all studies examining reasoning training compared with no treatment (N=4)	4	8	50%	3	8	38%	3	8	38%	2	7	29%
Grand Total and Percents across all studies examining reasoning training compared with any active-control condition or no treatment (N=7)	6	13	46%	3	13	23%	3	12	25%	2	12	17%
Grand Total and Percents across all studies examining inhibitory control training compared with any active-control condition (N=2)	0	7	0%	0	7	0%	0	7	0%	0	7	0%
Grand Total and Percents across all studies examining inhibitory control training compared with any active-control condition or no treatment (N=3)	0	9	0%	0	9	0%	0	9	0%	0	9	0%
Grand Total and Percents across all studies examining attention training compared with any active-control condition (N=2)	4	9	46%	4	12	33%	4	12	33%	4	12	33%
Grand Total and Percents across all studies examining attention training compared with no treatment (N=2)	8	12	67%	5	16	31%	5	13	38%	2	16	13%
Grand Total and Percents across all studies examining attention training compared with any active-control condition or no treatment (N=4)	12	21	58%	9	28	32%	9	25	36%	6	28	21%

(continued)

Table 8.16. Continued

Study #	Study Name	Condition of Interest	Comparison Condition	Significantly Better Improvement			Significantly Better Posttest			Significantly Better Posttest Only Including Measures Where This Was Looked at			Both Significantly Better Change and Posttest		
				# Sign.	# of Measures	% Sign.	# Sign.	# of Measures	% Sign.	# Sign.	# of Measures	% Sign.	# Sign.	# of Measures	% Sign.
	Grand Total and Percents across all studies examining WM training compared with any active-control condition ($N = 6$)			10	17	59%	5	17	29%	5	16	31%	5	17	29%
	Grand Total and Percents across all studies examining WM training compared with no treatment ($N = 9$)			13	47	28%	2	37	5%	2	33	6%	2	37	5%
	Grand Total and Percents across all studies examining WM training compared with any active-control condition or no treatment ($N = 15$)			23	64	36%	7	54	13%	7	49	14%	7	54	13%
	Grand Total and Percents across all studies examining training on multiple EFs compared with any active-control condition ($N = 4$)			2	8	25%	2	8	25%	2	7	29%	2	8	25%
	Grand Total and Percents across all studies examining training on multiple Efs compared with no treatment ($N = 4$)			11	26	42%	11	26	42%	11	26	42%	11	26	42%
	Grand Total and Percents across all studies examining training on multiple Efs compared with any active-control condition or no treatment ($N = 8$)			13	34	38%	13	34	38%	13	33	39%	13	34	38%

Grand Total and Percents across all studies compared with any active-control condition (N = 15)	18	45	40%	11	49	22%	11	46	24%	11	49	22%
Grand Total and Percents across all studies compared with no treatment (N = 17)	36	95	38%	21	89	24%	21	82	26%	17	88	19%
Grand Total and Percents across all studies compared with any active-control or no treatment condition (N = 32)	54	140	39%	32	138	23%	32	128	25%	28	137	20%

Note. For the condition of interest column: *Italic* font indicates a study that used noncomputerized cognitive training. Regular font indicates computerized cognitive training. An underline indicates training on "real-world" tasks. Red ink indicates reasoning training. Blue ink indicates inhibitory control training. Green ink indicates attention training. Brown ink indicates WM training. Violet ink indicates training on multiple EFs.

Borella et al. (2010) and Carretti et al. (2013) are included in this table because they used noncomputerized training. Their results are also presented in Tables 8.13 and 8.14, along with other studies of complex-span training.

[A] The WM training by Kyttälä, Kanerva, and Kroesbergen (2015) did not increase in difficulty (it was nonadaptive).

[B] The authors of this study did not include a correction for multiple comparisons. It is unclear which results would remain significant had they done that.

[C] The authors call this an experimental group, but, for the purpose of this table, we are considering it an active control.

[D] This study did not test the difference between posttest scores.

[E] The difference in rate of improvement between groups was not tested.

[F] Cheng et al. (2012) noted that when they combined the two cognitive training groups, the results for two of the three measures (Stroop and Trails B) are true only for the roughly 55% who attended ≥ 80% of the training sessions.

Of the 12 studies of noncomputerized training, an impressive 75% found at least suggestive evidence of EF benefits,[8] as did one study that used both computerized and noncomputerized training (Mackey et al., 2011). The other study that used both types of training (Blakey & Carroll, 2015) found little evidence of improvement and no evidence of better performance at posttest compared to the active control group.

For studies of training on reasoning or on multiple EFs, stronger evidence of EF benefits was found if an intervention was compared to a no-treatment group than to an active control group. However, for studies of miscellaneous training on WM or studies using noncomputerized training, stronger evidence of EF benefits was actually found when an intervention was compared to an active control group than when it was compared to no treatment.

The most disappointing results were found by Owen et al. (2010). In a study with over 10,000 participants (> 4,000 in each of two experimental groups and > 2,700 in the active control group; mean age = 40 years, range = 18–60 years), they found no better results from 6 weeks of online training in reasoning, planning, and problem-solving or online training in attention, memory, math, and visuospatial processing than from 6 weeks spent finding answers online to obscure knowledge questions. Their outcome measures were a grammatical reasoning test, a visuospatial WM test where participants had to remember which boxes had already been searched, and two non-EF measures (digit span and paired associate learning).

Table 8.17, which presents the percentage across all EF measures (except reasoning/fluid intelligence) on which persons who received other cognitive training showed more improvement and/or better posttest results than comparison groups across all studies and ages, broken down by study, appears online.

Five years later, Corbett et al. (2015) conducted a similar study with better outcomes. They enrolled > 6,500 participants (> 2,400 in each of two experimental groups and > 1,700 in the active control group; mean age = 59 years, age range not given). They found that those who trained on reasoning and problem-solving online and those who trained on attention, memory, math, and visuospatial processing online showed more improvement on a grammatical reasoning test, a verbal recognition memory test, and in self-reported daily living activities than active controls who worked on placing a series of statements in the correct order. Those who trained on reasoning also improved more on the visuospatial WM test where participants had to remember which boxes had already been searched and on paired associate learning than

---

[8] Those nine studies are: Ball et al. (2002), Rebok et al. (2014), and Willis et al. (2006)—all on the ACTIVE study; Blieszner, Willis, & Baltes (1981), Borella et al. (2010), Carretti et al. (2013), Kroesbergen, van't Noordende, & Kolkman (2014), Plemons et al. (1978), Röthlisberger et al. (2012), Semrud-Clikeman et al. (1999), and Tamm, Epstein, Peugh, Nakonezny, & Hughes (2013).

Table 8.17. Percentage of All EF Measures (Except Reasoning/Fluid Intelligence) on Which Persons Who Received Other Cognitive Training Showed More Improvement and/or Better Posttest Results Across All Studies and Ages, Broken Down by Study

Study #	Study Name	Condition of Interest	Comparison Condition	Significantly Better Improvement			Significantly Better Posttest			Significantly Better Posttest Only Including Measures Where This Was Looked at			Both Significantly Better Change and Posttest		
				# Sign.	# of Measures	% Sign.	# Sign.	# of Measures	% Sign.	# Sign.	# of Measures	% Sign.	# Sign.	# of Measures	% Sign.
			YOUNG CHILDREN (<7 YEARS OLD) WITH NO CLINICAL DIAGNOSIS												
1	Blakey & Carroll, 2015	Training in WM and inhibitory control	Training in perceptual judgments	1	4	25%	1	4	25%	1	4	25%	1	4	25%
2	Kroesbergen et al., 2014	Domain-general WM training	Domain-specific WM training	0	2	0%	0	2	0%	0	1	0%	0	2	0%
3	Rueda et al., 2005	Inhibition training: younger children	Watched children's videos	0	2	0%	0	2	0%	0	2	0%	0	2	0%
3	Rueda et al., 2005	Inhibition training: older children	Watched children's videos	0	2	0%	0	2	0%	0	2	0%	0	2	0%
4	Rueda et al., 2012	Inhibition training	Watched children's videos	0	3	0%	0	3	0%	0	3	0%	0	3	0%
5	Wass et al., 2011	Visual-attention training of infants	Viewed TV clips and images	3	4	75%	3	4	75%	3	4	75%	3	4	75%
2	Kroesbergen et al., 2014	Domain-general WM training	No treatment	1	2	50%	0	2	0%				0	2	0%
2	Kroesbergen et al., 2014	Domain-specific WM training	No treatment	1	2	50%	0	2	0%				0	2	0%
6	Kyttälä et al., 2015[A]	WM and counting training	No treatment	0	4	0%	0	4	0%	0	4	0%	0	4	0%

Table 8.17. Continued

Study #	Study Name	Condition of Interest	Comparison Condition	Significantly Better Improvement			Significantly Better Posttest			Significantly Better Posttest Only Including Measures Where This Was Looked at			Both Significantly Better Change and Posttest		
				# Sign.	# of Measures	% Sign.	# Sign.	# of Measures	% Sign.	# Sign.	# of Measures	% Sign.	# Sign.	# of Measures	% Sign.
7	Röthlisberger et al., 2012: 5-year-olds	*Training on EF laboratory tasks (including Stroop, Card Sort, Trail-Making, and Grass-Snow)*	No treatment	3	5	60%	3	5	60%	3	5	60%	3	5	60%
7	Röthlisberger et al., 2012: 6-year-olds	*Training on EF laboratory tasks (including Stroop, Card Sort, Trail-Making, and Grass-Snow)*	No treatment	1	5	20%	1	5	20%	1	5	20%	1	5	20%
			SCHOOL-AGE CHILDREN (7–15 YEARS OLD) WITH NO CLINICAL DIAGNOSIS												
8	Mackey et al., 2011	Computerized & noncomputerized reasoning training	Computerized & noncomputerized speed training	2	4	50%	0	4	0%	0	4	0%	0	4	0%
9	Johnstone et al., 2012[B]	Computerized training on self-ordered pointing & go/no-go tasks	No treatment	2	5	40%	2	5	40%	2	5	40%	2	5	40%
10	Wong et al., 2014[B]	Computerized WM training (Visuospatial & Auditory)	No treatment	3	11	27%	1	11	9%	1	11	9%	1	11	9%
			SCHOOL-AGE CHILDREN (7–15 YEARS OLD) WITH ADHD OR A LEARNING DISABILITY												
11	Alloway et al., 2013	High-frequency Jungle Memory™ WM training	Low-frequency Jungle Memory™ WM training	2	2	100%	0	2	0%	0	2	0%	0	2	0%

	Intervention	Control													
12	AixTent computerized attention training	Noncomputerized visual-perception training	1	8	13%	1	8	13%	1	8	13%	1	8	13%
11	High-frequency Jungle Memory™ WM training	No treatment	2	2	100%	0	2	0%	0	2	0%	0	2	0%
11	Low-frequency Jungle Memory™ WM training	No treatment	0	2	0%	0	2	0%	0	2	0%	0	2	0%
9	Computerized training on self-ordered pointing and go/no-go tasks	No treatment	4	5	80%	4	5	80%	4	5	80%	4	5	80%
13	*Training on visual and auditory attention tasks*	No treatment (ADHD control group)	2	2	100%	2	2	100%	2	2	100%	2	2	100%
13	*Training on visual and auditory attention tasks*	No treatment (typically developing)	2	2	100%	0	2	0%	0	2	0%	0	2	0%
14	*Pay Attention! noncomputerized intervention*	No treatment	12	27	44%	3	31	10%	3	19	16%	0	15	0%

ADULTS (18–55 YEARS OLD)

	Intervention	Control													
15	Web-based reasoning, planning, and problem-solving training	Web-based adaptive training in memory, attention, visuospatial processing, and math calculations	0	2	0%	0	2	0%	0	2	0%	0	2	0%
15	Web-based reasoning, planning, and problem-solving training	Answering obscure knowledge questions	0	2	0%	0	2	0%	0	2	0%	0	2	0%

(*continued*)

Table 8.17. Continued

Study #	Study Name	Condition of Interest	Comparison Condition	Significantly Better Improvement			Significantly Better Posttest			Significantly Better Posttest Only Including Measures Where This Was Looked at			Both Significantly Better Change and Posttest		
				# Sign.	# of Measures	% Sign.	# Sign.	# of Measures	% Sign.	# Sign.	# of Measures	% Sign.	# Sign.	# of Measures	% Sign.
15	Owen et al., 2010	Web-based training in memory, attention, visuospatial processing, and math calculations	Answering obscure knowledge questions	0	1	0%	0	1	0%	0	1	0%	0	1	0%
16	Penner et al., 2012[C]	Distributed-intensity BrainStim	High-intensity BrainStim WM training	3	8	38%	0	8	0%	0	8	0%	0	7	0%
17	Dahlin et al., 2008[B]	Letter memory, updating (verbal & nonverbal), and a complex-span task (Keep Track Task)	No treatment	1	5	20%	1	5	20%	1	5	20%	1	5	20%
16	Penner et al., 2012[B,C]	High-intensity BrainStim WM training	No treatment	0	8	0%	0	8	0%	0	8	0%	0	8	0%
16	Penner et al., 2012[B]	Distributed-intensity BrainStim WM training	No treatment	3	8	38%	0	8	0%	0	8	0%	0	8	0%
18	Schmiedek et al., 2010, 2014[B,D]	Nonadaptive tasks of perceptual speed, episodic memory, WM updating, and complex-span tasks	No treatment	0	5	0%									

#	Intervention	Control	OLDER ADULTS (OLDER THAN 55 YEARS)											
19	*Noncomputerized complex-span task*	Fill-in paper-and-pencil questionnaires	3	3	100%	3	3	100%	3	3	100%	3	3	100%
20	Working memory computerized training	Nonaerobic muscle training on recumbent bicycle	1	2	50%	1	2	50%	1	2	50%	1	2	50%
21	*Noncomputerized complex-span task (categorization WM span)*	Fill-in paper-and-pencil questionnaires	1	1	100%	1	1	100%	1	1	100%	1	1	100%
22	Web-based reasoning and problem-solving cognitive training (ReaCT)	Web-based game where statements needed to be put in correct order	2	2	100%	0	2	0%				0	2	0%
22	Web-based training in memory, attention, visuospatial processing, and math calculations	Web-based game where statements needed to be put in correct order	0	1	0%	0	1	0%				0	1	0%
23	<u>Reasoning and problem solving noncomputerized training including real-world tasks</u>	No treatment	1	1	100%									
24	*Inductive reasoning training (noncomputerized)*	No treatment							1	1	100%			

(continued)

Table 8.17. Continued

Study #	Study Name	Condition of Interest	Comparison Condition	Significantly Better Improvement			Significantly Better Posttest			Significantly Better Posttest Only Including Measures Where This Was Looked at			Both Significantly Better Change and Posttest		
				# Sign.	# of Measures	% Sign.	# Sign.	# of Measures	% Sign.	# Sign.	# of Measures	% Sign.	# Sign.	# of Measures	% Sign.
25	Cheng et al., 2012[F]	*Multidomain cognitive training (general EF training)*	No treatment	1	4	25%	1	4	25%	1	4	25%	1	4	25%
25	Cheng et al., 2012[F]	*Single-domain cognitive training (reasoning)*	No treatment	1	4	25%	1	4	25%	1	4	25%	1	4	25%
17	Dahlin, Nyberg, et al., 2008, Dahlin, Stigsdotter-Neely, et al. 2008[B]	Letter memory, updating (verbal & nonverbal), and a complex-span task (Keep Track Task)	No treatment	0	5	0%	0	5	0%	0	5	0%	0	5	0%
26	Plemons et al., 1978	*Visuospatial reasoning training (noncomputerized)*	No treatment	2	3	67%	1	3	33%	1	3	33%	1	3	33%
18	Schmiedek et al., 2010, 2014[B,D]	Nonadaptive tasks of perceptual speed, episodic memory, WM updating, and complex-span tasks	No treatment	2	5	40%									
27	Wang et al., 2011	Cooking task computerized training	Participated in other lab studies (no treatment)	1	2	50%	1	2	50%	1	2	50%	1	2	50%
28	Wilkinson & Yang, 2012	Stroop task computerized training	No treatment	0	3	0%	0	3	0%	0	3	0%	0	3	0%

		A			B			C			D		
29	Zinke et al., 2014 *WM and EF training*													No treatment
	Grand Totals and Percents across all studies examining noncomputerized training compared with any active-control condition (excluding Blakey & Carrol, 2015, and Mackey et al., 2011) (*N* = 3)	4	6	67%	4	6	67%	4	5	80%	4	6	67%
	Grand Totals and Percents across all studies examining noncomputerized training compared with any active-control condition (including Blakey & Carrol, 2015, and Mackey et al., 2011) (*N* = 5)	7	14	50%	5	14	36%	5	13	38%	5	14	36%
	Grand Totals and Percents across all studies examining noncomputerized training compared with no treatment (*N* = 10)	27	64	42%	13	68	19%	13	52	25%	9	51	18%
	Grand Totals and Percents across all studies examining noncomputerized training (excluding Blakey & Carrol, 2015, and Mackey et al., 2011) compared with any active-control condition or no treatment (*N* = 13)	31	70	44%	17	74	23%	17	57	30%	13	57	23%
	Grand Totals and Percents across all studies examining noncomputerized training (including Blakey & Carrol, 2015, and Mackey et al., 2011) compared with any active-control condition or no treatment (*N* = 15)	34	78	44%	18	82	22%	18	65	28%	14	65	22%
	Grand Totals and Percents across all studies examining computerized training compared with any active-control condition (*N* = 12)	16	45	36%	7	49	14%	7	46	15%	7	48	15%
	Grand Totals and Percents across all studies examining computerized training compared with no treatment (*N* = 7)	17	64	27%	8	54	15%	8	54	15%	8	54	15%
	Grand Totals and Percents across all studies examining computerized training compared with any active-control condition or no treatment (*N* = 19)	33	109	30%	15	103	15%	15	100	15%	15	102	15%
	Grand Totals and Percents across all studies for which participants trained on activities related to real-world activities compared with no treatment (*N* = 2)	2	3	67%	1	2	50%	1	2	50%	1	2	50%
	Grand Totals and Percents across all studies examining reasoning training compared with any active-control condition (*N* = 3)	4	10	40%	0	10	0%	0	8	0%	0	10	0%

(*continued*)

Table 8.17. Continued

Study #	Study Name	Condition of Interest	Comparison Condition	Significantly Better Improvement			Significantly Better Posttest			Significantly Better Posttest Only Including Measures Where This Was Looked at			Both Significantly Better Change and Posttest		
				# Sign.	# of Measures	% Sign.	# Sign.	# of Measures	% Sign.	# Sign.	# of Measures	% Sign.	# Sign.	# of Measures	% Sign.
			Grand Totals and Percents across all studies examining reasoning training compared with no treatment (N = 4)	4	8	50%	3	8	38%	3	8	38%	2	7	29%
			Grand Totals and Percents across all studies examining reasoning training compared with any active-control condition or no treatment (N = 7)	8	18	44%	3	18	17%	3	16	19%	2	17	12%
			Grand Totals and Percents across all studies examining inhibitory control training compared with any active-control condition (N = 2)	0	7	0%	0	7	0%	0	7	0%	0	7	0%
			Grand Totals and Percents across all studies examining inhibitory control training compared with any active-control condition or no treatment (N = 3)	0	10	0%	0	10	0%	0	10	0%	0	10	0%
			Grand Totals and Percents across all studies examining attention training compared with any active-control condition (N = 2)	4	9	46%	4	12	33%	4	12	33%	4	12	33%
			Grand Totals and Percents across all studies examining attention training compared with no treatment (N = 2)	16	31	52%	5	35	14%	5	23	22%	2	19	11%
			Grand Totals and Percents across all studies examining attention training compared with any active-control condition or no treatment (N = 4)	20	40	50%	9	47	19%	9	35	26%	6	31	19%
			Grand Totals and Percents across all studies examining WM training compared with any active-control condition (N = 6)	10	18	56%	5	18	28%	5	17	29%	5	17	29%
			Grand Totals and Percents across all studies examining WM training compared with no treatment (N = 9)	13	59	22%	2	49	4%	2	45	4%	2	49	4%

	Group 1			Group 2			Group 3			Group 4		
Grand Totals and Percents across all studies examining WM training compared with any active-control condition or no treatment ($N = 15$)	23	77	30%	7	67	10%	7	62	11%	7	66	11%
Grand Totals and Percents across all studies examining training on multiple EFs compared with any active-control condition ($N = 4$)	2	8	25%	2	8	25%	2	7	29%	2	8	25%
Grand Totals and Percents across all studies examining training on multiple EFs compared with no treatment ($N = 4$)	11	27	41%	11	27	41%	11	27	41%	11	27	41%
Grand Totals and Percents across all studies examining training on multiple EFs compared with any active-control condition or no treatment ($N = 8$)	13	35	37%	13	35	37%	13	34	38%	13	35	37%
Grand Totals and Percents across all studies compared with any active-control condition ($N = 15$)	20	51	39%	11	55	20%	11	51	22%	11	54	20%
Grand Totals and Percents across all studies compared with no treatment ($N = 17$)	44	128	34%	21	122	17%	21	106	20%	17	105	16%
Grand Totals and Percents across all studies compared with any active-control or no-treatment condition ($N = 32$)	64	179	36%	32	177	18%	32	157	20%	28	159	18%

Note. Results for reasoning/fluid intelligence (R/FL) are included here only for those studies that specifically trained people on reasoning (Ball et al., 2002, Blieszner et al., 1981, Cheng et al., 2012, Mackey et al., 2011, and Plemons et al., 1978). For studies that did not specifically train people on reasoning, results for R/FL are not included in Table 8.16 (although they are mentioned in the text), but results for all other EF measures are included.

For the condition of interest column: Italic font indicates a study that used noncomputerized cognitive training. Regular font indicates computerized cognitive training. An underline indicates training on real-world tasks. Red ink indicates reasoning training. Blue ink indicates inhibitory control training. Green ink indicates attention training. Brown ink indicates WM training. Violet ink indicates training on multiple EFs.

Borella et al. (2010) and Carretti et al. (2013) are included in this table because they used noncomputerized training. Their results are also presented in Tables 8.13 and 8.14, along with other studies of complex-span training.

[A] The WM training by Kyttälä, Kanerva, and Kroesbergen (2015) did not increase in difficulty (it was nonadaptive).

[B] The authors of this study did not include a correction for multiple comparisons. It is unclear which results would remain significant had they done that.

[C] The authors call this an experimental group, but for the purpose of this table, it is considered an active control.

[D] This study did not test the difference between posttest scores.

[E] The difference in rate of improvement between groups was not tested.

[F] Cheng et al. (2012) noted that when they combined the two cognitive training groups, the results for two of the three measures (Stroop and Trails B) are true only for the roughly 55% who attended ≥ 80% of the training sessions.

did the active control group, although those who trained on memory, attention, etc., did not.

Why were the results so much more positive in the study by Corbett et al. (2015) than in that by Owen et al. (2010)? One reason might be the longer and more intensive training. Although sessions in both studies were only about 10 min, in the Corbett et al.study they extended over 26 weeks, versus over only 6 weeks in the Owen et al. study. Indeed, Corbett et al. found that effects for the training of attention, memory, etc., were dose dependent: Those who completed 4–5 sessions per week (112 sessions total) showed better outcomes than those who completed fewer sessions per week. For the reasoning training (which showed more benefits), dose-response effects were not found. Another possible reason for the more positive results in the Corbett et al. study is that the older adults in that study seemed more highly motivated and exhibited a higher level of engagement than the younger adults in the Owen et al. study. During the first 6 weeks of training in the Corbett et al. study, participants completed over twice as many sessions (an average of 51) as did participants over the 6 weeks of the training in the Owen et al. study (an average of 25).

Three of four studies of attention training (75%) found at least suggestive evidence of benefits to attention (two were noncomputerized training: Semrud-Clikeman et al., 1999; Tamm et al., 2013; and one used computerized training: Wass, Porayska-Pomsta, & Johnson, 2011). The fourth study (which used computerized attention training: Tucha et al., 2011) found benefits on only one out of eight near-transfer measures. Tamm et al. and Tucha et al. had not corrected for multiple comparisons.

Five of the seven studies of reasoning training (71%) found at least suggestive evidence of improved reasoning. Those seven studies included three with noncomputerized training (the ACTIVE study—Ball et al., 2002, Rebok et al., 2014, Willis et al., 2006; Blieszner et al., 1981; Cheng et al., 2011; Plemons et al., 1978), two with computerized training (Corbett et al., 2015; Owen et al., 2010) and one with both computerized and noncomputerized training (Mackey et al., 2011).

The Two Studies That Used Noncomputerized Training of Complex-Span Tasks
The two studies that used noncomputerized training of complex-span tasks were done by many of the same people (Borella et al., 2010; Carretti et al., 2013), the second being essentially a replication and extension of the first. The training used in these studies was exceptionally brief (only three 60-min sessions over 2 weeks). Both were with older adults (65–75 years old). Borella et al. (2010) reported both more improvement and better posttest scores after the three training sessions than for the control group (which filled out questionnaires) on all EF outcome measures used (two that assessed WM—Backward Digit Span and Dot Matrix; one that assessed inhibition—Stroop; and one reasoning/fluid

intelligence—Cattell Culture Fair test). Benefits were still evident 8 months later on the Cattell test, although not on the other measures, compared to the control group. Carretti et al. (2013) found more improvement and better posttest performance on a near-transfer WM test among those who filled out questionnaires than among controls but no benefit for the Cattell test. The near-transfer WM benefit was still evident 6 months later. Such stellar results from so little training (only three 1-hour sessions over 2 weeks) calls out for replication attempts by others. No other complex-span training study has included follow-up testing. Note that these extremely impressive results were found with older adults.

The Nine Studies That Trained People on Miscellaneous WM Tasks
The most heroic and comprehensive training effort was conducted by Schmiedek, Lövdén, & Lindenberger (2010), who trained adults 7 days a week for 14 weeks, for a total of roughly 100 60-min sessions per participant. People were trained on 12 computerized tasks (six speed-of-processing tasks, plus a numerical memory updating task, a complex-span task [alpha span], a spatial N-back task, and one task each for memorizing lists of words, number-word pairs, and object positions in a grid). Schmiedek et al. intentionally varied content and procedures across tasks to emphasize learning cognitive skills rather than low-level strategies, although they did not dynamically increase difficulty.

They achieved impressive effect sizes. The effect sizes of benefits for young adults (20–31 years old) versus no-treatment controls on verbal episodic memory was > 0.50, numerical episodic memory ~ 0.45, N-back > 0.40, visuospatial reasoning > 0.35, and numerical reasoning > 0.30, although the training produced no significant benefits for young adults on complex-span tasks, memory updating, or Raven's Matrices. The effect sizes of benefits for older adults (65–81 years old) was 0.60 on rotation span, > 0.50 on Raven's Matrices, > 0.45 on episodic memory for word pairs, and > 0.40 on animal span, although the training produced no significant benefits for older adults on N-back, reading or counting span, memory updating, verbal, numerical, or visuospatial reasoning, or episodic memory.

Two years later, benefits to episodic memory and reasoning/fluid intelligence, but not WM, were still present for younger adults but no benefits were still evident for older adults (Schmiedek et al., 2014). The ages of 20–31 years are when most young people are living on their own for the first time, attending university, and/or starting careers and/or families. Perhaps younger adults continued to show benefits when older adults did not because their episodic memory, reasoning, and fluid intelligence continued to be challenged more than was true for older adults.

Buschkuehl et al. (2008) trained high-functioning 80-year-olds, many of whom had never used a computer before, on two computerized WM tasks twice a week for 45 min at a time over 12 weeks. The first task required remembering the order in which squares had been selected (the squares differed in color and spatial location). The second task required remembering the order in which stimuli (displayed one at a time) had been displayed, with an interposed task to indicate whether the displayed stimulus was right-side-up or upside-down. Those who went through the training improved more and performed better at posttest on visuospatial memory (Forward and Backward Spatial Span) but not on verbal memory (composite Forward and Backward Digit Span) compared to participants who did nonaerobic muscle training on a recumbent bicycle. Interestingly, the control group got worse from pre- to posttest.

Dahlin, Nyberg, Backman, and Neely (2008) and Dahlin, Stigsdotter-Neely, Larsson, Backman, and Nyberg (2008) trained younger and older adults (mean ages 24 and 68 years, respectively) on the Keep Track complex-span task and on five other tasks that required WM updating over a 5-week period (45-min sessions, three times a week). They found more improvement and better post-test performance by young adults who underwent this training than for no-treatment control subjects on the N-back test, but not on Backward Digit Span, Computation Span, Verbal Fluency, or fluid intelligence/reasoning (Raven's Matrices). The N-back benefit (the one benefit observed right after training) was no longer present 18 month later. Dahlin and colleagues did not find benefits from the training for older adults.

Wong, He, and Chan (2014) trained 6- to 12-year-old children with poor WM and ADHD (mean age = 8 years) on eight computerized WM tasks (five were visuospatial: indicate where objects had been in forward and backward order; three were auditory verbal: say back spoken letters or digits in forward and backward order) in 35- to 40-min sessions, 4 to 5 times a week over 5 weeks.[9] On very-near-transfer tasks (Span Board and Digit Span [forward + backward]), those who trained improved more than no-treatment controls. On the Span Board, trained children achieved higher scores than controls. On Digit Span, they simply caught up to controls. In follow-up testing 5 to 6 weeks later, the performance of trained children had deteriorated on Span Board so that it was no longer better than controls (but it was still better than their own pretest performance). On Digit Span, trained children maintained the gains they had achieved but still were no better than controls. There was no far transfer at posttest on any measure of inhibition or attention.

Kroesbergen et al. (2014) trained 6-year-old children who were poor at math on either verbal and visuospatial WM skills or on WM skills specific to numerical tasks. There were only eight 30-min sessions (two per week for 4 weeks) and the training was not computerized. Both groups improved more on visuospatial

---

[9] In analyzing their data, multiple comparisons were conducted, but no correction for that was made. It is unclear which, if any, results would remain significant had that correction been made.

WM than no-treatment controls but not on verbal WM. The WM benefit was comparable for both training groups. The group that trained on WM skills related to numeracy improved more in numeracy than controls.

Kyttälä et al. (2015) also used noncomputerized training with 6-year-olds. They trained some on counting and some on counting and WM, but without difficulty increasing. Here, too, there were only eight 30-min sessions over 4 weeks. No EF benefits in trained children were found compared to no-treatment controls.

Three studies varied characteristics of the training to see what is most helpful. Two varied frequency. Many studies over decades have documented that distributed or spaced practice usually yields better long-term results than massed practice (e.g., Landauer & Bjork, 1978; Rosenbaum et al., 2001; Shea & Morgan, 1979). Penner et al. (2012) showed this is also true for WM training. They had one group train 4 days a week over 4 weeks (massed practice) and one group train 2 days a week over 8 weeks (spaced practice). Sessions were 45 min long. On backward block span, Backward Digit Span, and the 3-back task, those who did spaced practice improved more than those who did massed practice and more than no-treatment controls. Improvement of the massed-practice and no-treatment groups did not differ. On other measures (Verbal Fluency and easier WM tasks: forward block span and 2-back) there were no group differences.

Alloway, Bibile, and Lau (2013) also varied frequency but kept duration constant so the spaced-practice group received fewer training sessions (one session a week for 8 weeks vs. four sessions a week for 8 weeks of online WM training [Jungle Memory*]). Whereas Penner et al. (2012) studied adults (mean age = 39 years), Alloway et al. studied children (mean age = 10.5 years). Alloway et al. found that the group with more sessions improved more on both WM measures (shape recall and a composite of Backward Digit Span and processing letter recall) compared to the group with less practice or no-treament controls, and those benefits were still evident 8 months later.

Prins, Dovis, Ponsioen, ten Brink, and van der Oord (2011) compared a modified version of Cogmed with elaborate gaming elements to the same modified version of Cogmed but without gaming elements (the way Cogmed is normally administered) for a very short time (once a week for 3 weeks). Their participants were 9½-year-olds with ADHD. Their only transfer measure was Corsi Blocks, on which there was no group difference, although those who trained with gaming elements improved significantly on that while those who trained without those elements did not. Children liked the version with gaming elements more and were more motivated to work at the modified Cogmed when those elements were present. Training sessions were 15 min, with the option of continuing for another 15 min. The group assigned to the version without gaming elements trained only for an additional 2.3 min on average; the group assigned to the version with gaming features trained for an extra 12.4 min on average.

The Four Studies That Trained People on Attention Tasks
Semrud-Clikeman et al. (1999) trained 10-year-old schoolchildren with ADHD of the inattentive type on a visual-attention task and an auditory-attention task in 60-min sessions twice a week for 18 weeks. On near-transfer measures (Brief Test of Attention and d2 Test of Attention), the children improved more than either no-treatment group (typically developing children and children with ADHD of the combined type). The intervention children performed better at posttest on both measures than the untrained ADHD children and, impressively, caught up to typically developing children on both measures.

Tamm et al. (2013) studied Pay Attention! noncomputerized training, which uses auditory stimuli and visual stimuli on cards to train sustained, selective, divided, and alternating attention, with difficulty increasing. They compared 30 min of this twice a week for 8 weeks to no intervention in children 7 to 15 years old (mean age = 9 years) with ADHD. (It appears that all, or most, participants were on medication for ADHD.) Nonblind observers (parents, clinicians, and the children themselves) reported some benefits, but teachers did not agree, and on objective neuropsychological tests the only benefit seen was on planning (which was probably the most difficult measure).

Tucha et al. (2011) studied possible benefits of AixTent computerized training for children 10 to 11½ years old with ADHD on medication compared to non-computerized visual perception training. AixTent aims to train four domains of attention (vigilance and alertness, which do not involve EFs, and selective and divided attention, which do) using computer tasks in everyday or gamelike situations, with difficulty increasing as performance improves. On the Zimmermann and Fimm Test of Attentional Performance (2002), children trained on AixTent improved more than children trained on the Frostig Developmental Program of Visual Perception in commission errors during divided attention, but not on omission errors or speed during divided attention, nor on any of those three measures for selective attention, and not in cognitive flexibility.

Wass et al. (2011) trained 11-month-old infants to sustain their attention (ignoring distractions) and to shift their attention. After just over 2 weeks, the trained infants were better at sustaining, switching, and disengaging attention than were infants who just watched TV clips and still images.

The Three Studies That Trained People on Inhibitory Control
In two independent studies, Rueda and colleagues (2005, 2012) trained 4½- to 6½-year-old children on computer games emphasizing inhibitory control (selective attention and response inhibition) five times a week for 45 min over 2 to 3

weeks. Unfortunately, in both studies they found neither more improvement on any EF measure nor better EF posttest performance compared with peers who simply watched videos.

At the other end of the age spectrum, Wilkinson and Yang (2012) trained older adults (60–84 years old, mean = 71) using the Stroop task in six 30-min sessions spread over 2 weeks. Participants improved on what they practiced but there was no transfer of benefits to untrained tasks, such as go/no-go (inhibitory control), Flanker (selective attention), or tests of task switching or reasoning. The type or presence/ absence of, feedback during training did not affect the outcome.

Two Studies That Trained People on WM and Inhibitory Control
Blakey and Carroll (2015) trained 4-year-olds on a noncomputerized WM task (Six Boxes) and computerized WM and inhibitory control tasks. There were only four 20-min training sessions (one per week over 4 weeks). They found benefits to WM, but not to inhibitory control or cognitive flexibility. On their only WM measure (backward word span), the children who trained on WM and inhibitory control improved more and performed better than their peers who trained on making perceptual judgments, and the benefit was still evident 3 months later. No benefits were found on any of their three measures of inhibitory control and cognitive flexibility. Three months later, reasoning was also tested. Those who had trained on WM and inhibitory control performed better at reasoning 3 months later than those who had trained on perceptual judgments, but there were no pretest scores for this.

Johnstone et al. (2012) devised a computerized training regime for children of 7 to 14 years (mean age = 9.7), almost half of whom had ADHD. Of those with ADHD, 87% were on medication. The children trained 5 days a week for 4 to 5 weeks (15–20 min per day) on two computer games. One was a self-ordered pointing task (which the researchers dubbed the "Feed the Monkey" game), where children needed to find which box contained a hidden object, trying not to repeat a choice. The other was a go/no-go task. Difficulty on each was incremented after error-free performance and reduced when five or more errors were made. One group got this training while their attention was passively monitored using EEG; the presence or absence of EEG monitoring had no effect on the results. The computerized training yielded more improvement and better posttest performance than no treatment on two inhibitory control (selective attention) measures (the Flanker and oddball tasks) for children with ADHD (though these findings might not have survived correction for multiple comparisons), but produced no benefits for typically developing children. The training did not improve performance even for children with ADHD on a go/no-go task (though training had specifically included that kind of task) or on the counting span task (which requires WM and other EFs) compared to no-treatment controls.

The Seven Studies That Trained People on Reasoning
The huge online studies by Owen et al. (2010) and Corbett et al. (2015) with over 10,000 and over 6,000 participants, respectively, were discussed in the begin-ning of this section (just after Table 8.16). The large ACTIVE study, with 2,832 older adults (mean age = 74 years), used noncomputerized cognitive training of problem-solving and reasoning (Group 1), verbal episodic memory (Group 2), and speed of processing (Group 3; Ball et al., 2002; Rebok et al., 2014; Willis et al., 2006). Participants were trained not just on laboratory measures but also using real-world activities (like food preparation and financial management for rea-soning; organizing and recalling shopping list items and remembering details on prescription labels for memory training). Only change scores were com-pared, not levels of posttest performance, and the authors didn't compare the performance of Groups 1, 2, and 3 to one another, but only compared each to no-treatment controls.

Reasoning training improved reasoning more than no treatment, and that remained true at follow-up testing 1, 2, and even 5 years later (that's impressive; indeed, the effect size 1 year later was still 0.40). Benefits were no longer evident 10 years later. No far transfer to reasoning was found for the other two types of training, nor did training on reasoning transfer to improved memory or speed. Those who trained on memory (or speed) improved more on memory (or speed, respectively) at posttest, and performed better than no-treatment controls 1, 2, and 5 years later (even 10 years later for speed, but not memory). The benefits noted here for each group were seen only on laboratory tests fairly similar to the laboratory measures on which participants had been trained, not on real-life measures. Not surprisingly, for each group, by 5 years after training, benefits were much reduced, but booster sessions helped slow the decline for reasoning.

Blieszner et al. (1981) trained older adults (mean age = 70, range = 60–85 years) over a period of 2 weeks (5 days a week, 60 min per day) on inductive reasoning, such as figuring out the next letter or number in a series. Compared to no treatment, training resulted in better posttest performance on one induc-tive reasoning measure but not on another, and degree of improvement was no better than among no-treatment controls because there were sizeable test–retest effects.

Cheng et al. (2012) trained older adults (mean age = 70 years, range = 65–75 years) for 12 weeks (60-min sessions, twice a week) on either multiple EF domains as well as real-world skills or specifically on reasoning. Both groups improved more and performed better at posttest on reasoning than no-treatment controls, although neither training group showed benefits relative to controls on Stroop or Trail-Making. This was true even though only slightly more than 50% completed either training. Those who completed ≥ 80% of either training performed better at posttest on both Stroop and Trails B than those who attended less training. Six months later, those who had trained more were still better on Trails B and now they were also better on reasoning than those who had trained less, although initially even those who had received less training showed a benefit on reasoning. By 1 after training, the differential benefit to reasoning was even greater, while the other relative benefits had disappeared.

Mackey et al. (2011) trained children 7 to 10 years old (mean age of 8½) on either reasoning or speed of processing, in both cases using computerized and noncomputerized games with two sessions a week for 60 min over 8 weeks. Those trained on reasoning improved more on reasoning (the TONI test) than those trained on speed, but posttest scores on reasoning were not significantly different between the two groups. That could be perhaps because training on speed of processing also aided reasoning somewhat, but without no-treatment controls we cannot know if that was the case or if the reasoning training produced no better posttest scores than simply taking the test twice (pre- and posttest).

Plemons et al. (1978) trained older adults (mean age = 70 years) on items designed to closely resemble, but not be identical to, those on the Figural Relations Diagnostic Test and Cattell-Horn measures for Figural Relations. Not surprisingly, people who practiced the training items improved more on both of those tests than no-treatment controls (although their posttest performance was only significantly better than controls for the Figural Relations Diagnostic Test, which the training resembled more closely than the Cattell-Horn measures). There was no generalization to a measure of inductive reasoning or a measure of crystallized intelligence.

Three Studies That Trained People on Multiple EF Skills
None of the studies found clear EF benefits. Wang et al. (2011) gave older adults (mean age = 66 years) only five sessions, one per week, each less than an hour long, of training on a computerized task related to real life—cooking a meal. This required planning, prioritizing, multitasking, and other challenging EF skills. On Letter-Number Sequencing (a difficult WM task), they found more improvement and better posttest performance than older adults who did other computerized training, but on a less difficult test of WM (Backward Digit Span), they found no difference between the groups. We would like to see follow-up of this study with more sessions of the intervention, testing on some of the high-levels skills involved in the training, and testing to see if and how long benefits last.

Röthlisberger et al. (2012) trained 5- and 6-year-olds on a host of EF laboratory tasks (including Stroop, Card Sort, Trail-Making, and Grass-Snow) for 6 weeks (30 min per day, 5 days a week). This noncomputerized training improved performance on a complex-span task (that taxes multiple EFs) in both age groups, and on task switching (switching from classic to reverse Flanker) in 5-year-olds. Both the degree of improvement and posttest scores on these two measures were better than for no-treatment controls. The only measure on which there was no benefit relative to controls was on the classic Flanker test (which assesses selective attention and is easier than the two measures on which group differences were found).

Zinke et al. (2014) trained older adults (mean age = 77 years) in only nine 30-min sessions (three per week for 3 weeks) on the spatial memory sub-test of the Kaufmann Assessment Battery for Children, a verbal WM task where participants heard a series of numbers and were to subtract two from each number and repeat the series of numbers, and the Tower of London. No benefits relative to no-treatment controls were found on the Tower of Hanoi or Corsi Block. Relative to controls, the training group improved more on Raven's Matrices, but there was no group difference in posttest scores. The training group improved more and performed better than controls at posttest on a measure of STM (Forward Letter Span) and that benefit was still evident 9 months later.

Neurofeedback

Neurofeedback uses scalp electrodes to monitor neural activity and gives participants continuous real-time feedback (e.g., by a visual-auditory display on a computer monitor) about whether they are getting closer to the goal for their neural activity. Participants typically have no clue how they are affecting their brain's electrical activity, but, despite that, are usually able to change their brainwave activity in the desired direction.

The neurofeedback training of Wang and Hsieh (2013) increased the amplitude of theta brain waves (4–7 Hz), especially over frontal-midline electrodes in older and younger adults (mean ages 65 and 22, respectively). Becerra et al. (2012) recruited older adults with abnormally high theta power and their neurofeedback training decreased their theta absolute power. Vollebregt, van Dongen-Boomsma, Buitelaar, and Slaats-Willemse (2014) used neurofeedback to normalize brain-wave abnormalities among children with ADHD.

Wang and Hsieh (2013) found that those who got the theta band uptraining for only 15 min, three times a week for only 4 weeks (only 12 sessions), performed better and improved more on the Flanker task (which requires selective attention) and, in the older group, the Sternberg task (which taxes WM and inhibitory control) than those who got sham neurofeedback training. A continuing question with neurofeedback is how long benefits last, since there is evidence they fade quickly. It would be interesting to see if the benefits can be replicated and if there is any evidence that they last even weeks or months.

The theta band downtraining by Becerra et al. (2012) over 10 to 12 weeks (two or three 30-min sessions per week) produced no EF benefits relative to the sham training. Both groups improved. Vollebregt et al.'s (2014) individually customized neurofeedback produced no EF benefits.

Two meta-analyses of the effectiveness of neurofeedback for children with ADHD report that, when the raters and testers were blinded and/or a sham or active control group was used (rather than just no treatment), no benefits of neurofeedback were significant (Cortese et al., 2016; Sonuga-Barke et al., 2013). The meta-analysis by Cortese and colleagues examined all eight RCTs reviewed by Sonuga-Barke et al. plus five more recent ones. Cortese et al. found no significant benefits of neurofeedback on laboratory measures of inhibition or sustained attention or on ratings of ADHD symptoms in general or inattentive symptoms, though nonblinded raters indicated a small but significant benefit for hyperactivity/impulsivity. A third meta-analysis of the effectiveness of neurofeedback for children with ADHD looked only at parent ratings (almost all of which were not blinded; van Doren et al., 2018). It looked at 10 RCTs, and consistent with the findings of the other two meta-analyses, found that nonblinded raters report that neurofeedback improves inattention. Van Doren et al. report that raters indicated that the benefits increased during the follow-up period after the neurofeedback sessions ended and that by the end of the follow-up period, the benefits reported for neurofeedback were equal to those reported for medication.

Physical-Activity Training to Improve EFs

Aerobic exercise with and without cognitive or motor-skill challenges is the most studied physical activity for improving EFs. Aerobic exercise is exercise that requires the consumption of substantially more oxygen than at rest. It involves expending energy that maintains an increased heart rate and increased oxygen uptake. Hence, its most proximal benefit is improved cardiorespiratory fitness. The next most studied physical activities to improve EFs are resistance training (also called strength training or weight training) and yoga.

A very impressive study by Sink et al. (2015), with eight sites and over 1,500 participants, randomly assigned sedentary, cognitively intact older adults 70 to 89 years old to moderate physical activity (walking, resistance training, and flexibility exercises) or health education (educational workshops and upper extremity stretching). After 2 full years of the intervention, they found no group difference on any cognitive measure, including EFs. The disappointing results do not seem to be due to the participants' being too old; indeed the authors found that benefits to EFs from physical activity were greater for those ≥ 80 years old and those with poorer baseline physical performance.

Aerobic Exercise with Fewer Cognitive Demands (Plain Aerobic Exercise)
Sixteen studies looked for EF benefits from aerobic exercise with little or no cognitive component.[10] Of those studies, 12 (75%) were with older adults, two were with children of 8 to 12 years, one was with 6-year-olds, and one was with youths and adults 17 to 47 years old. None was done with children younger than 6 years old. Only one study (Stroth et al., 2010) looked specifically at teens or young adults.

Over half the studies of plain aerobic exercise found no EF benefit at all (56%) and another found almost no EF benefit (Fisher et al. 2011; see Table 8.18). Two studies (13%; Kramer et al., 1999; Moul, Goldman, & Warren, 1995) found strong evidence of EF benefits from plain aerobic training.

In the Kramer et al. (1999) study, adults 60 to 75 years old were assigned to aerobic walking or flexibility training (stretching and toning) for 24 weeks (dose and frequency not given). Although on one measure of response inhibition (incongruent Simon task trials) the aerobic-walking group seems to have just caught up to the flexibility group, on another measure of response inhibition (the Stop-Signal task) and on task switching, those who did aerobic walking improved more and achieved better posttest performance than the flexibility-training group.

Moul et al. (1995) had adults 65 to 72 years old (mean age = 69) do 30 to 40 min of aerobic walking five times a week for 16 weeks and administered subscales of the Ross Information Processing Assessment (RIPA). On one of the two EF subscales (Organization: semantic categorization and word fluency) but not on the Problem-Solving and Abstract Reasoning subscale, those who did aerobic walking improved more and performed better at posttest than those who did resistance training or flexibility exercises. This is considered strong evidence by our criteria because we do not include performance on reasoning/fluid intelligence measures in our calculations (except for studies that specifically targeted that). Although omitting the reasoning subscale leaves only one dependent measure, according to our criteria, finding both more improvement and better posttest performance on that one measure qualifies as strong evidence. (Our criteria are better change and better outcome scores than a control group on > 67% of EF measures, excluding reasoning/fluid intelligence unless that was targeted in training).

Seven studies (44%) found suggestive evidence of EF benefits (i.e., more EF improvement or better EF posttest performance than a control group on > 50% of measures). On only 17% of the 70 measures where studies compared EF improvement across groups was there evidence that plain aerobic exercise

---

[10] Not included in any calculations discussed here are studies of aerobic exercise plus other forms of exercise or aerobic exercise plus other activities that did not involve physical activity and studies that compared aerobic exercise to medication, although they appear in Table 8.18.

Table 8.18. Percentage of EF Measures (Except Reasoning/Fluid Intelligence) on Which Persons Who Did Plain Aerobic Exercise Showed More Improvement and/or Better Posttest Results Than Comparison Groups Across All Studies and Ages, Broken Down by Study

Study #	Study Name	Condition of Interest	Comparison Condition	Significantly Better Improvement			Significantly Better Posttest			Significantly Better Posttest Only Including Measures Where This Was Looked at			Both Significantly Better Change and Posttest		
				# Sign.	# of Measures	% Sign.	# Sign.	# of Measures	% Sign.	# Sign.	# of Measures	% Sign.	# Sign.	# of Measures	% Sign.
				YOUNG CHILDREN (3–6 YEARS OLD)											
1	Fisher et al., 2011[A]	Aerobically intense PE	Standard PE	3	10	30%	2	10	20%	2	10	20%	2	10	20%
				SCHOOL-AGE CHILDREN (7–17 YEARS OLD)											
2	Schmidt et al., 2015	Aerobic exercise: High aerobic & low cognitive condition	Standard PE: low aerobic & low cognitive condition	0	3	0%	0	3	0%	0	3	0%	0	3	0%
3	Tuckman & Hinkle, 1986[B]	Aerobic running	Standard PE				1	2	50%	1	2	50%			
	Totals and Percents for children who did plain aerobic exercise compared with those who did standard PE			3	13	23%	3	15	20%	3	15	20%	2	13	15%
				ADULTS (17–47 YEARS OLD)											
4	Stroth et al., 2010	Running training	No treatment	2	4	50%	0	4	0%	0	4	0%	0	4	0%
				OLDER ADULTS (OLDER THAN 55 YEARS)											
5	Albinet et al., 2010	Aerobic walking, circuit training, and running	Stretching	1	1	100%	0	1	0%	0	1	0%	0	1	0%

(continued)

Table 8.18. Continued

Study #	Study Name	Condition of Interest	Comparison Condition	Significantly Better Improvement			Significantly Better Posttest			Significantly Better Posttest Only Including Measures Where This Was Looked at			Both Significantly Better Change and Posttest		
				# Sign.	# of Measures	% Sign.	# Sign.	# of Measures	% Sign.	# Sign.	# of Measures	% Sign.	# Sign.	# of Measures	% Sign.
6	Blumenthal et al., 1989[A]	Aerobic cycling, walking, and jogging	Yoga & flexibility	0	8	0%	0	8	0%	0	8	0%	0	8	0%
7	Erickson et al., 2011, Leckie et al., 2014, and McAuley et al., 2011[C]	Aerobic walking	Toning & stretching	0	3	0%									
8	Kramer et al., 1999	Aerobic walking	Toning & stretching	3	3	100%	2	3	67%	2	3	67%	2	3	67%
9	Moul et al., 1995	Aerobic walking	Mild range-of-motion & flexibility exercises	1	1	100%	1	1	100%	1	1	100%	1	1	100%
10	Smiley-Oyen et al., 2008	Aerobic exercise using equipment (treadmill, elliptical machines, etc.)	Flexibility exercises & resistance training	0	4	0%	0	4	0%	0	4	0%	0	4	0%
11	Voelcker-Rehage et al., 2011	Aerobic walking	Relaxation & stretching	0	2	0%	0	2	0%	0	2	0%	0	2	0%
	Totals and Percents for older adults who did plain aerobic exercise compared with toning & stretching			5	22	23%	3	19	16%	3	19	16%	3	19	16%
12	Dustman et al., 1984[A]	Aerobic walking	Resistance training	1	2	50%	1	2	50%	1	2	50%	1	2	50%

No.	Study	Control	Intervention												
				6	24	25%	4	21	19%	4	21	19%	4	21	19%
	Totals and Percents for older adults who did plain aerobic exercise compared with any physical-activity active control														
13				9	37	24%	7	36	19%	7	36	19%	6	34	18%
	Grand Totals and Percents for children and older adults who did plain aerobic exercise compared with any physical-activity active control														
	Fabre et al., 2002[D]	Leisure activities: Painting & choral singing	Aerobic training	0	1	0%	0	1	0%	0	1	0%	0	1	0%
14	Legault et al., 2011	Healthy aging education	Aerobic walking & flexibility training	0	6	0%	0	6	0%	0	6	0%	0	6	0%
	Totals and Percents for older adults who did plain aerobic exercise compared with any non-physical-activity active control			0	7	0%	0	7	0%	0	7	0%	0	7	0%
	Grand Totals and Percents for older adults who did plain aerobic exercise compared with any active control			6	31	19%	4	28	14%	4	28	14%	4	28	14%
6	Blumenthal et al., 1989[A]	No treatment	Aerobic cycling, walking, and jogging	0	8	0%	0	8	0%	0	8	0%	0	8	0%
12	Dustman et al., 1984	No treatment	Aerobic walking	1	2	50%	0	2	0%	0	2	0%	0	2	0%
15	Mortimer et al., 2012[AC]	No treatment	Aerobic walking	0	5	0%									
16	Oken et al., 2006[E]	No treatment	Aerobic walking and mild leg stretches	0	3	0%	0	3	0%	0	3	0%	0	3	0%
	Totals and Percents for older adults who did plain aerobic exercise compared with no treatment			1	18	6%	0	13	0%	0	13	0%	0	13	0%

(continued)

Table 8.18. Continued

Study #	Study Name	Condition of Interest	Comparison Condition	Significantly Better Improvement			Significantly Better Posttest			Significantly Better Posttest Only Including Measures Where This Was Looked at			Both Significantly Better Change and Posttest		
				# Sign.	# of Measures	% Sign.	# Sign.	# of Measures	% Sign.	# Sign.	# of Measures	% Sign.	# Sign.	# of Measures	% Sign.
		Totals and Percents for adults of all ages who did plain aerobic exercise compared with no treatment		3	22	14%	0	17	0%	0	17	0%	0	17	0%
		Totals and Percents for older adults who did plain aerobic exercise compared with any active-control condition or no treatment		7	49	14%	4	41	10%	4	41	10%	4	41	10%
		Grand Totals and Percents for children and older adults who did plain aerobic exercise compared with any active-control condition		9	44	20%	7	43	16%	7	43	16%	6	41	15%
		Grand Totals and Percents for children and all adults who did plain aerobic exercise compared with any active-control condition or no treatment		12	66	18%	7	60	12%	7	60	12%	6	58	10%
CHILDREN WHO TRAINED ON ENDURANCE, FLEXIBILITY, RESISTANCE, AND AEROBICS															
17	Gallotta et al., 2015[A,F]	Traditional PE Intervention	Coordinative PE Intervention	1	3	33%	0	3	0%	0	3	0%	0	3	0%
17	Gallotta et al., 2015[A,F]	Traditional PE Intervention	No treatment	3	3	100%	0	3	0%	0	3	0%	0	3	0%
		Totals and Percents for children who trained on endurance, flexibility, resistance, and aerobics compared with any active control or no treatment		4	6	67%	0	6	0%	0	6	0%	0	6	0%
ADULTS 40–72 YEARS OLD WITH CLINICALLY DIAGNOSED MAJOR DEPRESSIVE DISORDER (MDD)															
18	Hoffman et al., 2008[A]	Individualized aerobic training	Placebo pill	1	6	17%	1	6	17%	1	6	17%	1	6	17%

19	Langlois et al., 2013[c]	Aerobic exercise & resistance training using equipment (treadmill, elliptical machines, etc.)	No treatment	2	2	100%							0	1	0%
20	Williamson et al., 2009	Moderate-intensity physical activity: Aerobics, strength, balance, & flexibility	No treatment	0	1	0%	0	1	0%	0	1	0%	0	1	0%
21	Sink et al., 2015[c]	Moderate-intensity physical activity training (brisk walking), strength, flexibility, & balance training	Health education training (unlikely to increase physical activity)	0	3	0%									
20	Williamson et al., 2009	Moderate-intensity physical activity: Aerobics, strength, balance, & flexibility	Health education	0	1	0%	0	1	0%	0	1	0%	0	1	0%
	Totals and Percents for older adults who did plain aerobic activity plus resistance training compared with any active control or no treatment			2	7	29%	0	2	0%	0	2	0%	0	2	0%

(continued)

Table 8.18. Continued

Study #	Study Name	Condition of Interest	Comparison Condition	Significantly Better Improvement			Significantly Better Posttest			Significantly Better Posttest Only Including Measures Where This Was Looked at			Both Significantly Better Change and Posttest		
				# Sign.	# of Measures	% Sign.	# Sign.	# of Measures	% Sign.	# Sign.	# of Measures	% Sign.	# Sign.	# of Measures	% Sign.
ADULTS 40–72 YEARS OLD WITH CLINICALLY DIAGNOSED MAJOR DEPRESSIVE DISORDER (MDD)															
18	Hoffman et al., 2008[A]	Individualized aerobic training	Antidepressant medication: sertraline (Zoloft*)	0	6	0%	0	6	0%	0	6	0%	0	6	0%
22	Khatri et al., 2001	Aerobic walking and jogging	Antidepressant medication: sertraline (Zoloft*)	1	3	33%	0	3	0%				0	3	0%
	Grand Totals and Percents across all plain aerobic exercise studies (including studies where not just aerobic exercise was done) compared with any active control, medication, or no treatment			20	94	21%	8	83	10%	8	80	10%	7	81	9%

Note. Results for reasoning/fluid intelligence (R/FL) are not included in Table 8.18 (although they are mentioned in the text) but results for all other EF measures are included.

[A] The authors of this study did not include a correction for multiple comparisons. It is unclear which of their results would remain significant had they done that.

[B] Tuckman & Hinkle (1986) did not test the difference in rate of improvement between groups.

[C] This study did not test the difference between posttest scores.

[D] Fabre, Chamari, Mucci, Massé-Biron, & Préaut (2002) included only 8 participants per group and looked at outcomes after only 8 weeks.

[E] Oken et al. (2006) allowed people to enter the study—including the no-treatment group—who were doing ≤ 30 minutes of aerobic exercise a day.

[F] Gallotta et al. (2015) randomized by school but appear to have analyzed the data as if they randomized by individual children.

(aerobic exercise without motor skill or explicit EF demands) improved EFs more than any comparison condition. On only 11% of the 64 measures was EF posttest performance after weeks of plain aerobic exercise better than that of any comparison condition (see Table 8.18). Three studies compared plain aerobic exercise to standard physical education (PE; which presumably has some aerobic elements, hence potentially underestimating the benefits of aerobic exercise). Five compared plain aerobic exercise to no treatment (potentially overestimating the benefits of aerobic exercise). It ends up not mattering whether plain aerobic exercise was compared to standard PE, no treatment, or stretching and toning; in all cases, the mean percentage of measures on which the aerobic condition produced more EF improvement than the comparison condition was roughly 20% (23%, 14%, and 23%, respectively; see Table 8.18). The mean percentage of measures on which those who did plain aerobic exercises showed better EFs at posttest than a comparison condition varied from a low of 0% for the no-treatment control condition to 20% when standard PE was the comparison condition (16% when stretching and toning was the control condition). Fabre et al. (2002) included an excellent control condition (active, but not a physical activity; namely, leisure activities like painting and choral singing), but they included only eight participants per group (the minimum required to be included in this review) and looked at outcomes after only 8 weeks. They found no greater EF benefits from aerobic exercise than from the more sedentary leisure activities.

Studies of Plain Aerobic Exercise with ≥ 4 EF Measures That Found No EF Benefit

The most disappointing results come from five studies that looked at multiple EF measures and found no EF benefit at all from a training regimen of less cognitively demanding (i.e., plain) aerobic exercise. Blumenthal et al. (1989) had 60- to 83-year-olds (mean age = 67 years) do aerobic exercise for 16 weeks, in three 60-min sessions per week of which 45 min were aerobic, and found no EF benefits using design fluency, Verbal Fluency, Sternberg, Stroop, Backward Digit Span tasks, Trails B minus Trails A, and measures of selective attention and logical memory compared to active controls (who did yoga and flexibility exercises) or even no-treatment controls.

The same research group also looked at possible benefits of aerobic exercise (same frequency and duration as above) for sedentary, clinically depressed adults (ages 40 to 66, mean age of 52 years) compared with antidepressant medication (sertraline) or placebo pills (Hoffman et al., 2008). They found that those who did aerobic exercise improved no more than placebo controls on any of the same six neuropsychological EF measures used by Blumenthal et al. (1989). This was true even when looking only at the optimal exercisers. Exercisers also improved no more, and performed no better, than the medicated group on any neuropsy-chological measure except the Ruff 2 & 7 test (a measure of selective attention) and Trail-Making (a measure of cognitive flexibility), although the latter difference appears to be due to the medicated group inexplicably getting worse.[11]

The third study in this set of six had older adults free of dementia (mean age = 68, range = 60–79 years) do aerobic walking (Mortimer et al., 2012) for 30 min (during 50-min classes), three times a week for many weeks (40 weeks). Across all EF measures (Stroop, Trails B, Backward Digit Span, abstract verbal reasoning, category fluency, and an attention rating scale), the walking group showed zero benefits compared to no-treatment controls. When Mortimer and colleagues did a median split of the walking group, they found that the fast walkers improved more than the slow walkers on the Stroop task. They do not mention if the fast walkers improved more than the no-treatment group.

The fourth study (Legault et al., 2011) also had older adults (mean age = 76, range = 70–85 years) with normal cognitive functioning (and who had been exercising < 30 min a week) do aerobic walking (or stationary cycling for the few for whom walking was contraindicated). They, too, looked at Stroop and Trails B plus Flanker, task switching, self-ordered pointing, and a composite of all five. Participants did the aerobic exercise for 40 min (within 60-min classes) two times a week (plus 10–15 min each week on their own) for 17 weeks. No benefits were found on any EF measure, or the EF composite, for those who did aerobic exercise or those who did aerobic exercise plus recognition memory training compared to those who attended lectures on healthy aging.

The fifth study had adults 65 to 79 years old (mean age = 70) do 25–30 min of aerobic exercise (within 45-min classes) three times a week for 40 weeks and found no benefit to inhibitory control or cognitive flexibility as assessed by the Simon, Stroop, go/no-go, and Wisconsin Card Sort tests compared to other adults who did flexibility and resistance training exercises (Smiley-Oyen, Lowry, Francois, Kohut, & Ekkekakis, 2008).

Other Studies That Found Disappointing Results for EF Benefits From Aerobic Activities With Minimal Cognitive Demands
Six other studies, with fewer EF measures, also found little or no EF benefit from plain aerobic exercise. Oken et al. (2006) had healthy adults 65 to 85 years old

---

[11] Depression is often associated with poorer EFs. Hence, one might expect that anything that relieves depression (such as exercise or antidepressant medication) would improve EFs. However, depression is most likely to be associated with poorer EFs when the depression is severe (Mandelli, et al. 2006), recurrent (Paelecke-Habermann, Pohl, & Leplow, 2005), or treatment-resistant (Wroolie et al. 2006). In Hoffman et al.'s study, participants' depression tended to be mild to moderate (not severe), nonrecurrent, and responsive to treatment. Note this study and any other that compared an intervention program to medication are not included in any calculations we discuss.

(mean age = 72) do aerobic exercise in 60-min classes once a week and for an average of 56 min every day at home for 26 weeks. The researchers looked at the Stroop task, task switching, divided attention, and Letter-Number Sequencing. On none did the group that exercised improve more or perform better than no-treatment controls. However, people were allowed people to enter the study, including the no-treatment group, who were doing 30 min of aerobic exercise a day, and the authors speculate that there may have been ceiling effects on their cognitive measures.

Voelcker-Rehage, Godde, and Staudinger (2011) found no benefit in either speed or accuracy on the Flanker task for adults 63 to 79 years old (mean age = 70) from 35–50 min of aerobic walking three times a week for a full year (52 weeks) compared to stretching and relaxation exercises.

Another study compared 67-year-olds (age range: 55–80 years) assigned to aerobic walking to those assigned to stretching and toning on global and local switch costs (two EF measures). On neither was there a difference between groups in improvement or posttest performance (Erickson et al., 2011; Leckie et al., 2014; McAuley et al., 2011).

As mentioned above, Fabre et al. (2002) also found no EF benefit, although they included only one EF measure, their intervention was brief, and their number of subjects small. Their subjects were 60 to 76 years old (mean age = 66).

At the other end of the age spectrum, the one study of 6-year-olds (Fisher et al., 2011) included five behavioral measures of EFs plus the Conners Behavioral Rating Scale for Parents but found that 10 weeks of aerobically intense PE compared to standard PE yielded better improvement and better posttest performance on only two of the six EF tests (spatial span WM test in CANTAB*) and one subscale of Conners (inattentiveness). Fisher et al. found no benefit to planning, selective attention (on two different tests), memory of sequential order, or logical reasoning compared to standard PE.

The one study of adolescents and young adults (age range: 17–47; mean age = 23 years; Stroth et al., 2010) included three EF measures (Stroop, N-back, and the Hearts and Flowers task). The authors found no benefit from aerobic walking or running on the Stroop or N-back tasks, but more improvement (though not better posttest scores) than no-treatment controls on incongruent trials of the Hearts and Flowers task (which require inhibitory control).

Studies That Found Suggestive Evidence of EF Benefits From Aerobic Activities With Minimal Cognitive Demands
The studies that found suggestive evidence of EF benefits are Albinet et al. (2010), Dustman et al. (1984), Khatri et al. (2001), and Tuckman and Hinkle (1986). Albinet et al. used a challenging task (the Wisconsin Card Sort [WSCT]) to assess EF outcomes in sedentary seniors (65–78 years old, mean age of 71 years)

after the relatively short period of 12 weeks of 60-min classes three times a week of either plain aerobic exercise or stretching. Those who did aerobic exercise showed a greater decrease in WCST errors than those who stretched (the control group got worse at posttest), although posttest scores did not significantly differ between the groups.

Dustman et al. (1984) assigned adults 55 to 77 years old to 60-min classes of aerobic walking or resistance training that met three times a week for 17 weeks, or to no treatment. On the Stroop test, those who did aerobic walking improved more and performed better than both comparison groups. On digit span (forward and backward combined) there was no difference between those who did aerobic walking and those who did resistance training; aerobic walkers showed a tendency to improve more and perform better on this than those in the no-treatment group, but that was not significant. Fluid intelligence/reasoning (as assessed by the Cattell Culture Fair test) was no better (nor more improved) after 17 weeks of aerobic walking than after resistance training or no treatment. Since three groups were pairwise compared on each of three outcome measures, Dustman et al. should have included a correction for multiple comparisons.

The other study (besides Hoffman et al., 2008) that compared exercise to antidepressant medication was by Khatri et al. (2001). They compared 16 weeks of three weekly 45-min sessions of plain aerobic exercise to antidepressant medication for clinically depressed middle-aged adults (mean age = 57 years). They found that aerobic exercise and medication each succeeded in reducing participants' depression. Exercise improved performance on one of their three EF measures (Stroop) *more* than antidepressant medication did. Trails B performance tended to improve more from exercise than medication, but that was not significant. Exercise did not benefit Backward-Digit-Span performance, while medication did slightly, but that difference, too, was not significant. Unfortunately, a no-treatment group was not included, so we do not know how much performance might have improved simply from practice taking the tests.

In a study by Tuckman and Hinkle (1986), children of 8 to 12 years did 30 min of aerobic running three times a week. After the relatively short period of 12 weeks, no benefit on maze tracing was found, but those who did aerobic exercise had better posttest scores on a measure of cognitive flexibility (the alternative uses task) than those in standard PE. We do not know if the training group started out better because neither pretest nor change scores are given.

Comparing Studies of Plain Aerobic Exercise Where an EF Benefit Was Observed on at Least Half the EF Measures to Studies Where an EF Benefit Was Observed on 30% or Less of the Measures

It is not obvious why some studies found at least a suggestion of benefit to EFs (and two found strong evidence of EF benefits) while other studies found no evidence of this or very little. A slightly larger percentage of the studies with older adults that found at least a suggestion of EF benefits included brisk walking as at least one component of their aerobic exercise program (100%) than studies that found little or no EF benefit (88%). However, half the studies that found an EF benefit on ≥ 50% of measures used fast walking as their sole aerobic activity, but also half the studies that found no EF benefit on any measure used fast walking as the intervention. The meta-analysis by Scherder et al. (2014) found that walking improved the EFs of previously sedentary older persons if they were without cognitive impairment but not if they had cognitive impairment. One recent study reports better psychological and health benefits from physical activity done outside in nature than from the same physical activities done inside (Calogiuri et al., 2015). When aerobic walking aided EFs, was it usually done outside while the other aerobic activities were usually done inside? It would be interesting to follow up on this intriguing finding.

Those who found at least a suggestion of benefits did not study programs that lasted longer. The mean length of the interventions in the five studies to find at least a suggestive benefit to EFs was 16 weeks and the longest length was 24 weeks (see Table 8.7). The mean length of interventions in the 10 studies that found virtually no EF benefit was 27 weeks and the longest length was 52 weeks. Half of the studies that found no EF benefit were longer than any study that found at least a suggestion of EF benefits. Perhaps there's some reason why continuing a plain aerobic-exercise intervention longer is not beneficial for improving EF outcomes.

Those who found at least a suggestion of benefits did not study programs that had longer sessions. The mean duration of sessions in the five studies to find at least a suggestive benefit to EFs was 46 min, and for just the aerobic portion it was 35 min (see Table 8.7). The mean duration of sessions in the 10 studies that found virtually no EF benefit was 57 min and for just the aerobic portion it was 42 min. Perhaps continuing aerobic exercise sessions beyond 45 to 50 min or the aerobic portion of those sessions beyond 30 to 40 min yields no additional benefits and perhaps is counterproductive.

Similarly, Gomes-Osman, Cabral, Morris, McInerney, Cahalin et al. (2018), who recently systematically reviewed RCTs examining the potential benefits of exercise on cognition in older adults (≥ 60 years), also concluded that neither dose, duration, nor frequency of physical activity affected cognitive outcomes.

They also looked at total time exercising. They report that a total of at least 52 hours of physical activity (aerobic, resistance training, or mind-body) was associated with improved speed of processing and EFs, but not working memory, in older adults with and without cognitive impairment. We calculated total time from Table 7 and find that more total time exercising conferred no EF advantage in the studies reviewed here. Indeed, among the studies reviewed here, those finding clear or suggestive evidence of EF benefits had 1/3–1/2 as many total hours as studies that found no or almost no EF benefit. That may be because, unlike Gomes-Osman et al., we included both children and adults (not just older adults), excluded most studies of adults with cognitive impairment, included quasi-experimental designs as well as RCTs, included working memory under EFs, and included only papers published through 2015. Note also that Gomes-Osman et al.'s recommendation of 52 hours of moderate-vigorous exercise is based on total session time; if the portion of the exercise session devoted to aerobic activity is used instead, the number of hours goes down to about 18–25 hours.

There was little difference in the number of participants per condition. Those who found at least a suggestion of EF benefit had a mean of 35 participants per group (range = 10–77). Those who found virtually no evidence of a benefit had a mean of 34 participants per group (range = 8–65; see Table 8.7). *There was little difference in the mean age of older adults in studies that found at least a sugges-tion of benefit to EFs and studies that found little or no evidence of an EF benefit* (67 vs. 70 years, respectively; age ranges were 55–78 and 55–85, respectively; see Table 8.7).

Two thirds (67%) of the studies using standard PE as the control condition found little or no benefit to EFs; 57% of the studies using stretching and toning as the control condition found little or no benefit; 60% of the studies using a no-treatment control group found little or no benefit. As mentioned above, the per-centage of measures on which a greater EF improvement was found from plain aerobic exercise than in a control group was roughly 20%, regardless of whether the control condition was standard PE, stretching and toning, or no treatment. See Table 8.18.

Our earlier comment about monitoring and reporting compliance bears emphasizing here. How often did each participant attend his or her assigned sessions? Only 35% of plain aerobic activity studies reported this. How vigor-ously did each person participate in those sessions? Of course, people who rarely attended or rarely exerted themselves would not be expected to show much benefit. Similarly, if a person in a sedentary control group exercised vigorously during another part of the day, that too would minimize differences between groups. It would also complicate interpretation of results if people in a physical-activity condition exercised outside of (in addition to) that program. Very rarely did a study ask about, control for, or report monitoring participants' activities outside the intervention (this is less important for studies of sedentary adults, perhaps, but it is especially important for studies of children).

It could be that the critical difference between the studies where more or fewer EF benefits were found has to do with variables that few studies have reported, such as whether the group of participants developed significant camaraderie or not, whether the atmosphere created was one that fostered risking making a mistake versus one where participants worried about being embarrassed, characteristics of the activity leader (such as being supportive and not being punitive, having a strong conviction that EF benefits would be seen), or the physical environment in which the activity was done (e.g., outdoors in nature, outside in a city, or indoors). See the section entitled "Our Predictions About How to Most Effectively Improve EFs" for a fuller discussion of this.

Studies With Other Comparison Conditions or Additional Components to the Intervention Besides Plain Aerobic Exercise
Excellent results were achieved by Langlois et al. (2013). They studied adults 61 to 89 years old (mean age of 72 years, half of whom were frail) assigned to either business as usual or 60 min of physical exercise (aerobic exercise for 10–30 min plus resistance training for 10 min) three times a week for 12 weeks. They report only percent change and do not report posttest scores on any measures, so we do not know if there were any group differences in posttest performance. However, on both composite indices, one for inhibitory control (Stroop & Trails B minus Trails A) and one for WM (Backward Digit Span & Letter-Number Sequencing), those in the physical exercise group improved more. There was no difference on their measure of reasoning.

Two studies that looked at EF outcomes from aerobic exercise plus resistance, balance, and flexibility training found no evidence of any EF benefit at all compared to no treatment or health education (Sink et al., 2015; Williamson et al., 2009). See Table 8.18.

Masley et al. (2009) combined 30 to 45 min of aerobic exercise with training in stress management and a dietary intervention (so the benefits of aerobic exercise plus two additional components were investigated, not aerobic exercise alone) for adults 18 to 70 years old (mean age = 47). One group was assigned to aerobic exercise five to six times a week, another group to aerobic exercise three to four times a week, and a third group to no treatment. After 10 weeks, neither intervention group showed any benefit relative to no treatment for inhibitory control (Stroop test). Those who did aerobic exercise more often showed more improvement than controls on both composite EF indices. Those who exercised less often showed more improvement than controls on one of the composite EF indices. On no measure, however, was the posttest performance of either exercise group better than that of controls.

Relating the Results of This Review of EF Benefits From Plain Aerobic Activity to the Findings and Conclusions of Other Reviews
Consistent with the disappointing effects of plain, i.e., less cognitively demanding, aerobic exercise on EFs is the consistent finding that improvement in EFs and improvement in aerobic fitness are uncorrelated (for meta-analyses, see

Etnier et al., 2006; Young et al., 2015; for review, see Kramer & Erickson, 2007; also see Blumenthal et al., 1989; Davis et al., 2011; Smiley-Oyen et al., 2008).

Also consistent with our conclusion of disappointing EF benefits from aerobic exercise, a Cochran Review meta-analysis of 12 RCTs in older, cognitively healthy adults concluded that: "Overall *none* of our analyses showed a cognitive benefit from aerobic exercise even when the intervention was shown to lead to improved cardiorespiratory fitness. . . . Our analyses comparing aerobic exercise to any active intervention showed *no* evidence of benefit from aerobic exercise in any cognitive domain. This was also true of our analyses comparing aerobic exercise to no intervention" (Young et al., 2015, p.1, emphases added).

Older reviews have researched similar conclusions. Van Uffelen et al. (2008) reviewed five RCTs done with cognitively healthy older adults that looked at effects of aerobic exercise on EFs. Only one of those five studies (20%) found any benefit to EFs from aerobic exercise compared with control participants.

Kelly et al. (2014) reviewed 25 RCTs involving healthy older adults with no known cognitive impairment or any significant medical, psychiatric, or neurological problems. They concluded that "There is a lack of consistent evidence to show that aerobic interventions . . . result in improved performance on cognitive tasks for older adults without known cognitive impairment" (p.28). They found that on only 5% of EF measures did studies report significantly more EF improvement from aerobic exercise than from stretching/toning (positive results on only two out of 40 EF measures) or than from no-exercise active control conditions (positive results on only two out of 38 EF measures). Results were little better for aerobic exercise versus no treatment: Kelly et al. report that on only 12% of EF measures (five out of 12) did studies find more EF improvement from aerobic exercise than from no treatment.

Gates et al. (2013) reviewed 14 RCTs involving older adults with mild cognitive impairment. They concluded that only "trivial, nonsignificant effects were found for executive function" (p.1093) in their meta-analysis of aerobic exercise interventions.

Verburgh et al. (2014) conducted a meta-analysis of acute and chronic effects of exercise. Included in their analyses were three RCTs looking at the effects of chronic exercise in children and young adults (6 to 35 years old; Davis et al., 2011; Fisher et al., 2011; Stroth et al., 2010); they found "no significant overall effect of chronic physical exercise ($d = 0.14$, 95% CI −0.04 to 0.32, $p = 0.19$) on executive functions" (p. 973). Fisher et al. and Stroth et al. looked at plain aerobic exercise; Davis et al. looked at enriched.

Smith et al.'s (2010) meta-analysis in older adults found minimal benefits to EFs and none to WM. They included 16 RCTs that looked at the effect of aerobic exercise on EFs in cognitively healthy adults (in most, but not all cases, over 60 years old). Only one of those 16 studies found an effect size that was significant at $p < 0.05$ and that effect was significant at $p = 0.049$. Including the three studies on persons with mild cognitive decline brings the total number of studies to 19, but still only one had an effect size significant at $p < 0.05$. Two of the three studies showing the largest effects were not really studies of the benefits of aerobic exercise: Scherder et al. (2005) looked at slow walking (which is not aerobic) and Masley et al. (2009) looked at the benefits of stress management plus a dietary intervention plus aerobic exercise (which does not permit conclusions about the benefits of aerobic exercise per se). It's unclear what Smith et al.'s mean effect size for aerobic exercise benefits to EFs would have been with those two studies omitted, but it would certainly have been smaller. Twelve RCTs examined by Smith et al. looked at possible benefits of aerobic exercise to WM in adults (one of the 12 was with persons with mild cognitive decline); there were no WM benefits. The mean effect size across studies for the effect of aerobic exercise on WM was $g = 0.03$, ns. We, and most EF researchers, consider WM to be a component of EFs. If the studies Smith et al. included under WM had been combined with the studies they grouped under EFs, the mean effect size for that combined set of EF studies would not have been significant.

Consistent with these reviews are two cohort studies that followed people engaged in different activities. They found no protective role of physical activity in preventing cognitive decline. Wang et al. (2013) followed 1,463 adults in China without cognitive or physical impairment at baseline, age 65 at study entry, for 2.4 years. They found that cognitive activities, such as reading, playing cards, chess, or majiang, were associated with better EFs and less cognitive decline; social activities, such as visiting friends or family, were also associated with less cognitive decline. However, physical activities, whether they were walking or attending group exercise, were not associated at all with any reduced risk of cognitive decline.

Verghese et al. (2003) followed 469 adults in the United States living in the community and dementia-free at baseline, over age 75 at study entry, for 5.1 years. They found that reading, playing board games, playing a musical instrument, and especially social ballroom dance, were associated with a lower risk of dementia. However, physical activities such as walking, biking, swimming, or participating in group exercise, were not associated with any reduced risk of dementia.

One review concluded that aerobic activity does improve EFs of older, sedentary adults (Colcombe & Kramer, 2003). They included five studies of plain aerobic exercise reviewed here (Blumenthal et al. 1989; Dustman et al., 1984; Khatri et al., 2001; Kramer et al., 1999; Moul et al., 1995); we, too, found that four of those five studies found at least suggestive evidence of EF benefit. They also included a study by Williams and Lord (1997; which we discuss below because, in addition to aerobic exercise and resistance training, their exercise program included visuomotor coordination); we concluded that they, too, found suggestive evidence of EF benefits. On the small number of studies reviewed by Colcombe and Kramer (2003), we do not disagree with their conclusions. It is simply that a great many studies have been published since 2003.

Colcombe and Kramer also included two other studies that looked at EF outcomes that we have not included here (Emery, Schein, Hauck, & MacIntyre, 1998; Powell, 1974) because they both included clinical populations and we have included only very few such studies. Emery et al. looked at performance on Trails B and Verbal Fluency after 10 weeks of (Group A) aerobic exercise, education, and stress management, (Group B) just education and stress management, or (Group C) no treatment, among 67-year-olds with chronic obstructive pulmonary disease. They found more improvement in Verbal Fluency among those in Group A than those in no treatment, although there was no group difference in posttest performance or in improvement for Trails B. (Group B differed from neither Group A nor Group C on either measure in either improvement or posttest performance.)

Powell (1974) looked at fluid intelligence/reasoning (as assessed by Raven's Matrices) after 12 weeks of (Group A) what they termed "mild exercise" (brisk walking, calisthenics, and rhythmic movements), (Group B) a "social interaction" control condition (where participants played board games and did arts and crafts and music therapy together with others), and (Group C) business as usual among "institutionalized geriatric mental patients." They found more improvement on Raven's Matrices for Group A than Group C, although there was no significant difference at posttest. (Group B differed from neither Group A nor Group C in either improvement or posttest performance.) These two studies provide some support for an EF benefit from exercising.

In another review, Donnelly et al. (2016) included cross-sectional and longitudinal studies and studies of acute effects from a one-time exposure, which are not relevant to the present discussion. They also reviewed 10 reports of RCTs. All of those studies evaluated enriched aerobic exercise programs and are discussed in that section.

A Conundrum Concerning Aerobic Exercise and EFs

On the one hand, as just discussed, aerobic interventions do not seem to improve EFs. Many studies have found that whether or notEFs improve seems unrelated to whether aerobic fitness improves. On the other hand, people who are more physically active and have better aerobic fitness have been found repeatedly to have better EFs than those who are more sedentary (in children: Fedewa and Ahn, 2011; Gapin & Etnier, 2010; Hillman, Castelli, & Buck 2005; Scudder et al., 2014; Sibley & Etnier, 2003; in older adults: Boucard et al., 2012; Colcombe & Kramer, 2003; Voelcker-Rehage et al., 2011; at all ages: Etnier et al., 2006; Prakash, Voss, Erickson, & Kramer, 2015). Indeed, a computer simulation has estimated that a 5% reduction in physical inactivity among adults 45 or older in Australia would reduce dementia there by 11% (Nepal, Brown, & Ranmuthugala, 2010).

Perhaps people need to do aerobic activity for longer than has been investigated thus far in intervention studies (perhaps years vs. months). Many who are drawn to exercise have been physically active much of their lives, not just for the months or the year of a study.

People who freely choose to do aerobic activities probably enjoy them more than people who are randomly assigned to them. Evidence indicates that any cognitive benefit from physical activity may be proportional to how much the physical activity was enjoyed (Heyman et al., 2012; Raichlen, Foster, Gerdeman, Seillier, & Giuffrida, 2012). There is a biological reason why the ability of an activity to improve EFs may be proportional to how much joy the activity evokes (see the section "Our Predictions About How to Most Effectively Improve EFs" and Figure 8.7 in that section for elaboration of this point).

It may be that many who maintain better fitness do so by participating in physical activities that involve cognitive challenges and complex motor skills (such as ultimate Frisbee, squash, tennis, rock climbing, soccer, beach volleyball, social dance, or martial arts). Indeed, for people who regularly do what we have dubbed plain aerobic exercise, it is often not plain aerobic exercise for them. For committed runners or joggers, for instance, these activities are ripe with cognitive challenges as they strategically plan how, or if, they want to trade off speed and distance, minimize extra steps, etc. These activities can become exercises in mindfulness for them or provide the opportunity for exercising mindfulness. That is unlikely to be true for first-time exercisers assigned to an intervention. Thus, those who maintain a regular running regime by choice may do so more planfully or mindfully than those new to running (assigned to do it in some study). As we discuss under enriched aerobic exercise, however, interventions that have tried to specifically add cognitive and/or motor skill challenges have found results almost as disappointing as have studies of plain aerobic exercise.

For those who regularly engage in physical activities, these activities may be an important part of their social lives and/or an important source of pride and personal identity for them. The cognitive challenges, feelings of belonging, social benefits, feelings of pride and joy, and deep commitment to the activity and fellow teammates or exercise buddies may be critical to whether EFs benefit from these activities.

To the extent that aerobic exercise aids EFs, that might be because aerobic exercise improves mood (Khatri et al., 2001; Lane & Lovejoy, 2001; Williamson, Dewey, & Steinberg, 2001) and/or helps people sleep better (Foti, Eaton, Lowry, & McKnight-Ely, 2011; Loprinzi & Cardinal, 2011), given that EFs tend to be better when someone is happier and better rested (Borges et al., 2013; Hirt, Devers, & McCrea, 2008). As far as we know, those possibilities have not yet been investigated, except for mood in clinically depressed adults (Hoffman et al., 2008: Khatri et al., 2001).

Perhaps the correlation between better physical and cognitive fitness is due to one or more other variables and not to better fitness per se. Perhaps people who are more physically fit tend to eat better, get more sleep, or are healthier in general. Perhaps causality goes in the opposite direction: An individual probably needs good EFs, especially good inhibitory control and discipline, to maintain a regular exercise regimen.

In any case, the evidence so far seems to indicate that it is not aerobic fitness by itself that causes the cognitive benefit. Aerobic exercise interventions almost always improve aerobic fitness but less than half the time improve EFs. It is very possible, and we think probable, that engaging in physical activity does help EFs, but why that happens is not being consistently captured by physical-activity intervention studies. The section "Our Predictions About How to Most Effectively Improve EFs" discusses the possible aspects of physical activities important for improving EFs that have not be addressed in most intervention studies.

Physical Activity With More Cognitive and/or Motor Skill Demands (Enriched Aerobic Exercise)

General Comments

There have been a great many calls to move beyond an almost exclusive focus on plain aerobic exercise and resistance training to physical activities that tax EFs and motor coordination more (Best, 2010; Diamond, 2015; Ericsson, 2017; Ericsson & Karlsson, 2014, Moreau, 2015; Moreau & Conway, 2013; Myer et al., 2015; Pesce, 2012; Pesce, Leone, Motta, Marchetti, & Tomporowski, 2016; Pesce, Masci, et al., 2016; Sibley & Etnier, 2003; Tomporowski, Davis, Miller, & Naglieri, 2008; Tomporowski et al., 2015).

We had predicted that aerobic activity with EF demands would improve EFs more than aerobic activities with few EF demands (Diamond, 2015; Diamond & Ling, 2016; see also Best, 2010; Moreau, 2015; Pesce, 2012; Tomporowski et al., 2015). That prediction appears to have been confirmed. More studies of enriched aerobic exercise have yielded suggestive evidence of EF benefits than studies of plain aerobic exercise (see Figure 8.4a and Tables 8.1 and 8.2). In studies of enriched versus studies of plain aerobic exercise, more EF improvement than the control group was found on about twice as many measures (see Tables 8.1 and 8.2).

Three studies (Chuang, Hung, Huang, Chang, & Hung, 2015; Moreau, Morrison, & Conway, 2015; Schmidt et al., 2015) directly compared EF benefits from more versus less cognitively demanding aerobic activity (enriched vs. plain aerobic activity). Across the three studies, the evidence shows only a slight trend in the predicted direction. On only 57% of the seven measures across the three studies did enriched aerobic exercise improve EFs more than plain aerobic exercise. When Moreau et al. (2015) compared enriched aerobic exercise to WM training, they found comparable improvement and posttest scores; there were no significant differences in EF outcomes. That is quite impressive for enriched aerobic exercise. A study too new to be included in Tables 8.3 or 8.4 found more improvement on EFs from enriched than from plain aerobic exercise (Koutsandréou, Wegner, Niemann, & Budde, 2016).

Only two studies of enriched aerobic exercise (Chang et al., 2014; Williams & Lord, 1997) published by our cutoff date of 2015 found clear evidence (in both cases, more improvement and better posttest performance than the control group on 100% of their EF measures). Both studies, however, included only one EF measure. Williams and Lord also included a measure of reasoning/fluid intelligence (which we are considering far transfer for the WM training studies, so to be fair are not including it in calculations for other types of interventions). On that measure, they found neither more improvement nor better posttest performance among exercisers. These studies are discussed further in the section "Studies of Enriched Aerobic Exercise That Found Encouraging Results." (The study by Chang et al. is discussed under studies with children and the Williams and Lord study is discussed under studies with adults.)

Across all studies of enriched aerobic exercise, more improvement was found compared to control conditions on 36% of the 74 EF measures (twice as good as 18% of 66 EF measures for plain aerobic exercise) and better posttest performance than controls was found on 15% of the 41 EF measures for which data were reported (vs. 12% of the 60 measures for plain aerobic exercise). (See Tables 8.1, 8.18 and 8.19.)

Before proceeding further, we can mention one thing that is clear: Despite its widespread adoption by schools, there are no independent studies that have looked at whether Brain Gym* improves cognition in general or EFs in particular. That is, right now there is no evidence that Brain Gym improves either. An absence of evidence does not mean that Brain Gym might not, in fact, improve cognition, but it does mean that claims that such benefits have been established are untrue.

The percentage of measures on which more EF improvement was found in school-age children from enriched aerobic exercise than from the comparison condition was the same whether the comparison condition was PE or no-treatment (33%; see Table 8.19). For older adults, enriched aerobic exercise produced more improvement than any active control condition on 13% of EF measures. However, when enriched aerobic exercise was compared with no treatment for older adults, more benefit to EFs from enriched aerobic exercise was found on 45% of measures.

Weaker evidence of EF benefits from enriched aerobic activity was generally found for children than adults. For example, among children with no clinical diagnosis, enriched aerobic exercise resulted in more improvement on only 33% of 43 EF measures investigated and better posttest scores on only 8% of 25

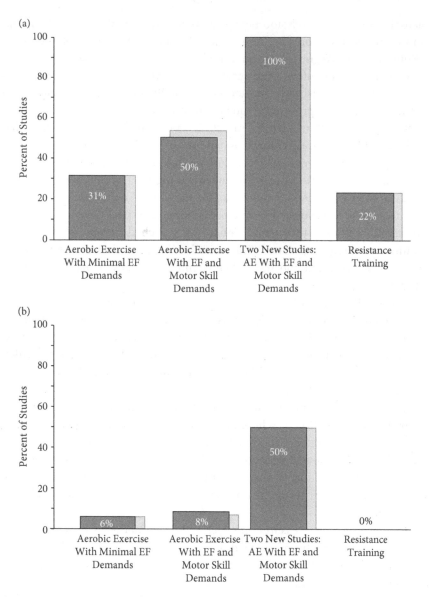

Figure 8.4. Success rates of plain and enriched aerobic exercise and resistance training for improving EFs. AE = aerobic exercise. The darker bars in the foreground present the results omitting studies with possibly spurious positive results. The lighter bars in the background present the results for all studies. Studies omitted for having positive results that might not have held up were those that had not corrected for multiple comparisons or had not conducted data analyses reflecting the level at which they randomized. Figure 8.4a: Percentage of studies finding at least suggestive evidence of physical activity benefiting any EFs, including reasoning (i.e., studies

measures, compared to control conditions. For adults with no clinical diagnosis, however, enriched aerobic exercise resulted in more improvement on 40% of 30 EF measures and better posttest scores on 15% of 13 measures compared to control conditions. (See Table 8.19.)

---

where the experimental group showed either more improvement or better posttest performance than a comparison group on > 50% of the EF measures). Plain aerobic exercise (N = 16): Albinet et al. (2010), Kramer et al. (1999), Moul et al. (1995), Stroth et al. (2010), and Tuckman and Hinkle (1986). Enriched aerobic exercise (N = 18): Chang et al.(2014) Chuang et al. (2015), Kim et al. (2011), Maillot et al. (2012) Moreau et al. (2015), Predovan et al. (2012), Staiano et al. (2012), Williams and Lord (1997). Newer studies (N = 2): Alesi et al. (2016) and Koutsandréou et al. (2016). Resistance training (N = 9): Liu-Ambrose et al. (2008) and Molloy, Beerschoten, Borrie, Crilly, and Cape (1988). For all studies: Plain aerobic exercise (N = 16): Albinet et al. (2010), Dustman et al. (1984), Kramer et al. (1999), Moul et al. (1995), Stroth et al. (2010), Tuckman and Hinkle (1986). Enriched aerobic exercise (N = 19): Chang, Hung, Huang, Hatfield, and Hung (2014), Chuang et al. (2015), Gallotta et al. (2015), Kim et al. (2011), Maillot, Perrot, and Hartley (2012), Moreau et al. (2015), Predovan et al. (2012), Staiano, Abraham, and Calvert (2012), Williams and Lord (1997). Newer studies (N = 2): Alesi et al. (2016) and Koutsandréou et al. (2016). Resistance training (N = 9): Liu-Ambrose et al. (2008) and Molloy, Beerschoten, Borrie, Crilly, and Cape (1988). Figure 8.4b: Percentage of studies finding clear evidence that physical activity benefits any EFs, including reasoning (i.e., studies where there was both more improvement and better posttest performance by the experimental group than by a comparison group on ≥ 67% of the EF measures used). Whenever a study reported > 67% of measures showing positive results for improvement or posttest and did not provide any data on the other, that study is not included in calculations of strong evidence because it is possible the results of that study might have met our criteria had the results not reported been included. For studies with the needed statistical analyses: Plain aerobics (N = 16): Kramer et al. (1999). Enriched aerobic exercise (N = 13): Chang et al. (2014) and Williams and Lord (1997). Newer studies (N = 2): Koutsandréou et al. (2016). Resistance training (N = 8): none. For all studies: Plain aerobics (N = 16): Kramer et al. (1999). Enriched aerobic exercise (N = 14): Chang et al. (2014) and Williams and Lord (1997). Newer studies (N = 2): Koutsandréou et al. (2016). Resistance training (N = 8): none. If the FitKids studies are counted as three separate, independent studies, then for enriched aerobic exercise, 47% of 19 enriched aerobics studies showed suggestive evidence, and 13% of 15 studies showed clear evidence. A caveat about Hillman et al. (2014) and Kamijo et al. (2011): Their suggestive evidence for enriched aerobic exercise was due to greater improvement in the enriched aerobic group on ≥ 50% of their measures, but in all cases for those two studies that greater improvement might be an artifact of the group's starting out with worse EFs, since posttest scores were quite similar across groups.

Table 8.19. Percentage of Measures on Which Persons Who Were Trained on Aerobic Activity With Cognitive and/or Motor Skill Demands Showed More Improvement and/or Better Posttest Results Than a Comparison Group on Measures of Executive Functions, Except Reasoning/Fluid Intelligence, Across All Studies and Ages, Broken Down by Study

Study #	Study Name	Condition of Interest	Comparison Condition	Significantly Better Improvement			Significantly Better Posttest			Significantly Better Posttest Only Including Measures Where This Was Looked at			Both Significantly Better Change and Posttest		
				# Sign.	# of Measures	% Sign.	# Sign.	# of Measures	% Sign.	# Sign.	# of Measures	% Sign.	# Sign.	# of Measures	% Sign.
			SCHOOL-AGE CHILDREN (5–17 YEARS OLD)												
1	Gallotta et al., 2015[A,B]	Coordinative-training PE intervention	Traditional PE intervention	2	3	67%	0	3	0%	0	3	0%	0	3	0%
2	Krafft, Pierce, et al., 2014, Krafft, Schaeffer, et al., 2014[C]	Jump rope, tag basketball, and soccer skills	Sedentary activities: Art and board games	2	10	20%	0	10	0%	0	4	0%	0	10	0%
3	Pesce et al., 2013[D]	PE with more cognitive demand taught by specialists	Standard PE (business as usual)	1	3	33%									
3	Pesce et al., 2013[D]	PE with more cognitive demand taught by specialists	Standard PE taught by specialists	1	3	33%									
4	Schmidt et al., 2015[E]	Team games: High aerobic & high cognitive condition	Aerobic exercise: High aerobic & low cognitive condition	1	3	33%	0	3	0%				0	3	0%

#	Study	Treatment	Control	n	n	%	n	n	%	n	n	%	n	n	%
4	Schmidt et al., 2015	Team games: High aerobic & high cognitive condition	Standard PE: Low aerobic & low cognitive condition (business as usual)	1	3	33%	0	3	0%				0	3	0%
	Totals and Percents for children who did aerobic exercise with cognitive and/or motor skill demands compared with any standard PE			3	9	33%	0	3	0%				0	3	0%
	Totals and Percents for children who did enriched aerobic exercise compared with any physical-activity active-control condition			6	15	40%	0	9	0%	0	3	0%	0	9	0%
	Totals and Percents for children who did enriched aerobic exercise compared with any active-control condition			8	25	32%	0	19	0%	0	7	0%	0	19	0%
	SCHOOL-AGE CHILDREN (7–17 YEARS OLD)														
5	Chaddock-Heyman et al., 2013, Hillman et al., 2014, and Kamijo et al., 2011[F]	FITKids (aerobic exercise & motor skills)	No treatment	3	8	38%	1	8	13%	1	8	13%	1	8	13%
	Chaddock-Heyman et al., 2013[G]	FITKids (aerobic exercise & motor skills)	No treatment	0	3	0%	0	3	0%	0	3	0%	0	3	0%
	Hillman et al., 2014[G]	FITKids (aerobic exercise & motor skills)	No treatment	1	4	25%	0	4	0%	0	4	0%	0	4	0%
	Kamijo et al., 2011[E,G]	FITKids (aerobic exercise & motor skills)	No treatment	2	3	67%	1	3	33%	1	3	33%	1	3	33%

(continued)

Table 8.19. Continued

Study #	Study Name	Condition of Interest	Comparison Condition	Significantly Better Improvement			Significantly Better Posttest			Significantly Better Posttest Only Including Measures Where This Was Looked at			Both Significantly Better Change and Posttest		
				# Sign.	# of Measures	% Sign.	# Sign.	# of Measures	% Sign.	# Sign.	# of Measures	% Sign.	# Sign.	# of Measures	% Sign.
6	Dalziell et al., 2015	Better Movers and Thinkers (BMT): Physical literacy, personal qualities, and thinking skills = performance	No treatment	0	1	0%	0	1	0%	0	1	0%	0	1	0%
7	Davis et al., 2007, 2011[C]	Jump rope, basketball, and soccer skills: 40-min sessions	No treatment	1	3	33%	1	3	33%	1	3	33%	1	3	33%
7	Davis et al., 2007, 2011[C]	Jump rope, basketball, and soccer skills: 20-min sessions	No treatment	0	3	0%	0	3	0%	0	3	0%	0	3	0%
1	Gallotta et al., 2015[A,B]	Coordinative-training PE intervention	No treatment	2	3	67%	0	3	0%	0	3	0%	0	3	0%
8	Reed et al., 2010[I]	Integrated physical activity into the teaching of language arts, math, and social studies	No treatment												

#	Study	Intervention	Comparison	n	Total	%	n	Total	%	n	Total	%	n	Total	%
	Totals and Percents for children who did enriched aerobic exercise compared with no treatment			6	18	33%	2	18	11%	2	18	11%	2	18	11%
	Grand Totals and Percents for typically developing children who did enriched aerobic exercise compared with any active-control condition or no treatment			14	43	33%	2	37	5%	2	25	8%	2	37	5%

CHILDREN (5–10 YEARS OLD) WITH A LEARNING DISORDER OR ADHD

#	Study	Intervention	Comparison	n	Total	%	n	Total	%	n	Total	%	n	Total	%
9	Chang et al., 2014	Aquatic coordination exercise program	No treatment	1	1	100%	1	1	100%	1	1	100%	1	1	100%
	Grand Totals and Percents for children (with or without a clinical diagnosis) who did aerobic exercise with cognitive and/or motor skill demands compared with no treatment			7	19	37%	3	19	16%	3	19	16%	3	19	16%

SCHOOL-AGE CHILDREN (7–17 YEARS OLD) WITH A LEARNING DISORDER OR ADHD

#	Study	Intervention	Comparison	n	Total	%	n	Total	%	n	Total	%	n	Total	%
	Westendorp et al., 2014[j]	Dynamic ball skills	Standard PE	0	2	0%	0	2	0%	0	2	0%	0	2	0%

ADOLESCENTS (15–19 YEARS OLD)

#	Study	Intervention	Comparison	n	Total	%	n	Total	%	n	Total	%	n	Total	%
10	Staiano et al., 2012[C,K]	Competitive exergames	Business as usual	1	1	100%	1	1	100%	1	1	100%			
10	Staiano et al., 2012[C,K]	Cooperative exergames	Business as usual	0	1	0%	0	1	0%	0	1	0%			
	Totals and Percents for adolescents 15–19 years old who did enriched aerobic exercise compared with no treatment			1	2	50%	1	2	50%	1	2	50%			
	Totals and Percents for participants less than 18 years old who did enriched aerobic exercise compared with no treatment			7	21	37%	4	21	19%	4	21	19%	3	19	16%
	Totals and Percents for participants less than 18 years old who did enriched aerobic exercise compared with any active-control condition or no treatment			15	44	34%	4	40	10%	4	28	14%	3	38	8%

(continued)

Table 8.19. Continued

Study #	Study Name	Condition of Interest	Comparison Condition	Significantly Better Improvement			Significantly Better Posttest			Significantly Better Posttest Only Including Measures Where This Was Looked at			Both Significantly Better Change and Posttest		
				# Sign.	# of Measures	% Sign.	# Sign.	# of Measures	% Sign.	# Sign.	# of Measures	% Sign.	# Sign.	# of Measures	% Sign.
				ADULTS (18–52 YEARS OLD)											
11	Moreau et al., 2015[E]	Designed wrestling sport with EF demands	Aerobic exercise	2	2	100%	1	2	50%	1	2	50%	1	2	50%
				OLDER ADULTS (OLDER THAN 55 YEARS)											
12	Chuang et al., 2015	Dance Dance Revolution	Brisk walking	1	2	50%	0	2	0%	0	2	0%	0	2	0%
13	Legault et al., 2011	Aerobic walking and flexibility training plus computerized cognitive training	Healthy aging education	0	6	0%	0	6	0%	0	6	0%	0	6	0%
	Totals and Percents for older adults who did enriched aerobic exercise compared with any active-control condition			1	8	13%	0	8	0%	0	6	0%	0	8	0%
	Grand Totals and Percents for adults (any age) who did enriched aerobic exercise compared with a vigorous physical-activity active-control condition			3	4	75%	1	4	25%	1	2	50%	1	4	25%
	Grand Totals and Percents for adults (any age) who did enriched aerobic exercise compared with any active-control condition			3	10	30%	1	10	10%	1	8	13%	1	10	10%

#	Study	Intervention	Control									
12	Chuang et al., 2015	Dance Dance Revolution	No treatment	1	2	50%	0	2	0%	0	2	0%
14	Kim et al., 2011	Latin dance—the Cha Cha (aerobic, social, partner dance)	Business as usual	1	2	50%	0	2	0%	0	2	0%
15	Klusmann et al., 2010[D]	Aerobic endurance, strength, coordination, flexibility, and balance training	No treatment	1	3	33%						
16	Maillot et al., 2012[D]	Exergames	Business as usual	3	4	75%						
17	Marmeleira et al., 2009	Aerobic activity plus task switching	Business as usual	0	4	0%	0	4	0%	0	4	0%
18	Predovan et al., 2012	Fast walking & aerobic dance exercise	No treatment	2	4	50%	0	4	0%	0	4	0%
19	Williams & Lord, 1997	Aerobic, balance, & coordination exercise	No treatment	1	1	100%	1	1	100%	1	1	100%
	Totals and Percents for older adults who did aerobic exercise with cognitive &/or motor skill demands compared with no treatment			9	20	45%	1	5	20%	1	13	8%
	Grand Totals and Percents for older adults who did aerobic exercise with cognitive &/or motor skill demands compared with any active-control condition or no treatment			10	28	36%	1	11	9%	1	21	5%
	Grand Totals and Percents for adults (any age) who did enriched aerobic exercise compared with any active-control condition or no treatment			12	30	40%	2	13	15%	2	23	9%

(continued)

Table 8.19. Continued

Study #	Study Name	Condition of Interest	Comparison Condition	Significantly Better Improvement			Significantly Better Posttest			Significantly Better Posttest Only Including Measures Where This Was Looked at			Both Significantly Better Change and Posttest		
				# Sign.	# of Measures	% Sign.	# Sign.	# of Measures	% Sign.	# Sign.	# of Measures	% Sign.	# Sign.	# of Measures	% Sign.
		Grand Totals and Percents across all ages comparing aerobic exercise with cognitive and/or motor skills demands to any physical-activity active-control condition		9	19	47%	1	13	8%	1	5	20%	1	13	8%
		Grand Totals and Percents for all ages of aerobic exercise with cognitive and/or motor skills demands compared with any non-physical-activity active-control condition		2	16	13%	0	16	0%	0	10	0%	0	16	0%
		Grand Totals and Percents across all ages of aerobic exercise with cognitive and/or motor skills demands compared with any active-control condition		11	35	31%	1	29	3%	1	15	7%	1	29	3%
		Grand Totals and Percents for all ages who did aerobic exercise with cognitive and/or motor skills demands compared with no treatment		16	39	41%	5	34	15%	5	26	19%	4	32	13%
		Grand Totals and Percents across all ages comparing aerobic exercise with cognitive and/or motor skills demands to any active-control condition or no treatment		27	74	36%	6	63	10%	6	41	15%	5	61	8%
COORDINATION TRAINING WITH MINOR AEROBIC COMPONENT (NOT AEROBIC EXERCISE + COORDINATION TRAINING)															
OLDER ADULTS (MEAN AGE OF 70 YEARS OLD)															
	Voelcker-Rehage et al., 2011	Coordination training	Relaxation & stretching	0	2	0%	0	2	0%	0	2	0%	0	2	0%

SCHOOL-AGE CHILDREN (7–17 YEARS OLD)

Koutsandréou et al., 2016	Coordinative exercise	Aerobic exercise	1	1	100%	0	1	0%	0	1	0%	0	1	0%	0	1	0%
Alesi et al., 2016	Soccer program	No treatment	1	2	50%	1	2	50%	1	2	50%	1	2	50%	1	2	50%
Koutsandréou et al., 2016	Coordinative exercise	No treatment	1	1	100%	1	1	100%	1	1	100%	1	1	100%	1	1	100%
Totals and Percents for newer studies with children who did aerobic exercise with cognitive and/or motor skill demands compared with any active-control or no treatment			3	4	75%	2	4	50%	2	4	50%	2	4	50%	2	4	50%

Note. Results for reasoning/fluid intelligence (R/FL) are not included in Table 8.19 (although they are mentioned in the text) but results for all other EF measures are included.

A Gallotta et al. (2015) randomized by school but appear to have analyzed the data as if they randomized by individual children.

B The authors of this study did not include a correction for multiple comparisons. It is unclear which of their results would remain significant had they done that.

C Participants were overweight.

D This study did not test the difference between posttest scores.

E One might plausibly expect EF benefits from aerobic exercise, so a failure to find a difference here might be due to both interventions' being beneficial, but instead results here were at least as promising as for other comparison conditions.

F In Kamijo et al. (2011) on the 3-Letter condition of the Sternberg test, the exercise group performed more poorly at pretest than controls. At posttest, both groups performed comparably. It's unclear whether the greater improvement by the children who exercised than controls reflects simply normal individual variation in developmental timetables, regression to the mean, or a benefit from the experimental condition. On the super-easy 1-Letter condition of the Sternberg test, the wait-list controls inexplicably got worse, making the difference in posttest scores between those who exercised and controls significant.

G These are the results if the FITKids studies are counted as three separate, independent studies. They are not included in the calculations for this table.

H The authors of this study did not conduct the needed multilevel data analysis. It is unclear how many of their results would remain significant had they done that.

I Reed et al. (2010) only included R/FL measures.

J Westendorp et al. (2014) randomized by class and had only three classes in each group; sample sizes of three gave them extremely low power to find any effect. All children had learning disorders and were attending special needs schools. It is unclear how much of an aerobic component this coordination training had, especially for the children who did not progress to the dynamic ball skills component. It is not included in our calculations.

K The difference in rate of improvement between groups was not tested.

The percentage of studies that involved children was only 18% for studies of plain aerobic exercise but 47% for enriched aerobic exercise. Only four of the nine studies that investigated enriched aerobic exercise in children less than 18 years old included an active control group. Only one study of possible EF benefits of enriched aerobic exercise has been done with adolescents and only one with adults younger than 55 years old. No study of enriched aerobic exercise has been done with children 3 to 6 years old.

Studies of Enriched Aerobic Exercise With Children That Found
Encouraging Results

Chang et al. (2014) found that children with ADHD (5–10 years old, mean age = 8½) who did aquatic exercise (aerobic exercise plus motor coordination) for 40 min in a 90-min class, two times a week, for only 8 weeks improved more and achieved better posttest performance on the one EF test administered (an inhibitory control measure: go/no-go) than did wait-list controls.

Staiano et al. (2012) randomly assigned overweight, low-income African American adolescents (mean age: 16.5 years, range: 15–19 years) to competitive or cooperative exergames (e.g., Wii Sports) or no treatment. Those who did exergames did them for 30 min, an average of once a week (the sessions were offered every day of the school week), for 10 weeks. On their EF measure (a composite of Trail-Making and Design Fluency), those who did competitive exergames improved more than those who did cooperative exergames or no treatment; those in the cooperative condition did not improve significantly more than the no-treatment group. Whether there were group differences in posttest performance was not reported.

Three papers reported on outcomes of the FITKids intervention. Two of the three studies (Hillman et al., 2014; Kamijo et al., 2011) reported somewhat encouraging results. The FITKids afterschool program includes aerobic exercise, resistance, endurance, motor skills training (e.g., practice dribbling), games requiring cooperating with teammates, healthy snacks, and health education. It takes place for 120 min (of which 75–95 min is aerobic exercise) five times a week. Children were evaluated after 36 weeks in the program. On average, children attended 82% of the 180 FITKids sessions. All three FITKids' studies were with children roughly 8 to 9 years old from the same program; the Hillman et al. study included children also reported in the other two studies.

Hillman et al. reported more improvement in accuracy, although not speed, on both selective attention (Flanker) and cognitive flexibility (task switching) in children who had been through FITKids versus wait-list controls. There were no group differences in posttest scores on either measure, however. Wait-list controls had started out with better accuracy and by the end of the intervention the accuracy of both groups on both tasks was comparable. Kamijo et al. reported on another EF outcome measure, the Sternberg task. Controls started out better; the children in FITKids caught up. Most of the catching-up was due to the wait-list controls' inexplicably getting worse on the easiest condition (see Figure 8.3 in Diamond & Ling, 2016). Because the EF benefits reported in the two papers appear to reflect catch-up, we are concerned that they could be due to differences in developmental timetables rather to the FITKids program per se.

Studies of Enriched Aerobic Exercise With Adults That Found Encouraging Results
Williams and Lord (1997) reported both more improvement and better posttest performance among exercisers on their one EF measure. Women, whose mean age was 72, who did aerobic exercise plus exercises for balance and for eye–hand and eye–foot coordination plus resistance training for 42 weeks (twice a week for 50–55 min) improved more and performed better at posttest on a composite WM-STM measure (Forward and Backward Digit Span combined) than their peers in the no-treatment group. Attrition was high, however; 24% for the exercise group and 16% for the control group. No benefit to fluid intelligence/reasoning (Cattell Culture Fair Test) was found.

Chuang et al. (2015) found that women (mean age: 68 years) who did *Dance, Dance Revolution* (aerobic exercise + coordination + cognitive demands) for 30 min three times a week for 12 weeks improved more in speed but not accuracy on their one EF measure (Flanker task) than their peers who did brisk walking for the same amount of time (posttest scores were not given).

Maillot et al. (2012) had sedentary adults (mean age: 74 years; range: 65–78) do exergames (e.g., Wii Sports, Wii Fit) for 60 min two times a week for 12 weeks and compared them to no-treatment sedentary adults. On two measure of inhibitory control (Trails B and Stroop) and two measures of reasoning (matrix reasoning and verbal reasoning), those who did exergames improved more than those who remained sedentary (whether there was a group difference in posttest performance was not reported). The only EF measure on which there was no group difference in improvement was spatial span (a WM measure).

Moreau et al. (2015) randomly assigned adults (mean age: 30 years; range: 18–52 years) to (a) training on a "designed sport" based loosely on freestyle wrestling with added EF, sensory, and motor coordination demands, (b) WM training, or (c) aerobic exercise. All occurred three times a week for 60-min sessions for 8 weeks. Training on the designed sport produced outcomes as good as targeted WM training on WM (Backward Digit Span and Letter-Number Sequencing). As mentioned above, that is impressive. The designed-sport group also improved more on those two tasks than those who just did aerobic exercise. Their posttest scores were better than the aerobic-exercisers for Letter-Number Sequencing but not for Backward Digit Span. It appears that there were ceiling effects for Backward Digit Span, limiting its sensitivity to group differences.

Studies of Enriched Aerobic Exercise That Found Ambiguous Results
Two studies (Kim et al., 2011, and Klusmann et al., 2010) found EF benefits from enriched aerobic exercise on one EF measure but not on another. Kim et al. (2011) found that men and women (mean age: 68 years) who learned the Cha Cha (aerobic exercise + coordination + cognitive demands + social interaction [eye contact and touch] with partners) in 60-min classes (with 45 min of aerobic dance), two times a week for 26 weeks improved more in Verbal Fluency but not on Trails B than no-treatment controls.

Klusmann et al. (2010) randomly assigned older women (mean age: 74; range: 70–93) to (Group A) a physical exercise condition that included coordination, aerobic exercise, and work on balance, endurance, strength, and flexibility, (Group B) a mental exercise condition, where people did computer games that challenged memory and creativity, or (Group C) no treatment. Both the mental and physical exercise groups (Groups A and B) participated in 90-min sessions, three times a week, for 26 weeks. Both exercise groups (whether physical or cognitive) improved more than no treatment on Trail-Making (B-A) but not on Stroop or semantic Verbal Fluency (whether there was a group difference in posttest performance was not reported).

Predovan et al. (2012) had adults 57 to 80 years old (mean age = 68) do fast walking and aerobic dance exercise (within 60-min classes, where the aerobic portion gradually increased from 15 to 40 min) three times a week for a relatively short period (12 weeks). Aerobic exercise that included dance movements produced no relative benefits compared to no treatment for Stroop interference (saying the color of the ink of color words rather than reading them), which is the most commonly used Stroop outcome measure. However, on a more sensitive Stroop condition (where subjects had to switch between saying the ink color and reading the word), those who exercised improved more in both speed and accuracy. There were no posttest performance differences; indeed, the accuracy of aerobic exercisers at posttest was not as good as the pretest accuracy of controls: exercisers improved from a mean of 4.8 errors to 3.0, while controls improved from a mean of 2.5 errors to 2.2. The increase in aerobic capacity correlated with posttest RT on the more sensitive Stroop condition for the exercise group.

Davis et al. (2007, 2011) studied overweight, sedentary 7- to 11-year-olds who did aerobic games (which included basketball and soccer skills, jumping rope, and other activities that were mentally challenging and fun), 5 days a week for 13 weeks. When the aerobic games sessions were 40 min (but not when they were 20 min), children improved more in planning (their most demanding EF measure) and were better on planning at posttest than their peers who had not exercised, although there was no benefit to attention or sequence recall. The latter two may have been too easy to detect an effect. Those who got the 20-min sessions showed no significant cognitive benefit relative to the non-exercise group, yet they improved as much in physical fitness as did those assigned to 40-min exercise sessions (as assessed by a treadmill test of endurance).

Gallotta et al. (2015) studied children in Grades 3 to 5 randomly assigned to specially designed coordinative-exercise PE, traditional PE, or no PE. The two PE programs each consisted of two 60-min classes per week for 22 weeks. The middle 30 min of each class for both conditions consisted of moderate-to-vigorous activity. The traditional PE intervention focused on endurance, flexibility, resistance, and aerobic training. The coordinative-exercise PE intervention included skills used in sports games, rhythmic activities, and gymnastics. The one cognitive measure used was a measure of selective attention (the d2-R Test of Attention), for which three dependent measures were derived. Children in the coordinative-exercise group improved more than those in traditional PE and in the no-treatment group on both concentration (the number of letters correctly marked minus number of errors of commission [letters incorrectly marked]) and on percentage of errors (the number of errors of omission and commission divided by the total number of items). On neither measure was their posttest performance significantly better than either group. On percentage of errors, those in traditional PE started off extremely well and the coordinative-exercise group simply caught up.

Speed on the d2-R test (total number of items processed) improved more among those who did traditional PE than among children in either of the other two groups. The traditional PE group showed marginally better concentration at both pre- and posttest. Although their concentration score remained the best of all three groups at the end, the coordinative-exercise group showed the most improvement. Randomization for this study was done by school, but it appears the data were analyzed as if randomization had been done by child. It is unclear which, if any, of the findings would be significant if multilevel analyses had been done.

Westendorp et al. (2014) assigned children 7 to 10 years old with learning difficulties to either a ball skills intervention or standard PE (40-min sessions, two times a week for 16 weeks). The ball skills were first practiced in simple, static settings (e.g., playing catch with another child) in the hope of automatizing them. Once automatized, they were applied to dynamic sports settings (e.g., team games), where children needed to pay attention to teammates, opponents, game rules, and time, and where EF skills, including action planning, problem-solving, and cognitive flexibility, are thought to be critical, and where the aerobic component was greater.

Westendorp et al. found no group difference in improvement or posttest performance on any EF or academic skills measure (but they randomized by class and had only three classes in each group; sample sizes of three gave them extremely low power to find any effect). They did find that those who improved most in ball skills also showed the greatest improvement on the Tower of London task 6 months later. Westendorp's team speculates that those with weaker ball skills never progressed to practicing in dynamic sports settings where EFs are challenged; since their EFs were not challenged, they did not improve. Without progressing to the dynamic component, they were probably not aerobically challenged either, and either coordination exercise or aerobic exercise without the other might be less effective in improving EFs (see Marchetti et al., 2015).

Pesce et al. (2013) studied children 5 to 10 years old. Some were randomly assigned to PE that was intentionally more cognitively demanding and taught by a PE specialist, others were assigned to the standard PE curriculum taught by a PE specialist (the active control group), and still others were assigned to regular PE. These three PE conditions occurred for 60 min once a week for 26 weeks. They found no benefit to planning and no benefit from cognitively demanding PE on inhibition, as assessed by the Stroop task, but typically developing children who had been in cognitively demanding PE improved more than those in either of the other two groups in cognitive flexibility (task switching). Neither pre- nor posttest scores were reported, however, so we do not know if their task switching was better at posttest than the other two groups or if they simply caught up to the others. For children with developmental coordination challenges, the cognitively demanding PE condition was too challenging.

In a beautifully designed study by Schmidt et al. (2015), children 10 to 12 years old were randomly assigned to either exercise with high demands on both cognitive engagement and physical exertion (enriched aerobic exercise), exercise with high physical exertion demands but low cognitive demands (plain aerobic exercise), or exercise with low demands on both cognitive engagement and physical exertion (standard PE). That is the best study design of any of the extant investigations of the possible benefits of aerobic activity with cognitive demands. Unfortunately, the investigators found no differential benefits on the N-back or Flanker tests. The group that did enriched aerobic exercise improved more in cognitive flexibility (global switch costs) than the other two groups, but they started out worse than those groups and there was no significant difference in posttest switching performance (just in degree of change; see Figure 8.5). Thus, as with FITKids, the significant change reflects one group of children catching up

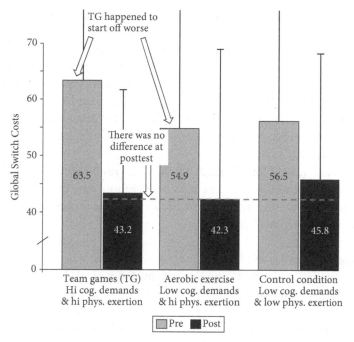

Figure 8.5. On one EF measure, Schmidt et al. (2015) found greater improvement from enriched aerobic exercise, but that seems to reflect catching up to the other children.

to the other children. It is important to note that Schmidt et al.'s (2015) intervention included only 12 sessions (two sessions per week for only 6 weeks). Perhaps if it had been more intensive or longer, greater differential benefits would have emerged.

Better Movers and Thinkers (BMT; Education Scotland Foghlam Alba, n.d.) is an approach to physical education developed in Scotland that specifically targets improving not only aerobic fitness, but also balance, gross motor coordination, rhythm, and timing, as well as EFs, confidence, determination, perseverance, and learning to work together with others in ways that children thoroughly enjoy. So far there is just an exploratory study on its feasibility (Dalziell, Boyle, & Mutrie, 2015). One class ($N = 25$) in one school received BMT and one class ($N = 21$) in another school received standard PE twice a week for 16 weeks. The children were 9 to 10 years old. Two EF measures were administered: nonverbal reasoning and visuospatial WM, both from the Lucid Assessment System for Schools (LASS). For analyses, the WM score was combined with the score on a STM measure (Forward Digit Span) also from LASS.

There were no significant group differences on either measure, although the BMT group improved significantly on the combined WM + STM composite and the control group did not. Phonological awareness improved more in the BMT group than among controls, and BMT children reported that they felt more confident and were better able to focus and concentrate when it came to their schoolwork. Since in this pilot study assignment to condition was by class and there was only one class per condition, if Dalziell et al. had analyzed their data reflecting that, they would have had no statistical power to detect any group difference.

Chang et al. (2013) found exactly the same EF results whether children did moderate- or low-intensity soccer practice. Since those were the only two groups in the study, we have no way of knowing if soccer practice benefitted both groups equally or benefitted neither.

Studies of Enriched Aerobic Exercise That Found No EF Benefits
Krafft, Schaeffer, et al. (2014) and Krafft, Schwarz, et al. (2014) had overweight 8- to 11-year-olds do 40 min of aerobic games every school day for 34 weeks. They found no improvement in inhibitory control (anti-saccade task) or selective attention (Flanker task) compared to peers assigned to sedentary activities, such as art and board games. For a small subset of participants (who also underwent neuroimaging), Krafft, Schaeffer, et al. also reported teachers' ratings on the BRIEF. Teachers rated those who exercised as more improved on the Global Executive Composite and Metacognition Indices (but not the Behavioral Regulation Index) than those who did sedentary activities (posttest scores for the BRIEF are not given, so we do not know if the exercisers were scored higher than the other group or simply caught up to them).

A third report on EF outcomes from FITKids (Chaddock-Heyman et al., 2013), in addition to the two reports discussed above, looked at two inhibitory control outcome measures (Flanker and go/no-go) and found neither more improvement nor better posttest scores on either in the FITKids group versus controls. Both groups made fewer errors on incongruent Flanker trials on posttest than they had on the pretest (i.e., both groups improved in inhibitory control). This improvement was significant for the FITKids group but not for controls. However, when the improvement of the two groups on incongruent Flanker trials was compared, there was no significant difference.

Marmeleira, Godinho, and Fernandes (2009) randomly assigned sedentary older adults (mean age: 68 years, range: 60–82 years) to aerobic exercise that intentionally included demands on inhibitory control, WM, planning, and processing speed (done three times a week for 60 min at a time for 12 weeks) or to no treatment. They found no EF benefits.

Two Studies Too Recent To Be Included in Tables 8.3 or 8.4 Or Our Tabulations, But That Deserve Mention

Alesi et al. (2016) found that children who practiced soccer showed more improvement and better posttest scores on the Tower of London than did sedentary control subjects, although on the easier Corsi Block spatial WM test there was no group difference. The mean age in both groups was 9 years.

Koutsandréou et al. (2016) found that children 9 to 10 years old who did coordinative exercise improved more in WM (as assessed by digit span) than those who did aerobic exercise, but again that might be only because the former group started behind (posttest scores were comparable for both groups). However, compared to a control group that did homework assignments while the other two groups were exercising, the coordinative-exercise children both improved more and had better posttest scores. That was not true for the plain-aerobic-exercise group.

In addition, it is worth mentioning the study by Ishihara et al. (2017), which looked at the EF benefits for 6- to 12-year-old children of learning to play tennis the traditional way (by practicing individual skills, such as the forehand swing) versus learning to play by actually playing tennis (in a modified, age-appropriate way).[12] They found greater EF benefits from playing tennis than from practicing tennis skills in isolation. Those who learned tennis by playing tennis improved on all three EF measures and on all three core EF skills (inhibitory control [Stroop task], WM [2-back task], and cognitive flexibility [task switching on a global-local task]). Those who learned tennis by practicing individual skills improved only on WM (2-back task). On a composite of the three EF measures, the tennis-playing group showed more improvement than the skill-practicing group. On posttest scores, however, there were no group differences. Children were not randomly assigned, so the children in the two groups might have differed a priori in ways that affected the results. Also, although the children in each group had been receiving tennis instruction for some time, pretesting was done before one tennis lesson and posttesting was done after that one lesson.

A Study of Coordination Training With Less of an Aerobic Component

Voelcker-Rehage et al. (2011), as mentioned above, found no greater benefit on the Flanker task in either speed or accuracy from aerobic walking versus stretching exercises. They also found no greater benefit on the Flanker task from

---

[12] The authors' rationale for their study was that since playing tennis requires more top-down cognitive control and puts greater demands on the abilities to override automatic response tendencies (inhibitory control) and to flexibly adjust in real time (cognitive flexibility) than doing repetitive exercises to work on tennis techniques (e.g., ball feeding), playing tennis should improve EFs more than repetitive exercises.

coordination exercises that addressed fine and gross motor skills, including balance, eye–hand coordination, leg–arm coordination, spatial orientation, and reacting to moving objects. The coordination group showed more improvement than those who did stretching exercises on the only non-EF cognitive measure (a visual-search task). The participants in the study were older adults (mean age = 70 years) and their exercise sessions were 35–50 min, three times a week. These results are consistent with neither coordinative exercise nor aerobic exercise being very effective in improving EFs without the other. Of course, the critical evidence missing here are results from a combined aerobic exercise + coordination training group.

Discussion of Results for Studies of Enriched Aerobic Exercise

Just as with studies of plain aerobic exercise, it is *not* that the studies that found clear or suggestive evidence of EF benefits from enriched aerobic exercise investigated programs that lasted longer, had longer sessions, were more frequent, or had more participants than studies that found few if any EF benefits (see Table 8.7 above).

Also as mentioned for plain aerobic exercise, perhaps studies have not found more EF benefits because participants do not do the activities for long enough. That might be even more critical for enriched aerobic exercise because it takes time to develop motor skills to the point where serious EF challenges can be added in a dynamic context. For example, Westendorp et al. (2014) speculate that participants with weaker ball skills never progressed to practicing in dynamic sports settings where EFs were challenged; since their EFs were not challenged, they did not improve.

Both studies of exergames (Maillot et al., 2012; Staiano et al., 2012) found encouraging results. This deserves further investigation.

While combining aerobic exercise with coordination training does not guarantee EF benefits, since several studies of enriched AE have failed to find EF improvements, it does seem clear that either coordination training or aerobic exercise alone, without the other, is even less likely to improve EFs. No benefit to EFs have been found from coordination training with less of an aerobic component (Voelcker-Rehage et al., 2011; Westendorp et al., 2014). Fewer EF benefits are found from plain aerobic exercise than from aerobic exercise with more cognitive and motor skill demands (see Table 8.7).

Too many studies of enriched aerobic exercise included only one EF measure. To seriously investigate EF benefits, more than one or even two measures should be used.

It may be that the participants need to engage in a sport, or an activity like dance, rather than do exercises drawn from that sport or activity done out of context. Certainly, correlational studies consistently show that athletes show better

EFs than non-athletes. For example, basketball players show particularly good selective attention (Kioumourtzoglou, Kourtessis, Michalopoulou, & Derri, 1998) and baseball players are particularly good at response inhibition (Kida, Oda, & Matsumura, 2005).

This hypothesis is consistent with the findings of Pesce, Masci, et al. (2016), who found that when children used the ball skills on which they were trained in a lot of their own outdoor play, more of an EF benefit was seen. The ball skills became something the children needed for something they wanted to do. If people learn something far better when they need it for something they want to do, as has repeatedly been shown (e.g., Cordova & Lepper, 1996; Freeman et al., 2014; Olson, 1964), training them in skills isolated from their use in a real-world activity seems ill-advised.

Participants are more likely to be emotionally invested in a sport than in decontextualized exercises, and their emotional investment may be key to whether that activity, even if it challenges EFs, ends up improving EFs (see the section "Our Predictions About How to Most Effectively Improve EFs"). One way to increase emotional investment is to give participants even a small decision-making role in the training activity, giving them some say in how the activity is organized or done (Ackerlund Brandt, Dozier, Juanico, Laudont, & Mick, 2015; Khan, Nelson, & Whyte, 2014; Williams, Cox, Kouides, & Deci, 1999).

Moderate- or high-intensity aerobic exercise may be more beneficial for EFs than low-intensity exercise (Hsu, Best et al., 2018). If so, it is not clear that high intensity yields better EF outcomes than moderate intensity (e.g., high- and lower-intensity soccer practice were found to produce similar EF gains by Chang et al., 2013). On the other hand, Coe, Pivarnik, Womack, Reeves, and Malina (2006) report that children who engaged in vigorous physical activity outside of school did better academically than those who exercised at only a moderate level (EFs were not assessed). Marchetti et al. (2015) report that motor-skill training improved inhibitory control only of those teenagers who were stronger and more aerobically fit. The role of intensity in directly affecting EF improvement or in moderating the effect of other features of an intervention on EFs might be worth further study.

An alternative perspective is provided by Tomporowski and Pesce (personal communication, September 23, 2017), who argue that while moderate-to-vigorous exercise yields more physical fitness benefits than lower-intensity exercise, aerobic intensity might not be central to cognitive benefits. They argue that what is driving the EF benefits from enriched aerobic exercise, such as sports, are the cognitive demands of the activity and that, during much of the skills training, aerobic demands are relatively low.

Based on their systematic review of cross-sectional, longitudinal, cohort, RCT, and acute-effects (from a one-time exposure) studies in children, Donnelly and colleagues (2016) concluded, "Based on the evidence available . . . PA has a positive influence on cognition" (p. 1197). Relevant here are the 10 reports on RCTs they reviewed. They counted each report as an independent datapoint (repeatedly saying that the 10 papers represent 10 studies). However, the two papers by Davis et al. (2007, 2011) were about the same RCT (the first paper contained a subset of the participants included in the second), the three papers by Krafft were about the same RCT (Krafft, Pierce et al., 2014; Krafft, Schaeffer, et al., 2014; Krafft, Schwarz, et al., 2014; with the first two papers reporting on a subset of the participants included in the third paper), and the four papers on FITKids were about the same implementation of that program (Chaddock-Heyman et al., 2013; Hillman et al., 2014; Kamijo et al., 2011; Monti, Hillman, & Cohen, 2012; the paper by Hillman et al. included participants included in the three other papers). The tenth paper was the soccer study by Chang et al. (2013), which, as already mentioned, compared more- to less-intensive soccer practice and found no difference in EF outcomes.

Eight of the 10 papers included in Donnelly et al.'s (2016) review are included in the present review. The study by Krafft, Pierce, et al. (2014) is not included here because it reported brain imaging outcomes, not cognitive outcomes. The paper by Monti et al. (2012) is not included because the only cognitive outcome they looked at was relational memory, and it is not clear that falls under EFs. In any case, Monti et al. found no benefits to relational memory, so that would only further reinforce our point that there are few EF benefits from enriched or plain aerobic exercise interventions.

Donnelly et al. (2016) wrote that multiple measures of cognition were included in all 10 of the papers on RCTs they reviewed. That is incorrect. Chang et al. (2013) looked at just the Flanker test, Kamijo et al. (2011) at just the Sternberg test, and Monti et al. (2012) at just a single measure of relational memory. Krafft, Pierce, et al. (2014) included no posttest measure of cognition. Donnelly et al. counted Kamijo et al. as providing evidence of the cognitive benefits of physical activity, but that is based on the wait-list control group inexplicably getting worse at posttest on the easiest condition. Chaddock-Heyman et al. found a benefit to speed of processing but none to EFs. Donnelly et al. note that, on incongruent trials on the Flanker task (which assess the EF ability inhibitory control), Chaddock-Heyman et al. found significant improvements in accuracy for those in FITKids but not for wait-list controls, but Donnelly et al. neglected to note that when Chaddock-Heyman et al. directly compared those change scores, they were not significantly different between the two groups. Donnelly et al. also do not mention that Chang et al. found no difference in cognitive outcomes for their two groups, making it impossible to draw a conclusion about any benefits. In sum, we find Donnelly et al.'s conclusion that, "Overall, the results of studies using RCT designs have consistently demonstrated significant improvements in the treatment groups, particularly for EF tasks" (p. 1204) to be unjustified.

In their systematic review of physical-activity intervention studies in children (most of which involved enriched aerobic exercise), Singh, Saliasi, et al. (2018) found five studies that they rated as high in methodological quality and that looked at the effects of physical activity on cognitive performance. They found that only 10 out of the total of 21 analyses (48%) conducted by the five studies showed a significant beneficial effect of physical activity on cognition. The authors concluded, "There is currently inconclusive evidence for beneficial effects of PA interventions on cognitive and overall academic performance in children."

Northey, Cherbuin, Pumpa, Smee, and Rattray (2017) do not report statistical analyses for EFs but only for "global cognition" (which included attention and memory in addition to EFs), so it is not possible to draw conclusions about possible benefits of physical activity specifically for EFs from that review.

The review by Vazou, Pesce, Lakes, and Smiley-Oyen (2019) included many studies that looked at other aspects of cognition, not EFs. Since studies with and without EF outcome measures were combined in their analyses, as well as EF and non-EF outcomes within a study, it is not possible to draw any conclusion about possible benefits of physical activity specifically for EFs from that review.

Resistance Training

Resistance training (also known as strength training or weight training) involves moving your limbs against resistance (anything that makes movement more difficult). Such resistance can be provided by your body weight, gravity, stretch bands, or weights. The most proximal benefits expected from resistance training are improved muscle strength and tone and improved endurance. There is little evidence that resistance training improves EFs. No study has found strong evidence of resistance training's aiding EFs. Only two studies of resistance training (Liu-Ambrose et al., 2008, and Molloy et al., 1988) have found EF benefits on 50% of assessment measures and none on more than that (see Figure 8.4 above and Table 8.20 below). Only one study of resistance training (Liu-Ambrose et al., 2008) has found better posttest performance on any EF measure when comparing those who did resistance training to any control group (and that study found better posttest scores on only two of its four measures [50%] using the low bar of a no-treatment control group; see Table 8.20).

Across the four studies that compared resistance training to toning and stretching, resistance training improved EFs more on four of 14 EF measures (29%) and produced no instances of better posttest performance. Across the four studies that compared resistance training to no treatment, resistance training

Table 8.20. Percentage of EF Measures (Except Reasoning/Fluid Intelligence) on Which Persons Who Did Resistance Training Showed More Improvement and/or Better Posttest Results Than Comparison Groups Across All Studies and Ages, Broken Down by Study

Study #	Study Name	Condition of Interest	Comparison Condition	Significantly Better Improvement			Significantly Better Posttest			Significantly Better Posttest Only Including Measures Where This Was Looked at			Both Significantly Better Change and Posttest		
				# Sign.	# of Measures	% Sign.	# Sign.	# of Measures	% Sign.	# Sign.	# of Measures	% Sign.	# Sign.	# of Measures	% Sign.
			OLDER ADULTS (OLDER THAN 55 YEARS)												
1	Cassilhas et al., 2007[A]	Resistance training	Toning & stretching	1	3	33%									
2	Liu-Ambrose et al., 2010, 2012	Resistance training + balance (1 time per week)	Toning & stretching	1	5	20%	0	5	0%	0	5	0%	0	5	0%
2	Liu-Ambrose et al., 2010, 2012	Resistance training + balance (2 times per week)	Toning & stretching	2	5	40%	0	5	0%	0	5	0%	0	5	0%
3	Moul et al., 1995	Resistance training	Mild range-of-motion & flexibility exercises	0	1	0%	0	1	0%	0	1	0%	0	1	0%
	Totals and Percents for older adults who did resistance training compared with toning and stretching			4	14	29%	0	11	0%	0	11	0%	0	11	0%
4	Kimura et al., 2010	Resistance training	Health education	0	2	0%	0	2	0%	0	2	0%	0	2	0%

#	Study	Treatment	Control												
5	Smiley-Oyen et al., 2008	Resistance training + flexibility exercises	Aerobic exercise using equipment	0	4	0%	0	4	0%	0	4	0%	0	4	0%
6	Brown et al., 2009	Resistance training + balance	No treatment	0	3	0%	0	3	0%	0	3	0%	0	3	0%
7	Dustman et al., 1984	Resistance training	No treatment	0	2	0%	0	2	0%	0	2	0%	0	2	0%
8	Liu-Ambrose et al., 2008	Resistance training	No treatment	2	4	50%	2	4	50%	2	4	50%	2	4	50%
9	Molloy et al., 1988	Resistance training	No treatment	1	2	50%	0	2	0%	0	2	0%	0	2	0%
	Totals and Percents for older adults who did resistance training compared with no treatment			3	11	27%	2	11	18%	2	11	18%	2	11	18%
	Grand Totals and Percents for older adults who did resistance training compared with any active-control condition			4	20	20%	0	17	0%	0	17	0%	0	17	0%
	Grand Totals and Percents across all resistance training studies compared with any active-control condition or no treatment			7	31	23%	2	28	7%	2	28	7%	2	28	7%

Note. Results for reasoning/fluid intelligence are not included in Table 8.20 (although they are mentioned in the text) but results for all other EF measures are included.
[A] This study did not report the difference between posttest scores.

improved EFs more on three of 11 measures (27%) and produced better posttest performance on two of 11 measures (18%). It should be noted, though, that three of the eight studies that looked at EF benefits from resistance training had hoped not to find them, since resistance training was the active control condition, not the condition of primary interest.

Similar conclusions have been reached by Fedewa and Ahn (2011), Snowden et al. (2011), Gates et al. (2013), and Kelly et al. (2014) from their meta-analyses.

Fedewa and Ahn concluded, "No significant effects of physical-activity program [on children´s academic achievement or general cognitive outcomes (not EF-specific)] were found when resistance training or combined training was applied" (p. 527). Snowden et al., who looked at studies of community-dwelling older adults, concluded, "None of the intervention categories had sufficient or strong evidence of effectiveness in maintaining or improving cognition. . . . [Two of these categories consisted of] studies of strength training on general cognition and executive function" (p. 706).

Gates et al. stated: "Resistance training was provided in two trials and produced . . . nonsignificant results on executive function" (p.1093). Kelly et al. reported finding no EF benefits in their meta-abalysis from resistance training versus no-exercise active controls (3 RCTs), although they found significantly more improvement in reasoning, but not in WM, when comparing resistance training to stretching/toning (3 RCTs). Studies included in Kelly et al.'s review reported significantly more improvement from resistance training than no-exercise active controls on only two out of 11 separate EF measures (18%) and significantly more improvement from resistance training than stretching/toning on only four out of 18 separate EF measures (22%).

For the present review, we found nine studies that examined EF performance before and after a regimen of resistance training. All nine were with older adults. In two studies (Dustman et al., 1984; Smiley-Oyen et al., 2008), resistance training was the active control condition, not the activity of primary interest; no EF benefits were found from resistance training in these studies.

In the study by Moul et al. (1995), both aerobic walking and resistance training were the activities of interest. On neither subscale of RIPA, nor on problem-solving and abstract reasoning, nor on word fluency and semantic categorization, did those who did resistance training outperform those who did flexibility training.

The most encouraging findings for resistance training were found by Liu-Ambrose et al. (2010) and Liu-Ambrose, Nagamatsu, Voss, Khan, and Handy (2012). They found that after a year of resistance training (52 weeks), women (mean age = 69 years) improved more on two different measures of inhibitory control than their peers who had done balance and toning exercises 60 min twice a week for the same 52 weeks. The two measures were Stroop interference (RT difference on a block of incongruent trials vs. a block of colored x's) and the Flanker Effect ([RT difference on incongruent minus congruent trials] divided by RT on congruent trials). The benefit on Stroop was true whether participants did resistance training once or twice a week. The benefit on Flanker was only true for those who did resistance training two times weekly. Posttest performance did not differ across groups on either the Stroop or Flanker task; only change scores differed. On EF measures that less often show treatment effects (RT on incongruent Flanker trials, Backward Digit Span minus Forward Digit Span, and Trails B minus Trails A), there were no group differences in either improvement or posttest scores.

Cassilhas et al. (2007) found greater improvement on visuospatial WM (Backward Corsi Blocks) and verbal STM (Forward Digit Span) after men (mean age = 68 years) did 24 weeks of resistance training (whether high or low intensity) than after 24 weeks of stretching and warm-up exercises. Posttest scores were not given. No benefit to verbal WM (Backward Digit Span) or spatial STM (Forward Corsi Blocks) was found from resistance training.

With older adults (mean age = 82 years) who had fallen and were at risk of further falls, Liu-Ambrose et al. (2008) investigated the benefits of 26 weeks of a home-based balance and strength retraining program. Note that this involved not just resistance training but also training in balance. They found more improvement and better posttest performance in speed and accuracy on the Stroop test from this program than business as usual, but no group difference in either change or posttest on Backward Digit Span or Trails B.

Molloy et al. (1988) found less decline in Verbal Fluency in 82-year-old women who did 13 weeks of resistance training compared with no-treatment controls— but posttest scores on Verbal Fluency did not differ between the groups (controls started out somewhat better but by posttest had declined to almost the level the exercise group was at). On digit span, again the controls started out slightly better, but both groups held their own, so that pre- and posttest scores were comparable within group and between groups.

Kimura, Obuchi et al. (2010) failed to find a benefit on task switching from 3 months of resistance training. They randomly assigned participants to either resistance training (combination of leg press, knee extension, hip abduction, and rowing) or a health-education course. There were no significant between-group differences in RT or in accuracy. Kimura et al. speculated that this result might be due to an insufficiently long or intense resistance-training program.

Mindfulness Training (Including More Sedentary Mindfulness As Well As More Physically-Active Mindfulness, Such as Yoga or Taekwondo)

All mindfulness approaches have in common quieting and focusing the mind, inhibiting internal and external distraction, so that one stays fully present to the current moment. Thirty-nine (39) studies looked at possible EF benefits from training involving some form of mindfulness. After one initial study each of Taekwondo and yoga showed promising results, follow-up studies of Taekwondo and yoga have found more disappointing outcomes (with the notable exception of a study of yoga by Gothe, Kramer, & McAuley, 2014, 2017). Very promising results have been found for two different mindfulness training methods both based on Chinese mind–body practices (Chan et al., 2013; Tang et al., 2007) and for a school program called MindUP. Quadrato Motor Training holds some promise and is worth further study. Indeed, mindfulness practices involving movement have yielded extremely promising results for improving EFs, better than for mindfulness practices primarily done seated and considerably better than for many movement activities without a mindfulness component.

More EF benefits from sitting mindfulness might be found if initially stressed individuals were the study participants. Mindfulness practices reduce stress and stress impairs EFs. Thus, helping severely stressed people feel calmer and less stressed should improve their EFs. Four studies discussed below report results consistent with this prediction (one by Bilderbeck, Farias, Brazil, Jakobowitz, & Wikholm [2013]; one by Gothe, Keswani, & McAuley [2016] and Gothe et al., [2014, 2017], and one each by Jensen, Vangkilde, Frokjaer, & Hasselbalch [2012] and Jha et al.[2015]).

Yoga

Yoga might be considered a dynamic form of mindfulness or a mindful physical activity. We decided to tabulate the results separately for yoga and other mindfulness activities involving movement before looking at the results, based solely on there being enough studies of yoga for it to merit its own table. There are many forms of yoga, but all involve physical movement and postures that emphasize flexibility and balance (asanas), breathing (pranayama), and meditative exercises (dhyana). Unfortunately, yoga intervention studies rarely report the amount or proportion of time spent doing each of the individual components.

Also, it is difficult to discern from written reports how faithful any program has been to the centuries-long mindfulness tradition behind yoga. Indeed, Sullivan et al. (2018, p. 6) noted that "Much of modern yoga practice focuses primarily on physical postures and movement sequences, [but] the traditional roots are centered on a philosophical path towards understanding the causes of suffering and its alleviation (Mallinson & Singleton, 2017; Stoler-Miller, 1998)." That is quite important, for as Trulson (1986) found for taekwondo, when the traditional mindfulness practice is emphasized, positive outcomes are found, but when just the physical exercise aspect is emphasized, positive outcomes are not found. The best EF outcomes from yoga are reported by the most recent and the earliest studies.

The most recent study (Gothe et al., 2014) had an excellent experimental design. Older adults with a mean age of 62 years (range = 55–79) were randomized to either hatha yoga with poses like warrior pose and sun salutations plus deep breathing exercises and meditative exercises or to stretching and resistance training without yoga poses or breathing or meditative exercises. Both conditions were delivered by certified trainers and included comparable levels of social interaction with the trainer, research staff, and others in the group. The conditions only lasted a short time (8 weeks) but both were given in 60-min classes three times each week. It is not known if the trainers for the control group knew the experimental hypotheses or if the testers were blind to the hypotheses or group assigments of the participants.

It is exciting that benefits from yoga were found on all three of their EF outcome measures: task switching (which assesses cognitive flexibility), N-back (which requires WM and inhibitory control), and the running memory complex-span task (which also requires WM and inhibitory control).[13] Impressively, for all three tasks, although the group that did yoga performed better on the experimental condition, both groups performed comparably on the control condition. For task switching there was no difference between the groups on the single task block (the control condition). Those who did yoga needed to slow down less to preserve their accuracy on both kinds of trials in the mixed block (switch trials and repeat trials). On both of those trial types, their speed improved more and was faster at posttest than those who did stretching and resistance training. Accuracy was high on both trial types in both groups, so it was not that the two groups chose different speed–accuracy trade-offs. The percentage of correct responses on both types of trials was marginally higher and showed marginally more improvement among those who did yoga; the differences in accuracy were probably not significantly different between the two groups because of ceiling effects.

Again, on the N-back task, there was no difference between the groups on the control condition, 1-back. In the 2-back condition, those who had done yoga improved more than those who had done stretching and resistance training. Indeed, the latter group showed no posttest improvement on the task. For the running span task, on the trials where there were no distractors, the two groups

---

[13] Enthusiasm should be tempered a bit, however, because they did not correct for multiple comparisons. Not all their positive findings might still be significant had they done that.

performed comparably. On the key trials where one or two distractors appeared at the beginning of the string of numbers to be recalled, accuracy improved more among those who had done yoga than among control subjects. Indeed, the latter group showed no posttest improvement on the task.

In a follow-up paper too recent to be included in Table 8.3 or our calculations, Gothe et al. (2017) report on two other EF outcome measures from their study. Their results for Trail-Making follow the same pattern as above—no difference between groups on the control condition (Trails A) but noticeable improvement by the yoga group on the EF-demanding condition (Trails B) and on the difference (Trails B minus Trails A), with no improvement on either by the control group, hence significantly more improvement by the yoga practitioners. On the Flanker task, those who had done yoga improved more in their speed on all types of trials, the EF-demanding ones (incongruent trials, that tax inhibitory control) and control trial types (neutral and congruent trials), than did those who had done stretching and resistance training. Results for the Flanker Effect ([RT difference on incongruent minus congruent trials] divided by RT on congruent trials) were not reported. The positive results across all five EF measures reported in the two papers is impressive indeed. Importantly, the yoga group showed more reduced stress and anxiety according to both self-report and cortisol measures than did the control group (Gothe et al., 2016).

In a small study lasting only 4 weeks, Manjunath and Telles (2001) found that 12-year-old girls assigned to yoga (which included relaxation and awareness training in addition to exercises) improved more, and performed better at posttest, on a measure of planning and inhibitory control (the Tower of London, their only EF measure) than did girls who had been assigned to regular physical training.

A follow-up study by Telles, Singh, Bhardwaj, Kumar, and Balkrishna (2013) with almost five times as many children, both boys and girls (a bit younger; mean age of 10½ years), and lasting three times longer, found more disappointing results (see Table 8.21). On the only EF outcome measure (Stroop) there was no difference between those who did yoga (including breathing techniques, postures, guided relaxation, and chanting) and those who did standard PE. Indeed, they found no significant benefit of any kind from the yoga.

With incarcerated youth and adult women, Bilderbeck et al. (2013) found that only 10 weeks of hatha yoga once a week for 2 hours reduced stress and psychological distress and increased positive affect. Their one cognitive measure was a go/no-go task (which assesses inhibitory control). They found that those who did yoga were more accurate on go trials of the go/no-go task at posttest than no-treatment controls. Fewer commission errors (fewer incorrect presses on no-go

Table 8.21. Percentage of Measures on Which Persons Who Did Yoga Showed More Improvement and/or Better Posttest Results Than Comparison Groups on Measures of EFs, Except Reasoning/Fluid Intelligence, Across All Studies and Ages, Broken Down by Study

Study #	Study Name	Condition of Interest	Comparison Condition	Significantly Better Improvement			Significantly Better Posttest			Significantly Better Posttest Only Including Measures Where This Was Looked at			Both Significantly Better Change and Posttest		
				# Sign.	# of Measures	% Sign.	# Sign.	# of Measures	% Sign.	# Sign.	# of Measures	% Sign.	# Sign.	# of Measures	% Sign.
				SCHOOL-AGE CHILDREN (7–15 YEARS OLD)											
1	Manjunath & Telles, 2001	Yoga: Physical training + relaxation + awareness training	Physical training	3	3	100%	2	3	67%	2	3	67%	2	3	67%
2	Telles et al., 2013 [A]	Yoga: Physical training + relaxation + awareness training	Physical exercise	0	2	0%	0	2	0%	0	2	0%	0	2	0%
	Totals and Percents for school-age children who did yoga compared with a physical-activity active-control condition			3	5	60%	2	5	40%	2	5	40%	2	5	40%
	Purohit & Pradhan, 2017 [B]	Yoga	No treatment	2	3	67%	0	3	0%	0	3	0%	0	3	0%
				ADULTS (21–68 YEARS OLD) IN PRISON											
3	Bilderbeck et al., 2013 [C]	Hatha yoga	No treatment	1	3	33%	1	3	33%	1	3	33%			
				OLDER ADULTS (OLDER THAN 55 YEARS) WITH NO CLINICAL DIAGNOSIS											
4	Gothe et al., 2014 [D]	Hatha yoga	Stretching and resistance training	5	6	83%	2	6	33%	2	6	33%	2	6	33%

(continued)

Table 8.21. Continued

Study #	Study Name	Condition of Interest	Comparison Condition	Significantly Better Improvement			Significantly Better Posttest			Significantly Better Posttest Only Including Measures Where This Was Looked at			Both Significantly Better Change and Posttest		
				# Sign.	# of Measures	% Sign.	# Sign.	# of Measures	% Sign.	# Sign.	# of Measures	% Sign.	# Sign.	# of Measures	% Sign.
	Gothe et al., 2017[B]	Hatha yoga	Stretching and resistance training	5	5	100%	5	5	100%	5	5	100%	5	5	100%
5	Oken et al., 2006 [A]	Iyengar yoga	Aerobic exercise	0	3	0%	0	3	0%	0	3	0%	0	3	0%
6	Blumenthal et al., 1989	Yoga and flexibility	No treatment	0	8	0%	0	8	0%	0	8	0%	0	8	0%
7	Hariprasad et al., 2013[D,E]	Yoga	No treatment	4	7	57%	3	7	0%	3	7	43%	3	7	0%
5	Oken et al., 2006	Iyengar yoga	No treatment	0	3	0%	0	3	0%	0	3	0%	0	3	0%
	Totals and Percents for older adults who did yoga compared with an active-control condition			5	9	56%	2	9	22%	2	9	22%	2	9	22%
	Totals and Percents for older adults who did yoga compared with no treatment			4	18	22%	3	18	17%	3	18	17%	3	18	17%
	Grand Totals and Percents for older adults who did yoga compared with an active-control condition or no treatment			9	27	33%	5	27	19%	5	27	19%	5	27	19%
	Grand Totals and Percents for adults (any age) who did yoga compared with no treatment			4	18	22%	4	21	19%	4	21	19%	3	18	17%

	Yoga group											
Grand Totals and Percents for adults (any age) who did yoga compared with any active-control condition or no treatment	14	32	44%	11	35	31%	11	35	31%	10	32	31%
Grand Totals and Percents across all ages comparing yoga to any active-control condition	13	19	68%	9	19	47%	9	19	47%	9	19	47%
Grand Totals and Percents across all ages comparing yoga to any active-control condition (except aerobic exercise) or no treatment	12	29	41%	8	32	25%	8	32	25%	7	29	24%
Grand Totals and Percents across all ages comparing yoga to any active-control condition or no treatment	12	32	38%	8	35	23%	8	35	23%	7	32	22%

ADULTS 18–55 YEARS OLD WITH CLINICALLY DIAGNOSED WITH MAJOR DEPRESSIVE DISORDER (MDD)

8	Sharma et al., 2006	Sahaj yoga + antidepressant medication	Antidepressant medication	0	3	0%	0	3	0%	0	3	0%	0	3	0%
Grand Totals and Percents across all ages compared with any active control, medication, or no treatment			12	35	34%	8	38	21%	8	38	21%	7	35	20%

Note. Results for reasoning/fluid intelligence (R/FL) are not included in Table 8.21 (although they are mentioned in the text) but results for all other EF measures are included.

[A] One might plausibly expect EF benefits from physical exercise or aerobic exercise, so a failure to find a difference here might be due to both interventions' being beneficial, rather than practicing yoga's being ineffectual, thus we have not included the null findings here when calculating totals or percentages, except where otherwise noted.

[B] This study was published after the 2015 cutoff date. We include it here because we think it is important, but it is not included in any calculations of percentages or totals.

[C] Bilderbeck et al. (2013) did not do pretesting. Participants were incarcerated.

[D] The authors of this study did not correct for multiple comparisons. It is unclear which of their results would remain significant had they done that.

[E] The authors of this study did not conduct the needed multilevel data analysis. It is unclear how many of their results would remain significant had they done that.

trials) would have indicated better inhibitory control. Although on average those who did yoga were considerably better at not pressing on no-go trials than were controls, there was so much variability among participants that that difference was not significant. Also, it should be noted that Bilderbeck et al. administered the go/no-test only after the intervention, not before.

At the other end of the age spectrum, yoga was the control condition in the Blumenthal et al. (1989) study of 67-year-olds discussed above under aerobic exercise. Their EF findings for yoga were as disappointing as their EF findings for aerobic exercise (see Table 8.21). Yoga was the primary condition of interest in the Oken et al. (2006) study of 72-year-olds also discussed above under aerobic exercise. Here, too, their EF findings for yoga were as disappointing as their EF findings for aerobic exercise, although, as mentioned above, there may have been ceiling effects that limited the ability to find group differences, and the researchers allowed people to enter the study, including the no-treatment group, who were doing up to 30 min of exercise a day.

Yoga was also the primary condition of interest in the study by Hariprasad et al. (2013) of 75-year-olds. Hariprasad and colleagues used seven EF tasks and found more improvement on four—Backward and Forward Spatial Span, Verbal Fluency (controlled oral word association), and response time on the Stroop test—among those who did yoga than no-treatment controls, but no difference in improvement on Backward Digit Span, accuracy on the Stroop test, or Trails B, and no difference in posttest scores except on Forward Spatial Span and Verbal Fluency. The study had a high dropout rate of over 25%, and the investigators did not correct for multiple comparisons and did not do multilevel data analyses, although they had done block randomization of elderly homes. Had their data analyses taken into account their method of randomization and had they corrected for their many group comparisons on EF and non-EF measures, their four significant EF results would probably not have held up.

With adults suffering from major depression, Sharma, Pomeroy, and Baron (2006) found that Sahaj yoga in addition to antidepressant medication produced no greater benefits on any of their three EF measures compared to medication alone. These results are reminiscent of those reported by Hoffman et al. (2008), who found that clinically depressed adults who did aerobic exercise showed no greater EF benefits than those who took placebo pills.

In a study too recent to be included in Table 8.4 or our calculations, Purohit and Pradhan (2017) randomized adolescents in the Bangalore area of India who had been orphaned (and hence stressed) to a yoga program that ran for a long time with frequent sessions (3 months of 90-min sessions 4 days a week) or to a waiting list. The yoga group improved more on Stroop and Backward Digit Span than wait-list controls, although not more on Trails B. Posttest scores did not differ between groups.

There is considerable evidence that stress impairs EFs (Arnsten, 2015; Arnsten & Goldman-Rakic 1998; Arnsten, Mazure, & Sinha, 2012; Liston, McEwen, & Casey, 2009; Morgan, Doran, Steffian, Hazlett, & Southwick, 2006). Yoga reduces feelings of stress and anxiety and decreases physiological indices of stress (Li & Goldsmith, 2012; Pascoe & Bauer, 2015; Tyagi & Cohen, 2016; West, Otte, Geher, Johnson, & Mohr, 2004). Thus, it makes sense that one mechanism by which yoga might improve EFs is via reducing stress, which would be consistent with benefits found by Bilderbeck et al. (2013), Gothe et al. (2014, 2016, 2017), and Purohit and Pradhan (2017). However, there is also evidence that other forms of physical activity, including aerobic exercise, reduce stress and anxiety (Herring, O'Connor, & Dishman, 2010; Salmon, 2001; von Haaren et al., 2016), yet aerobic exercise interventions have had disappointing results for benefiting EFs. Yoga is also a mindfulness exercise and as such should train attention and cognitive inhibition. It is likely that the direct training of EFs together with reducing stress might account for EF benefits from yoga for people who are experiencing stress. Note that most studies did not select for stressed participants and did not assess stress levels before and after, and four out of the seven studies found weak evidence of EF benefits (see Table 8.21).

In their review that included 15 RCTs that Gothe and McAuley (2015) judged to have examined the effect of yoga on EFs, the authors concluded there was a moderate effect on EFs ($g = 0.27$, $p = .001$). They included studies excluded here either because participants were likely to have brain damage (e.g., patients with multiple sclerosis) or because no dependent measure qualified as an EF measure in our judgement.

Chinese Mind–Body Practices

Both methods based on Chinese mind–body practices eschew struggling or trying to control one's thoughts; instead they see this coming naturally, without much effort, once stress is reduced and one's mind and body are more relaxed and in balance. Trying to force things is seen as counterproductive. Both studies were done in China; it is unclear if such promising results from such short exposure would be found in the West, but both studies found clear evidence of EF benefits, so the methods deserve follow-up study.

Chan et al. (2013) studied Nei Yang Gong, which thought only recently developed is based on traditional Chinese Chan-based mind-body exercises (from the *Chan* tradition named *Chanwuyi* from Sanhuang monastery; Chan, 2010). Like

Tai Chi and Qijong, Nei Yang Gong involves sets of slow, smooth, gentle, and calm movements.[14] It has two primary purposes:

First, it aims to foster self-awareness and mental self-control to help restore a calm and relaxed state. Second, it helps to reduce stress, increase flexibility of the limbs, and improve the circulation of Qi and blood. . . . Nei Yang Gong has been developed on the basis of the Chan medical model, which emphasizes the maintenance of a natural and relaxed attitude to achieve smooth circulation of Qi and blood. In this way, Nei Yang Gong differs from some of the other mind–body techniques, such as mindfulness and meditation, which require a high degree of conscious mental awareness and self-control. (Chan et al., 2013, p. 2).

The control condition was progressive muscle relaxation (PMR), a well-established behavioral technique selected because of well-replicated empirical evidence that it alleviates anxiety and physiological arousal (Lohaus & Klein-Hessling, 2003; Omizo, Loffredo, & Hammett, 1982; Singh, Rao, Prem, Sahoo, & Keshav, 2009).

Participants attended sessions where they practiced for 1 hour in 5-min chunks, twice weekly for 4 weeks (i.e., only a short time), but also practiced at home roughly 20 min a day, 4 days a week on average (according to parents' logs). Participants were children with Autism Spectrum Disorder 6 to 17 years old (mean = 11.9 years) who were grouped into closely matched pairs, with one member of each pair randomly assigned to the experimental condition and one to the control condition.

Results are summarized in Table 8.22. Children in the Nei Yang Gong group improved more and performed better at posttest on the Tower of London (which requires planning and inhibitory control) than children in the PMR group; indeed, the reduction in rule violations was four times greater in the experimental group than in the control group. On another challenging EF measure (the Color Trails test), however, there were no differences between groups. Parents rated children in the Nei Yang Gong group has having fewer temper tantrums at

---

[14] For the purposes of their study, five types of movements were used: "tranquil stand, shoulder relaxation, nasal bridge massage, Qi-circulating movement, and passive Dan Tian breathing. The movements were arranged in a fixed sequence and incorporated with specific pieces of music to facilitate the children's mastery of the technique and to keep them engaged. . . . To foster self-awareness and self-control, the children were also encouraged to practice some forms of Nei Yang Gong that served as self-guided massages for relaxing and calming oneself whenever they feel distressed and frustrated, e.g., rolling their hands slowly up and down between the chest and the abdomen, resting their hands on their abdomen while quietly observing their breathing. The selected Nei Yang Gong movements involved simple bodily actions (e.g., moving hands/fingers up and down, and bending the knees) and the children were only asked to perform the movements in a relaxed and natural manner" (Chan et al., 2013, p. 5).

Table 8.22. Percentage of Measures on Which Persons Who Practiced Mindfulness Practices Involving Movement (Other Than Yoga) Showed More Improvement and/or Better Posttest Results on Near-Transfer Measures of EFs, Except Reasoning/Fluid Intelligence, Across All Studies and Ages, Broken Down by Study

	Study			Significantly Better Improvement			Significantly Better Posttest			Significantly Better Posttest only including measures where this was looked at			Both Significantly Better Change and Posttest		
Study #	Study Name	Condition of Interest	Comparison Condition	# Sign.	# of Measures	% Sign.	# Sign.	# of Measures	% Sign.	# Sign.	# of Measures	% Sign.	# Sign.	# of Measures	% Sign.
				SCHOOL-AGE CHILDREN (7–15 YEARS OLD)											
1	Lakes et al., 2013[A]	Leadership Ed. thru Athletic Development (LEAD): Taekwondo with added features	Standard PE	1	2	50%	0	6	0%	0	6	0%	0	6	0%
2	Lakes & Hoyt, 2004[A,B,C]	Taekwondo martial arts	Standard PE	3	5	60%	4	5	80%	4	5	80%	3	5	60%
	Totals and Percents for school-age children who practiced Taekwondo compared to standard PE			4	7	57%	4	11	36%	4	11	36%	3	11	27%
				OLDER ADULTS (OLDER THAN 55 YEARS)											
3	Taylor-Piliae et al., 2010[D]	Tai Chi	Aerobic, resistance + flexibility training	1	2	50%									
3	Taylor-Piliae et al., 2010[D]	Tai Chi	Course on healthy aging	1	2	50%									
4	Mortimer et al., 2012[C,D]	Tai Chi	No treatment	3	5	60%									

Table 8.22. Continued

Study				Significantly Better Improvement			Significantly Better Posttest			Significantly Better Posttest only including measures where this was looked at			Both Significantly Better Change and Posttest		
Study #	Study Name	Condition of Interest	Comparison Condition	# Sign.	# of Measures	% Sign.	# Sign.	# of Measures	% Sign.	# Sign.	# of Measures	% Sign.	# Sign.	# of Measures	% Sign.
5	Nguyen & Kruse, 2012	Tai Chi	No treatment	1	1	100%	1	1	100%	1	1	100%	1	1	100%
	Totals and Percents for older adults who practiced Tai Chi compared to any active-control condition			2	4	50%									
	Totals and Percents for older adults who practiced Tai Chi compared to no treatment			4	6	67%	1	1	100%	1	1	100%	1	1	100%
	Totals and Percents for older adults who practiced Tai Chi compared to any active-control condition or no treatment			6	10	60%	1	1	100%	1	1	100%	1	1	100%
	Grand Totals and Percents across all ages comparing Taekwondo or Tai Chi to any active-control condition or no treatment			10	17	59%	5	12	42%	5	12	42%	4	12	33%
CHILDREN (6–17 YEARS OLD) WITH AUTISM SPECTRUM DISORDER															
6	Chan et al., 2013	Nei Yang Gong (based on traditional Chinese mind–body practices)	Progressive muscle relaxation	2	2	100%	1	2	50%	1	2	50%	1	2	50%
	Totals and Percents for children who practiced mindfulness involving movement (other than yoga) compared to any active-control condition			6	10	60%	1	1	100%	1	1	100%	1	1	100%

ADULTS (18–55 YEARS OLD)

#	Study	Intervention	Comparison	n	tot	%	n	tot	%	n	tot	%	n	tot	%
7	Ben-Soussan et al., 2015[D]	Quadrato Motor Training	Simple motor and verbal training	2	2	100%									
8	Venditti et al., 2015	Quadrato Motor Training	Walking	2	2	100%	2	2	100%	2	2	0%	2	2	100%
	Totals and Percents for adults of all ages who practiced mindfulness involving movement (other than yoga) compared to any active-control condition			6	8	75%	2	2	100%	2	2	100%	2	2	100%
	Grand Totals and Percents across all ages comparing mindfulness involving movement (other than yoga) to any active-control condition			12	17	71%	7	15	47%	7	15	47%	6	15	40%
	Grand Totals and Percents across all ages comparing mindfulness involving movement (other than yoga) to any active-control condition or no treatment			16	23	70%	8	16	50%	8	16	50%	7	16	44%

Note. Results for reasoning/fluid intelligence (R/FL) are not included in Table 8.22 (although they are mentioned in the text) but results for all other EF measures are included.

[A] The authors of this study did not conduct the needed multilevel data analysis. It is unclear how many of their results would remain significant had they done that.

[B] Lakes & Hoyt (2004) used participants who were 5 to 11 years old.

[C] The authors of this study did not correct for multiple comparisons. It is unclear which of their results would remain significant had they done that.

[D] This study did not report the difference between posttest scores.

posttest than control children and as having a greater reduction in the intensity of such outbursts than control children.

After a short exposure to integrative body–mind training (IBMT) of only five 20-min sessions, university students who did IBMT performed better and had improved more on the only near-transfer EF measure used (Flanker) than those who did relaxation training. Far transfer to Raven's Matrices was not significantly better than for controls, but those in the IBMT group improved significantly on Raven's, whereas the improvement of the relaxation-training group was not significant. (These results are summarized below in Table 8.23 for more sedentary mindfulness interventions.)

In a follow-up study published only in Chinese (and so not included in our tables or our tabulations) middle- and high-school students in China were trained on IBMT in 30-min sessions daily over 6 weeks (Tang, 2005, 2009). According to Tang et al. (2012) what they found was that those who did IBMT improved more in sustained and selective attention, Raven's Matrices, positive emotions, and academic test scores than an unspecified active control group.

Taekwondo

Lakes and Hoyt (2004) found that children 5 to 11 years old assigned to 16 weeks (45-min sessions 2–3 times per week) of a variant of the traditional martial art, Taekwondo, showed benefits (more improvement and better posttest scores) on all dimensions of EFs studied (e.g., cognitive [focused vs. distractible] and affective [persevere vs. quit] and emotion regulation) compared to their peers assigned to standard PE. Better posttest performance but not more improvement was also seen on the freedom from distraction subscale of the WISC-III and on the arithmetic subtest of that, although not on the digit span subtest component. These are encouraging results, but the data were not analyzed taking into account that cluster randomization was used nor correcting for multiple comparisons. It is unclear which results if any would have been significant had the data analyses reflected the method of randomization and corrected for multiple comparisons. As usual, greater benefits were seen for the children starting out with worse EFs. Greater benefits were also seen in older children than in younger ones; that should be further investigated to see if there is a lower limit to the age when Taekwondo might be practiced with the goal of improving EFs.

The martial arts program Lakes and Hoyt (2004) investigated is called Leadership Education through Athletic Development (LEAD). Developed by Pasquinilli (2001), LEAD is a program born out of the Korean Moogong Ryu martial arts tradition and incorporates not only Taekwondo, but also Hapkido and Gumdo. Pasquinilli delivered the in-school program studied by Lakes and Hoyt; it is unclear if someone else's delivering the program would produce the same impressive gains. A primary goal of LEAD (and most traditional martial arts) is self-improvement and character development. In LEAD this is achieved through iteratively evaluating one's thoughts and actions and then working to improve them. At the start of each class, students spend a few minutes sitting in meditation. They are instructed to clear their minds of thoughts and worries and to focus solely on their breathing. Deep-breathing techniques are taught and reinforced during meditation. Then each student is to ask him- or herself three questions that emphasize self-monitoring and planning: (a) Where am I? (i.e., focus on the present moment), (b) What am I doing? and (c) What should I be doing? The latter two questions direct children to select specific behaviors, compare their current behavior to their goal, and generate concrete plans to improve their behavior. The values of respect, humility, responsibility, honor, perseverance, discipline, focus, and self-control are emphasized.

Lakes and colleagues (2013) next studied a Taekwondo program (not LEAD) with seventh and eighth graders (vs. K–5 in Lakes & Hoyt, 2004), in a public school (Lakes & Hoyt had been in a private school), with larger classes (50 per class vs. 16 in Lakes & Hoyt), and over a longer period (9 months vs. 3 in Lakes & Hoyt), but with assessments on far fewer participants per group (30 per condition vs. 104 in Lakes & Hoyt). PE classes were five times a week (vs. three in Lakes & Hoyt), but martial arts was still done in only two of those PE periods. Results were distinctly less positive than in Lakes and Hoyt. On the only behavioral EF outcome measure (Hearts and Flowers), the Taekwondo group performed no better than those who did standard PE on either block of the task requiring EFs. Parents rated the behavior, although not the attention, of those who did Taekwondo as more improved than those who did standard PE.

T'ai Chi

Three studies of t'ai chi met criteria for inclusion here. The best of the three is by Taylor-Piliae et al. (2010), who randomly assigned healthy older adults (mean age = 69, range = 60–84 years) to 6 months (26 weeks) of t'ai chi, "Western exercise" (WE), or a "healthy aging" curriculum. The t'ai chi classes were taught by a t'ai chi grand master. Twelve postures were taught over the first 12 weeks and were practiced with continuous movement from one to the next, in a slow and rhythmic motion. In addition to the t'ai chi postures, participants were taught other elements of t'ai chi, including breathing, relaxation, attention to feeling, inattention to thoughts, upright and relaxed posture, and a slow and relaxed pace. WE consisted of aerobic, resistance, and flexibility training. Classes in t'ai chi and WE were twice a week for 60 min. Participants in t'ai chi were encouraged to practice on their own every day and at least three times a week. Those in WE were encouraged to practice 3 days a week (each time, at least 30 min of walking followed by 10–25 min of resistance and flexibility training). The healthy aging classes met once a week for 90 min and covered a variety of topics, such as healthy eating and medical and legal advice, and included visits to markets to learn about reading food labels and selecting produce.

The study by Taylor-Piliae et al. (2010) included two EF measures (semantic fluency and Backward Digit Span). On the latter, those who did t'ai chi improved more than those who did WE or attended classes in healthy aging. It is not possible to tell from the report if posttest scores differed between groups. At follow-up 6-months after the classes ended, the t'ai chi group not only had maintained its superiority on the Backward Digit Span but also demonstrated further improvements. (Those in the t'ai chi and WE groups had been instructed to continue to attend one class a week and do three home-based sessions each week during the 6 months between the end of the interventions and when the follow-up assessments occurred.) People seemed to enjoy the t'ai chi more than the other two conditions, because attendance was higher for the t'ai chi classes than for WE or healthy aging during the 6 months of the intervention and higher than for WE during the 6 months following that (healthy aging classes were not offered after the intervention period).

Mortimer et al. (2012) compared 40 weeks of t'ai chi (50 min per day, three times a week) to no treatment for older adults (mean age = 68, range = 60–79 years). Those who did t'ai chi improved more than their peers who did not receive an intervention on Trails B, category fluency, and a rating scale of attention, although with correction for multiple comparisons these results might not have remained significant. There were no group differences on Stroop, Backward Digit Span, or abstract verbal reasoning. A control group that met just to engage in social interaction (to control for possible social benefits of t'ai chi) improved more than the no-treatment group on only one EF measure (category fluency). They were never compared to the Tai Chi group, but the change score for Tai Chi appears to be significantly better on Trails B than for the social group.

Nguyen and Kruse (2012) compared 26 weeks of t'ai chi to no treatment for older adults (mean age = 69 years). The t'ai chi lessons were twice a week for 60 min each (of which 15 min was warm-up and 15 min was cool-down). The only EF measure was Trails B. Those who did t'ai chi showed more improvement and better posttest scores than control participants.

Wayne, Walsh, et al. (2014) conducted a meta-analysis with these three studies plus one not included here because its participants suffered from depression (Lavretsky et al., 2011) and they included a measure under EFs (Forward Digit Span) that we consider a measure of STM, not EFs. Nonetheless, they concluded that *compared to no-treatment*, the effect size of t'ai chi's benefits for EFs is a whopping 0.90 (Hedges' *g*) and *compared to active control groups*, its effect size is $g = 0.51$. All these studies were conducted with older adults. It would be wonderful to see t'ai chi studies with children and younger adults.

Kelly et al. (2014) included two studies of t'ai chi in their meta-analysis. Those studies were Mortimer et al. (2012) and Nguyen and Kruse (2012). They concluded that t'ai chi produced significant benefits to attention compared to no-exercise control groups and particularly impressive benefits to speed of processing, a non-EF ability. They report, however, better results for the t'ai chi participants versus controls on only five out of the 15 measures (33%) that they considered to be EF measures across the two studies.

Quadrato

Quadrato Motor Training, developed by Patrizio Paoletti, is a mindful movement activity where a person starts in one corner of a 0.5-m × 0.5-m square and is to keep moving to corners of the square in response to recorded verbal instructions calling out the number associated with the corner to move, for example, from Position 2 to 3, from 3 to 1, 1 to 4, 4 to 4 (i.e., don't move), 4 to 2, 2 to 4, 4 to 3, etc., while looking straight ahead, rather than down at the square. Young women (mean age 24 years) who did Quadrato every day for only 7 min for only 4 weeks improved more than did others who did verbal training or simple motor training on both outcome measures from the alternative uses task (the only EF task administered), which assesses cognitive flexibility (Ben-Soussan, Berkovich-Ohana, Piervincenzi, Glicksohn, & Carducci, 2015; for a review, see Ben-Soussan et al., 2015). Those who did Quadrato improved more in ideational fluency (number of uses generated) and ideational flexibility (number of different categories from which answers were generated). Whether posttest scores differed is not indicated. The same benefits were found in a second study with young women (mean age 30 years) who did Quadrato with the same dose, duration, and frequency—except here both more improvement and better posttest scores were reported for both outcomes measures of the alternative uses task (Venditti et al., 2015). This is worth following up. Quadrato is simple, takes very little time, and can be done anywhere.

Mindfulness-Based Stress Reduction (MBSR)

MBSR was developed by Kabat-Zinn (1990, 1994) beginning in the 1970s. It includes elements of sitting meditation focused on one's breath, awareness of sensations in one's body (body-scan), bringing a relaxed calm to one's mind and body, and simple yoga movements, with an emphasis on being nonjudgmental, trying to adopt a beginner's mind and stay in the present moment, and being kind to oneself and others. There is some evidence that MBSR (Kabat-Zinn, 1990, 1994) can perhaps improve selective attention but there is little evidence of other EF benefits. Six studies have looked at MBSR (see Table 8.23); however, one study looked at only a week's worth of training, another trained novices

Table 8.23. Percentage of Measures on Which Persons Who Practiced Relatively Sedentary Mindfulness Showed More Improvement and/or Better Posttest Results Than Comparison Groups on Measures of EFs, Except Reasoning/Fluid Intelligence, Across All Studies and Ages, Broken Down by Study

Study #	Study Name	Condition of Interest	Comparison Condition	Significantly Better Improvement			Significantly Better Posttest			Significantly Better Posttest Only Including Measures Where This Was Looked at			Both Significantly Better Change and Posttest		
				# Sign.	# of Measures	% Sign.	# Sign.	# of Measures	% Sign.	# Sign.	# of Measures	% Sign.	# Sign.	# of Measures	% Sign.
				YOUNG CHILDREN (3–6 YEARS OLD)											
1	Flook et al., 2015	Mindfulness-based kindness curriculum	Business as usual: standard curriculum (no treatment)	0	4	0%	0	4	0%	0	4	0%	0	4	0%
				SCHOOL-AGE CHILDREN (7–15 YEARS OLD)											
2	Napoli et al., 2005	Attention Academy program	Reading or other activities	2	3	67%	1	3	33%	1	1	100%	1	3	33%
3	Schonert-Reichl et al., 2015	MindUp	Business as usual: standard curriculum (no treatment)	6	9	67%	6	9	67%	6	9	67%	6	9	67%
	Totals and Percents for children of any age who practiced relatively sedentary mindfulness compared with no treatment			6	13	46%	6	13	46%	6	13	46%	6	13	46%
	Grand Totals and Percents for school-age children who practiced relatively sedentary mindfulness compared with any active-control condition or no treatment			8	12	67%	7	12	58%	7	10	70%	7	12	58%

	Mindfulness intervention	Active-control / comparison	8	16	50%	7	16	44%	7	14	50%	7	16	44%
	Grand Totals and Percents for children of all ages who practiced relatively sedentary mindfulness compared with any active-control condition or no treatment		8	16	50%	7	16	44%	7	14	50%	7	16	44%
	**ADOLESCENTS (15–18 YEARS OLD) IN PRISON**													
4	Leonard et al., 2013 "Power Source": Group-based cognitive-behavioral/mindfulness meditation	Evidence-based cognitive-perception	0	3	0%	1	4	25%	1	4	25%	0	4	0%
	**ADULTS (18–75 YEARS OLD): MISCELLANEOUS PROGRAMS**													
5	Ainsworth et al., 2013 Focused attention meditation training	Relaxation training	0	1	0%	0	1	0%	0	1	0%	0	1	0%
5	Ainsworth et al., 2013 Open-monitoring meditation training	Relaxation training	0	1	0%	0	1	0%	0	1	0%	0	1	0%
6	Allen et al., 2012 1-Hour focused-attention + open-monitoring meditation & 1-hour developing fullness of feeling	Reading aloud followed by discussion	1	3	33%	0	3	0%	0	3	0%	0	3	0%
7	Jha et al., 2010 Mindfulness training similar to MBSR + additional content relevant to military deployment & stress resilience: high practice	Mindfulness training similar to MBSR + additional content relevant to military deployment & stress resilience: low practice	1	1	100%	1	1	100%	1	1	100%	1	1	100%

(continued)

Table 8.23. Continued

Study				Significantly Better Improvement			Significantly Better Posttest			Significantly Better Posttest Only Including Measures Where This Was Looked at			Both Significantly Better Change and Posttest		
Study #	Study Name	Condition of Interest	Comparison Condition	# Sign.	# of Measures	% Sign.	# Sign.	# of Measures	% Sign.	# Sign.	# of Measures	% Sign.	# Sign.	# of Measures	% Sign.
8	Jha et al., 2015[A]	Training-focused mindfulness-based mind fitness training	Didactic-focused mindfulness-based mind fitness training	0	2	0%	1	3	33%	1	3	33%	0	3	0%
9	Josefsson et al., 2014	Mindfulness meditation (notice whatever arises with a nonjudg-mental accepting attitude)	Relaxation training	0	4	0%	0	4	0%	0	4	0%	0	4	0%
10	Mrazek et al., 2013	Mindfulness training course that emphasized physical posture + focused-attention meditation	Nutrition course	4	4	100%	2	4	50%	2	4	50%	2	4	50%
11	Tang et al., 2007	Integrative body-mind training (IBMT; based on traditional Chinese mind–body practices)	Western-based relaxation training	1	1	100%	1	1	100%	1	1	100%	1	1	100%

	Study	Intervention	Comparison												
5	Ainsworth et al., 2013	Focused-attention meditation training	No treatment	1	1	100%	1	1	100%	1	1	100%	1	1	100%
5	Ainsworth et al., 2013	Open-monitoring meditation training	No treatment	1	1	100%	1	1	100%	1	1	100%	1	1	100%
12	Greenberg et al., 2012, 2013	Mindfulness-based cognitive therapy	No treatment	0	4	0%	2	5	40%	2	5	40%	0	4	0%
7	Jha et al., 2010	Mindfulness training similar to MBSR + additional content relevant to military deployment & stress resilience: low practice	No treatment	0	1	0%	0	1	0%	0	1	0%	0	1	0%
7	Jha et al., 2010	Mindfulness training similar to MBSR + additional content relevant to military deployment & stress resilience: High practice	No treatment	1	1	100%	1	1	100%	1	1	100%	1	1	100%
8	Jha et al., 2015[A]	Training-focused mindfulness-based mind fitness training	No treatment	0	1	0%	2	3	67%	2	3	67%	0	3	0%

(continued)

Table 8.23. Continued

Study #	Study Name	Condition of Interest	Comparison Condition	Significantly Better Improvement			Significantly Better Posttest			Significantly Better Posttest Only Including Measures Where This Was Looked at			Both Significantly Better Change and Posttest		
				# Sign.	# of Measures	% Sign.	# Sign.	# of Measures	% Sign.	# Sign.	# of Measures	% Sign.	# Sign.	# of Measures	% Sign.
8	Jha et al., 2015[A]	Didactic-focused mindfulness-based mind fitness training	No treatment	0	1	0%	1	3	33%	1	3	33%	0	3	0%
9	Josefsson et al., 2014	Mindfulness meditation (notice whatever arises with a nonjudg-mental accepting attitude)	No treatment	0	4	0%	0	4	0%	0	4	0%	0	4	0%
13	Morrison et al., 2014	Short-form mindfulness training	No treatment	1	5	20%	1	5	20%	1	5	20%	1	5	20%
	Totals and percents for adults who practiced relatively sedentary mindfulness (not including MBSR or those who received an intensive exposure to mindfulness) compared to any active-control condition			7	17	41%	5	18	28%	5	18	28%	4	18	22%
	Totals and percents for adults who practiced relatively sedentary mindfulness (not including MBSR or those who received an intensive exposure to mindfulness) compared to no treatment			4	19	21%	9	24	38%	9	24	38%	4	23	17%

	Totals and percents for adults who practiced relatively sedentary mindfulness (not including MBSR or those who received an intensive exposure to mindfulness) compared to any active-control condition or no treatment		11	36	31%	14	42	33%	14	42	33%	8	41	20%

ADULTS (18–75 YEARS OLD): INTENSIVE EXPOSURE

14	Chambers et al., 2008	Intensive Vispassana meditation course	No treatment	1	2	50%	0	2	0%	0	2	0%	0	2	0%
15	Heeren et al., 2009	Mindfulness-based cognition training (MBCT)	No treatment	2	5	40%	2	5	40%	2	5	40%	2	5	40%
16	MacLean et al., 2010; Sahdra et al., 2011	Intensive Shamatha meditation retreat	No treatment	1	1	100%	1	1	100%	1	1	100%	1	1	100%
17	Zanesco et al., 2013	Vipassana meditation retreat	No treatment	0	1	0%	1	1	100%	1	1	100%	0	1	0%
	Totals and Percents for those who received an intensive exposure to mindfulness compared to no treatment			4	9	44%	4	9	44%	4	9	44%	3	9	33%
	Totals and Percents for adults who practiced any relatively sedentary mindfulness (other than MBSR) compared with no treatment			8	28	29%	13	33	39%	13	33	39%	7	32	22%
	Totals and Percents for adults who practiced any relatively sedentary mindfulness (other than MBSR) compared with any active-control condition or no treatment			15	45	33%	18	51	35%	18	51	35%	11	50	22%

(continued)

Table 8.23. Continued

ADULTS (18–75 YEARS OLD): MINDFULNESS-BASED STRESS REDUCTION (MBSR)

Study #	Study Name	Condition of Interest	Comparison Condition	Significantly Better Improvement			Significantly Better Posttest			Significantly Better Posttest Only Including Measures Where This Was Looked at			Both Significantly Better Change and Posttest		
				# Sign.	# of Measures	% Sign.	# Sign.	# of Measures	% Sign.	# Sign.	# of Measures	% Sign.	# Sign.	# of Measures	% Sign.
18	Jensen et al., 2012	MBSR	Nonmindfulness stress reduction	1	5	20%	0	5	0%	0	5	0%	0	5	0%
19	MacCoon et al., 2014	MBSR	Health enhancement program	0	1	0%	0	1	0%	0	1	0%	0	1	0%
20	Zeidan et al., 2010	MBSR	Book listening: Tolkein's *The Hobbit*	2	3	67%	0	3	0%	0	3	0%	0	3	0%
21	Anderson et al., 2007	MBSR	No treatment	0	3	0%	0	3	0%	0	3	0%	0	3	0%
18	Jensen et al., 2012	MBSR	No treatment	2	5	40%	0	5	0%	0	5	0%	0	5	0%
22	Meland et al., 2015	MBSR	No treatment	2	4	50%	0	4	0%	0	4	0%	0	4	0%
23	Moynihan et al., 2015	MBSR	No treatment	1	2	50%	1	2	50%	1	2	50%	1	2	50%
	Totals and Percents for adults who practiced MBSR compared with any active-control condition	MBSR		3	9	33%	0	9	0%	0	9	0%	0	9	0%

Description												
Totals and Percents for adults who practiced MBSR compared with no treatment	5	14	36%	1	14	7%	1	14	7%	1	14	7%
Grand Totals and Percents for adults who practiced MBSR compared with any active-control condition or no treatment	8	23	35%	1	23	4%	1	23	4%	1	23	4%
Grand Totals and Percents for adults who practiced any form of relatively sedentary mindfulness compared with any active-control condition	10	26	38%	5	27	19%	5	27	19%	4	27	15%
Grand Totals and Percents for adults who practiced any form of relatively sedentary mindfulness compared with no treatment	13	42	31%	14	47	30%	14	47	30%	8	46	17%
Grand Totals and Percents for adults who practiced relatively sedentary mindfulness compared with any active-control condition or no treatment	23	68	34%	19	74	26%	19	74	26%	12	73	16%
Grand Totals and Percents across all ages for those who practiced any form of relatively sedentary mindfulness compared with any active-control condition	12	32	38%	7	34	21%	7	32	22%	5	34	15%
Grand Totals and Percents across all ages for those who practiced any form of relatively sedentary mindfulness compared with no treatment	19	55	35%	20	60	33%	20	60	33%	14	59	24%
Grand Totals and Percents across all ages for those who practiced any form of relatively sedentary mindfulness compared with any active-control or no-treatment condition	31	87	36%	27	94	29%	27	92	29%	19	93	20%

Note. Results for reasoning/fluid intelligence (R/FL) are not included in Table 8.23 (although they are mentioned in the text) but results for all other EF measures are included.

[A] Jha et al. (2015) did not conduct the needed multilevel data analyses. It is unclear how many of their results would remain significant had they done that. They also did not report differences in the degree of improvement.

on sessions that were extremely long, and another used only one EF outcome measure and one unlikely to be sensitive to EF benefits from MBSR.

Anderson, Lau, Segal, and Bishop (2007) found no EF benefits on a CPT test (sustained attention) or the Stroop test (inhibitory control), but they had participants do MBSR only once a week (much less often than other studies, and much longer sessions than other studies: 120 min).

MacCoon, MacLean, Davidson, Saron, and Lutz (2014) found no EF benefit (compared to a health enhancement program) but only included one EF measure (CPT), and arguably selective attention is more relevant to EFs than sustained attention, which CPTs assess.

Zeidan, Johnson, Diamond, David, and Goolkasian (2010) looked at only one week of MBSR (four sessions total). They included two EF outcome measures. On neither of them did the MBSR group perform better at posttest than those who listened to a reading of *The Hobbit*, but on Verbal Fluency and on one outcome variable for the N-back test (hit rate), the MBSR group showed more improvement than controls. That the MBSR group showed more improvement in the number of correct responses in a row on N-back suggests an improvement in attention.

The possible benefits of MBSR for improving EFs are probably underestimated by the three studies above. It is not surprising that 1 week of MBSR produced little benefit (Zeidan et al., 2010). It is not that surprising that doing MBSR only once a week did not produce much EF benefit, and 2-hour sessions would be extremely long for novices (Anderson et al., 2007).

An MBSR training study by Jensen et al. (2012) suggested that daily MBSR for 8 weeks may be able to improve selective attention. Jensen's team compared 8 weeks of daily MBSR to non-mindful stress reduction, and for half the no-treatment group offered monetary incentives to improve their posttest performance. Jensen and colleagues looked at five EF outcome measures. On one measure of selective attention (the d2 Test of Attention), the MBSR group improved more than active controls and more than no-treatment controls (whether they were given monetary incentives or not). On the other selective attention measure (CombiTVA), the MBSR group improved more than both no-treatment groups, although not more than active controls. Benefits were limited to selective attention; the MBSR group did not outperform any control group in cognitive flexibility, inhibitory control, or sustained attention. Importantly, Jensen et al. assessed stress levels and found that at posttest, the MBSR group had the most reduced levels and lowest levels on physiological indices of stress and on reported perceived stress.

Moynihan et al. (2013) also looked at 8 weeks of MBSR but included no measure of attention. Their one EF outcome measure was Trail-Making. On the ratio of Trails B to Trails A, those who did MBSR improved more and performed better at posttest than no-treatment controls, although this benefit was gone 11 weeks later.

The primary benefit of MBSR is to reduce stress, and reducing stress has been shown repeatedly to improve EFs (Arnsten et al., 2012; Liston et al., 2009). Participants in the six MBSR studies cited above were mentally and physically healthy adults. We predict that when MBSR is studied with persons feeling significant stress or pain, stronger evidence that MBSR improves EFs will be found (as well as marked reductions in stress or pain, which has been demonstrated many times). Indeed, the study that found selective attention benefits (Jensen et al., 2012) is also the study that found reductions in stress. Also, EF benefits have been studied after only short courses of MBSR—only 8 weeks, except in Zeidan et al. (2014) where it was even shorter (only 1 week). There is also little evidence yet on whether (or how long) benefits last after one stops doing MBSR, or whether benefits are maintained with continued MBSR practice.

Meland et al. (2015) studied people who they assumed would be under stress because of an excessive workload (personnel at military helicopter units during a prolonged period of high workload). Their MBSR intervention lasted a long time (4 months with 3-hour sessions twice a week). As it turns out, all the personnel had been in the military for an average of 15 to 20 years and neither the MBSR group nor the no-treatment control group was very stressed at pre- or posttest. On anxiety, worry, and depression scales of 0 (not at all) to 3 (very much), the mean for both groups at both time points was under 1 (and for depression under 0.5). MBSR did, however, reduce the self-reported score on both the worry and depression scales more than shown in the control group, although the posttest scores were not significantly different between groups.

On the SART (a go/no-go sustained attention response task), Meland et al. (2015) found neither better outcomes nor better improvement in the group that went through the mindfulness training versus the no-treatment group. However, on a shortened version of the Attention Capture Task (ACT), they found more improvement in selective attention at both Distances 1 and 2, although not at Distance 3, compared to no-treatment controls. That is, across four EF attention measures (one from SART and three from ACT), they found more improvement on two (50%) and no better posttest performance on any. Our prediction of better outcomes with a longer course of mindfulness practice is not supported by this study. The jury is still out on our prediction of better outcomes with more stressed individuals, because the participants in this study did not feel particularly stressed.

Two other studies, both from Jha's lab, have also looked at possible EF benefits of mindfulness training for persons under stress, although unfortunately neither study assessed stress levels, so we do not know if the training helped to relieve stress. Jha, Stanley, Kiyonaga, Wong, and Gelfand (2010) looked at 8 weeks of mindfulness training for individuals under stress (U.S. Army Marine Corps reservists preparing for deployment to Iraq). The training was similar to MBSR but included additional content relevant to military deployment and stress resilience. Training sessions were 2 hours once a week for 8 weeks, supplemented by an 8-hour silent retreat near the end and instructions to practice 30 min daily.

Overall, participants who practiced mindfulness did not perform better on the Operation Span EF task than no-treatment controls. However, those who practiced more ($\geq$ 400 min over 8 weeks, an average of 634 min) improved more and performed better on Operation Span at posttest than did those who got the same training but practiced less (an average of only 151 min over 8 weeks) or no-treatment controls. In particular, civilians' scores on Operation Span remained stable over the 8 weeks, but for Marines in the control group or who practiced mindfulness only a little, Operation Span scores deteriorated over the 8 weeks, presumably due to the stress of their pending war-zone deployment. On the other hand, Operation Span scores for Marines who spent more hours practicing mindfulness improved modestly. Their positive affect (enthusiasm and energy), as assessed by the PANAS scale, improved.

In a later study, Jha et al. (2015) looked at 8 weeks of mindfulness training versus no treatment for active-duty U.S. Army soldiers 8 to 10 months prior to their deployment to Afghanistan. This mindfulness intervention departed even more from MBSR and was designed specifically for individuals who had had prolonged exposure to severe stress. There were two training conditions; one involved only didactic instruction during the 2-hour classes, whereas the other involved not only instruction, but practice in the mindfulness exercises during each class. Both groups attended four training sessions (once per week in the first 4 weeks). Throughout the 8 weeks, both groups were to practice mindfulness 30 min a day on their own.

The EF outcome measure was SART (a go/no-go sustained attention response task). Those who practiced mindfulness in class and on their own were more accurate on the task at posttest than those who received verbal instruction on mindfulness in class and practiced on their own or no-treatment controls. Those who practiced in class and on their own also made fewer commission errors at posttest than no-treatment controls. Those who received only verbal instruction in class and practiced on their own showed commission error outcomes intermediate between the other two groups—neither significantly different from those who practiced in class and on their own or from no-treatment controls. Although group differences were found in accuracy and commission errors, there were no group differences in RT. The unit of randomization was units of soldiers, but the data were analyzed as if randomization had been at the level of individual participants. It does not appear that any EF group differences would have been significant had the data analysis reflected the method of randomization.

Other Mindfulness Interventions with Adults
Short Mindfulness Interventions

Mrazek, Franklin, Phillips, Baird, and Schooler (2013) looked at the benefits of a 2-week introductory mindfulness course. College students (mean age = 21 years) were randomly assigned to a class on mindfulness or on nutrition. Each met four times a week for 45 min. The mindfulness training emphasized physical posture and 10 to 20 min of focused-attention meditation. Participants were supposed to integrate mindfulness into their daily activities and to meditate for 10 min each day outside of class. The mindfulness training reduced mind-wandering on three different measures compared to active controls. It also improved EFs as indexed by the Operation Span task, although posttest scores were not significantly better than for those who studied nutrition. Performance on reading comprehension items from the GRE improved more and was better at posttest than for the control group. That improvement (in reading comprehension) and the improvement on the Operation Span task were mediated by reduced mind-wandering (i.e., better attention).

In contrast to Chinese mind–body interventions that produced more EF benefits than did relaxation training (see above), a sitting mindfulness meditation intervention that emphasized noticing whatever thoughts, sensations, or perceptions arose in a nonjudgmental and accepting way did not produce any greater EF benefits (as assessed by the Stroop task) than did relaxation training or business as usual (large test–retest effects for all groups; Josefsson, Lindwall, & Broberg, 2014). No differential outcomes on any of the diverse psychological well-being measures were found between the meditation and relaxation groups. Participants in the study were employees at local companies (mean ages per group = 49 and 50 years, respectively) randomly assigned to 4 weeks of training in sitting meditation or relaxation (two 45-min sessions per week).

Ainsworth, Eddershaw, Meron, Baldwin, and Garner (2013) compared a variant of open-monitoring meditation different from that used by Josefsson et al. (2014) to focused-attention meditation. For the former, participants were instructed to allow a sense of awareness of the breath and physical sensations to gradually expand, including sights, sounds, smells, and emotions, allowing any sense of comfort or discomfort to become part of their awareness, noticing whatever changes occurred. The latter group was instructed to find a salient sensation and keep their focus on that in as detailed and exactly pinpointed a way as possible; gently but firmly bringing their attention back whenever it wandered. The Ainsworth group found that both mindfulness conditions produced more improved and better posttest selective attention performance than was found for the no-treatment group (as indexed by RTs on incongruent Flanker trials), mirroring positive results for Chinese mind–body versus relaxation training. This benefit was specific to focused attention; alerting and orienting did not show any group difference. The authors noted that each open-monitoring meditation session began with focused attention before widening the field of attention, and they speculated that, especially since all participants were novice meditators, the open-monitoring condition may have had more aspects of focused attention than had been intended. Participants' mean age was 20 years. The interventions were extremely brief—just three 1-hour sessions, spread over 8 days, plus instructions to practice 10 min each day on their own.

Greenberg, Reiner, and Meiren (2012) looked at cognitive flexibility and Greenberg, Reiner, Meiren (2013) looked at backward inhibition (BI) with healthy young adults (mean age = 26 years) before and after 6 weeks of mindfulness-based cognitive therapy (MBCT) modified to include handling stress. There were seven 120-min sessions over the first 5 weeks, a half-day retreat in Week 6, plus instructions to do 20 min of daily practice on their own. Cognitive flexibility (reduced rigidity) was better in those who did MBCT than in wait-list controls at posttest. For BI, the MBCT group was able to increase speed without sacrificing accuracy, whereas controls markedly improved their accuracy but at the cost of a modest decrease in speed.

Allen et al. (2012) compared mindfulness training in focused-attention and open-monitoring meditation and in developing fullness of feeling to reading aloud in a group followed by discussion in young adults (mean age = 27 years) naïve to meditation. Each group met for 6 weeks, once a week for 120 min. There were no group differences in go/no-go performance or error awareness, but on the Counting Stroop task, those who went through the mindfulness training showed a greater reduction in the RT difference between congruent and incongruent trials (more improvement in inhibitory control) than those who were in the reading group.

Interventions Involving Intensive Immersion in Mindfulness
Four studies have looked at possible EF benefits from intensive immersion in mindfulness. Two were done with persons who had not meditated before:

One and a half weeks of 11 hours of mindfulness meditation each day yielded no benefit on task switching, but it improved WM more than no treatment. The meditation group started off slightly worse in WM than controls on the Backward Digit Span (mean scores of 8.0 vs. 8.35) and ended up slightly better (9.8 vs. 8.4), although the between-group differences were not significant at pre- or posttest (Chambers, Lo, & Allen, 2008).

Heeren, van Broeck, and Philippot (2009) enrolled participants with a wide age span (27–75 years, with a mean of 54 years) and looked at the benefits of 8 weeks of daily mindfulness-based cognitive therapy (MBCT). Heeren et al. found more improvement and better posttest performance on three independent measures of Verbal Fluency (semantic, phonemic, and verb fluency) and on the Haylings test (which assesses inhibitory control) compared with no-treatment controls, though they found neither better performance nor more improvement on a measure of cognitive flexibility (Trail-Making) or another measure of inhibitory control (go/no-go). If the EF benefit from mindfulness training is primarily to attention, then a benefit to cognitive inhibition (Haylings test) would be expected more than a benefit to response inhibition (go/no-go test), since selective attention and cognitive inhibition are both aspects of interference control.

All participants in the study by MacLean et al. (2010) were highly experienced meditators, who had each attended at least three 5- to 10-day mindfulness retreats over the preceding 10 years. Only 2 hours a day of Shamatha meditation over a relatively long period (a 12-week retreat; i.e., spaced rather than massed practice) resulted in better and more improved sustained attention (the only EF outcome measure used) compared with those who did not attend the retreat (MacLean et al., 2010). Thus, even for such experienced meditators, the 12-week retreat resulted in sustained attention benefits. Sahdra et al. (2011) further reported that the retreat attendees improved on response inhibition more than those who did not attend the retreat, and the response inhibition benefit was still evident 5 months later. When the control group later attended the same retreat, they too improved in response inhibition, but there was no additional control group with whom to compare their gains.

A 4-week retreat of 9.8 hours per day of Vipassana meditation did not improve inhibition (go/no-go) more than no treatment, though the meditators performed better at posttest than non-meditators (seemingly they started better and stayed better; Zanesco, King, MacLean, & Saron, 2013). No other EF measure was administered. Retreat participants had been meditating for 14 years on average, and those in the control group had been meditating on average for 10 years.

Mindfulness Interventions with Youths 16 to 18 Years Old

Leonard et al. (2013) looked at the benefits of mindfulness combined with cognitive-behavioral therapy (a program called Power Source) for individuals under stress (incarcerated adolescents). The young men (mean age = 17, range = 16–18 years old) were assigned to Power Source or an evidence-based cognitive-perception intervention for 10 sessions, 75 min each, over 3 to 5 weeks. Their only EF outcome measure was the Flanker test, and there were no group differences on that, or on the alerting or orienting components of the ANT test. However, across all three components of the ANT (Flanker and the two non-EF components), the Power Source group showed less reduction in accuracy over time than the control group (worsening EFs over time is expected in those in highly stressful situations), though no savings in RT. The authors used a group randomized design but did not analyze their data taking that into account. It is doubtful that their one significant result would have remained significant taking that into account.

Morrison, Goolsarran, Rogers, and Jha (2014) looked at the benefits of short-form mindfulness training compared to no treatment for 18-year-old university students, done once a week, for 1 hour, over 7 weeks. They used SART (a go/no-go sustained attention response task) and an Operation Span task (which assesses WM plus other EFs) to assess EF outcomes. There were no group differences on Operation Span in speed or accuracy. There was also no group difference in speed on SART. However, on the ability to correctly withhold responses to non-target items, those who practiced mindfulness improved more and were more accurate at posttest.

In-School Mindfulness Interventions With Young Children

Very promising results have been found for the MindUP program, developed by the Hawn Foundation for elementary school classrooms (Hawn Foundation, 2008). It consists of three daily sessions of 3 min of mindfulness (focusing on breathing and attentive listening to a resonant sound), twelve 45-min lessons designed to improve EFs, social-emotional understanding, and a positive mood, plus practicing those skills throughout the school day, scaffolded if necessary by the teacher. In addition, it includes instruction on the brain bases of EFs and a social-responsibility component that includes acts of kindness in school every week and working together on a community-service activity.

Fourth-graders (9 to 11 years old) who had a year of MindUP rather than the regular curriculum, which also includes a social-responsibility component, showed both better posttest performance and more improvement on all RT measures of inhibitory control used (both on the Flanker task and the Hearts and Flowers task) plus better emotion regulation (Schonert-Reichl et al., 2015). MindUP benefitted cognitive flexibility (task switching, which was assessed on the Flanker/Reverse Flanker task). These results are strong. Replication studies and examination of how long benefits last seem warranted.

Napoli, Krech, & Holley (2005) looked at a 24-week mindfulness program for children in Grades 1 to 3, which, like MBSR, included elements of sitting meditation focused on one's breath, body-scan, awareness of sensations, and simple yoga movements, encouraging students to be nonjudgmental, focus on the present moment, and adopt a beginner's mind. Sessions were 45 min, only once every 2 weeks (i.e., spaced vs. massed practice). Children randomly assigned to this improved more in selective (but not sustained) attention and decreased more in test anxiety than other children randomly assigned to reading or other quiet activities.

Flook et al. (2010, 2015) conducted the only other studies where EF outcomes of in-school mindfulness programs were investigated. The EF results were weak.

Flook and colleagues (2015) investigated a composite mindfulness and kindness preschool program for children 4 to 5 years old consisting of two 20–30 min lessons a week for 12 weeks (~10 hours total). The mindfulness practice aimed at cultivating attentional and emotional regulation; the kindness practices were aimed at empathy, gratitude, and sharing. On none of the four EF outcome measures (Flanker [accuracy and speed], delay of gratification, or card sort) were there any group differences between children who had received the program and wait-list controls.

An earlier study led by Flook (Flook et al., 2010) that evaluated a program called mindful awareness practices (MAPs) does not appear in our tables or tabulations because the only outcome measures were parent and teacher ratings. No behavioral measure of EFs was used. MAPs consisted of exercises designed to promote heightened and receptive attention to moment-by-moment experience that were done twice a week for 8 weeks by children 7 to 9 years old. Parents and teachers filled out a rating scale (the BRIEF). On none of the three indices of the BRIEF (metacognition, behavior regulation, or global executive composite) did parents or teachers rate the children assigned to MAPs as more improved than children assigned to do reading. Both parents and teachers rated the children who had done MAPs as better on inhibitory control (behavioral regulation) than children who had been assigned to silent reading instead, but they had also rated the MAPs children as better on this at pretest; there was no difference in change scores. However, the children who most needed help on EFs (the children worse on EFs at the outset) improved more from MAPs than from reading on all three indices of the BRIEF, according to both parent and teacher ratings.

Concluding Remarks Concerning EF Benefits from Mindfulness

The results presented in Table 8.22 are striking! Every study of a mindfulness activity involving movement (i.e., dynamic mindfulness, other than yoga) found either clear or suggestive evidence of EF benefits. No other program or intervention can make that claim for 100% of its studies. Looking at the percentage of EF measures on which benefits were found, again mindfulness activities involving movement (other than yoga) show the best results. However, when one looks only at near-transfer measures, Cogmed is close behind (although many of the EF measures used to evaluate Cogmed closely resembled the training tasks, whereas that is not true for mindful movement activities). On an impressive 70% of the 23 EF outcome measures, people who did a mindful movement activity (other than yoga) improved more than controls. Omitting the three studies that did not conduct the needed data analyses, this percentage becomes even more impressive. On 82% of the 11 EF outcome measures, those who participated in a mindful movement activity improved more than controls. All the activities included in Table 8.22 (Taekwondo, Tai Chi, traditional Chinese mind–body practices, and Quadrato Motor Training) deserve further study.

People who did mindful movement performed better at posttest than controls on only 50% of the 16 EF outcome measures for which data are available. These numbers change when studies with possibly spurious positive results are removed: Then the results show that those who did mindful movement performed better at posttest than controls on 80% of the five EF outcome measures for which data are available. That is better than the EF results for any other program.

Largely because of studies' failure to report whether there was a group difference in posttest performance, we rated only two studies of mindful movement as providing clear evidence. Suggestive evidence (better change or better outcome on only 50% or more of the measures) is a low bar to pass. Also, bear in mind that it is not uncommon for initial findings to look strong but then for those promising findings to not hold up in subsequent studies. No mindfulness practice involving movement in Table 8.22 has more than three studies evaluating its EF benefits.

Some studies of relatively sedentary mindfulness report promising results as well. Across all 23 studies, 57% report at least suggestive evidence of EF benefits (see Figure 8.6). That is better than most of the methods for improving EFs reviewed here (see Table 8.1). Similarly, omitting the one study with possibly spurious positive results, 55% report at least suggestive evidence of EF benefits. Across the 84 EF measures in these 22 studies, relatively sedentary mindfulness improved performance more than did the control condition on 38% of the measures. That percent is not very high, but it still puts relatively sedentary mindfulness interventions in the top half (see Tables 8.1 and 8.2).

Mak et al. (2018), who reviewed the results for mindfulness interventions with children and adolescents, found the results disappointing. Of the 13 RCTs in their review, only five (38%) found a significant effect on an EF measure. Of the 28 EF outcome variables across those studies, only 11 (39%) found a positive effect on EFs from mindfulness. We looked at almost twice as many studies and looked at results for both children and adults. Looking only at the studies of more sedentary mindfulness with children that we reviewed, 67% of three studies found at least suggestive evidence of EF benefits, and on 50% of the 16 EF measures across the three studies, improvement was better from the mindfulness training than from the control condition. More sedentary mindfulness might not be a great idea for improving EFs in young children.

Results from evaluations of relatively sedentary mindfulness practices generally do not look good for cognitive flexibility, although Greenberg et al. (2012) found better cognitive flexibility after MBCT than was shown by wait-list controls at posttest. The MBSR studies by Anderson et al. (2007) and Jensen et al. (2012) found no benefit to cognitive flexibility, and the mindfulness intervention studies of Chambers et al. (2008) and Flook et al. (2015) found none either. Schonert-Reichl et al. (2015) found better switching on the Flanker/Reverse Flanker test after MindUP. Moynihan et al. (2013) found a benefit to cognitive flexibility (i.e., switching, as indicated by results for the Trails B:Trails A ratio) from MBSR, although not to cognitive flexibility, as indicated by Trails B performance by itself, consistent with Heeren et al.'s (2009) finding of no benefit from MBCT on Trail-Making.

In contrast, the mindful-movement program Quadrato has twice been found to reduce cognitive rigidity and improve flexibility (Ben-Soussan et al., 2015; Venditti et al., 2015), though the measure of cognitive flexibility used (the alternative uses task) has never been used in studies of more sedentary mindfulness practices to our knowledge.

Mindfulness practices generally target attention or interference control, and the EF results from studies of mindfulness look best for selective attention and cognitive inhibition. The clearest example of this is the study of daily MBSR for 2 months by Jensen et al. (2012), who found strong results for selective attention on multiple measures but no benefits for cognitive flexibility, inhibitory control, or sustained attention relative to no-treatment or active controls. Heeren et al. (2009) found both more improvement and better posttest performance on a measure of cognitive interference control (Haylings test) after 2 months of MBCT every day, but no benefits to cognitive flexibility or inhibitory control on Trail-Making or go/no-go. Across two of three measures of selective attention, Meland et al. (2015) found benefits from MBSR. Ainsworth et al. (2013) found that, whether young adults were trained on open-monitoring or focused-attention meditation, they showed more improvement and better posttest performance on the Flanker test than did those who received relaxation training. Practicing mindfulness was found to help preserve performance on the Flanker test in the face of stress (Leonard et al., 2013). Mrazek et al. (2013) found reduced mind-wandering on three different measures after a 2-week introductory mindfulness course for college students, and they found that improvements in WM and reading comprehension after the 2-week course were mediated by reduced mind-wandering (i.e., better attention). Among fourth graders, Schonert-Reichl et al. (2015) found more improvement and better posttest performance on all three RT measures of selective attention in the Flanker/Reverse Flanker task. Among first to third graders, Napoli et al. (2005) found more improvement in selective attention but not sustained attention in those who practiced mindfulness versus those who did reading instead.

Benefits to selective attention have also been reported for yoga, Chinese mind–body training, and taekwondo. Gothe et al. (2017) for yoga and Tang et al. (2007) for IBMT report benefits evident on the Flanker task for yoga and for IBMT, respectively. Lakes & Hoyt (2004) report that those who were trained in taekwondo showed less distractibility at posttest than those in standard PE.

Benefits from mindfulness are clearer for selective attention than for sustained attention. Results from relatively sedentary mindfulness practices have generally been negative on CPTs (Anderson et al., 2007; MacCoon et al. 2014) and on SART (Meland et al., 2015). There are exceptions, though: MacLean et al. (2010) found benefits on a CPT. Jha et al. (2015) and Morrison et al. (2014) found benefits on SART.

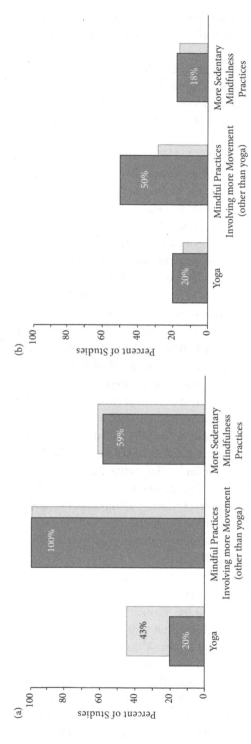

Figure 8.6. Success rates of mindfulness practices for improving EFs. Figure 8.6a: Percentage of studies finding even suggestive evidence that a mindfulness activity benefits EFs (including reasoning). Figure 8.6b: Percentage of studies finding clear evidence that a mindfulness activity benefits EFs (including reasoning). The darker bars in the foreground present the results for all studies. Studies omitted for having positive results that might not have held up were those that had not corrected for multiple comparisons or had not conducted data analyses reflecting the level at which they randomized. Whenever a study reported > 67% of measures showing positive results for improvement or posttest and did not provide any data on the other, that study is not included in calculations of strong evidence because it is possible the results of that study might have met our criteria had the results not reported been included. Given that, it is possible for the percentage of studies showing strong evidence to occasionally be higher than the percentage showing suggestive evidence, as is the case here for yoga. Figure 8.6a: Percentage of studies finding even suggestive evidence that a mindfulness activity benefits EFs, including reasoning (i.e., studies where the experimental group showed either more improvement or better posttest performance than a comparison group on ≥ 50% of the EF measures). Yoga (N = 5): Gothe et al.

(2014), Hariprasad et al. (2013), and Manjunath and Telles (2001). Mindful practices involving movement (N = 5): Ben-Soussan et al. (2015), Chan et al. (2013), Nguyen and Kruse (2012), Taylor-Piliae et al. (2010), and Venditti et al. (2015). More sedentary mindfulness practices (N = 22): Ainsworth et al. (2013), Chambers et al. (2008), Heeren et al. (2009), Jha et al. (2010), MacLean et al. (2010), Meland et al. (2015), Moynihan et al. (2013), Mrazek et al. (2013), Napoli et al. (2005), Sahdra et al. (2011), Schonert-Reichl et al. (2015), Tang et al. (2007), Zanesco et al. (2013), and Zeidan et al. (2010). For all studies: Yoga (N = 7): Gothe et al. (2014), Hariprasad et al. (2013), and Manjunath and Telles (2001). Mindful practices involving movement (N = 8): Ben-Soussan et al. (2015), Chan et al. (2013), Lakes et al. (2013), Lakes and Hoyt (2004), Mortimer et al. (2012), Nguyen and Kruse (2012), Taylor-Piliae et al. (2010), and Venditti et al. (2015). More sedentary mindfulness practices (N = 23): Ainsworth et al. (2013), Chambers et al. (2008), Heeren et al. (2009), Jha et al. (2010, 2015), MacLean et al. (2010), Meland et al. (2015), Moynihan et al. (2013), Mrazek et al. (2013), Napoli et al. (2005), Sahdra et al. (2011), Schonert-Reichl et al. (2015), Tang et al. (2007), Zanesco et al. (2013), and Zeidan et al. (2010). Figure 8.6b: Percentage of studies finding clear evidence that a mindfulness activity benefits EFs, including reasoning (i.e., studies where there was both more improvement and better posttest performance by the experimental group than by a comparison group on ≥ 67% of the EF measures used). Whenever a study reported ≥ 67% of measures showing positive results for improvement or post-test and did not provide any data on the other, that study is not included in calculations of strong evidence because it is possible the results of that study might have met our criteria had the results not reported been included. Given that, it is possible for the percentage of studies showing evidence to occasionally be higher than the percentage showing suggestive evidence, as is the case here for yoga. For studies with the needed statistical analyses: Yoga (N = 5): Manjunath and Telles (2001). Mindful practices involving movement (N = 4): Nguyen and Kruse (2012) and Venditti et al. (2015). More sedentary mindfulness practices (N = 22): Ainsworth et al. (2013), Jha et al. (2010), MacLean et al. (2010), Sahdra et al. (2011), and Schonert-Reichl et al. (2015). For all studies: Yoga (N = 7): Manjunath and Telles (2001). Mindful practices involving movement (N = 7): Lakes and Hoyt (2004), Nguyen and Kruse (2012), and Venditti et al. (2015). More sedentary mindfulness practices (N = 23): Ainsworth et al. (2013), Jha et al. (2010), MacLean et al. (2010), Sahdra et al. (2011), and Schonert-Reichl et al. (2015).

Benefits from mindfulness are clearer for interference control (inhibition at the level of attention [selective attention] and at the level of cognition) than for response inhibition (inhibition at the level of action). As mentioned above, Heeren et al. (2009) found benefits from mindfulness on the Haylings test (cognitive inhibition) but not on a go/no-go test (response inhibition). Allen et al. (2012) and Zanesco et al. (2013) also found no benefit on go/no-go.

There is evidence that one of the main problems in cognitive aging is the reduced ability to ignore or inhibit distractions (i.e., poorer interference control, including selective attention; Gazzaley, Clapp, McEvoy, Knight, & D'Esposito, 2008; Gazzaley, Cooney, Rissman, & D'Esposito, 2005; Gazzaley & D'Esposito, 2007; Hasher & Zacks, 1988). To the extent that that is the case, and to the extent that a mindfulness practice can help with that, mindfulness training might be an effective approach to help curb cognitive decline in older adults. Practicing mindfulness might aid WM because of its benefits to interference control (Jha et al., 2010; Mrazek et al., 2013).

Mindfulness practices reduce stress. They may improve EFs because they reduce stress, and perhaps to the extent that they succeed in reducing stress. Findings consistent with that hypothesis have been found for studies of yoga (Bilderbeck et al., 2013; Gothe et al., 2016; Purohit & Pradhan, 2017) and more sedentary meditation (Jha et al., 2010, 2015; Leonard et al., 2013; Napoli et al., 2005).

School Programs Intended to Benefit EFs

Some of the most encouraging findings for improving EFs come from studies of school programs (see Table 8.24).[15]

The PATHS (Promoting Alternative THinking Strategies; Kusché & Greenberg, 1994) program is an add-on to the school curriculum delivered 3 days a week for 20 to 30 min, but what is covered there is intended to be practiced throughout the school day, scaffolded if necessary by the teacher. The PATHS program promotes prosocial behavior, emotional understanding and other emotional and social competencies, self-control, and social problem-solving. Riggs, Greenberg, Kusché, and Pentz (2006) found that second and third graders who had experienced PATHS were better at posttest and more improved on both EF measures they used (Stroop and Verbal Fluency) than their peers who received the regular curriculum only. This deserves follow-up. An important caveat is that the data were analyzed as if individuals (rather than classes) had been assigned to condition; it's unclear which results, if any, would be significant with proper

[15] The MindUP school program, Attention Academy Program, and Flook's two in-school programs were discussed in the section on mindfulness.

Table 8.24. Percentage of Measures on Which Persons Trained With Various School Programs Showed More Improvement and/or Better Posttest Results on Measures of Executive Functions, Except Reasoning/Fluid Intelligence Across All Studies and Ages, Broken Down by Study

Study #	Study Name	Condition of Interest	Comparison Condition	Significantly Better Improvement			Significantly Better Posttest			Significantly Better Posttest Only Including Measures Where This Was Looked at			Both Significantly Better Change and Posttest		
				# Sign.	# of Measures	% Sign.	# Sign.	# of Measures	% Sign.	# Sign.	# of Measures	% Sign.	# Sign.	# of Measures	% Sign.
			YOUNG CHILDREN (3–6 YEARS OLD)												
1	Diamond et al., 2007[A]	Tools of the Mind	Another curriculum newly developed by the school district				1	4	25%	1	4	25%			
2	Domitrovich et al., 2007[B]	Preschool PATHS	Business as usual: standard curriculum	0	4	0%	0	4	0%	0	4	0%	0	4	0%
3	Flook et al., 2015	Composite mindfulness and kindness curriculum	Business as usual: standard curriculum	0	4	0%	0	4	0%	0	4	0%	0	4	0%
4	Lillard & Else-Quest, 2006[A]	Montessori	Business as usual: standard curriculum				2	3	67%	2	3	67%			
5	Raver et al., 2008, 2011	Chicago School Readiness Project	Business as usual: standard curriculum	2	3	57%	2	3	67%	2	3	67%	2	3	67%
	Solomon et al., 2017[C]	Tools of the Mind	Playing to Learn: play-based school program	1	1	100%									

(continued)

Table 8.24. Continued

Study #	Study Name	Condition of Interest	Comparison Condition	Significantly Better Improvement			Significantly Better Posttest			Significantly Better Posttest Only Including Measures Where This Was Looked at			Both Significantly Better Change and Posttest		
				# Sign.	# of Measures	% Sign.	# Sign.	# of Measures	% Sign.	# Sign.	# of Measures	% Sign.	# Sign.	# of Measures	% Sign.
		Totals and Percents for young children trained using various school programs (excluding ones involving mindfulness included in Table 8.24) compared with no treatment		2	7	29%	4	10	40%	4	10	40%	2	7	29%
		Totals and Percents for young children trained using any school program compared with no treatment		2	11	18%	4	14	29%	4	14	29%	2	11	18%
		Grand Totals and Percents for young children trained using a school program (excluding ones involving mindfulness included in Table 8.24) compared with any active-control condition or no treatment		2	7	29%	5	14	36%	5	14	36%	2	7	29%
		Grand Totals and Percents for young children trained using any school program compared with any active-control condition or no treatment		2	11	18%	5	18	28%	5	18	28%	2	11	18%
CHILDREN (5–7 YEARS OLD)															
6	Napoli et al., 2005	Attention Academy program	Reading or other activities	2	3	67%	1	3	33%	1	3	33%	1	3	33%
7	Blair & Raver, 2014[D]	Tools of the Mind: all children	Business as usual: standard curriculum	1	3	33%	1	6	17%	1	6	17%	1	6	17%
7	Blair & Raver, 2014[D]	Tools of the Mind: lower SES	Business as usual: standard curriculum	3	3	100%	3	3	100%	3	3	100%	3	3	100%

#	Intervention	Comparison	n	N	%	n	N	%	n	N	%	n	N	%
	Intervention Programme for Self-Regulation and Executive Functions (PIAFEx).	Business as usual: standard curriculum	4	7	57%	1	14	7%	1	1	100%	1	14	7%
8	Play-based approach to training EFs: Helping story characters overcome challenges	Business as usual: standard curriculum	7	14	50%	1	14	7%	1	1	100%	1	14	7%
Totals and Percents for children trained using a school program compared with no treatment			11	20	55%	5	10	50%	5	23	22%	5	23	22%
Totals and Percents for children trained using a school program compared with an active-control condition or no treatment			13	23	57%	6	11	55%	6	26	23%	6	26	23%
SCHOOL-AGE CHILDREN (7–15 YEARS OLD)														
9	Metacognitive WM training using a strategy game	General cognitive activities	0	3	25%	0	3	0%	0	3	0%	0	3	0%
10	Embedded EF training in reading comprehension lessons	Business as usual: standard curriculum	0	1	0%	0	1	0%	0	1	0%	0	1	0%
4	Montessori curriculum: older children	Business as usual: standard curriculum				1	2	50%	1	1	100%			
11	PATHS	Business as usual: standard curriculum	2	2	100%	2	2	100%	2	2	100%	2	2	100%

(continued)

Table 8.24. Continued

Study #	Study Name	Condition of Interest	Comparison Condition	Significantly Better Improvement			Significantly Better Posttest			Significantly Better Posttest Only Including Measures Where This Was Looked at			Both Significantly Better Change and Posttest		
				# Sign.	# of Measures	% Sign.	# Sign.	# of Measures	% Sign.	# Sign.	# of Measures	% Sign.	# Sign.	# of Measures	% Sign.
12	Schonert-Reichl et al., 2015	MindUp	Business as usual: standard curriculum	6	9	67%	6	9	67%	6	9	67%	6	9	67%
			Totals and Percents for school-age children trained using a school program (excluding ones involving mindfulness) compared with no treatment	2	3	67%	3	5	60%	3	4	75%	2	3	67%
			Totals and Percents for school-age children trained on any school program compared with no treatment	8	12	67%	9	14	64%	9	13	69%	8	12	67%
			Totals and Percents for school-age children trained using a school program (excluding ones involving mindfulness) compared with any active-control condition or no treatment	2	6	33%	3	8	38%	3	7	43%	2	6	33%
			Totals and Percents for school-age children trained on any school program compared with any active-control condition or no treatment	8	15	53%	9	17	53%	9	16	56%	8	15	53%
			Totals and Percents for children of all ages trained using a school program (excluding ones involving mindfulness) compared with any active-control condition	0	3	0%	1	7	14%	1	7	14%	0	3	0%
			Totals and Percents for children of all ages trained on any school program compared with any active-control condition	2	6	33%	2	10	20%	2	8	25%	1	6	17%

	Totals and Percents for children of all ages trained using a school program (excluding ones involving mindfulness) compared with no treatment	15	30	50%	12	38	32%	12	24	50%	9	33	27%
	Totals and Percents for children of all ages trained on any school program compared with no treatment	21	47	45%	18	55	33%	18	41	44%	15	50	30%
	Totals and Percents for children of all ages trained using a school program (excluding ones involving mindfulness) compared with any active control condition or no treatment	15	33	45%	13	45	29%	13	31	42%	9	36	25%
	Totals and Percents for children of all ages trained on any school program compared with any active control condition or no treatment	23	49	47%	20	61	33%	20	45	44%	16	52	31%

SCHOOL-AGE CHILDREN (7–15 YEARS OLD) WITH ADHD

13	Menezes et al., 2015[D]	Intervention Programme for Self-Regulation and Executive Functions (PIAFEx) Social skills intervention	1	3	33%	3	18	17%	3	18	17%	1	18	6%
	Grand Totals and Percents for all children, with or without ADHD, trained using a school program (excluding ones involving mindfulness) compared with any active-control condition	1	6	17%	4	25	16%	4	25	16%	1	21	5%
	Grand Totals and Percents for all children, with or without ADHD, trained using any school program compared with any active-control condition	3	9	33%	5	28	18%	5	26	19%	2	24	8%

(continued)

Table 8.24. Continued

Study #	Study Name	Condition of Interest	Comparison Condition	Significantly Better Improvement			Significantly Better Posttest			Significantly Better Posttest Only Including Measures Where This Was Looked at			Both Significantly Better Change and Posttest		
				# Sign.	# of Measures	% Sign.	# Sign.	# of Measures	% Sign.	# Sign.	# of Measures	% Sign.	# Sign.	# of Measures	% Sign.
		Grand Totals and Percents across all ages trained using a school program (excluding ones involving mindfulness) compared with any active-control condition or no treatment		16	36	44%	16	63	25%	16	49	33%	10	54	19%
		Grand Totals and Percents across all ages trained using any school program compared with any active control condition or no treatment		24	52	46%	23	79	29%	23	63	37%	17	70	24%

Note. Results for reasoning/fluid intelligence (R/FL) are not included in Table 8.24 (although they are mentioned in the text) but results for all other EF measures are included.

[A] This study did not collect pretest data.

[B] The author of this study did not conduct the needed multilevel data analysis. It is unclear how many of their results would remain significant had they done that.

[C] This study was published after the 2015 cutoff date. It is included here because it is important, but it is not included in any calculations of percentages or totals.

[D] The authors of this study did not include a correction for multiple comparisons. It is unclear which of their results would remain significant if they had.

multilevel data analyses. PATHS delivered to Head Start preschoolers 1 day a week produces marked social and emotional benefits but has yet to demonstrate EF benefits (Domitrovich, Cortes, & Greenberg, 2007).

The Chicago School Readiness Project (CSRP) was an add-on for Head Start preschool classrooms. It emphasized developing verbally skilled strategies for emotion regulation. Stress-reduction workshops were conducted for teachers all year. Children with the worst externalizing behavior received one-on-one counseling. CSRP is more focused on behavior management than PATHS but otherwise had similar goals (see Table 8.1 in Diamond & Lee, 2011). Four-year-olds who experienced CSRP in Head Start improved more and performed better at posttest on two of the three measures of inhibitory control used (Tapping and Balance Beam composite, and experimenters' ratings of impulse control) than their peers in regular Head Start classes (Raver et al., 2008, 2011). The CSRP children didn't show a benefit in delaying gratification, but only one study we've reviewed has found improvement on the delay of gratification paradigm from any program. CSRP children also improved more in vocabulary, letter naming, and math than did controls. CSRP's improvement of academic skills was mediated largely via its improvement of EFs. EFs in the Spring of preschool predicted achievement 3 years later in both math and reading (Li-Grining et al., 2011). These results, too, are most encouraging.

Results from three independent evaluations have been published on the Tools of the Mind (Tools) curriculum for preschool and kindergarten. Tools (Bodrova & Leong, 2007) is based on the work of Vygotsky (1978) and is a full curriculum. It emphasizes improving EFs (especially self-control), social and emotional skills, and building a sense of community as much as academics. Diamond et al. (2007) found better inhibitory control on the Reverse Flanker task in 5-year-old children who had been through Tools than in their peers who had been in another new curriculum of which the district was quite proud. There were no significant group differences, however, on the standard Flanker task or on Blocks 1 or 2 of the Hearts and Flowers task (all of which had ceiling effects) or on Block 3 of Hearts and Flowers (which suffered from a floor effect). Whether children were in Tools or not accounted for more variance in EFs than did age or gender. The children were not evaluated before the intervention, so it is possible that children in Tools of the Mind had better inhibitory control to begin with, though the groups were closely matched on a great many demographic variables.

Blair and Raver (2014) found better and more improved emotion regulation on the dot-probe task, but they did not find better or more improved inhibitory control and cognitive flexibility on the Hearts and Flowers task, card sorting, or the Flanker task (which included reverse and mixed Flanker trials), in kindergarten children who received Tools versus the regular school curriculum.[16] Benefits to reading, math, and vocabulary were even larger the following year (in Grade 1) compared with controls. Effects were about eight times larger in low-income schools (see Figures 8.3a and 8.3b). Low-income children who had experienced Tools were better and more improved on fluid intelligence/reasoning (Raven's Matrices) than their peers who had received the standard curriculum, but this was not significant for more economically advantaged children.

In another study of Tools, published after our cutoff date and so not included in our tables or analyses (Solomon et al., 2018), a daycare-based Tools program for 3- to 4-year-olds was compared to a high-quality, existing play-based program. Children whose parents had rated them as highly hyperactive and/or inattentive at the outset of the year showed greater gains on an inhibitory control task ("Touch your toes when I say touch your head" and "Touch your head when I say touch your toes") than children in the existing program. The authors concluded that "Tools may be advantageous in classrooms with children experiencing greater challenges with self-regulation, at no apparent cost to those less challenged in this regard" (p. 2). In sum, the results for Tools are encouraging and deserve more longitudinal follow-up.

Other work not included in our tables or analyses because the first report was published in Portuguese (Dias & Seabra, 2013) and a later one was published after our 2015 cutoff date (Dias & Seabra, 2016) also deserves mention. Dias and Seabra developed an Intervention Programme for Self-Regulation and Executive Functions (PIAFEx) for schoolchildren in Brazil that intentionally borrows some principles from Tools. In their study, which is included in our tables and analyses, Menezes et al. (2015) looked at whether this program could help 7- to 13-year-old children (mean age = 10 years) with ADHD when delivered twice a week for 1 hour over 35 weeks. They found a benefit on Stroop and a verbal WM test compared to no-treatment controls, but no benefits on a visuospatial WM Test, Trails B, the Wisconsin Card Sort, Verbal Fluency, or other measures, and it is not clear if the benefits they found would remain significant if corrections for multiple comparisons had been done.

Those disappointing results are consistent with other studies of Tools itself as an add-on to existing curricula. Leong and Bodrova originally tried Tools as an add-on, with Tools activities done for roughly an hour a day. Benefits were narrow and specific to the context in which the skills were practiced. Clements et al. (2012) replicated the limited benefits from Tools as an add-on. Diamond et al. (2007), Blair and Raver (2014), and Diamond, Lee, Senften, Lam, and Abbott (accepted) replicated the marked benefits from Tools as

---

[16] Blair and Raver (2014) conducted multiple comparisons without correcting for that in their significance testing. Not all their positive findings might still be significant had they done that.

an all-day curriculum originally reported more informally by Bodrova and Leong (2001).

In a study with first graders (mean age = 6 years), Dias and Seabra (2016) looked at PIAFEx implemented as the school curriculum (all day, 5 days a week, for 15 weeks). Compared to those in regular first-grade classes, children in PIAFEx improved more on some measures of inhibitory control (Trails B and the Simon task) and cognitive flexibility (Trails B and errors on a cancellation attention task [CAT] that required switching attention) but not on other measures of inhibitory control (Stroop and go/no-go) or cognitive flexibility (score on the CAT). Information is not provided on whether there was any group difference in posttest scores and no correction for multiple comparisons was made.

Two other play-based school programs have been investigated. Traverso, Viterbori, and Usai (2015) looked at EF benefits from a program where children were asked to help two story characters overcome various challenges. The program was short (only 1 month and only three 30-min sessions per week). Those who participated in the program showed more improvement in, and better posttest performance on, RT on the Flanker test than those who did not participate. There were no group differences in accuracy. Children in the program also improved more on other measures of inhibitory control (delaying gratification and circle drawing) as well as measures of WM (backward word span and Keep Track span), attention (matching familiar figures), and cognitive flexibility (accuracy in the mixed block of the Hearts and Flowers test, although there were no differences in RT for that block). That is quite impressive. No differences were found on another delay of gratification measure or go/no-go. Often when people improve on a measure, they improve on either speed or accuracy, so we would say that this program helped children perform better on most measures, including the Flanker and Hearts and Flowers tests, but not on go/no-go. Enthusiasm needs to be tempered a bit, however, because Traverso et al. conducted multiple comparisons without correcting for that in their significance testing.

The other play-based program involved a story protagonist who had certain things to do (Caviola, Mammarella, Cornoldi, & Lucangeli, 2009). The children (fourth graders) and teacher discussed what strategy to use to achieve the protagonist's objective. This program too was very brief (only 1 month and only two 50-min sessions per week). The only EF outcome measures were three WM tests, and no benefits compared to business-as-usual controls were found on any of them.

Garcia-Madruga et al. (2013) investigated EF and reading benefits from embedding training in EFs (WM, attention, inhibition, and switching) within work on reading comprehension. This was done with 8- to 9-year-olds and also for only a very short time (only 1 month; 50-min sessions three times per week). On neither WM (reading span) nor reasoning was any benefit found relative to no-treatment controls, but a benefit to reading comprehension was found.

Lillard and Else-Quest (2006) looked at benefits from Montessori education but included very few EF outcome measures and did not collect any pre-intervention data. Parents of all children in the study had wanted their children to attend public Montessori instruction; choice of who got in was made by lottery. Children who attended Montessori showed more cognitive flexibility than children in the standard curriculum both at 6 years of age on card sorting and at 12 years on a creativity measure. There was no difference in delay of gratification at age 6.

Of the 13 school programs (including those studied by Flook et al., 2015, Napoli, 2005, and Schonert-Reichl et al., 2015), 54% found at least suggestive evidence of EF benefits. The following school programs show promising evidence of EF benefits: Attention Academy, CSRP, MindUP, Montessori, PATHS, and Tools. Of the seven studies investigating those promising school programs, 75% found at least suggestive evidence of EF benefits (Table 8.1). When the two studies whose positive findings might not have held up if the requisite statistical analyses were conducted are omitted, the results show 67% of studies of the promising school programs found at least suggestive evidence of EF benefits (Table 8.2). Only mindful movement practices have found better results than promising school programs. In Table 8.1, this difference is small; promising school programs show the best results on two of the metrics and show second best on the other two. When studies with potentially spurious results are omitted, however, promising school programs drop to third place on two metrics and second on the other two—they are still second only to mindfulness practices involving movement but now by a wider margin. These school programs share the goals of promoting social and emotional outcomes as well as academic excellence and minimizing stress. These promising school programs deserve further study.

EF Outcomes From Other Programs

Results for EF benefits are summarized here for Experience Corps, theater, piano instruction, and learning digital photography or quilting, and El Sistema orchestra.

Experience Corps (Glass et al., 2004) is a program that brings older adults into schools with the goals of improving the mental and physical health of the senior volunteers who participate and helping students feel more at home and less alienated in school, as well as improving their academic outcomes. Older African American women who participated improved more on selective attention (Flanker) than their peers who did not participate. Posttest scores are not given. No other EF outcome measure was used (Carlson et al., 2009). We'd like to see this promising program receive more study.

Park et al. (2014) conducted an extremely careful, very well-designed study with disappointing results. Older adults were randomly assigned to spend 14 weeks learning digital photography, quilting, or both, or to either of two controls conditions: participating in structured activities with others in a social club or alone. No differential benefit to any EF skill was found (not on Flanker, Stockings of Cambridge, or WM) or to fluid intelligence (Raven's). Episodic, recognition, and recall memory improved in those who did photography but not in the quilters.

Bugos, Perlstein, McCrae, Brophy, & Bedenbaugh (2007) used random assignment to examine the potential of 6 months of individualized piano instruction for mitigating age-related cognitive decline in healthy adults 60 to 85 years old. They found more improvement on one EF measure (Trials B) than in no-treatment controls and that benefit was still maintained 3 months later. On their other EF measure (a composite of Forward Digit Span [an STM measure] and Backward Digit Span), though, they found no group difference.

Noice, Noice, and Staines (2004) reported that older adults who were randomly assigned to theater (training in acting; seven 90-min sessions over the course of a month) improved more and were better at posttest in problem-solving (the Means-End Problem-Solving Procedure by Platt & Spivack, 1975) than their peers assigned to either visual arts or no treatment. On the other EF measure, the Listening Span task, those who received acting training also improved more than the other groups but this just missed being statistically significant ($p = .056$). Participants assigned to visual arts appreciation did not differ from those assigned to no treatment in improvement or posttest on either EF measure. Those who did theater improved more in feelings of psychological well-being and reported more well-being at posttest than either control group (visual arts or no treatment), although there were no differences in self-esteem. The gains by those who had trained in theater persisted undiminished for 4 months after the training ended.

There are many reasons why theater might be an excellent avenue for improving EFs. One factor might be its positive impact on emotional and social factors, such as Noice et al. (2004) found for psychological well-being in their study. Noice and Noice (2006) offer other possible factors, such as practice in processing material at a deep level (so that it is remembered better) and practice in staying in the present moment (inhibiting attentional or cognitive interference):

> Actors . . . determine the goal of every utterance of the character, breaking down scripts into what they call "beats" (the smallest goal-directed chunks of dialogue). . . . A link is forged between almost every word or phrase and the goal that caused the character to utter it (Noice & Noice, 1997, 2004). A consistent

finding in the text-comprehension literature is that goal statements are better recalled than nongoal statements (e.g., Trabasso & van den Broek, 1985). Processing the script at such depth produces a great deal of verbatim retention without rote memorization. . . .

During rehearsal, they try to devote all their conscious awareness to remaining in the present moment by attending to the other actors, only glancing down at the script when necessary. . . .

[In a subsequent study (Noice & Noice, 2004)] we specifically told [participants] not to try to remember the words but to put all their concentration on meaning them (i.e., actively using them to gain a specific end such as warning a friend). . . . Meaning the words produced greater retention than memorizing them did. . . . This finding has been replicated repeatedly using different populations and procedures and various types of materials. (Noice & Noice, 2006, p. 15)

Further study of the potential EF benefits of theater with children, young adults, and elders would be most welcome, especially when training in generalizing the cognitive skills learned in theater to other contexts is provided.

An excellent study of El Sistema (Holochwost et al., 2017) came out too late to be included in Table 8.4 or our calculations, but it deserves mention. El Sistema is an orchestral music program developed by Jose Antonio Abreu in Venezuela to rescue poor children through music (Booth & Tunstall, 2016). It emphasizes playing together in ensemble from the start, the joy of making music, not embarrassing anyone over a mistake, building community, learning to work together and learning from one another (child teaching children), and demanding daily practice and training. A predominantly African American parochial school in Philadelphia offers El Sistema and decides who gets in by lottery. The parents of all children in this study wanted their children to get El Sistema; half got randomly selected for it. None of the 265 children in the study were classified as special education and most were lower income. El Sistema meets for 120 min every school day (39 weeks/year). Forty minutes of that is instruction in a small-group setting and 40 min is rehearsal in an ensemble. The drop-out rate from El Sistema over the 3 years of the study was very low (only 10%). Holochwost and colleagues used an intent-to-treat data analysis, which is the most conservative and most rigorous. Testers were blind to condition.

Holochwost and colleagues found standardized test scores, academic grades in English and math, and performance on seven out of nine (78%) EF measures improved more (and were better at posttest) for children in El Sistema than for children in the control group (see Table 8.25.) Effect sizes on the Flanker and Stroop tests were quite large (0.5 or greater). Some effects were not evident, however, until children had been in El Sistema for 3 years.

Table 8.25. Results in the Holochwost et al. (2017) Study of EI Sistema

Variable	Difference*	Significance	Effect Size
Standardized test scores (Terra Nova)	13.5 points higher	$p < .007$	$d = 0.24$
Grades in English	2.5 points higher	$p < .001$	$d = 0.30$
Grades in Math	3.9 points higher	$p < .01$	$d = 0.42$
EF Measures:			
1) Go/No-go: better accuracy	3.5% higher	$p < .004$	$d = 0.40$
2) Flanker Test: better accuracy	6.6% higher	$p < .01$	$d = 0.35$
more efficient (based on RT and percent correct)	12.3 better	$p < .001$	$d = 0.50$
smaller Flanker effect (RT difference on incongruent & congruent trials)		$p < .001$	$d = 0.50$
3) Stroop Test: more efficient (based on RT and percent correct)		$p < .001$	$d = 0.50$
faster RT	313 msec faster	$p < .0001$	$d = 0.57$
(for the bottom 25%)	620 msec faster		$d = 1.13$
4) Wisconsin Card Sorting Test: fewer perserverative errors		$p < .02$	$d = 0.18$
5) Tower of London	NS (too difficult)		
6) Trail-Making Test	NS (too insensitive)		

*Those in the program for 3 years vs. controls.

Across All Approaches to Improving EFs, Which Are the *Most* Promising Thus Far?

The approach that has been most successful thus far at improving EFs is **mindfulness practices involving movement** (such as t'ai chi, taekwondo, Chinese mind–body, and Quadrato). Every single study that has investigated whether training in a mindful movement practice can improve EFs has found at least suggestive evidence that it can. That is not true of any other approach we have examined. The EF results for mindfulness practices involving movement are far better than those for other movement activities without a mindfulness component and better than those for mindfulness practices primarily done seated, although that is not true if the results for yoga are included with the other mindful movement practices. The superiority of the results for mindful movement practices is especially evident when studies are omitted that had positive results that might not have held up had they corrected for multiple comparisons or had not taken into account when analyzing their data that they had randomized at the group level (Table 8.2). Here, mindful movement practices show the best EF outcomes of all approaches on all four of our indices. The difference between the percentage of studies showing at least suggestive evidence of EF benefits for mindful movement practices and the approach with the second-best results is a whopping 33%. When looking at the percentage of studies showing clear evidence of EF benefits, a 10% difference between the results for mindfulness practices involving movement and the next-most-successful approach is found. These results for mindful movement should be followed up.

More studies of the mindfulness practices involving movement that have already been studied are needed, with more EF outcome measures and more longitudinal follow-up. We would also encourage research of possible EF benefits from other mindful movement practices, such as aikido, judo, qigong, or the Niroga program (Frank et al., 2012; Frank, Kohler, Peal, & Bose, 2017). We are perplexed that the results have been so mixed for yoga, with two studies finding outstanding EF results (Gothe et al., 2014, 2016, 2017; Manjunath & Telles, 2001) but the other six studies finding less. Research exploring the possible reasons for this or better identifying the conditions under which yoga is most beneficial for EFs would be most welcome.

The second most successful approach for improving EFs is **promising school programs**. They show consistently better results than any cognitive training approach (computerized or noncomputerized) for improving EFs across all indices we used (see Tables 8.1 and 8.2).

The best results for improving inhibitory control from any approach have been found for school programs with children in preschool through Grade 4 (MindUP, PATHS, Tools of the Mind, and CSRP). To our knowledge, no work has been published of a school program that tried to improve inhibitory control in children beyond Grade 4. That school programs have been the most successful of all approaches at improving inhibitory control matters because inhibitory control seems to be the EF most predictive of long-term outcomes (Miller et al., 2011; Moffitt et al., 2011).

Expectations of individuals who deliver the programs and interventions are likewise important. Although teachers in the control group for the Tools school program study by Diamond et al. (2007) were as enthusiastic and optimistic about the prospects of the new program they were delivering as were the Tools teachers, for many other studies of noncomputerized approaches, the possibility exists that it was the expectations of teachers or trainers that drove the results rather than the program itself. It is critical to have a control condition for which there is great excitement and for which expectations are as high as for the experimental condition.

Programs that are part of a public school's curriculum have several critical advantages. They can reach the most children, in the most economical way, and in the fairest way, in the sense that all schoolchildren can be reached (not just the privileged few who can afford to pay for outside programs). School programs are also able to provide greater doses, frequency, and duration than most other interventions or approaches to improving EFs. That is especially true when EF training is embedded in activities throughout the school day (as is done in Tools and Montessori and to some extent in PATHS and CSRP). Also, school programs can train diverse EFs under very diverse circumstances. Training diverse EFs makes it more likely that more EFs will improve, since transfer is narrow and people generally only improve on what they train on. Training under diverse circumstances makes transfer to other contexts, especially novel ones, more likely. The results suggest that the combination of a lot of training and practice under diverse circumstances appears to be particularly effective. We hope school systems and others will take note of this.

The third most successful approach at improving EFs is **noncomputerized cognitive training** (see Tables 8.1 and 8.2). It falls in the top half of all approaches for improving EFs on all four indices in Tables 8.1 and 8.2. Across all studies of noncomputerized cognitive training, 67% report at least suggestive evidence of EF benefits. EF outcomes from noncomputerized cognitive training are better than those for any type of computerized cognitive training. The higher levels of in-person interaction may account for the encouraging results in comparison to most computerized approaches, especially if the teacher or trainer is supportive and has great confidence that the trainee will succeed.

While for school programs those administering the outcome measures have generally been blind as to who received the intervention, for noncomputerized cognitive training, the norm has unfortunately been the reverse: Those administering the assessment measures have generally not been blind to who was in which condition—notable exceptions being the ACTIVE study (Ball et al., 2002; Rebok et al., 2014; Willis et al., 2006), Cheng et al. (2012), and Mackey et al. (2011). Without blinding, it is possible for tester expectations to affect the results.

Outstanding results for WM have been found in two studies from a group that used noncomputerized complex-span training (Borella et al., 2010; Carretti et al., 2013). It will be interesting to see if these results hold up in other studies by other groups, especially since it is surprising to see such good results from such a minimal amount of training (only three 1-hour sessions over 2 weeks).

The fourth most-successful approach for improving EFs is **Cogmed**. It (like the approaches ranked first, second, and third) was ranked in the top half of all approaches for improving EFs on all four indices in both Table 8.1 and 8.2. No other approaches can claim this. No other computerized cognitive training approach shows results as good as those for Cogmed. No other computerized cognitive training approach ranks in the top half of all approaches. Cogmed certainly succeeds in improving the aspects of WM it trains. It is the only method of computerized training to consistently show sustained near-transfer benefits. Benefits to WM from Cogmed have been shown to last for 3 to 6 months and even for 1 year. It may also improve other aspects of WM and attention. More studies are needed to see if Cogmed improves WM and perhaps attention in school situations and in other arenas of life. Selective attention (indeed, interference control in general of both internal and external distractions) is closely tied to WM. We would not be surprised if WM training improved interference control, including selective attention, but convincing evidence in support of that does not presently exist. Cogmed is marketed as beneficial to children with ADHD, yet its generalization to ADHD symptomatology has not been confirmed by blinded observers. Ideally, WM and attention in the real world should be assessed in objective ways; no one who administers or scores the measures should be aware of which children are in the experimental group and which are not.

WM training, whether using Cogmed or N-back tasks, may be a promising approach for older adults beginning to suffer from selective WM deficits. We recommend more study of that. Age-related cognitive decline is often specifically in WM (Hedden & Park, 2001; Park & Payer, 2000; Wang et al., 2011). The one study that tried Cogmed training with older adults (mean age of 64; Brehmer et al., 2012) found those who trained on Cogmed showed more improvement on all four (100%) of their EF near-transfer measures and on the Cognitive Failures Questionnaire than controls who did nonincrementing Cogmed; all improvements were still evident 3 months later. Older adults (mean age of 68) who trained on N-back tasks in the study by Stepankova et al. (2014) improved more and performed better at posttest than no-treatment controls on both of their WM measures (Letter-Number Sequencing and Forward + Backward Digit Span task), visuospatial processing (block design), and visuospatial reasoning (matrix design). Li et al. (2008) found only very narrow transfer in their N-back training study that included older adults (mean age of 74), but their only other outcome measures were complex-span tasks, which one might expect would be insensitive to N-back training. Older adults might well comprise a population in whom computerized WM training could be especially beneficial.

Karbach and Verhaeghen (2014) similarly concluded from their meta-analysis that WM training might be highly effective for older adults with WM decline. A word of caution is warranted here, however, in that the Cogmed study with older adults (Brehmer et al., 2012) and one of the N-back studies with older adults (Stepankova et al., 2014) conducted multiple comparisons without correcting for that in their significance testing. Their results might not look so rosy had they done that.

Do not give up on older folks. EFs can be improved even in those more than 70 years old. Sink et al. (2015) found more EF improvement from their physical-activity training program in those 80 or older than in those younger. Williams and Lord (1997) found EF benefits from enriched aerobic exercise among participants whose mean age was 72. Noncomputerized reasoning and problem-solving training that included real-world tasks improved the reasoning and problem-solving of seniors whose mean age was 74 years, and those benefits were still evident 1, 2, and even 5 years later (the ACTIVE study: Ball et al., 2002; Rebok et al., 2014; Willis et al., 2006). In Karbach and Verhaeghen's (2014) meta-analysis, prolonged practice with computerized WM training showed gains as large for older adults as those for younger adults.

Across all the approaches reviewed here (except aerobic exercise interventions), generally more weeks has produced better results than fewer weeks, within the range of durations studied. Cogmed training has generally been more successful at improving EFs than N-back training. One reason for that might be that the duration of Cogmed training is usually longer (5–8 weeks, vs. 2–5 weeks for N-back). Similarly, Basak et al. (2008) found that 5 weeks of training using Rise of Nations produced better EF benefits than 2½ weeks. Of three mindfulness retreats, the one that lasted longest (13 weeks vs. 1.5 or 4 weeks) and had the most spaced practice (2 hours a day vs. 10 or 11 hours per day) produced the best EF results (MacLean et al., 2010, vs. Chambers et al., 2008, & Zanesco et al., 2013).

There are exceptions, however. For example, two of the studies with the best EF outcomes (Tang et al., 2007, which used IBMT, and Green et al., 2012, which used Cogmed) lasted only 1 week and 4 weeks, respectively. Perhaps studies of aerobic exercise interventions have not found better EF outcomes from longer interventions because most of the interventions (whether more or less successful) have generally lasted far longer than cognitive interventions (for both plain and enriched aerobic exercise programs, the more successful ones lasted on average 16 and 17 weeks, respectively, whereas less successful ones lasted on average 27 and 20 weeks, respectively).

In general, better results have been found with training sessions that lasted 30 min or more than with shorter sessions, though the results for Quadrato Motor Training with only 7-min sessions is a marked exception (Ben-Soussan et al., 2015). Cogmed sessions have generally lasted 30 to 45 min, whereas N-back sessions have generally been shorter (lasting only 15–30 min). Perhaps that is one reason why EF outcomes have generally been better for Cogmed than N-back training. Cogmed has yielded better results with children 7 to 14 years old than with children 4 to 5 years old (the former practiced 30–45 min at a time, the latter only 15 min). Mawjee et al. (2014, 2015) found, however, that, at least for adults, the benefits from 45 min of Cogmed a day were no greater than the benefits from 15 min per day. Davis et al. (2007, 2011) found better EF outcomes from 40-min sessions of enriched aerobic exercise than from 20-min sessions. The benefits from Tools as an add-on to existing curricula are markedly less than when the Tools training in EFs is embedded in all activities throughout the school day (Blair & Raver, 2014; Bodrova & Leong, 2007; Clements et al., 2012; Diamond et al., 2007; Miller et al., 2014).

Perhaps studies of aerobic exercise interventions have generally not found better results from longer sessions because *all* sessions (whether for the more or less successful programs) have lasted more than 30 min, and even the aerobic portion has, in general, (both for the more and less successful programs) been over 30 min (see Table 8.7). On average, sessions in more successful plain and enriched aerobic exercise programs lasted 46 and 56 min, respectively, whereas sessions for plain and enriched aerobic exercise programs less successful at improving EFs were a little longer (57 and 64 min, respectively).[17] Perhaps a session length of about 45 to 55 min is better than one of more than 55 min; at some point, sessions might get too long and produce diminishing returns. The aerobic portion of sessions across more and less successful plain and enriched aerobic exercise interventions has varied from a mean of 35 min (more successful plain aerobic exercise programs) to a mean of 48 min (less successful enriched aerobic exercise programs). Perhaps once a threshold of 30 to 40 min for the aerobic portion is reached, there are no further EF benefits, or even diminishing returns, from going longer than that. Similarly, a study with very long sessions of MBSR (2 hours) found no EF benefits from that (Anderson et al., 2007); the sessions were likely too long (especially for novices).

Most studies have focused on training WM. There is some evidence that training attention or reasoning might produce better results. Two of four studies of attention training (50%; one noncomputerized: Semrud-Clikeman et al., 1999; one computerized: Wass, Porayska-Pomsta, & Johnson, 2011) report at least suggestive evidence of EF benefits. Five of the seven studies of reasoning training (71%; three noncomputerized ones: the ACTIVE study [Ball et al., 2002; Rebok et al., 2014; Willis et al., 2006]; Blieszner et al., 1981; Plemons et al., 1978; one computerized: Corbett et al., 2015; one of both computerized and noncomputerized training: Mackey et al., 2011) found at least suggestive evidence of improved reasoning.

---

[17] FITKids was an outlier here, lasting 120 minutes. Without FITKids, the mean duration of sessions of less successful enriched aerobic programs reviewed here was 57 minutes.

In the ACTIVE study, the better reasoning of those trained on reasoning versus no-treatment controls remained true 1, 2, and even 5 years later (and participants here were older adults). The probably reason why the fourth study of reasoning training (Cheng et al., 2012) did not find suggestive evidence of benefits is because so many participants did not complete the training. Those who completed at least 80% of the training showed better and more improved reasoning 6 months later than those who had trained less, and by 1 year after training, that difference was even greater.

Both studies of exergames (Maillot et al., 2012; Staiano et al., 2012) found suggestive evidence of EF benefits. That merits further investigation.

We would also like to point out the success in improving EF outcomes of targeted training that involves real-world activities—the ACTIVE study (Ball et al., 2002; Rebok et al., 2014; Willis et al., 2006; where training included not only laboratory tasks but also real-world activities such as food preparation and managing a budget), Experience Corps (Carlson et al., 2009), El Sistema music (Holochwost et al., 2017), theater (Noice et al., 2004), and the tantalizing results from three recent studies of sports (Alesi et al., 2014, 2016; Ishihara et al., 2017; Koutsandréou et al., 2016). These, too, merit further investigation.

Across All Approaches, Which Have Been Least Successful Thus Far in Improving EFs?

EF results have been worse for **resistance training** than any other method for improving EFs reviewed here. Resistance training comes in last on three of the four indices for assessing program efficacy in improving EFs in both Tables 8.1 and 8.2. It falls in the bottom half on all four indices in both tables. No study of resistance training found strong evidence of EF benefits; only 22% found even suggestive evidence. Across 30 EF measures investigated across nine studies, resistance training failed to produce better posttest EF performance than the comparison condition on 93% of the measures investigated. A caveat is that resistance training was the active control condition in four of the nine studies reviewed here; in those four studies, investigators had hoped not to find EF benefits for resistance training.

The next-worst results for EF benefits come from studies of aerobic exercise interventions without explicit EF challenges or motor skill demands (**plain aerobic exercise**), like brisk walking or running. That, too, falls in the bottom half on all four metrics in both Tables 8.1 and 8.2.

The third-worst results for broad EF benefits (near and far transfer, including reasoning) come from studies of **computerized complex-span** training.

Although several computerized cognitive training approaches claim to aid ADHD, there is a lack of objective evidence that such training improves ADHD symptoms or academic performance, although unblinded, subjective parent ratings often indicate benefits. Our conclusion here is consistent with those of others (Cortese et al., 2015; Rapport et al., 2013; Sonuga-Barke et al., 2013). For example, Cortese et al. concluded that, "For trials implementing working memory training . . . effects on ADHD were negligible even considering most proximal measures. This suggests that this form of training, which has been widely promoted for use with patients with ADHD, has little or no efficacy for core ADHD symptoms. . . . Crucially, there was . . . no evidence that these effects generalized to important areas of everyday functioning, which themselves are influenced by working memory ability, such as reading and arithmetic" (pp.171–172).

When potentially spurious positive results are excluded, EF benefits from **yoga** are quite disappointing (Table 8.2). A few studies have found outstanding EF results from yoga, but most studies have not. It is unclear why there is such a stark discrepancy across studies. Perhaps the critical difference is how the yoga was presented (were the mindful, spiritual aspects front and center, or was it the physical exercise component?) and/or characteristics of the instructor.

Our prediction that aerobic exercise that trains and challenges EF skills (**enriched aerobic exercise**) would improve EFs more than plain aerobic exercise was supported, but the EF results for enriched aerobic exercise still fall among the bottom half of all approaches investigated. The EF results for enriched aerobic exercise are better than for plain aerobic exercise on all four of our indices, although two of the differences are slight (see Tables 8.1 and 8.2). Thus, EF outcomes for enriched aerobic exercise are better than for plain aerobic exercise, but the results are still relatively poor compared with other methods for improving EFs.

The newer studies of sports (Alesi et al., 2016; Ishihara et al., 2017; Koutsandréou et al., 2016) provide more encouraging results than studies of enriched aerobic exercise that only included sports elements. We think participating in a sport is a more promising approach; this is discussed further below in the section "Our Predictions About How to Most Effectively Improve EFs."

Results for far transfer are generally poor, regardless of the method of training—and all results for resistance training and plain aerobic exercise are far transfer—consistent with the generally disappointing EF benefits from those activities. Of the four Cogmed studies that looked at far transfer to reasoning/fluid intelligence, only one study (25%) found it; Klingberg et al. (2005) found both more improvement and better final test scores. Across all six studies of complex-span training (computerized and noncomputerized), only one study (Borella et al., 2010) found far-transfer benefits to reasoning/fluid intelligence, although all six studies looked. Across all 11 studies of N-back training that looked for benefits to reasoning/fluid intelligence, only four reported even suggestive evidence of that (36%). Three (27%) reported clear evidence (Jaeggi et al., 2010; Rudebeck et al., 2012; Stepankova et al., 2014). Of the five N-back studies with only a no-treatment control group, four (80%) reported at least suggestive evidence of far transfer to reasoning/fluid intelligence, but only one study out of seven (14%) with an active control group found that.

One might say that it is unfair to compare cognitive benefits from cognitive training to those from physical-activity training. Yet, if aerobic exercise and resistance training are promoted specifically as ways to improve EFs, then there should be evidence that they do that. The results for EF benefits are pretty poor for resistance training and plain aerobic exercise even if no comparison to other training methods is made (e.g., not even suggestive evidence of EF benefits in 78% of resistance-training studies and in 67%–69% of aerobic exercise studies; see Tables 8.1 and 8.2). Also, resistance training and aerobic exercise interventions last far longer than do cognitive training programs. For example, the length of Cogmed training is generally 5 to 8 weeks, while the length of complex-span or N-back training is only 2 to 5 weeks. On the other hand, the average length of resistance-training interventions has been 26 weeks, of plain aerobic exercise, 23 weeks, and of enriched aerobic exercise, 19 weeks—on average over 400% longer than cognitive training interventions.

It is highly likely that a great many studies reviewed here were underpowered to find significant EF effects and many did not choose their outcome measures well. On the other hand, surely some studies that failed to find significant EF benefits were never published. That is particularly likely for studies of WM or attention training, since an EF benefit would have been the primary focus of such studies. For only roughly 50% of the approaches reviewed here have at least half the studies reported at least suggestive evidence of EF benefits (see Tables 8.1 and 8.2).

Limitations of the Present Systematic Review and a Call to Others to Analyze the Extant Literature in Ways Other Than We Have

There is no one right way to analyze results across studies. We encourage others to use the rich information provided here on each study in the text and in Tables 8.3 and 8.4 to try different ways to make sense of the results across studies.

For example, others might choose to exclude studies with only one EF outcome measure. Should a study with only one EF outcome measure be said to provide clear or strong evidence, even if the study found more improvement and better posttest scores on that one measure? Should a study with only one EF out-come measure that found no benefits on that be considered as providing negative evidence when, if more EF measures had been included, some EF benefits might have been found?

What about studies that choose several insensitive outcome measures or sev-eral outcome measures insensitive to the kind of approach they were evaluating? Should those studies be considered as providing weak evidence or should the results for their weak measures be discounted? Should outcome measures carry more or less weight depending on their difficulty? We encourage others to come up with alternative perspectives on what constitutes strong or clear evidence and/or what constitutes at least suggestive evidence.

Certainly, studies should be required to report results for all their outcome variables. To the extent that some studies have not done that, it diminishes the validity of conclusions that can be drawn from the literature. The same goes for publishing reports of negative results; valid conclusions cannot be drawn if studies with positive results continue to be published more than those with neg-ative findings.

The same cognitive task is often administered differently or analyzed differ-ently by different researchers, complicating conclusions that can be drawn from the literature. For example, the critical Stroop condition can be a single-task con-dition (say the color of the ink of color words) or a mixed-task condition (say the color of the ink of color words except when the word is in a box, then read the word) and the dependent measure can be the percentage of correct responses on incongruent trials in the critical block, or on all trials in that block, or the difference in accuracy on that block and an easier block. The proportion of in-congruent trials in the critical Stroop block (trials where a color word appears in the ink of another color) can vary from 100% to 33%. Similarly, the proportion of no-go trials in a go/no-go task can vary from 50% to 20% across studies. While Trail-Making is usually administered the same way by all, some researchers use performance on Trails B as their EF measure, whereas others use performance on Trails B minus performance on Trails A. The number of trials administered can also vary widely across studies for tasks that are called by the same name.

The "same" program or intervention can be administered differently by dif-ferent individuals. Too rarely have studies checked or reported fidelity in implementing an intervention, and almost never has consistency across different individuals nominally implementing the same program been checked.

For example, Zeidan et al. (2010) found no EF benefits from MBSR but had participants do it for only 1 week (four sessions total), far shorter than other studies. Anderson et al. (2007) also found no EF benefits from MBSR, but had participants do it only once a week, much less often than other studies, and in far longer sessions than other studies (2 hours). Is it fair to count those studies when evaluating the efficacy of MBSR for improving EFs? The same question applies to a third MBSR study. MacCoon et al. (2014) found no EF benefit from MBSR training but included only one EF measure (CPT) and that assesses sustained attention. Arguably it is selective attention rather than sustained attention that one would predict MBSR might improve. This is an example of using only one EF outcome measure and likely not a sensitive one.

We feel strongly that it is important to show both more improvement and better posttest outcomes than controls. Others may feel it suffices to show better improvement alone.

Others may want to exclude outcome measures that bear a close similarity to the cognitive tasks on which participants were trained.

Some may disagree with what we consider to be EF tests or tests of other cognitive abilities. For example, we have categorized mental rotation tasks as assessing spatial ability rather than EFs, and we have characterized cancellation measures (except those with particularly demanding selective attention demands, such as the d2 Test of Attention) as assessing speed of processing rather than EFs; others might want to categorize these as EF measures.

Similarly, some may disagree with what EF ability we consider a test to be assessing. For example, there is considerable debate about what the Stroop test assesses; we consider it a measure of inhibition. Most people consider N-back tasks to be WM measures, but because the presence of lures puts demands on inhibitory control, we consider N-back tasks to be measures of WM + inhibitory control.

Although we noted where studies did not correct for multiple comparisons or did not analyze their data in accord with group randomization although they had randomized at the level of group (not individuals), we still recorded their findings as they reported them. Others might want to take a more stringent approach toward those studies by omitting them, by asking the authors to re-analyze their data, or by adjusting significance levels or re-analyzing the data themselves. For improvements that reflect the experimental group simply catching up to the control group, we counted those improvements but indicated our skepticism about whether they really reflect genuine benefits from the experimental condition. Others might discount those improvements altogether.

It was very difficult for us to extract whether the experimental group had truly improved more than the control group or whether the experimental group truly performed better at posttest than the control group for a great many studies. It is certainly possible, indeed likely, that despite our best efforts, we have made mistakes in interpreting reported results. Far too few studies reported effect sizes. Some reported just means and standard deviations without giving p values for between-group comparisons. A more enterprising individual might contact the study authors and ask if they might be willing to provide effect sizes and any other missing data analyses, or at least to provide enough information for effect sizes to be calculated.

Certainly, a shortcoming of our review is that we counted each significant result equally without taking effect size into account. We counted the number of significant effects without taking into account the size of any because for a number of studies, we could not determine effect sizes. We strongly encourage others to make a determination of whether strong or clear evidence exists for a program by taking into account the size of the effects.

We included all studies meeting our criteria that we could find regardless of how old they were. Others might want to exclude studies published before a certain date. We excluded studies with fewer than eight participants per group; someone else might want a higher, perhaps much higher, cutoff. One might want to exclude studies where the intervention occurred for fewer than x number of sessions or x number of weeks (our only exclusionary criterion relevant to this was to exclude studies that looked at only a single instance or session). Others might choose to exclude studies where the attrition rate was too high, compliance was too low, evidence was lacking for even the most proximal benefits from the intervention (i.e., no improvement on exactly what was trained, which was exceedingly rare), or where raters or testers were not blind to which condition subjects had participated in.

We did not include in our calculations studies published after 2015, although we noted some such studies when the delay in publication of this volume allowed more time. Conclusions might change with the addition of newer studies. We included three studies that had not done pretesting; others might choose to exclude them. We included some studies that had not used random assignment; others might choose to exclude those.

Others might want to exclude all studies that included only a no-treatment or business-as-usual control group because that provides only a very low bar to pass. Note, however, that for Cogmed, miscellaneous other cognitive training, plain aerobic exercise, and yoga, stronger results for the condition of interest were found when it was compared to an active control condition than when it was compared to no treatment (see Tables 8.9, 8.16, 8.18 and 8.21).

We encourage others to look at the data differently from the way we have and to see how their conclusions confirm or differ from ours. We do not mean our systematic review to be the final say, but simply one credible way to look at the evidence available through 2015.

We offer a caution, though, about the seeming mathematical precision of meta-analyses, except for studies of Cogmed or N-back training. The interventions are simply too diverse; their methods, content, dose, frequency, and duration are too different; and their outcomes measures are too different from one study to another for a meta-analysis to be meaningful. Only for Cogmed and N-back is there sufficient comparability across a number of studies. While several studies of Cogmed and of N-back training have looked at EF improvements using quite similar trainings and similar outcome measures, the same cannot be said for any other approach to improving EFs. There have been several studies of aerobic exercise, resistance training, yoga, and more sedentary mindfulness; however, few have looked at benefits of the same activities and few have used the same outcome measures. Most approaches to improving EFs (except Cogmed and N-back) have had only one or two studies investigating them.

A Call to Researchers to Consider Additional Analyses of Their Data

It might be worthwhile for studies to analyze their results separately for participants initially most behind on EFs, since everything suggests that those individuals are likely to show the largest benefits from the experimental condition. It might also be worthwhile to analyze results separately for those who attended a large percentage (perhaps ≥ 90%) of the sessions for the experimental condition or who showed the most direct benefit from the experimental condition (whether that is most improvement on the cognitive tasks on which they trained or most improvement on fitness or skill measures directly linked to the physical activity on which they trained). Studies might want to analyze results with all participants and a second time excluding participants who attended only a few sessions of the experimental condition. It might be worthwhile to assess mood and/or feelings of efficacy, pride, and/or self-confidence before and after, and do an analysis of the EF results once including only those whose mood or self-confidence improved, since we predict those are the individuals most likely to show the largest benefits from the experimental condition.

There has been much debate about what makes an adequate control condition and what makes an optimal control condition. As Simons et al. (2016, p. 116) wrote, "Just because a control group is active does not mean it is adequate." Time actually spent in the control condition should be comparable to that for the experimental condition. Expectation of benefits should ideally be as high for the control condition as for the experimental one, since we know that expectations can play a large role in any effect (Boot et al., 2013; Rosenthal & Jacobsen, 1968). Since expectations for, and excitement about, something new are usually high, in part simply because it is new, ideally the control condition should be something new.

Klingberg had reasoned that an excellent control for Cogmed would be the same Cogmed games, just without difficulty increasing. Many have criticized this control condition as being too unlikely that participants would expect similar benefits and too boring (potentially affecting participants' motivation to play the training games, the amount of time they spent playing them, and the length of time they remained in the study). Klingberg and his colleagues, to their credit, have tried to collect empirical data on this. They have evidence that participants have not reported feeling significantly more bored by the nonincrementing version of Cogmed than by regular Cogmed (e.g., Bergman Nutley et al., 2011) nor have they dropped out at higher rates. Nevertheless, Klingberg and colleagues have little evidence that participants found either version to be of passionate interest or that participants felt the nonincrementing version to be as exciting or deeply engaging as the regular version.

Mackey et al. (2011) had one group train on reasoning and another group train on speed of processing. These were equally interesting to participants, but the two abilities are not independent. A smaller difference might have been found between groups than if one of the conditions had been more unrelated to EFs, such as recall or recognition memory or perceptual discrimination. Any new skill will require EFs initially to acquire the skill, and that can potentially reduce between-group differences in outcomes.

Matching experimental and control conditions extremely well, where efforts are made to vary only one variable—as Schmidt et al. (2015) attempted to do when they had a high physical demand and high cognitive demand condition and a high physical demand and low cognitive demand condition—may match conditions so well that it is difficult to find significantly stronger benefits from the condition of interest.

We are impressed by the use of visual search (cognitively demanding but not requiring EFs) as the control condition for N-back training by Redick et al. (2013), single-task training as the control condition in many studies of task-switching training (Dörrenbächer et al., 2014; Karbach & Kray, 2009; Kray et al., 2012; Pereg et al., 2013), visual-search training as two control conditons for complex-span training (Harrison et al., 2013), visual-perception training as a control condition for attention training (Tucha et al., 2011), watching children's videos or TV clips and images as the control condition for training young children on computer games requiring inhibition and for training infants on visual attention (Rueda et al., 2005, 2012; Wass et al., 2011), stretching and toning as the control condition for aerobic exercise or resistance training (see Tables 8.18–8.20), and sedentary activities, such as painting or other visual arts or board games, as control conditions for aerobic exercise (Fabre et al., 2002; Krafft, Pierce, et al., 2014; Krafft, Schaeffer, et al., 2014; Krafft, Schwarz, et al., 2014) or theater (Noice et al., 2004).

As the control conditions for learning digital photography or quilting, Park et al. (2014) included one control condition that had similar social group interactions (a social club with common structured activities) and one that specifically omitted that (structured activities done alone that relied on existing knowledge or not empirically shown to improve cognition). That was an admirable study design indeed! The control condition in one of our studies was also good because it pitted two newly introduced curricula against one another (Tools of the Mind and a new curriculum the district itself had developed, for which the district had very high hopes; Diamond et al., 2007).

Finally, we need to be assessing the outcomes we really care about, i.e., the ability to use EFs in real life. Objective, real-world measures of EFs are desperately needed. Right now, by and large, the choice is between objective but arbitrary laboratory tests or subjective questionnaires asking about use of EFs in real-life situations. Perhaps virtual reality technology will provide ways to objectively assess the ability to use EFs in real-life situations.

A Call to Researchers to Study Factors Affecting How Long Benefits Last

Does it matter which method is used to try to sustain benefits (e.g., booster sessions or embedding challenges to EFs in daily activities)? If refresher or booster sessions are used, at what intervals should they be given, and for what duration? Do the answers to these questions differ by type of program, EF component (e.g., WM or response inhibition), the age, gender, or cultural group of the participants, or other variables?

There is a desperate need for more studies that look at benefits months and years after an intervention has ended. How long do benefits last? What affects how long, or if, benefits last? Little is known about whether the length of time that benefits last differs by any characteristic of the participants or mentors or by type of program or activity. What do participants do after an intervention ends—do they continue doing that activity on their own? Do they find other ways to challenge the EFs that were challenged during their training? For school programs, do teachers and programs in subsequent years reinforce the EF-enhancing aspects of a program that produced EF benefits?

No one has looked at whether EF benefits last from any form of physical activity (plain aerobic exercise, resistance training, enhanced aerobic exercise, yoga, martial arts, or anything else) except Taylor-Piliae et al. (2010) and Oswald, Gunzelmann, Rupprecht, and Hagen (2006). Longitudinal follow-up should be done for the most promising physical-activity approaches.

Two studies report benefits still evident 5 years later (Oswald et al., 2006, for balance, coordination, and flexibility training + cognitive training sessions, and less so for cognitive training alone; [18] and the ACTIVE study: Ball et al., 2002; Rebok et al., 2014; Willis et al., 2006, for noncomputerized reasoning training that included some real-world situations).

There were age differences in the longevity of benefits from the Schmiedek et al. (2014) study of intensive computerized training in WM and speed of processing. Younger adults still showed benefits in episodic memory and reasoning/fluid intelligence (though not WM) 2 years later. For older adults, no benefits were evident 2 years later.

Academic benefits from CSRP were still evident 3 years later and were mediated by improved EFs (Li-Grining et al., 2011). The benefit to reading from Tools that had not been significant at the end of kindergarten was significant in the Fall of Grade 1 (Blair & Raver, 2014). The benefit to vocabulary was sustained and expanded to include all children who had received Tools, not just those in high-poverty schools. The effect on mathematics was somewhat reduced by Grade 1. Notably, the benefits to reading, vocabulary, and math relative to controls in Grade 1 were present despite controlling for both pre- and posttest kindergarten results, indicating that benefits continued to accrue over and above those seen at the end of kindergarten. EF benefits from school programs in the early years need replication studies and should try to follow children for several years.

WM benefits from Cogmed have been found 6 months later (Bigorra et al., 2015; Holmes et al., 2009; van der Donk et al., 2015) and 1 year, but not 2 years, later (Roberts et al., 2016). Far-transfer benefits from Cogmed were no longer present 3 months later in Klingberg et al. (2005) but were even larger 6 months later in Bigorra et al. and were present for the first time 6 months later in Holmes

---

[18]   Among older adults, 75 to 93 years old (mean age = 80) at the study's start, Oswald et al. (2006) found that 5 years later those in the no-treatment, psychoeducational, physical activity, or combined psychoeducation + physical activity conditions showed declines on the study's cognitive measures. Those who had received combined cognitive and physical training or cognitive training alone showed significant cognitive preservation that was still evident 5 years later. The scores for the combined cognitive and physical training group were the highest of any group 5 years later on reasoning (WAIS similarities), several memory tasks, and speed of processing. (The term *highest scores* here means showing the least decline.) The physical training in the Oswald et al. (2006) study involved no resistance training and little aerobic exercise. It concentrated instead on balance, eye–hand coordination, motor coordination, and flexibility, including movements from gymnastics, dance, and yoga, although it also included playing tennis and table tennis. The cognitive training included practice on visual-search tasks, a maze task, and a Stroop word-color task (with an emphasis on speed), and lots of memory tasks (e.g., remembering phone numbers, shopping lists, and names) where memory strategies were taught. The psychoeducational intervention involved lectures, group discussions, exercises, and role play on everyday problems (e.g., avoiding falls, dealing with the death of a loved one, nutrition, and understanding prescription labels). Trainings were administered to small groups of 15 to 20 persons every week or two for a total of 30 sessions. The cognitive and psychoeducational sessions were 90 min; physical exercise was 45 min; the combined cognitive training plus physical training was 90 + 45 min (135 min) as was the combined psychoeducation and physical training. (This study was not included in our tables of calculations because pre- to post-test change scores are only reported averaged across multiple EF and non-EF domains. We mention it here because for the 5-year follow-up results, they report outcomes for individual measures, including measures of EFs.)

et al. Parental ratings on the BRIEF in the Bigorra et al. study did not significantly differ between the Cogmed and control group right after training, but 6 months later the parents of those who had done Cogmed saw more improvements than the parents of controls. Teachers saw some benefits right away, but they saw more and larger benefits 6 months later compared with their ratings of children in the control group. Holmes et al. found a benefit to mathematical reasoning 6 months after Cogmed that had not been evident immediately after Cogmed training. Roberts et al. reported only on math performance 2 years after training, but they found that those who had trained on Cogmed performed worse in math 2 years later than others who had received regular classroom instruction rather than CogMed.

Benefits from N-back training have been shown to last 2 to 5 months for N-back performance itself but not for other EF measures (Pugin et al., 2014). Benefits for reasoning from reasoning training have been found even 5 years later (Ball et al., 2002; Willis et al., 2006). Benefits from 100 sessions of processing speed and WM training were not present for WM 2 years later, but benefits to episodic memory and fluid intelligence (present immediately after training for younger, but not older, adults) were still evident 2 years later (Schmiedek et al., 2014). Benefits from complex-span training were not still evident 18 months later (Dahlin, Nyberg, et al., 2008). Benefits to inhibitory control from BrainGame Brian were still evident 3 months later, though the benefit to visuo-spatial WM relative to controls was slightly reduced 3 months later (Dovis et al., 2015). Benefits from a 3-month Shamatha mindfulness retreat were still evident 5 months later (MacLean et al., 2010).

The ACTIVE study and Taylor-Piliae et al. (2010) are the only studies reviewed here that looked at whether continued practice or booster sessions could help after an intervention had ended. Taylor-Piliae et al. asked participants to continue doing what they had been assigned to (t'ai chi or aerobic exercise + resistance training & flexibility training)—one class a week and three sessions at home per week—during the 6 months between when the programs ended and follow-up assessment. They found that at 6-month follow-up assessment, the t'ai chi group had not only maintained its superiority in WM but had improved even more in WM. The ACTIVE study (Ball et al., 2002; Rebok et al., 2014; Willis et al., 2006) found that booster sessions seemed to help preserve the reasoning gains longer from their training. We know very little about when to give booster sessions or what the best durations or frequencies might be.

As already mentioned, it is unrealistic to expect benefits, much less lasting ones, if participants do not attend training sessions. Attendance (compliance) should be monitored and reported. As also mentioned, it is unrealistic to expect benefits to last indefinitely if one does not continue using and challenging the trained skills and continue doing the kind of activity that led to the improvement in the first place. No study of EF benefits has looked at whether or how long participants continue to do the activity they were trained on or continue to challenge the EF skills on which they were trained. Presumably using and challenging the skills needs to be embedded in one's regular routine, or at least periodic booster or refresher sessions should be offered. Research is sorely needed on whether these assumptions are correct.

What About Training People in Strategies to Minimize the Need for EFs, so That People Do Not Have to Expend So Much Effort Trying to Exercise EFs?

Most EF tasks assess the ability to exercise EFs, but on Mischel's delay-of-gratification task (Mischel, Ayduk, & Mendoza-Denton, 2003), the children who succeed usually do so by finding ways to minimize the EF demands (e.g., by looking away or finding something else to do so they are not so tempted to eat the treat). Much of Baumeister's work, too, finds better self-control outcomes in adults who find ways to reduce the need for self-control is not taxed as much (Baumeister & Alquist, 2009; Muraven & Baumeister, 2000).

It makes good sense to avoid expending effort needlessly. Thus, in addition to helping people improve their EFs, also teaching them how to minimize the demands on their EFs (e.g., by writing themselves notes, thereby reducing WM demands, or by placing unhealthy foods out of sight, so that less willpower is required to avoid eating them) might end up being one of the best ways for people to achieve optimal outcomes.

It would be of great interest to see training studies move beyond *only* trying to improve EFs to start trying to help people be more efficient in their deployment of EFs, learning ways to minimize or circumvent the need for EFs wherever possible. Why expend all the effort to exercise EFs when you could achieve the same excellent result without using EFs, thus saving your finite energy resources for when you really need them?

What About Looking at the EF Benefits of Being Outside in Nature?

There has been very little study of the benefits for EFs of being in nature. This might be well worth looking into. Perhaps the EF benefits from brisk walking have generally been better than for any other form of plain aerobic exercise or resistance training because walking was more often done outdoors than other types of exercise in those studies. Some mindful movement activities, such as t'ai chi, are usually done outdoors, but others like taekwondo or Quadrato are not.

One study found that children with ADHD concentrated better after walking in a park (Faber Taylor & Kuo, 2009). Indeed, the effect sizes were so impressive that the authors suggested that "doses of nature" might serve as a safe and inexpensive way to manage ADHD symptoms. Other researchers have fatigued the attention of participants and then had them spend 40 min walking in the natural environment, walking in an urban environment, or listening to soft music and relaxing (Hartig, Mang, & Evans, 1991). They found that those who walked in nature performed better on proofreading (requiring focused concentration) than those in the other two conditions. In a follow-up study, Hartig, Evans, Jamner, Davis, and Gärling (2003) found that a walk in a nature reserve improved performance on an attention test, reduced stress, and increased positive feelings.

Pesce, Masci, et al. (2016) conducted an RCT with 460 children 5 to 10 years old that contrasted 6 months of weekly physical education games with more cognitive and motor skill demands with traditional PE. Children in the enriched PE exhibited more improvement in inhibitory control than children in traditional PE. Importantly, time playing outdoors seemed to be critical to this effect. Only when the training in ball skills was paralleled by a medium-to-high level of outdoor play was this effect evident. The authors concluded, "Outdoor play appears to offer the natural ground for the stimulation by designed PA games to take root in children's mind" (Pesce, Masci, et al., 2016, p. 1).

Another recent study found greater psychological and health benefits from physical activity done outside in nature than from the same activities done inside (Calogiuri et al., 2015). Atchley, Strayer, and Atchley (2012) report an impressive 50% improvement in EFs after participants had taken part in a 4- to 6-day wilderness hiking trip. Kaplan (1995) and Atchley et al. (2012) have theorized that natural environments help to restore attention because people do not have to work so hard to concentrate in nature, there are fewer distractions; that "rest," they theorize, helps to restore attention and the ability to concentrate and focus. Ulrich (1983) has theorized that, because of our evolutionary past, the visual and aesthetic properties of nature produce an automatic response that can reduce stress and evoke positive emotions.

Our Predictions About How to Most Effectively Improve EFs

We predict that the activities that will most successfully improve EFs will include each of the following elements:

- They will tax EFs, continually challenging them in new and different ways.
- They will be personally meaningful and relevant, inspiring a deep commitment and emotional investment on the part of participants to the activity and perhaps also to one another.

- They will have a mentor or guide who firmly believes in the efficacy of the activity and is supportive (sincerely cares about and believes steadfastly in the individual participants).
- They will provide joy, reduce feelings of stress and loneliness and inspire self-confidence and pride.

What activities are most likely to have those characteristics? We propose the answer is *real-world activities*, as studies of El Sistema music (Holochwost et al., 2017), Experience Corps (Carlson et al., 2009), theater (Noice et al., 2004), and sports (Ishihara et al., 2017; Koutsandréou et al., 2016) suggest. We predict that a great many activities not yet studied for their possible EF benefits might well improve EFs, including *group musical activities* (such as band, choral singing, or a drumming circle; Ho, Tsao, Bloch, & Zeltzer, 2011; Metzler-Baddeley et al., 2014; Smith, Viljoen, & McGeachie, 2014), *mindful movement activities* (such as aikido, judo, jiujitsu, qigong, and taekkyeon), *sports* (such as basketball, synchronized swimming, rock climbing, or rowing crew), *other physical activities* (such as orienteering, wilderness survival, or youth circus; Bolton, 2004; Davis & Agans, 2013), *communal dance forms* (such as contradance, hip hop, and rueda (Gill, 2009), *other creative activities* (such as filmmaking), *social-service activities* (such as "Free the Children"—now called "WE"; Kielburger & Kielburger, 2008; Kielburger & Major, 1999), participating in the *Boy Scouts* or *4-H* (Gestsdóttir & Lerner, 2007; Lerner, Lerner, Bowers, Lewin-Bizan, & von Eye, 2011), *caring for an animal* (Ling, Kelly, & Diamond, 2016; Raina, Waltner-Toews, Bonnet, Woodward, & Abernathy, 1999), musical or physical activities that are less communal, or any number of other activities that tax EFs, engender a strong emotional commitment on the part of participants, have inspiring leaders, bring joy, and build self-confidence.

Continually Challenge EFs in New and Different Ways

Real-world activities train diverse EF skills in diverse situations. Rarely does exactly the same situation occur twice in real life. When EFs are always trained in the same few contexts, the training is less likely to generalize outside those contexts (for a similar arguments, see Moreau & Conway, 2014, and Pesce, Croce, et al., 2016).

It has been known for decades that variable training (or varied practice), where participants are continually presented with novel situations in which to practice a skill, leads to better long-term performance than constant or blocked training with the same materials (Ahissar & Hochstein, 2004; Bransford et al., 1977; Rosenbaum et al., 2001; Schmidt & Bjork, 1992; Shapiro & Schmidt, 1982). A problem with many computerized cognitive training and physical-activity training regimens is there is a fair bit of repetition within a very limited set of contexts. People then become very good in those specific contexts, but such training does not provide a good basis for generalizing to other contexts. People only come to generalize by being presented with lots of different contexts where that skill is needed. School programs that embed training in, and challenges to, EF skills in diverse activities, such as reading, math, and play, capitalize on this principle.

Training regimens typically focus on training one skill at a time, eliminating demands on all other skills for the moment. However, the real world is inherently complex, often requiring multiple EF skills at once or in close succession.[19] Perhaps that is one reason why Ishihara et al. (2017) found better results when youngsters practiced a simplified form of tennis from the start rather than starting the usual way with practicing individual tennis skills in isolation.

The training strategies that produce better long-term results, that encourage learning at a deeper rather than at a more superficial level, generally take longer to show benefits. For example, benefits from constant practice are evident earlier than benefits from variable practice, but the transfer from constant practice is exceedingly narrow and task-specific. Similarly, massed practice (practicing a lot over a short period) produces better immediate gains than does spaced practice, but spaced practice (where practice is distributed over time) leads to better long-term gains (Landauer & Bjork, 1978; Rea & Modigliani, 1985; Rosenbaum et al., 2001; Shea & Morgan, 1979).

If we want lasting benefits, we need to be patient. We will likely need to continue the training for longer and will likely need to wait until longer after the training ends to see the full benefit. These assumptions should be tested.

Training diverse skills in parallel takes longer to show gains than training just one skill. For example, Bergman Nutley et al. (2011) found greater immediate

---

[19] Diamond and Ling (2016, p. 40) stated:

Most sports place demands on each of the EFs. Participants need to remember complex movement sequences, mentally work with lots of information, processing in real-time cues such as people's positions and where they will likely go next (for ball sports, cues about the ball's location and trajectory), mentally compare the present situation with past ones, and use that to predict what is likely to happen next or down the line (i.e., they must use WM). Participants need to inhibit attending to distractions and keep their attention focused; they must inhibit a planned action when that is suddenly no longer a good idea and inhibit what might be their first inclination, such as the temptation to try to score oneself rather than passing (i.e., they must use inhibitory control). And, they must use cognitive flexibility: The situation is constantly changing. Participants must quickly and accurately evaluate and respond to those changes, flexibly switching plans in real time, adjusting to the unexpected, adapting to complex and rapidly changing conditions. The situation they are faced with at any moment is often different from anything they have faced before. They can never know for sure what someone else will do; at best they can only predict. Some of this can become automatized and no longer require top-down control, but (a) that is less true for people relatively new to a sport and (b) typically the difficulty of what one is facing keeps increasing. As other players or opponents get better at the sport, the inherent difficulty of what one is faced with increases, providing constant challenge.

benefits to WM and reasoning, respectively, when children were trained only on WM or only on reasoning than when children were trained on both. If we want robust, lasting benefits to diverse EF skills, longer training may be needed and assessment at longer intervals after training might be needed if we are to see the full gains.

Relevant here is a hypothesis championed by both Pesce and Vazou that the difference between physical-activity programs that are successful in improving EFs and those that are not lies in the presence of skilled instructors who use effective teaching methods to create challenging learning contexts that promote mental engagement and the motivation to push oneself and master new skills (Pesce, Masci, et al., 2016; Vazou & Smiley-Oyen, 2014).

Deep Commitment, Passionate Interest, Emotional Investment

We predict that whether participants are emotionally invested in an activity that requires EFs may be key to whether that activity improves EFs. EFs should improve most when people are engaged in activities they care deeply about and for which improving EFs improves performance. Aspects of activities that may lead people to deeply care about the activity (become emotionally invested in it) include feeling the activity matters, having a say in how the activity is done, forming strong personal bonds with others doing or teaching the activity, thoroughly enjoying the activity, gaining feelings of pride, self-confidence, or improved self-esteem from doing the activity, and feeling challenged in a good way by it.

Emotional investment matters because, if someone is deeply committed to an activity, that person will devote great time and effort to it. When doing something you thoroughly enjoy, 'work' feels like 'play.' If that activity happens to train and challenge EFs, then sizeable EF improvements should be seen, because it is the time spent practicing, pushing oneself to improve that drives the benefit (Diamond & Lee, 2011; Ericsson, 2006; Ericsson et al., 2009; Ericsson & Towne, 2010). Few of the scores of attempts to improve EFs have looked at participants engaged in anything they deeply care about.

An exception is perhaps Prins et al.'s (2011) study of Cogmed with gaming elements. They found that the gaming elements really sparked the children's interest and the extra time children spent doing Cogmed, beyond that required, was six times greater for the version with gaming elements versus normal Cogmed.

If participants really enjoyed the activity that was studied, they would be more likely to continue doing it after the study. Doing the activity itself and continuing to derive joy from it should help to extend the duration of EF benefits. Research shows that we have better working memory and selective attention when we're happy (Csikszentmihalyi, Abuhamdeh, & Nakamura 2005; Von Hecker & Meiser, 2005; Wendt, Tuckey, & Prosser, 2011; Yang, Yang, & Isen, 2013). The strongest effect is on cognitive flexibility (Hirt et al., 2008). People are able to work more flexibly (Murray, Sujan, Hirt, & Sujan, 1990) and more readily see connections among unusual and atypical members of categories (Isen, Daubman, & Nowicki, 1987; Isen, Johnson, Mertz, & Robinson, 1985) when they are happy.

There is some evidence that EF benefits from any activity may be proportional to how much joy that activity instills (El Haj, Postal, & Allain, 2012; Heyman et al., 2012; Lee, Chan, & Mok, 2010; Raichlen et al., 2012). A possible biological mechanism underlying that could be: When people are enjoying themselves, endocannabinoids (endogenous cannabinoids in the brain) activate the dopamine neurons that project to prefrontal cortex and the nucleus accumbens (Okon-Singer, Hendler, Pessoa, & Shackman, 2015; Wang & Lupica, 2014). The projection to prefrontal cortex aids EFs and the projection to the nucleus accumbens embellishes the experience of pleasure and the willingness to stay on task, endure countless hours of hard work and boring practice, and push oneself to keep improving, all in service of achieving one's goal (Floresco, 2015; see Figure 8.7).

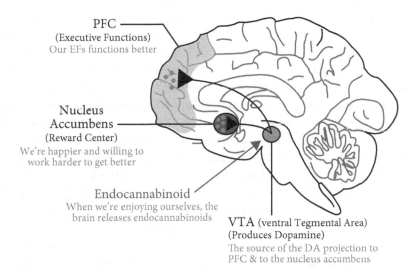

PFC
(Executive Functions)
Our EFs functions better

Nucleus
Accumbens
(Reward Center)
We're happier and willing to
work harder to get better

Endocannabinoid
When we're enjoying ourselves, the
brain releases endocannabinoids

VTA (ventral Tegmental Area)
(Produces Dopamine)
The source of the DA projection to
PFC & to the nucleus accumbens

Figure 8.7. How the brain's release of endocannabinoids might help EFs and the willingness to persevere to achieve a goal. When an individual is happy, endocannabinoids activate the ventral tegmental area (the VTA), the source of the dopamine projection to prefrontal cortex (PFC; central for EFs) and the nucleus accumbens (central to the experience of pleasure and willingness to persevere in the service of a goal).

Real-World Activities Versus Practicing Isolated, Decontextualized Skills

People learn something when it is relevant to (when they need it for) what they want to do (Cordova & Lepper, 1996; Olson, 1964). Training decontextualized skills, isolated from their use in a real-world activity, is unlikely to engender deep personal commitment. We predict that training real-world activities will help EFs more than training isolated skills.

Training people on arbitrary computerized tasks or on skills abstracted from a sport, without ever playing the sport, have thus far produced minimal and/or extremely narrow EF benefits. We often train people on decontexualized, component skills first, such as learning technique on a musical instrument or learning individual sports skills, instead of training them by having them play with others from the start in an orchestra or by having them play a simplified version of the sport from the start. The intriguing results from the study by Ishihara et al. (2017) that if tennis is taught by playing a simplified version of tennis from the outset, children improve more and faster at tennis and EFs, than if tennis is taught the traditional way by first learning and practicing the forehand stroke, then backhand, etc., deserve to be followed up. The same principle applies to El Sistema (playing in an orchestra from the outset) versus traditional music instruction (Booth & Turnstall, 2016).

Empowering Participants by Giving Them a Say

Letting participants have a say in how an activity is organized or conducted increases their commitment to it. When people have a say, they experience more ownership of the activity. It is theirs, rather than something imposed on them. Having input (even about something as trivial as the order in which things are done) has been consistently shown to produce more engagement in the activity and more improvement (Cordova & Lepper, 1996; Hooyman, Wulf, & Lewthwaite, 2014; Iyengar & Lepper, 1999; Khan et al., 2014; Williams et al., 1999), even when participants were instructed to do exactly what they would have chosen to do anyway (Ackerlund Brandt et al., 2015). It is empowering to feel that your opinion and ideas count (Eisman et al., 2016; Eitam, Kennedy, & Higgins, 2013; Larson, 2000; Ryan & Deci, 2000). We predict that people will be more invested in EF training (and experience greater EF gains) if they have even a small say in shaping the training activity. Giving people a voice in shaping an EF-training activity has yet to be tested. A fundamental problem with RCTs is that people randomized to do something usually do not have the same commitment to it as people who chose to do it.

Interpersonal Components

The character and quality of interpersonal aspects of an activity are likely far more important than most EF researchers have appreciated.

Positive Relationship Between the Trainer or Mentor and the Participants
Multiple lines of evidence strongly suggest that personal characteristics of those leading a program have a major impact on how beneficial a program is. This has received too little attention in the EF-training literature and deserves more study. The beneficial effects of someone who believes in and cares about you can be huge. A supportive mentor, who believes in the program and the ability of participants to succeed, who genuinely cares about each individual participating, and who helps build the self-confidence and self-esteem of participants, can be critical to a program's success (Frank, 1961; Freedman, 1993).

Lakes and Hoyt (2004) found tremendous EF benefits when the developer of a Taekwondo program was the person administering it. Not only did he believe in his program, but also by all accounts he is a remarkable human being. Whether the person leading a program is committed to its success, believes firmly in its efficacy, and believes in the participants in the program, and whether the local community is supportive of an intervention and has had a say in crafting it, are just some of the many factors that might be key to why or whether a program is successful.

Cogmed includes a one-on-one in-person mentoring component with a supportive, encouraging adult. One study suggested that that component might be even more decisive for the benefits from Cogmed training than the computerized component that is emphasized (de Jong, 2014). The two times Cogmed has been compared to other programs with significant trainer–participant interaction, the benefits from Cogmed and the other programs have not differed much (Gray et al., 2012; van der Donk et al., 2015).

A deeply caring relationship between the trainer and the children produces the best outcomes. After reviewing copious amounts of data from all over the world, Melhuish concluded that what matters most for early-childhood-education outcomes is not the adult to child ratio, class size, instructional style, or quality of materials. What matters most is the adult–child relationship (Melhuish, 2004; Melhuish, Ereky-Stevens, et al., 2015).

Smith and Smoll have repeatedly found that win–loss records bear little relation to youths' self-esteem, enjoyment of a sport, performance anxiety, or feelings about their coach. Indeed, "Virtually all the systematic variance in outcome was accounted for by differences in coaching behaviors" (Smith & Smoll, 1997, p. 17). The most positive outcomes occurred with coaches who conveyed that they genuinely cared about the youths, were generous in giving praise and in giving encouragement in the face of mistakes, minimized stress (in particular, were never hurtful or mean and never embarrassed a team member), fostered camaraderie, and emphasized the importance of having fun while doing the activity (Smith, Smoll, & Barnett, 1995; Smoll, Smith, Barnett, & Everett, 1993).

The founder of Communities in Schools, Bill Milliken, has famously said, "It's relationships, not programs, that change children. A great program simply creates the environment for healthy relationships to form between adults and children. Young people thrive when adults care about them . . . and when they also have a sense of belonging to a caring community" (https://www. communitiesinschools.org/about-us/).

None of this proves that the relationship between the trainer or mentor and the trainees will prove decisive for EF outcomes, but we predict it will.

Building Social Connections and a Sense of Camaraderie

Humans are fundamentally social (Baumeister & Leary, 1995; Cacioppo & Patrick, 2008). We need to feel liked and accepted. We need to feel we're not alone. Feeling socially excluded not only is painful subjectively, it also activates the same brain network as that for physical pain (Eisenberger, 2012). We thrive when we know beyond a shadow of a doubt that there are people who care about us, believe in us, and are there for us. There is evidence that people tend to be far more invested in an activity if they are working together with others toward an important shared goal (Michael, Sebanz, & Knoblich, 2016). We are often happiest when we feel part of a group working toward a common goal (Putnam, 2000).

Many real-world activities, such as sports, involve working together with others toward a common goal. Some of the best results for improving EFs have come from programs that build feelings of community and connections with others (e.g., Experience Corps; Carlson et al., 2009). It is interesting that in the study by Verghese et al. (2003) that followed almost 500 adults who showed no sign of dementia at age 75 for 5 years, the researchers found that those who did social ballroom dance showed the least signs of dementia, while other physical activities, such as walking, biking, swimming, or participating in group exercise, were not associated with any reduced risk of dementia.

Results of three different meta-analyses indicate that people show greater adherence to an exercise program (fewer missed sessions, longer participation) when they participate in groups (especially cohesive ones without major differences in ability) rather than on their own (Burke, Carron, Ets, Mtoumanis, & Estabrooks, 2006; Carron, Hausenblas, & Mack, 1996; Dishman & Buckworth, 1996). Thus, positive social elements might aid EF benefits in part just by increasing exposure to the activity.

When we're lonely, our EFs suffer (Cacioppo & Patrick, 2008; Campbell et al., 2006). When we feel socially supported, we show better EFs (Cacioppo & Patrick, 2008; Tangney, Baumeister, & Boone, 2004). Feeling alone, without social support, has been shown to impair selective attention, self-control, and reasoning (Baumeister, DeWall, Ciarocco, & Twenge, 2005; Campbell et al., 2006; Twenge, Catanese, & Baumeister, 2002). Even anticipating being alone in the future has been shown to impair logical reasoning (a higher-order EF), although not simple memorization (which does not require EFs; Baumeister, Twenge, & Nuss, 2002).

That was true even in comparison with anticipating other negative experiences. Conversely, simple getting-to-know-you interactions with strangers (without any cooperative goal) was found in one study to boost EFs as much as doing cognitive activities (Ybarra, Winkielman, Yeh, Burnstein, & Kavanagh, 2011).

Minimize Stress and Avoid Negative Experiences

Studies of various mindfulness practices provide evidence in support of our hypothesis that programs that reduce stress will be more effective in improving EFs (for yoga: Bilderbeck et al., 2013; Gothe et al., 2016; Purohit & Pradhan, 2017; for more sedentary meditation: Jha et al., 2010, 2015; Leonard et al., 2013; Napoli et al., 2005). For example, Gothe et al. (2014, 2016, 2017) found that yoga resulted in more reduced stress and anxiety according to both self-report and cortisol measures than was found for the control group, and that yoga produced impressive EF outcomes across all five measures compared to the control group. Similarly, Napoli et al. (2005) found that children in a mindfulness program decreased more in test anxiety and improved more in EFs than other children randomly assigned to reading or other quiet activities.

We would like to underline a finding from Curtis, Smith, & Smoll (1979) and Smith et al. (1983). In both studies, the investigators found that although baseball and basketball coaches rarely engaged in punitive or hostile actions toward the youths they were coaching, those rare behaviors had devastating and disproportionate impacts. It is not enough to usually be supportive. It is an important principle that one negative act, such as humiliating someone, can override the benefit of scores of positive ones.

Montessori (1989) was adamant that one should never embarrass a child. Mentors and program leaders need to create an environment where participants feel safe to take risks and try. That means that participants feel it is okay if they make mistakes. Treating errors and failed attempts as learning opportunities, or as simply what happens when you venture beyond what you are already confident of, has been demonstrated to be important for improving at diverse skills (Blackwell, Trzesniewski, & Dweck, 2007; Dweck, 2002, 2006). We predict that will also be key for improving EFs.

The biological mechanisms by which even mild stress disproportionately affects prefrontal cortex, the brain region that plays a key role in subserving EFs, have been well described. There are more receptors for the stress hormone cortisol in prefrontal cortex than in any other region of the primate brain (Sánchez, Young, Plotsky, & Insel, 2000). Thus, prefrontal cortex is especially sensitive to increases in cortisol. Mild stress markedly increases the amount of the neurotransmitters dopamine and norepinephrine in prefrontal cortex but not elsewhere in the brain (Deutch & Roth, 1990; Finlay, Zigmond, & Abercrombie., 1995). These levels of dopamine and norepinephrine are too high for prefrontal cortex to function properly. Higher levels of dopamine in prefrontal cortex during stress correlate with the degree of EF impairment (Murphy, Arnsten, Goldman-Rakic, & Roth, 1996). High levels of dopamine and norepinephrine interfere with signal transfer from the dendrites to the cell body, impairing signal-to-noise in prefrontal cortex (Marek & Aghajanian, 1999; Yang & Seamans, 1996). High levels of norepinephrine during stress also engage low-affinity norepinephrine receptors (alpha-1 receptors; Arnsten, 2000; Ramos et al., 2005) that impair prefrontal cortex function by reducing neuronal firing (Birnbaum et al., 2004; Mao, Arnsten, & Li, 1999). Indeed, scientists have worked out the intracellular signaling events that open ion channels and weaken prefrontal cortex network connections (Arnsten, 2009). Even mild stress impairs the communication between prefrontal cortex and other brain regions, which impairs EFs (Liston, McEwen, & Casey, 2009).

It is no accident that stress increases both cortisol and catecholamine neurotransmitter levels in prefrontal cortex. In part, prefrontal cortisol receptors regulate prefrontal dopamine and norepinephrine levels. During stress, the higher levels of cortisol block the transporters that would normally clear dopamine and norepinephrine, allowing levels of those neurotransmitters to increase (Grundemann, Schechinger, Rappold, & Schömig, 1998)

Improve Self-Confidence and Increase Feelings of Self-Efficacy

When people feel confident that they are capable of succeeding and believe that through effort theycan improve has been shown, in multiple arenas, to be pivotal in affecting whether people do succeed (Bandura, 1994, 2006; Blackwell et al., 2007; Dweck, 2002, 2006; Murphy & Dweck, 2010). Our expectations about whether or not we can do something have a huge effect on whether we succeed (Aronson et al., 1999; Good, Aronson, & Harder, 2008; Steele & Aronson, 1998). We do not know of any data specifically on the importance of believing in yourself or expecting that you can succeed for whether or how much your EFs improve. However, we predict that these attitudes will be as important for improving EFs as they are for improving on anything else.

It helps people to believe in themselves and to feel proud and self-confident if they are given challenges that are do-able but push their limits (so they can see for themselves that they are capable). It also helps if the trainer, mentor, or teacher shows that he or she firmly believes that the trainee or student will succeed (Rosenthal & Jacobsen, 1968). One way to show faith in someone is to give them an important responsibility. For example, when students who were major discipline problems and poor readers were asked to take on the responsibility of tutoring students several years younger who were struggling with reading, both groups improved significantly in reading, the tutors' school attendance, grades, attitudes toward school, and self-concept improved, and their discipline problems disappeared (Cardenas, Montecel, Supik, & Harris, 1992).

Many of the real-world activities reviewed here take participants repeatedly through a cycle of what had looked impossible becoming easy after hours and hours of practice. Participants see themselves accomplishing things they had never thought possible. That builds confidence.

Final Thoughts

We predict that if a program challenges EFs and brings joy, builds self-confidence, and enhances social well-being, EFs should improve more than if the program focuses only on challenging EFs. That is, supporting the other aspects of an individual (emotional, social, and physical) that support optimal EF performance may be key to seeing benefits and seeing them last.

That prediction is consistent with a theory advanced by Diamond (2013, 2014; Diamond & Ling, 2016; Ling et al., 2016), illustrated in Figure 8.8, which holds that activities that will most successfully improve EFs will not only directly train and challenge EFs, but also indirectly support EFs by lessening things that impair them (like sadness or stress) and by enhancing things that support them (like joy or feeling socially supported or self-confident). People show better EFs when they are happy, feel socially supported, and are healthy and physically fit (Etnier et al., 2006; Hirt et al., 2008; Isen et al., 1987). These are not independent factors. For example, when feelings of being socially supported improve, people also feel happier. The different parts of a person are fundamentally interrelated (Diamond, 2007). Similarly, when people are sad, stressed, lonely, or not physically fit, those conditions impair prefrontal cortex functioning and hence EFs. Indeed, prefrontal cortex and EFs show earlier and greater impairments from sadness, stress, or loneliness than any other brain region or skill or ability (Arnsten, 2015; Baumeister, DeWall, Ciarocco, & Twenge, 2005). Thus, if someone is stressed, sad, or lonely, the very EF skills a program is trying to improve will suffer.

This is a markedly different perspective from that of most EF researchers. Most EF-training studies have focused only on directly training EFs (or improving aerobic fitness to improve EFs), ignoring powerful emotional and social factors that affect EFs. Most EF-training studies have not trained participants on anything they care deeply about.

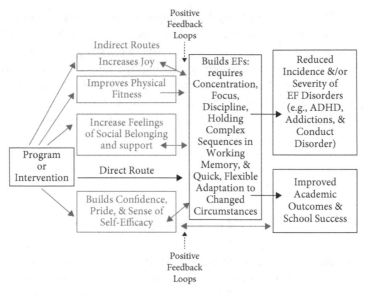

Figure 8.8. Model of direct and indirect routes by which a program or intervention might improve EFs, thereby leading to better school outcomes and the reduced incidence or severity of mental health disorders reflecting poor EFs. While training and challenging EFs are needed for them to improve, that alone may not be enough to achieve the best results. Reprinted with permission from Diamond, A. (2012). Activities and programs that improve children's executive functions. *Current Directions in Psychological Science, 21,* 335–341.

More studies should look at affect, mood, stress levels, and feelings of being socially supported before and after an intervention. Interventions that focus on reducing stress levels should enroll stressed individuals if they want to see sizeable EF benefits.

It could be that the critical difference between the studies where more or fewer EF benefits were found has to do with variables that few studies have reported, such as participants' emotional investment in the training activity, the physical environment in which the activity was done (e.g., outdoors in nature, outside in a city, or indoors), whether the group of participants developed significant camaraderie or not, whether the atmosphere created was one that fostered risk-taking (including risking making a mistake) versus one where participants worried about being embarrassed, and whether the activity leader had a strong conviction that EF benefits would be seen and was supportive rather than punitive.

If a Real-Life Activity Improves EFs (Be It Theatre, Martial Arts, Soccer, a School Curriculum, or Something Else), What Aspect(s) of the Program Are Responsible for That? Why Did the Program Improve EFs?

Our hunch is that the most beneficial programs are gestalts and trying to study just one aspect in isolation will not prove a fruitful endeavor. Beneficial programs work, we hypothesize, because they not only train and challenge EF skills, but also bring joy, pride, and self-confidence, engender a deep commitment, and provide a sense of social belonging (e.g., team membership). For a similar perspective, see Pesce (2012).

For example, soccer is not just aerobic; it requires and builds many fine and gross motor skills, such as eye–hand coordination and balance, requires and builds EF skills, including focused attention, WM, and cognitive flexibility, can build self-confidence and provide great joy, and is social, with all members of a team striving toward a common goal.

The Taekwondo program studied by Lakes and Hoyt (2004) not only worked on physical fitness and motor skills but also trained and challenged EFs and addressed children's social and emotional needs.

Smith and Smoll (1997) found the best outcomes from sports programs where participants helped and supported one another, where they felt the coach genuinely cared about them, where self-confidence was built up and youths' voices were honored, and where participants were not worried about being shamed or embarrassed.

Similarly, the school programs found to improve EFs not only train and challenge diverse EF skills but also address social and emotional needs that support performing at one's best.[20] They build feelings of community and pride, reduce

---

[20] For example, both the Tools program and Montessori curricula for young children (a) embed training in, and challenges to, EFs in all aspects of the school day, (b) provide supports (scaffolds) for weak EFs so all children experience success and can practice trying to exercise EFs, (c) gradually remove supports as children improve (thus progressively increasing difficulty), (d) go to great lengths to avoid having any child feel embarrassed, (e) imbue the attitude that mistakes are learning opportunities, (f) make it quite clear that they have faith in each child and that each and every one will succeed, (g) give children important responsibilities (conveying the message that each is needed and each is capable), (h) give children a say in planning their day and what skills they work on (encouraging feelings of autonomy and empowerment), (i) provide no extrinsic rewards (such as stickers; in Montessori programs there are not even grades; the intrinsic reward from learning is considered sufficient motivation), (j) nurture a feeling of community, where the children help one another, (k) place a strong emphasis on oral language, (l) have the children engage in active, hands-on learning much of the day singly or with one, two, or a few other children, which enables the teacher to (m) provide individual attention (observing carefully and listening with total attention to what a child has to say), (n) provide individual instruction, and (o) permit each child to progress at his or her own rate (individualized pacing). We expect that the whole package is critical to producing the benefits.

interpersonal conflict, try not to embarrass any child, and build in training on, and progressively greater challenges to, self-control, selective attention, and WM into most school activities. Isolating individual aspects to try to determine which is the critical component will likely risk losing the benefits (for a similar perspective, see Park et al., 2007).

Which Will Matter More, the Type of Program or the Way It Is Done?

We predict that the way a program it is done will prove to be more decisive. An example of "Tain't What You Do (It's the Way That You Do It)" is provided by Trulson (1986). He studied two martial arts programs. One was traditional Taekwondo emphasizing self-control and character development. The other was martial arts presented only as a physical activity and competitive sport. The first produced benefits (e.g., less aggression and anxiety and improved self-esteem). The latter produced deficits (e.g., more aggressiveness and diminished self-esteem).

Similarly, a sports program can be destructive if it tears down participants' self-esteem, is relentlessly competitive emphasizing being better than someone else rather than better than one's own past best, abdicates character-building aspects of the activity, or forgets that first and foremost the activity should be fun. Indeed, Smith, Smoll, and their colleagues have repeatedly found that sports programs high in supportiveness produce major benefits to youths' self-esteem and willingness to persist in the face of adversity, whereas sports programs nominally the same (the same sport, with youths of the same age) where supportiveness was low produce the opposite effects on self-esteem and perseverance (Smith & Smoll, 1997). Programs high in supportiveness had four features: Instead of emphasizing competing against others, the coaches emphasized "giving maximum effort and making improvement. The explicit and primary focus [was] on having fun, deriving satisfaction from being on the team, learning sport skills, and increasing self-esteem" (p. 18). Second, the coaches gave a lot of positive reinforcement, encouragement, and sound technical instruction, and avoided responding hostilely or punitively. They specifically tried to reduce youths' fear of failing. Third, through modeling supportive behaviors and praising actions that promoted team unity, the coaches established norms on their teams that emphasized "mutual obligation to help and support one another" (p.18). Fourth, the coaches involved the youths in decisions regarding team rules.

Smith and Smoll (1997, p. 17) concluded, "The most important factor determining outcomes is the manner in which this important social learning situation [i.e., the physical activity] is structured and supervised." We don't know that the effects of program characteristics that Trulson (1986) and Smith and Smoll (1997) reported will be found for EF outcomes, but we predict they will.

Almost any activity could probably be the means for improving EFs as long as it has the elements mentioned above—(1) it keeps taxing EFs in new and different ways, (2) it is personally meaningful, inspiring a deep commitment and emotional investment, (3) it has a mentor who firmly believes in the activity and in the trainees, and (4) it provides joy and camaraderie, reduces feelings of stress, and inspires self-confidence and pride. The way an activity is done will prove, we predict, to be more critical than what the activity is.

References

Ackerlund Brandt, J. A., Dozier, C. L., Juanico, J. F., Laudont, C. L., & Mick, B. R. (2015). The value of choice as a reinforcer for typically developing children. *Journal of Applied Behavior Analysis, 48,* 344–362.

Ackerman, P. L., Kanfer, R., & Calderwood, C. (2010). Use it or lose it? Wii brain exercise practice and reading for domain knowledge. *Psychology and Aging, 25*(4), 753–766.

Ahissar, M., & Hochstein, S. (2004). The reverse hierarchy theory of visual perceptual learning. *Trends in Cognitive Sciences, 8*(10), 457–464.

Ainsworth, B., Eddershaw, R., Meron, D., Baldwin, D. S., & Garner, M. (2013). The effect of focused attention and open monitoring meditation on attention network function in healthy volunteers. *Psychiatry Research, 210*(3), 1226–1231.

Albinet, C. T., Boucard, G., Bouquet, C., & Audiffren, M. (2010). Increased heart rate variability and executive performance after aerobic training in the elderly. *European Journal of Applied Physiology, 109,* 617–624.

Alesi, M., Bianco, A., Luppina, G., Palma, A., & Pepi, A. (2016). Improving children's coordinative skills and executive functions: The effects of a football exercise program. *Perceptual and Motor Skills, 122*(1), 27–46.

Alesi, M., Bianco, A., Padulo, J., Vella, F. P., Petrucci, M., Paoli, A., . . . Pepi, A. (2014). Motor and cognitive development: The role of karate. *Muscle, Ligaments and Tendons Journal, 4*(2), 114–120.

Allen, M., Dietz, M., Blair, K. S., van Beek, M., Rees, G., Vestergaard-Poulsen, P., . . . Roepstorff, A. (2012). Cognitive-affective neural plasticity following active-controlled mindfulness intervention. *Journal of Neuroscience, 32*(44), 15601–15610.

Alloway, T. P., & Alloway, R. G. (2010). Investigating the predictive roles of working memory and IQ in academic attainment. *Journal of Experimental Child Psychology, 106*(1), 20–29.

Alloway, T. P., Bibile, V., & Lau, G. (2013). Computerized working memory training: Can it lead to gains in cognitive skills in students? *Computers in Human Behavior, 29,* 632–638.

Alloway, T. P., Gathercole, S. E., & Pickering, S. J. (2006). Verbal and visuospatial short-term and working memory in children: Are they separable? *Child Development, 77*(6), 1698–1716.

Alloway, T. P., Gathercole, S. E., Willis, C., & Adams, A. M. (2004). A structural analysis of working memory and related cognitive skills in young children. *Journal of Experimental Child Psychology, 87*(2), 85–106.

Anderson, M. C., & Levy, B. (2009). Suppressing unwanted memories. *Current Directions in Psychological Science, 18*(4), 189–194.

Anderson, N. D., Lau, M. A., Segal, Z. V., & Bishop, S. R. (2007). Mindfulness-based stress reduction and attentional control. *Clinical Psychology and Psychotherapy, 14*(6), 449–463.

Angevaren, M., Aufdemkampe, G., Verhaar, H. J. J., Aleman, A., & Vanhees, L. (2008). Physical activity and enhanced fitness to improve cognitive function in older people without known cognitive impairment. *Cochrane Database of Systematic Reviews, 3*, CD005381.

Anguera, J. A., Boccanfuso, J., Rintoul, J. L., Al-Hashimi, O., Faraji, F., Janowich, J., . . . Gazzaley, A. (2013). Video game training enhances cognitive control in older adults. *Nature, 501*, 97–101.

Archontaki, D., Lewis, G. J., & Bates, T. C. (2013). Genetic influences on psychological well-being: A nationally representative twin study. *Journal of Personality, 81*(2), 221–230.

Arnsten, A. F. T. (2000). Stress impairs prefrontal cortical function in rats and monkeys: Role of dopamine D1 and norepinephrine a-1 receptor mechanisms. *Progress in Brain Research, 126*, 183-192.

Arnsten, A. F. T. (2009). Stress signalling pathways that impair prefrontal cortex structure and function. *Nature Reviews Neuroscience, 10*(6), 410–422.

Arnsten, A. F. T. (2015). Stress weakens prefrontal networks: Molecular insults to higher cognition. *Nature Neuroscience, 18*, 1376–1385.

Arnsten, A. F. T., & Goldman-Rakic, P. S. (1998). Noise stress impairs prefrontal cortical cognitive function in monkeys: Evidence for a hyperdopaminergic mechanism. *Archives of General Psychiatry, 55*, 362–368.

Arnsten, A. F. T., Mazure, C. M., & Sinha, R. (2012). This is your brain on meltdown. *Scientific American, 306*(4), 48–53.

Aron, A. R., Behrens, T. E., Smith, S., Frank, M. J., & Poldrack, R. A. (2007). Triangulating a cognitive control network using diffusion-weighted magnetic resonance imaging (MRI) and functional MRI. *Journal of Neuroscience, 27*(14), 3743–3752.

Aronson, E. (1999). The power of self-persuasion. *American Psychologist, 54*(11), 875–884.

Atchley, R. A., Strayer, D. L., & Atchley, P. (2012). Creativity in the wild: Improving creative reasoning through immersion in natural settings. *PLoS One, 7*(12), e51474.

Au, J., Buschkuehl, M., Duncan, G. J., & Jaeggi, S. M. (2016). There is no convincing evidence that working memory training is NOT effective: A reply to Melby-Lervåg and Hulme (2015). *Psychonomic Bulletin & Review, 23*(1), 331–337.

Au, J., Sheehan, E., Tsai, N., Duncan, G. J., Buschkuehl, M., & Jaeggi, S. M. (2015). Improving fluid intelligence with training on working memory: A meta-analysis. *Psychonomic Bulletin & Review, 22*(2), 366–377.

Avtzon, S. A. (2012). Effect of neuroscience-based cognitive skill training on growth of cognitive deficits associated with learning disabilities in children grades 2–4. *Learning Disabilities: A Multidisciplinary Journal, 18*(3), 111–122.

Baddeley, A. (1992). Working memory. *Science, 255*(5044), 556–559.

Baddeley, A. D., & Hitch, G. J. (1994). Developments in the concept of working memory. *Neuropsychology, 8*(5), 485–493.

Bailey, C. E. (2007). Cognitive accuracy and intelligent executive function in the brain and in business. *Annals of New York Academy of Sciences, 1118,* 122–141.

Ball, K., Berch, D. B., Helmers, K. F., Jobe, J. B., Leveck, M. D., Marsiske, M., . . . Willis, S. L. (2002). Effects of cognitive training interventions with older adults: A randomized controlled trial. *Journal of the American Medical Association, 288*(18), 2271–2281.

Baltes, P. B., & Lindenberger, U. (1988). On the range of cognitive plasticity in old age as a function of experience: 15 years of intervention research. *Behavior Therapy, 19,* 283–300.

Bandura, A. (1994). Self-efficacy. In R. J. Corsini (Ed.), *Encyclopedia of Psychology* (2nd edition, Vol. 3, pp. 368–369). New York, NY: Wiley.

Barenberg, J., Berse, T., & Dutke, S. (2011). Executive functions in learning processes: Do they benefit from physical activity? *Educational Research Review, 6*(3), 208–222.

Barnes, D. E., Santos-Modesitt, W., Poelke, G., Kramer, A. F., Castro, C., Middleton, L. E., & Yaffe, K. (2013). The Mental Activity and eXercise (MAX) trial: A randomized controlled trial to enhance cognitive function in older adults. *Journal of the American Medical Association (Internal Medicine), 173,* 797–804.

Basak, C., Boot, W. R., Voss, M. W., & Kramer, A. F. (2008). Can training in a real-time strategy video game attenuate cognitive decline in older adults? *Psychology and Aging, 23*(4), 765–777.

Baumeister, R. F., & Alquist, J. L. (2009). Is there a downside to good self-control? *Self and Identity, 8*(2–3), 115–130.

Baumeister, R. F., DeWall, C. N., Ciarocco, N. J., & Twenge, J. M. (2005). Social exclusion impairs self-regulation. *Journal of Personality and Social Psychology, 88*(4), 589–604.

Baumeister, R. F., & Leary, M. R. (1995). The need to belong: Desire for interpersonal attachments as a fundamental human motivation. *Psychological Bulletin, 117,* 497–529.

Baumeister, R. F., Twenge, J. M., & Nuss, C. K. (2002). Effects of social exclusion on cognitive processes: Anticipated aloneness reduces intelligent thought. *Journal of Personality and Social Psychology, 83,* 817–827.

Becerra, J., Fernández, T., Roca-Stappung, M., Díaz-Comas, L., Galán, L., Bosch, J., . . . Harmony, T. (2012). Neurofeedback in healthy elderly human subjects with electroencephalographic risk for cognitive disorder. *Journal of Alzheimer's Disease, 28,* 357–367.

Beck, S. J., Hanson, C. A., Puffenberger, S. S., Benninger, K. L., & Benninger, W. B. (2010). A controlled trial of working memory training for children and adolescents with ADHD. *Journal of Clinical Child & Adolescent Psychology, 39*(6), 825–836.

Ben-Soussan, T. D., Berkovich-Ohana, A., Piervincenzi, C., Glicksohn, J., & Carducci, F. (2015). Embodied cognitive flexibility and neuroplasticity following Quadrato Motor Training. *Frontiers in Psychology, 6,* 1021.

Ben-Soussan, T. D., Glicksohn, J., & Berkovich-Ohana, A. (2015). From cerebellar activation and connectivity to cognition: A review of the Quadrato Motor Training. *Biomed Research International, 2015,* 954901–954911.

Bergman-Nutley, S., & Klingberg, T. (2014). Effect of working memory training on working memory, arithmetic and following instructions. *Psychological Research, 78,* 869–877.

Bergman Nutley, S., Söderqvist, S., Bryde, S., Thorell, L. B., Humphreys, K., & Klingberg, T. (2011). Gains in fluid intelligence after training non-verbal reasoning in 4-year-old children: A controlled, randomized study. *Developmental Science, 14*(3), 591–601.

Best, J. R. (2010). Effects of physical activity on children's executive function: Contributions of experimental research on aerobic exercise. *Developmental Review, 30*(4), 331–351.

Bigorra, A., Garolera, M., Guijarro, S., & Hervás, A. (2015). Long-term far-transfer effects of working memory training in children with ADHD: A randomized controlled trial. *European Child & Adolescent Psychiatry, 25,* 853–867.

Bilderbeck, A. C., Farias, M., Brazil, I. A., Jakobowitz, S., & Wikholm, C. (2013). Participation in a 10-week course of yoga improves behavioural control and decreases psychological distress in a prison population. *Journal of Psychiatric Research, 47,* 1438–1445.

Birnbaum, S. B., Yuan, P., Wang, M., Vijayraghavan, S., Bloom, A., Davis, D. J., . . . Arnsten, A. F. T. (2004). Protein kinase C overactivity impairs prefrontal cortical regulation of working memory. *Science, 306*(5697), 882–884.

Blair, C. (2002). School readiness: Integrating cognition and emotion in a neurobiological conceptualization of children's functioning at school entry. *American Psychologist, 57*(2), 111–127.

Blair, C., & Raver, C. (2014). Closing the achievement gap through modification of neurocognitive and neuroendocrine function: Results from a cluster randomized controlled trial of an innovative approach to the education of children in kindergarten. *PLoS One, 9*(11), e112393.

Blair, C., & Razza, R. P. (2007). Relating effortful control, executive function, and false-belief understanding to emerging math and literacy ability in kindergarten. *Child Development, 78*(2), 647–663.

Blakey, E., & Carroll, D. J. (2015). A short executive function training program improves preschoolers' working memory. *Frontiers in Psychology, 6,* 1827.

Blieszner, R., Willis, S. L., & Baltes, P. B. (1981). Training research in aging on the fluid ability of inductive reasoning. *Journal of Applied Developmental Psychology*, 2, 247–265.

Blumenthal, J. A., Emery, C. F., Madden, D. J., George, L. K., Coleman, R. E., Riddle, M. W., . . . Williams, R. S. (1989). Cardiovascular and behavioral effects of aerobic exercise training in healthy older men and women. *Journal of Gerontology*, 44(5), M147–M157.

Bodrova, E., & Leong, D. J. (2007). *Tools of the mind: The Vygotskian approach to early childhood education* (2nd ed.). New York, NY: Pearson.

Bolton, R. (2004). *Why circus works: How the values and structures of circus make it a significant developmental experience for young people* (Doctoral dissertation). Murdoch University, Perth, Australia.

Boot, W. R., Simons, D. J., Stothart, C., & Stutts, C. (2013). The pervasive problem with placebos in psychology: Why active control groups are not sufficient to rule out placebo effects. *Perspectives in Psychological Science*, 8(4), 445–454.

Booth, E., & Tunstall, T. (2016). *Playing for their lives: The global El Sistema movement for social change through music*. New York, NY: Norton.

Borella, E., Carretti, B., & Pelgrina, S. (2010). The specific role of inhibition in reading comprehension in good and poor comprehenders. *Journal of Learning Disabilities*, 43(6), 541–552.

Borges, J. G., Ginani, G. E., Hachul, H., Cintra, F. D., Tufik, S., & Pompéia, S. (2013). Executive functioning in obstructive sleep apnea syndrome patients without comorbidities: Focus on the fractionation of executive functions. *Journal of Clinical and Experimental Neuropsychology*, 35(10), 1094–1107.

Boucard, G. K., Albinet, C. T., Bugaiska, A., Bouquet, C. A., Clarys, D., & Audiffren, M. (2012). Impact of physical activity on executive functions in aging: A selective effect on inhibition among old adults. *Journal of Sport & Exercise Psychology*, 34(6), 808–827.

Bransford, J. D., Franks, J. J., Morris, C. D., & Stein, B. S. (1977). Some general constraints on learning and memory research. In L. S. Cermak & F. I. M. Craik (Eds.), *Levels of processing in human memory* (pp. 331–354). Hillsdale, NJ: Erlbaum.

Brehmer, Y., Westerberg, H., & Bäckman, L. (2012). Working-memory training in younger and older adults: Training gains, transfer, and maintenance. *Frontiers in Human Neuroscience*, 6(63), 63.

Broidy, L. M., Nagin, D. S., Tremblay, R. E., Brame, B., Dodge, K. A., Fergusson, D. E., . . . Vitaro, F. (2003). Developmental trajectories of childhood disruptive behaviors and adolescent delinquency: A six-site cross-national study. *Developmental Psychology*, 39(2), 222–245.

Brown, A. K., Liu-Ambrose, T., Tate, R., & Lord, S. R. (2009). The effect of group-based exercise on cognitive performance and mood in seniors residing in intermediate care and self-care retirement facilities: A randomised controlled trial. *British Journal of Sports Medicine*, 43, 608–614.

Brown, T. E., & Landgraf, J. M. (2010). Improvements in executive function correlate with enhanced performance and functioning and health-related quality of life: Evidence from 2 large, double-blind, randomized, placebo-controlled trials in ADHD. *Postgraduate Medicine, 122*(5), 42–51.

Bryck, R. L., & Fisher, P. A. (2012). Training the brain: Practical applications of neural plasticity from the intersection of cognitive neuroscience, developmental psychology, and prevention science. *American Psychologist, 67*(2), 87–100.

Bugos, J. A., Perlstein, W. M., McCrae, C. S., Brophy, T. S., & Bedenbaugh, P. (2007). Individualized piano instruction enhances executive functioning and working memory in older adults. *Aging and Mental Health, 11*, 464–471.

Buitenweg, J. I. V., Murre, J. M. J., & Ridderinkhof, K. R. (2012). Brain training in progress: A review of trainability in healthy seniors. *Frontiers in Human Neuroscience, 6*(eCollection 2012), 183.

Burke, C. A. (2010). Mindfulness-based approaches with children and adolescents: A preliminary review of current research in an emergent field. *Journal of Child and Family Studies, 19*(2), 133–144.

Burke, S. M., Carron, A. V., Eys, M. A., Ntoumanis, N., & Estabrooks, P. A. (2006). Group versus individual approach? A meta-analysis of the effectiveness of interventions to promote physical activity. *Sport and Exercise Psychology Review, 2*(1), 19–35.

Buschkuehl, M. (2011). *N-back training: Benefits and mechanisms.* First International Workshop on Cognitive and Working Memory Training, University of Maryland Center for Advanced Study of Language (CASL), August 24, 2011, College Park, MD.

Buschkuehl, M., Jaeggi, S. M., Hutchison, S., Perrig-Chiello, P., Dapp, C., Muller, M., . . . Perrig, W. J. (2008). Impact of working memory training on memory performance in old-old adults. *Psychology and Aging, 23*(4), 743–753.

Bustamante, E. E., Williams, C. F., & Davis, C. L. (2016). Physical activity interventions for neurocognitive and academic performance in overweight and obese youth: A systematic review. *Pediatric Clinics of North America, 63*, 459–480.

Cacioppo, J., & Patrick, W. (2008). *Loneliness: Human nature and the need for social connection.* New York, NY: W. W. Norton & Co., Inc.

Calogiuri, G., Evensen, K., Weydahl, A., Andersson, K., Patil, G., Ihlebaek, C., & Raanaas, R. K. (2015). Green exercise as a workplace intervention to reduce job stress. Results from a pilot study. *Work, 53*, 99–111.

Campbell, W. K., Krusemark, E. A., Dyckman, K. A., Brunell, A. B., McDowell, J. E., Twenge, J. M., & Clementz, B. A. (2006). A magnetoencephalography investigation of neural correlates for social exclusion and self-control. *Social Neuroscience, 1*, 124–134.

Cárdenas, J. A., Robledo Montecel, M., Supik, J. D., & Harris, R. J. (1992). Coca-Cola Valued Youth Program: Dropout prevention strategies for at-risk students. *Texas Researcher, 3*, 111–130.

Carlson, M. C., Erickson, K. I., Kramer, A. F., Voss, M. W., Bolea, N., Mielke, M., . . . Fried, L. P. (2009). Evidence for neurocognitive plasticity in at-risk older adults: The experience corps program. *The Journals of Gerontology*, *64A*(12), 1275–1282.

Carlson, S. M., & Moses, L. J. (2001). Individual differences in inhibitory control and children's theory of mind. *Child Development*, *72*(4), 1032–1053.

Carretti, B., Borella, E., Zavagnin, M., & Beni, R. (2013). Gains in language comprehension relating to working memory training in healthy older adults. *International Journal of Geriatric Psychiatry*, *28*, 539–546.

Carron, A. V., Hausenblas, H. A., & Mack, D. E. (1996). Social influence and exercise: A meta-analysis. *Journal of Sport and Exercise Psychology*, *18*(1), 1–16.

Cassilhas, R. C., Viana, V. A., Grassmann, V., Santos, R. T., Santos, R. F., Tufik, S., & Mello, M. T. (2007). The impact of resistance exercise on the cognitive function of the elderly. *Medicine and Science in Sports and Exercise*, *39*(8), 1401–1407.

Caviola, S., Mammarella, I. C., Cornoldi, C., & Lucangeli, D. (2009). A metacognitive visuospatial working memory training for children. *International Electronic Journal of Elementary Education*, *2*(1), 122–136.

Cepeda, N. J., Pashler, H., Vul, E., Wixted, J. T., & Rohrer, D. (2006). Distributed practice in verbal recall tasks: A review and quantitative synthesis. *Psychological Bulletin*, *132*, 354–380.

Chacko, A., Bedard, A. C., Marks, D. J., Feirsen, N., Uderman, J. Z., Chimiklis, A., . . . Ramon, M. (2014). A randomized clinical trial of Cogmed working memory training in school-age children with ADHD: A replication in a diverse sample using a control condition. *Journal of Child Psychology and Psychiatry*, *42*, 769–783.

Chaddock, L., Pontifex, M. B., Hillman, C. H., & Kramer, A. F. (2011). A review of the relation of aerobic fitness and physical activity to brain structure and function in children. *International Neuropsychological Society*, *17*(6), 975–985.

Chaddock-Heyman, L., Erickson, K. I., Voss, M. W., Knecht, A. M., Pontifex, M. B., Castelli, D. M., . . . Kramer, A. F. (2013). The effects of physical activity on functional MRI activation associated with cognitive control in children: A randomized controlled intervention. *Frontiers in Human Neuroscience*, *7*(72), 1–12.

Chambers, R., Lo, B. C. Y., & Allen, N. B. (2008). The impact of intensive mindfulness training on attentional control, cognitive style, and affect. *Cognitive Therapy and Research*, *32*(3), 303–322.

Chan, A. S. (2010). *The Shaolin Chanwuyi: A Chinese Chan Buddhism*. Hong Kong: Chanwuyi Publishing.

Chan, A. S., Sze, S. L., Siu, N. Y., Lau, E. M., & Cheung, M.-C. (2013). A Chinese mind-body exercise improves self-control of children with autism: A randomized controlled trial. *PLoS One*, *8*(7), e68184.

Chang, Y. K., Hung, C. L., Huang, C. J., Hatfield, B. D., & Hung, T. M. (2014). Effects of an aquatic exercise program on inhibitory control in children with ADHD: A preliminary study. *Archives of Clinical Neuropsychology*, *29*(3), 217–223.

Chang, Y. K., Pan, C. Y., Chen, F. T., Tsai, C. L., & Huang, C. C. (2012). Effect of resistance-exercise training on cognitive function in healthy older adults: A review. *Journal of Aging and Physical Activity*, *20*, 497–517.

Chang, Y. K., Tsai, Y. J., Chen, T. T., & Hung, T. M. (2013). The impacts of coordinative exercise on executive function in kindergarten children: An ERP study. *Experimental Brain Research, 225*(2), 187–196.

Chein, J. M., & Morrison, A. B. (2010). Expanding the mind's workspace: Training and transfer effects with a complex working memory span task. *Psychonomic Bulletin & Review, 17*(2), 193–199.

Cheng, Y., Wu, W., Feng, W., Wang, J., Chen, Y., Shen, Y., . . . Li, C. (2012). The effects of multi-domain versus single-domain cognitive training in non-demented older people: A randomized controlled trial. *BMC Medicine, 10*(1), 30.

Chooi, W. T., & Thompson, L. A. (2012). Working memory training does not improve intelligence in healthy young adults. *Intelligence, 40*, 531–542.

Chuang, L. Y., Hung, H. Y., Huang, C. J., Chang, Y. K., & Hung, T. M. (2015). A 3-month intervention of Dance Dance Revolution improves interference control in elderly females: A preliminary investigation. *Experimental Brain Research, 233*(4), 1181–1188.

Clements, D. H., Sarama, J., & Layzer, C. (2012). *The efficacy of an intervention synthesizing scaffolding designed to promote self-regulation with an early mathematics curriculum: Effects on executive function.* Paper presented at Society for Research on Educational Effectiveness (SREE) Conference on Understanding Variations in Treatment Effects, March 7, 2012, Washington, DC.

Coe, D. P., Pivarnik, J. M., Womack, C. J., Reeves, M. J., & Malina, R. M. (2006). Effect of physical education and activity levels on academic achievement in children. *Medicine and Science in Sports and Exercise, 38*, 1515–1519.

Cohen, J. D., Perlstein, W. M., Braver, T. S., Nystrom, L. E., Noll, D. C., Jonides, J., & Smith, E. E. (1997). Temporal dynamics of brain activation during a working memory task. *Nature, 386*(6625), 604–608.

Colcombe, S. J., & Kramer, A. F. (2003). Fitness effects on the cognitive function of older adults: A meta-analytic study. *Psychological Science, 14*(2), 125–130.

Cole, M. W., & Schneider, W. (2007). The cognitive control network: Integrated cortical regions with dissociable functions. *NeuroImage, 37*(1), 343–360.

Collins, A., & Koechlin, E. (2012). Reasoning, learning, and creativity: Frontal lobe function and human decision-making. *PLoS Biology, 10*(3), e1001293.

Conway, A. R. A., Kane, M. J., & Engle, R. W. (2003). Working memory capacity and its relation to general intelligence. *Trends in Cognitive Science, 7*(12), 547–552.

Corbett, A., Owen, A., Hampshire, A., Grahn, J., Stenton, R., Dajani, S., . . . Williams, G. (2015). The effect of an online cognitive training package in healthy older adults: An online randomized controlled trial. *Journal of the American Medical Directors Association, 16*, 990–997.

Cordova, D. I., & Lepper, M. R. (1996). Intrinsic motivation and the process of learning: Beneficial effects of contextualization, personalization, and choice. *Journal of Educational Psychology, 88*, 715–730.

Cortese, S., Ferrin, M., Brandeis, D., Buitelaar, J., Daley, D., Dittmann, R. W., . . . Sonuga-Barke, E. J. (2015). Cognitive training for attention-deficit/hyperactivity disorder: Meta-analysis of clinical and neuropsychological outcomes from randomized controlled trials. *Journal of the American Academy of Child and Adolescent Psychiatry, 54*, 164–174.

Cortese, S., Ferrin, M., Brandeis, D., Holtmann, M., Aggensteiner, P., Daley, D., . . . European ADHD Guidelines Group (EAGG). (2016). Neurofeedback for attention-deficit/hyperactivity disorder: Meta-analysis of clinical and neuropsychological outcomes from randomized controlled trials. *Journal of the American Academy of Child & Adolescent Psychiatry, 55*, 444–455.

Crescioni, A. W., Ehrlinger, J., Alquist, J. L., Conlon, K. E., Baumeister, R. F., Schatschneider, C., & Dutton, G. R. (2011). High trait self-control predicts positive health behaviors and success in weight loss. *Journal of Health Psychology, 16*(5), 750–759.

Cserjési, R., Luminet, O., Poncelet, A. S., & Schafer, J. (2009). Altered executive function in obesity. Exploration of the role of affective states on cognitive abilities. *Appetite, 52*(2), 535–539.

Csikszentmihalyi, M., Abuhamdeh, S., & Nakamura, J. (2005). Flow. In A. J. Elliot & C. S. Dweck (Eds.), *Handbook of competence and motivation* (pp. 598–608). New York, NY, US: Guilford Publications.

Curtis, B., Smith, R. E., & Smoll, F. L. (1979). Scrutinizing the skipper: A study of leadership behaviors in the dugout. *Journal of Applied Psychology, 64*, 391–400.

Dahlin, E., Nyberg, L., Backman, L., & Neely, A. S. (2008). Plasticity of executive functioning in young and older adults: Immediate training gains, transfer, and long-term maintenance. *Psychology and Aging, 23*, 720–730.

Dahlin, E., Stigsdotter-Neely, A., Larsson, A., Backman, L., & Nyberg, L. (2008). Transfer of learning after updating training mediated by the striatum. *Science, 320*(5882), 1510–1512.

Dalziell, A., Boyle, J., & Mutrie, N. (2015). Better Movers and Thinkers (BMT): An exploratory study of an innovative approach to physical education. *Europe's Journal of Psychology, 11*, 722–741.

Davidson, M. C., Amso, D., Anderson, L. C., & Diamond, A. (2006). Development of cognitive control and executive functions from 4–13 years: Evidence from manipulations of memory, inhibition, and task switching. *Neuropsychologia, 44*(11), 2037–2078.

Davis, C. L., Tomporowski, P. D., Boyle, C. A., Waller, J. L., Miller, P. H., Naglieri, J. A., & Gregoski, M. (2007). Effects of aerobic exercise on overweight children's cognitive functioning: A randomized controlled trial. *Research Quarterly for Exercise and Sport, 78*(5), 510–519.

Davis, C. L., Tomporowski, P. D., McDowell, J. E., Austin, B. P, Miller, P. H., Yanasak, N. E., . . . Naglieri, J. A. (2011). Exercise improves executive functions and achievement and alters brain activation in overweight children: A randomized, controlled trial. *Health Psychology, 30*(1), 91–98.

Davis, J. C., Marra, C. A., Najafzadeh, M., & Liu-Ambrose, T. (2010). The independent contribution of executive functions to health related quality of life in older women. *BMC Geriatrics, 10*(1), 16–23.

Davis, J. L., & Agans, J. P. (2013). *Community circus: Fostering autonomy, belonging, and competence in youth.* Poster presented at the 5th International Conference on Self-Determination Theory, June 28, 2013, Rochester, NY.

de Jong, P. (2014). *Effects of training working memory in adolescents with a below average IQ.* Paper presented at Workshop on Enhancing Executive Functions in Education, May 20, 2014, Nijmegen, The Netherlands.

De Luca, C. R., & Leventer, R. J. (2008). Developmental trajectories of executive functions across the lifespan. In V. Anderson, R. Jacobs, & P. J. Anderson (Eds.), *Executive functions and the frontal lobes: A lifespan perspective* (pp. 24–56). New York, NY: Taylor & Francis Group.

D'Esposito, M., Aguirre, G. K., Zarahn, E., Ballard, D., Shin, R. K., & Lease, J. (1998). Functional MRI studies of spatial and nonspatial working memory. *Cognitive Brain Research, 7*(1), 1–13.

D'Esposito, M., Detre, J. A., Alsop, D. C., Shin, R. K., Atlas, S., & Grossman, M. (1995). The neural basis of the central executive system of working memory. *Nature, 378*(6554), 279–281.

D'Esposito, M., Postle, B. R., Ballard, D., & Lease, J. (1999). Maintenance versus manipulation of information held in working memory: An event related fMRI study. *Brain and Cognition, 41*(1), 66–86.

Deutch, A. Y., & Roth, R. H. (1990). The determinants of stress-induced activation of the prefrontal cortical dopamine system. *Progress in Brain Research, 85,* 367–403.

de Vries, M., Prins, P. J. M., Schmand, B. A., & Geurts, H. M. (2014). Working memory and cognitive flexibility-training for children with an autism spectrum disorder: A randomized controlled trial. *Journal of Child and Adolescent Psychiatry, 56,* 566–576.

Diamond, A. (1995). Evidence of robust recognition memory early in life even when assessed by reaching behavior. *Journal of Experimental Child Psychology, 59,* 419–456.

Diamond, A. (2006). The early development of executive functions. In E. Bialystok & F. I. M. Craik (Eds.), *Lifespan cognition: Mechanisms of change* (pp. 70–95). New York, NY: Oxford University Press.

Diamond, A. (2007). Interrelated and interdependent. *Developmental Science, 10,* 152–158.

Diamond, A. (2013). Executive functions. *Annual Review of Psychology, 64*(1), 135–168.

Diamond, A. (2014). Want to optimize executive functions and academic outcomes? Simple, just nourish the human spirit. *Minnesota Symposia on Child Psychology, 37,* 205–232.

Diamond, A. (2015). Effects of physical exercise on executive functions: Going beyond simply moving to moving with thought. *Annals of Sport Medicine, 2*(1), 1–5.

Diamond, A., Barnett, W. S., Thomas, J., & Munro, S. (2007). Preschool program improves cognitive control. *Science, 318*(5855), 1387–1388.

Diamond, A., & Lee, K. (2011). Interventions and programs demonstrated to aid executive function development in children 4–12 years of age. *Science, 333*(6045), 959–964.

Diamond, A., Lee, C., Senften, P., Lam, A., & Abbott, D. (accepted). Randomized control trial of Tools of the Mind: Marked benefits to kindergarten children and their teachers. *PLoS One*.

Diamond, A., & Ling, D. S. (2016). Conclusions about interventions, programs, and approaches for improving executive functions that appear justified and those that, despite much hype, do not. *Developmental Cognitive Neuroscience, 18*, 34–48.

Dias, N. M., & Seabra, A. G. (2016). The promotion of executive functioning in a Brazilian public school: A pilot study. *Spanish Journal of Psychology, 18*(e8), 1–14.

Dishman, R. K., & Buckworth, J. (1996). Increasing physical activity: A quantitative synthesis. *Medicine and Science in Sports and Exercise, 28*, 706–719.

Domitrovich, C. E., Cortes, R. C., & Greenberg, M. T. (2007). Improving young children's social and emotional competence: A randomized trial of the preschool "PATHS" curriculum. *The Journal of Primary Prevention, 28*(2), 67–91.

Donnelly, J. E., Hillman, C. H., Castelli, D., Etnier, J. L., Lee, S., Tomporowski, P., . . . Szabo-Reed, A. N. (2016). Physical activity, fitness, cognitive function, and academic achievement in children: A systematic review. *Medicine & Science in Sports & Exercise, 48*(6), 1223–1224.

Dorbath, L., Hasselhorn, M., & Titz, C. (2011). Aging and executive functioning: A training study on focus-switching. *Frontiers in Psychology, 2*, 257.

Dörrenbächer, S., Müller, P. M., Tröger, J., & Kray, J. (2014). Dissociable effects of game elements on motivation and cognition in a task-switching training in middle childhood. *Frontiers in Psychology, 5*, 1275–1296.

Dovis, S., Van der Oord, S., Wiers, R. W., & Prins, P. J. M. (2015). Improving executive functioning in children with ADHD: Training multiple executive functions within the context of a computer game. A randomized double-blind placebo controlled trial. *PLoS One, 10*, e0121651.

Drollette, E. S., Scudder, M. R., Raine, L. B., Moore, R. D., Saliba, B. J., Pontifex, M. B., & Hillman, C. H. (2014). Acute exercise facilitates brain function and cognition in children who need it most: An ERP study of individual differences in inhibitory control capacity. *Developmental Cognitive Neuroscience, 7*, 53–64.

Duncan, G. J., Dowsett, C. J., Claessens, A., Magnuson, K., Huston, A. C., Klebanov, P., . . . Japel, C. (2007). School readiness and later achievement. *Developmental Psychology, 43*(6), 1428–1446.

Dunning, D. L., Holmes, J., & Gathercole, S. E. (2013). Does working memory training lead to generalized improvements in children with low working memory? A randomized controlled trial. *Developmental Science, 16*(6), 915–925.

Dustman, R. E., Ruhling, R. O., Russell, E. M., Shearer, D. E., Bonekat, W., Shigeoka, J. W., . . . Bradford, D. C. (1984). Aerobic exercise training and improved neuropsychological function of older individuals. *Neurobiology of Aging, 5*(1), 35–42.

Dweck, C. S. (2002). The development of ability conceptions. In A. Wigfield & J. S. Eccles (Eds.), *Development of achievement motivation* (pp. 57–88). New York, NY: Academic Press.

Dweck, C. S. (2006). *Mindset: The new psychology of success*. New York: Random House.

Eakin, L., Minde, K., Hechtman, L., Ochs, E., Krane, E., Bouffard, R., . . . Looper, K. (2004). The marital and family functioning of adults with ADHD and their spouses. *Journal of Attention Disorders, 8*(1), 1–10.

Education Scotland Foghlam Alba (n.d.). *Better movers and thinkers.* Retrieved from https://education.gov.scot/improvement/documents/hwb9-better-mover-and-thinkers.pdf)

Egeland, J., Aarlien, A. K., & Saunes, B.-K. (2013). Few effects of far transfer of working memory training in ADHD: A randomized controlled trial. *PLoS One, 8*(10), e75660.

Eisenberger, N. I. (2012). The neural bases of social pain: Evidence for shared representations with physical pain. *Psychosomatic Medicine, 74*, 126–135.

Eisenberg, D. P., & Berman, K. F. (2010). Executive function, neural circuitry, and genetic mechanisms in schizophrenia. *Neuropsychopharmacology, 35*(1), 258–277.

Eisman, A. B., Zimmerman, M. A., Kruger, D., Reischl, T. M., Miller, A. L., Franzen, S. P., & Morrel-Samuels, S. (2016). Psychological empowerment among urban youth: Measurement model and associations with youth outcomes. *American Journal of Community Psychology, 58*, 410–421.

Eitam, B., Kennedy, P. M., & Higgins, E. T. (2013). Motivation from control. *Experimental Brain Research, 229*, 475–484.

El Haj, M., Postal, V., & Allain, P. (2012). Music enhances autobiographical memory in mild Alzheimer's disease. *Educational Gerontology, 38*(1), 30–41.

Eldreth, D. A., Patterson, M. D., Porcelli, A. J., Biswal, B. B., Rebbechi, D., & Rypma, B. (2006). Evidence for multiple manipulation processes in prefrontal cortex. *Brain Research, 1123*(1), 145–156.

Emery, C. F., Schein, R. L., Hauck, E. R., & MacIntyre, N. R. (1998). Psychological and cognitive outcomes of a randomized trial of exercise among patients with chronic obstructive pulmonary disease. *Health Psychology, 17*, 232–240.

Engle, R. W., & Kane, M. J. (2004). Executive attention, working memory capacity, and a two-factor theory of cognitive control. In B. Ross (Ed.), *The psychology of learning and motivation,* Vol. 44 (pp. 145–199). New York, NY: Elsevier.

Engle, R. W., Tuholski, S. W., Laughlin, J. E., & Conway, A. R. (1999). Working memory, short-term memory, and general fluid intelligence: A latent-variable approach. *Journal of Experimental Psychology: General, 128*(3), 309–331.

Ericsson, I. (2017). Effects of physical activity and motor skills acquisition on executive functions and scholastic performance: A review. *Progress in Education, 43*, 71–104.

Ericsson, I., & Karlsson, M. K. (2014). Motor skills and school performance in children with daily physical education in school—A 9-year intervention study. *Scandinavian Journal of Medicine & Science in Sports, 24*, 273–278.

Ericsson, K. A. (2006). The influence of experience and deliberate practice on the development of superior expert performance. In K. A. Ericsson, N. Charness, P. Feltovich, & R. R. Hoffman (Eds.), *Cambridge handbook of expertise and expert performance* (pp. 685–706). Cambridge, UK: Cambridge University Press.

Ericsson, K. A., Nandagopal, K., & Roring, R. W. (2009). Toward a science of exceptional achievement: Attaining superior performance through deliberate practice. *Annals of New York Academy of Sciences, 1172,* 199–217.

Ericsson, K. A., & Towne, T. J. (2010). Expertise. *Wiley Interdisciplinary Reviews: Cognitive Science, 1*(3), 404–416.

Erickson, K. I., Voss, M. W., Prakash, R. S., Basak, C., Szabo, A., Chaddock, L., . . . Kramer, A. F. (2011). Exercise training increases size of hippocampus and improves memory. *Proceedings of the National Academy of Sciences of the United States of America, 108,* 3017–3022.

Espy, K. A. (2004). Using developmental, cognitive, and neuroscience approaches to understand executive control in young children. *Developmental Neuropsychology, 26*(1), 379–384.

Etnier, J. L., Nowell, P. M., Landers, D. M., & Sibley, B. A. (2006). A meta-regression to examine the relationship between aerobic fitness and cognitive performance. *Brain Research Reviews, 52*(1), 119–130.

Etnier, J. L., Salazar, W., Landers, D. M., Petruzzello, S. J., Han, M., & Nowell, P. M. (1997). The influence of physical fitness and exercise upon cognitive functioning: A meta-analysis. *Journal of Sport & Exercise Psychology, 19*(3), 249–277.

Faber Taylor, A., & Kuo, F. E. (2009). Children with attention deficits concentrate better after walk in the park. *Journal of Attention Disorders, 12*(5), 402–409.

Fabio, R. A., Castriciano, C., & Rondanini, A. (2015). ADHD: Auditory and visual stimuli in automatic and controlled processes. *Journal of Attention Disorders, 19*(9), 771–778.

Fabre, C., Chamari, K., Mucci, P., Massé-Biron, J., & Préfaut, C. (2002). Improvement of cognitive function by mental and/or individualized aerobic training in healthy elderly subjects. *International Journal of Sports Medicine, 23*(6), 415–421.

Farah, M. J., Shera, D. M., Savage, J. H., Betancourt, L., Giannetta, J. M., Brodsky, N. L., . . . Hurt, H. (2006). Childhood poverty: Specific associations with neurocognitive development. *Brain Research, 1100,* 166–174.

Fedewa, A. L., & Ahn, S. (2011). The effects of physical activity and physical fitness on children's achievement and cognitive outcomes: A meta-analysis. *Research Quarterly for Exercise and Sport, 82*(3), 521–535.

Fiebach, C. J., Ricker, B., Friederici, A. D., & Jacobs, A. M. (2007). Inhibition and facilitation in visual word recognition: Prefrontal contribution to the orthographic neighborhood size effect. *NeuroImage, 36*(3), 901–911.

Finlay, J. M., Zigmond, M. J., & Abercrombie, E. D. (1995). Increased dopamine and norepinephrine release in medial prefrontal cortex induced by acute and chronic stress: Effects of diazepam. *Neuroscience, 64,* 619–628.

Fisher, F., Boyle, J. M. E., Paton, J. Y., Tomporowski, P. D., Watson, C., McColl, J. H., & Reilly, J. J. (2011). Effects of a physical education intervention on cognitive function in young children: Randomized controlled pilot study. *BMC Pediatrics, 11*(1), 97–105.

Flook, L., Goldberg, S. B., Pinger, L., & Davidson, R. J. (2015). Promoting prosocial behavior and self-regulatory skills in preschool children through a mindfulness-based kindness curriculum. *Developmental Psychology, 51*(1), 44–51.

Flook, L., Smalley, S. L., Kitil, M. J., Galla, B. M., Kaiser-Greenland, S., Locke, J., . . . Kasari, C. (2010). Effects of mindful awareness practices on executive functions in elementary school children. *Journal of Applied School Psychology, 26*(1), 70–95.

Floresco, S. B. (2015). The nucleus accumbens: An interface between cognition, emotion, and action. *Annual Review of Psychology, 66*, 25–52.

Frank, J. L., Bose, B. K., Ancin, D., Hanson, A., Zelmer, V., & Dunlap, J. (2012). *Transformative Life Skills program and training curriculum.* Oakland, CA: Niroga Institute.

Foti, K. E., Eaton, D. K., Lowry, R., & McKnight-Ely, L. R. (2011). Sufficient sleep, physical activity, and sedentary behaviors. *American Journal of Preventive Medicine, 41*(6), 596–602.

Frank, J. (1961). *Persuasion and healing: A comparative study of psychotherapy.* Baltimore, MD: Johns Hopkins University Press.

Frank, J. L., Bose, B. K., Ancin, D., Hanson, A., Zelmer, V., & Dunlap, J. (2012). *Transformative Life Skills program and training curriculum.* Oakland, CA: Niroga Institute.

Frank, J. L., Kohler, K., Peal, A., & Bose, B. (2017). Effectiveness of a school-based yoga program on adolescent mental health and school performance: *Findings from a randomized controlled trial. Mindfulness, 8*, 544–553.

Freeman, S., Eddy, S. L., McDonough, M., Smith, M. K., Okoroafor, N., Jordt, H., & Wenderoth, M. P. (2014). Active learning increases student performance in science, engineering, and mathematics. *Proceedings of the National Academy of Sciences of the United States of America, 111*, 8410–8415.

Freedman, M. (1993). *The kindness of strangers.* Cambridge, UK: Cambridge University Press.

Friedman, N. P., Haberstick, B. C., Willcutt, E. G., Miyake, A., Young, S. E., Corley, R. P., & Hewitt, J. K. (2007). Greater attention problems during childhood predict poorer executive functioning in late adolescence. *Psychological Science, 18*(10), 893–900.

Friedman, N. P., Miyake, A., Robinson, J. L., & Hewitt, J. K. (2011). Developmental trajectories in toddlers' self-restraint predict individual differences in executive functions 14 years later: A behavioral genetic analysis. *Developmental Psychology, 47*(5), 1410–1430.

Gallotta, M. C., Emerenziani, G. P., Iazzoni, S., Meucci, M., Baldari, C., & Guidetti, L. (2015). Impacts of coordinative training on normal weight and overweight/obese children's attentional performance. *Frontiers in Human Neuroscience, 9*, 577.

Gapin, J., & Etnier, J. L. (2010). The relationship between physical activity and executive function performance in children with attention-deficit hyperactivity disorder. *Journal of Sport and Exercise Psychology, 32*, 753–763.

García-Madruga, J. A., Elosúa, M. R., Gil, L., Gómez-Veiga, I., Vila, J. Ó., Orjales, I., . . . Duque, G. (2013). Reading comprehension and working memory's executive processes: An intervention study in primary school students. *Reading Research Quarterly, 48*(2), 155–174.

Gates, N., Singh, M. A. F., Sachdev, P. S., & Valenzuela, M. (2013). The effect of exercise training on cognitive function in older adults with mild cognitive impairment: A meta-analysis of randomized controlled trials. *The American Journal of Geriatric Psychiatry, 21*, 1083–1097.

Gathercole, S., Dunning, D. L., & Holmes, J. (2012). Cogmed training: Let's be realistic about intervention research. *Journal of Applied Research in Memory and Cognition, 1*(3), 201–203.

Gathercole, S. E., Pickering, S. J., Knight, C., & Stegmann, Z. (2004). Working memory skills and educational attainment: Evidence from National Curriculum assessments at 7 and 14 years of age. *Applied Cognitive Psychology, 18*, 1–16.

Gazzaley, A., Clapp, W., McEvoy, K., Knight, R., & D'Esposito, M. (2008). Age-related top-down suppression deficit in the early stages of cortical visual memory processing. *Proceedings of the National Academy of Sciences of the United States of America, 105*(35), 13122–13126.

Gazzaley, A., Cooney, J. W., Rissman, J., & D'Esposito, M. (2005). Top-down suppression deficit underlies working memory impairment in normal aging. *Nature Neuroscience, 8*, 1298–1300.

Gazzeley, A., & D'Esposito, M. (2007). Top-down modulation and normal aging. *Annals of the New York Academy of Sciences, 1097*(1), 67–83.

Gestsdóttir, S., & Lerner, R. M. (2007). Intentional self-regulation and positive youth development in early adolescence: Findings from the 4-H Study of Positive Youth Development. *Developmental Psychology, 43*(2), 508–521.

Gill, S. (2009). *Dancing the rhythms of life: Toward appreciating dancing*. Retrieved from www.SalsAmigos.org

Glass, T. A., Freedman, M., Carlson, M. C., Hill, J., Frick, K. D., Ialongo, N., . . . Fried, L. P. (2004). Experience Corps: Design of an intergenerational program to boost social capital and promote the health of an aging society. *Journal of Urban Health, 81*(1), 94–105.

Gomes, H., Duff, M., Ramos, M., Molholm, S., Foxe, J. J., & Halperin, J. (2012). Auditory selective attention and processing in children with attention-deficit/hyperactivity disorder. *Clinical Neurophysiology, 123*(2), 293–302.

Gomes-Osman, J., Cabral, D. F., Morris, T. P., McInerney, K., Cahalin, L. P., Rundek, T., Oliveira, A., & Pascual-Leone, A. (2018). Exercise for cognitive brain health in aging: A systematic review for an evaluation of dose. *Neurology Clinical Practice, 8*, 257–265. doi:10.1212/CPJ.0000000000000460

Good, C., Aronson, J., & Harder, J. A. (2008). Problems in the pipeline: Stereotype threat and women's achievement in high-level math courses. *Journal of Applied Developmental Psychology, 29*, 17-28.

Gothe, N. P., Keswani, R. K., & McAuley, E. (2016). Yoga practice improves executive function by attenuating stress levels. *Biological Psychology, 121*, 109–116.

Gothe, N. P., Kramer, A. F., & McAuley, E. (2014). The effects of an 8-week Hatha yoga intervention on executive function in older adults. *The Journals of Gerontology Series A: Biological Sciences and Medical Sciences, 69*(9), 1109–1116.

Gothe, N. P., Kramer, A. F., & McAuley, E. (2017). Hatha yoga practice improves attention and processing speed in older adults: Results from an 8-week randomized control trial. *Journal of Alternative and Complementary Medicine, 23*(1), 3–40.

Gothe, N., & McAuley, E. (2015). Yoga and cognition: A meta-analysis of chronic and acute effects. *Psychosomatic Medicine, 77*, 784–797.

Gray, J. R., Chabris, C. F., & Braver, T. S. (2003). Neural mechanisms of general fluid intelligence. *Nature Neuroscience, 6*(3), 316–322.

Gray, S. A., Chaban, P., Martinussen, R., Goldberg, R., Gotlieb, H., Kronitz, R., & Tannock, R. (2012). Effects of a computerized working-memory training program on working memory, attention, and academics in adolescents with severe LD and comorbid ADHD: A randomized controlled trial. *Journal of Child Psychology and Psychiatry, 53*, 1277–1284.

Green, C. S., & Bavelier, D. (2008). Exercising your brain: A review of human brain plasticity and training-induced learning. *Psychology and Aging, 23*, 692–701.

Green, C. T., Long, D. L., Green, D., Losif, A.-M., Dixon, J. F., Miller, M. R., . . . Schweitzer, J. B. (2012). Will working memory training generalize to improve off-task behavior in children with attention-deficit/hyperactivity disorder? *Neurotherapeutics, 9*(3), 639–648.

Greenberg, M. T., & Harris, A. R. (2012). Nurturing mindfulness in children and youth: Current state of research. *Child Development Perspectives, 6*(2), 161–166.

Greenberg, J., Reiner, K., & Meiran, N. (2012). "Mind the trap": Mindfulness practice reduces cognitive rigidity. *PLoS One, 7*, e36206.

Greenberg, J., Reiner, K., & Meiran, N. (2013). "Off with the old": Mindfulness practice improves backward inhibition. *Frontiers in Psychology, 3*, 618–626. doi:10.3389/fpsyg.2012.00618

Gropper, R. J., Gotlieb, H., Kronitz, R., & Tannock, R. (2014). Working memory training in college students with ADHD or LD. *Journal of Attention Disorders, 18*, 331–345.

Gründemann, D., Schechinger, B., Rappold, G. A., & Schömig, E. (1998). Molecular identification of the corticosterone-sensitive extraneuronal catecholamine transporter. *Nature Neuroscience, 1*, 349–351.

Hall, P., Crossley, M., & D'Arcy, C. (2010). Executive function and survival in the context of chronic illness. *Annals of Behavioral Medicine, 39*(2), 119–127.

Hamre, B. K., & Pianta, R. C. (2001). Early teacher-child relationships and the trajectory of children's school outcomes through eighth grade. *Child Development, 72*(2), 625–638.

Hankivsky, O. (2008). *Cost estimates of dropping out of high school in Canada.* Ottawa, Canada: Canadian Council on Learning.

Hardy, J. L., Nelson, R. A., Thomason, M. E., Sternberg, D. A., Katovich, K., Farzin, F., & Scanlon, M. (2015). Enhancing cognitive abilities with comprehensive training: A large, online, randomized, active-controlled trial. *PLoS One, 10*(9), e0134467.

Hariprasad, V. R., Koparde, V., Sivakumar, P. T., Varambally, S., Thirthalli, J., Varghese, M., . . . Gangadhar, B. N. (2013). Randomized clinical trial of yoga-based intervention in residents from elderly homes: Effects on cognitive function. *Indian Journal of Psychiatry, 55*, S357.

Harrison, T. L., Shipstead, Z., Hicks, K. L., Hambrick, D. Z., Redick, T. S., & Engle, R. W. (2013). Working memory training may increase working memory capacity but not fluid intelligence. *Psychological Science, 24*(12), 2409–2419.

Hartig, T., Evans, G. W., Jamner, L. D., Davis, D. S., & Gärling, T. (2003). Tracking restoration in natural and urban field settings. *Journal of Environmental Psychology, 23*(2), 109–123.

Hartig, T., Mang, M., & Evans, G. W. (1991). Restorative effects of natural environment experiences. *Environment and Behavior, 23*(1), 3–26.

Hasher, L., & Zacks, R. T. (1988). Working memory, comprehension, and aging: A review and a new view. In G. H. Bower (Ed.), *The psychology of learning and motivation: Advances in research and theory*, Vol. 22 (pp. 193–225). San Diego, CA: Academic Press.

Hawkes, T., Manselle, W., & Woollacott, M. H. (2013). Cross-sectional comparison of executive attention function in normally aging long-term t'ai chi, meditation, and aerobic fitness practitioners versus sedentary adults. *The Journal of Alternative and Complementary Medicine, 20*, 178–184. 1–7.

Hawn Foundation. (2008). *Mindfulness education*. Miami Beach, FL: Author.

Hedden, T., & Park, D. (2001). Aging and interference in verbal working memory. *Psychology and Aging, 16*, 666–681.

Heeren, A., van Broeck, N., & Philippot, P. (2009). The effects of mindfulness on executive processes and autobiographical memory specificity. *Behaviour Research and Therapy, 47*(5), 403–409.

Helms, D., & Sawtelle, S. M. (2007). A study of the effectiveness of cognitive skill therapy delivered in a video-game format. *Optometry & Vision Development, 38*(1), 19–26.

Herring, M. P., O'Connor, P. J., & Dishman, R. K. (2010). The effect of exercise training on anxiety symptoms among patients: A systematic review. *Archives of Internal Medicine, 170*(4), 321–331.

Heyman, E., Gamelin, F. X., Goekint, M., Piscitelli, F., Roelands, B., Leclair, E., . . . Meeusen, R. (2012). Intense exercise increases circulating endocannabinoid and BDNF levels in humans: Possible implications for reward and depression. *Psychoneuroendocrinology, 37*, 844–851.

Hillman, C. H., Castelli, D. M., & Buck, S. M. (2005). Aerobic fitness and neurocognitive function in healthy preadolescent children. *Medicine and Science in Sports and Exercise, 37*(11), 1967–1974.

Hillman, C. H., Erickson, K. I., & Kramer, A. F. (2008). Be smart, exercise your heart: Exercise effects on brain and cognition. *Nature Reviews Neuroscience, 9*(1), 58–65.

Hillman, C. H., Pontifex, M. B., Castelli, D. M., Khan, N. A., Raine, L. B., Scudder, M. R., . . . Kamijo, K. (2014). Effects of the FITKids randomized controlled trial on executive control and brain function. *Pediatrics, 134*(4), e1063–1071.

Hindin, S. B., & Zelinski, E. M. (2012). Extended practice and aerobic exercise interventions benefit untrained cognitive outcomes in older adults: A meta-analysis. *Journal of the American Geriatrics Society, 60*(1), 136–141.

Hirt, E. R., Devers, E. E., & McCrea, S. M. (2008). I want to be creative: Exploring the role of hedonic contingency theory in the positive mood-cognitive flexibility link. *Journal of Personality and Social Psychology, 94*(2), 214–230.

Ho, P., Tsao, J. C., Bloch, L., & Zeltzer, L. K. (2011). The impact of group drumming on social-emotional behavior in low-income children. *Evidence-based Complementary and Alternative Medicine, 2011,* 250708–250714.

Hoffman, B. M., Blumenthal, J. A., Babyak, M. A., Smith, P. J., Rogers, S. D., Doraiswamy, P. M., & Sherwood, A. (2008). Exercise fails to improve neurocognition in depressed middle-aged and older adults. *Medicine and Science in Sports and Exercise, 40*(7), 1344–1352.

Holmes, J., Gathercole, S. E., & Dunning, D. L. (2009). Adaptive training leads to sustained enhancement of poor working memory in children. *Developmental Science, 12*(4), F9–F15.

Holmes, J., Gathercole, S. E., Place, M., Dunning, D. L., Hilton, K. A., & Elliott, J. G. (2010). Working memory deficits can be overcome: Impacts of training and medication on working memory in children with ADHD. *Applied Cognitive Psychology, 24*(6), 827–836.

Holochwost, S. J., Propper, C. B., Wolf, D. P., Willoughby, M. T., Fisher, K. R., Kolacz, J., . . . Jaffee, S. R. (2017). Music education, academic achievement, and executive functions. *Psychology of Aesthetics, Creativity, and the Arts, 11*(2), 147–166.

Hooyman, A., Wulf, G., & Lewthwaite, R. (2014). Impacts of autonomy-supportive versus controlling instructional language on motor learning. *Human Movement Science, 36,* 190–198.

Hovik, K. T., Saunes, B.-K., Aarlien, A. K., & Egeland, J. (2013). RCT of working memory training in ADHD: Long-term near-transfer effects. *PLoS One, 8*(12), e80561.

Howard-Jones, P. (2014). *Neuroscience and education: A review of educational interventions and approaches informed by neuroscience.* Millbank, UK: Education Endowment Foundation.

Hsu, C. L., Best, J. R., Davis, J. C., Nagamatsu, L. S., Wang, S., Boyd, L. A., . . . Liu-Ambrose, T. (2018). Aerobic exercise promotes executive functions and impacts functional neural activity among older adults with vascular cognitive impairment. *British Journal of Sports Medicine, 52,* 184–186.

Hughes, C. (2005). Executive function and development. In B. Hopkins (Ed.), *Cambridge encyclopedia of child development* (pp. 313–316). Canbridge, UK: Cambridge University Press.

Hughes, C., & Dunn, J. (1998). Understanding mind and emotion: Longitudinal associations with mental-state talk between young friends. *Developmental Psychology, 34*(5), 1026–1037.

Hughes, C., & Ensor, R. (2008). Does executive function matter for preschoolers' problem behaviors? *Journal of Abnormal Child Psychology, 36*(1), 1–14.

Isen, A. M., Daubman, K. A., & Nowicki, G. P. (1987). Positive affect facilitates creative problem solving. *Journal of Personality and Social Psychology, 52,* 1122–1131.

Isen, A. M., Johnson, M. M., Mertz, E., & Robinson, G. F. (1985). The influence of positive affect on the unusualness of word associations. *Journal of Personality and Social Psychology, 48*, 1413–1426.

Ishihara, T., Sugasawa, S., Matsuda, Y., & Mizuno, M. (2017). Improved executive functions in 6–12-year-old children following cognitively engaging tennis lessons. *Journal of Sports Sciences, 35*, 2014–2020.

Ishihara, T., Sugasawa, S., Matsuda, Y., & Mizuno, M. (2017). The beneficial effects of game-based exercise using age-appropriate tennis lessons on the executive functions of 6–12-year-old children. *Neuroscience Letters, 642*, 97–101.

Iyengar, S. S., & Lepper, M. R. (1999). Rethinking the value of choice: A cultural perspective on intrinsic motivation. *Journal of Personality and Social Psychology, 76*(3), 349–366.

Jacob, R., & Parkinson, J. (2015). The potential for school-based interventions that target executive function to improve academic achievement: A review. *Review of Educational Research, 85*(4), 512–552.

Jacques, S., & Marcovitch, S. (2010). Development of executive function across the life span. In W. F. Overton (Ed.), *Cognition, biology and methods across the life-span: Volume 1 of the handbook of life-span development* (pp. 431–466). Hoboken, NJ: Wiley.

Jaeggi, S. M., Buschkuehl, M., Jonides, J., & Perrig, W. J. (2008). Improving fluid intelligence with training on working memory. *Proceedings of the National Academy of Sciences of the United States of America, 105*(19), 6829–6833.

Jaeggi, S. M., Buschkuehl, M., Jonides, J., & Shah, P. (2011). Short- and long-term benefits of cognitive training. *Proceedings of the National Academy of Sciences of the United States of America, 108*(25), 10081–10086.

Jaeggi, S. M., Buschkuehl, M., Jonides, J., & Shah, P. (2012). Cogmed and working memory training—Current challenges and the search for underlying mechanisms. *Journal of Applied Research in Memory and Cognition, 1*, 211–213.

Jaeggi, S. M., Buschkuehl, M., Perrig, W. J., & Meier, B. (2010). The concurrent validity of the N-back task as a working memory measure. *Memory, 18*(4), 394–412.

Jenner, H. (1990). The Pygmalion effect: The importance of expectancies. *Alcoholism Treatment Quarterly, 7*(2), 127–133.

Jensen, C. G., Vangkilde, S., Frokjaer, V., & Hasselbalch, S. G. (2012). Mindfulness training affects attention—or is it attentional effort? *Journal of Experimental Psychology, 14*(1), 106–123.

Jha, A. P., Morrison, A. B., Dainer-Best, J., Parker, S., Rostrup, N., & Stanley, E. A. (2015). Minds "at attention": Mindfulness training curbs attentional lapses in military cohorts. *PLoS One, 10*(2), e0116889.

Jha, A. P., Stanley, E. A., Kiyonaga, A., Wong, L., & Gelfand, L. (2010). Examining the protective effects of mindfulness training on working memory capacity and affective experience. *Emotion, 10*(1), 54–64.

Johnstone, S. J., Roodenrys, S., Blackman, R., Johnston, E., Loveday, K., Mantz, S., & Barratt, M. F. (2012). Neurocognitive training for children with and without AD/HD. *ADHD Attention Deficit and Hyperactivity Disorders, 4*(1), 11–23.

Josefsson, T., Lindwall, M., & Broberg, A. G. (2014). The effects of a short-term mindfulness based intervention on self-reported mindfulness, decentering, executive attention, psychological health, and coping style: Examining unique mindfulness effects and mediators. *Mindfulness, 5*(1), 18–35.

Kabat-Zinn, J. (1990). *Full catastrophe living: Using the wisdom of your body and mind to face stress, pain, and illness*. New York, NY: Dell Publishing.

Kabat-Zinn, J. (1994). *Wherever you go, there you are: Mindfulness meditation in everyday life*. New York, NY: Hachette Books.

Kable, J. W., Calufield, K. M., Falcone, M., McConnell, M., Bernardo, L., Parthasarathi, T., . . . Lerman, C. (2017). No effect of commercial cognitive training on brain activity, choice behavior, or cognitive performance. *Journal of Neuroscience, 37*, 7390–7402.

Kamijo, K., Pontifex, M. B., O'Leary, K. C., Scudder, M. R., Wu, C. T., Castelli, D. M., & Hillman, C. H. (2011). The effects of an afterschool physical activity program on working memory in preadolescent children. *Developmental Science, 14*(5), 1046–1058.

Kane, M. J., Brown, L. H., McVay, J. C., Silvia, P. J., Myin-Germeys, I., & Kwapil, T. R. (2007). For whom the mind wanders, and when: An experience-sampling study of working memory and executive control in daily life. *Psychological Science, 18*(7), 614–621.

Kane, M. J., Conway, A. R., Miura, T. K., & Colflesh, G. J. (2007). Working memory, attention control, and the N-back task: A question of construct validity. *Journal of Experimental Psychology, 33*(3), 615–622.

Kaplan, S. (1995). The restorative benefits of nature: Toward an integrative framework. *Journal of Environmental Psychology, 15*(3), 169–182.

Karbach, J., & Kray, J. (2009). How useful is executive control training? Age differences in near and far transfer of task-switching training. *Developmental Science, 12*(6), 978–990.

Karbach, J., & Verhaeghen, P. (2014). Making working memory work: A meta-analysis of executive-control and working memory training in older adults. *Psychological Science, 25*(11), 2027–2037.

Kelly, M. E., Loughrey, D., Lawlor, B. A., Robertson, I. H., Walsh, C., & Brennan, S. (2014). The impact of exercise on the cognitive functioning of healthy older adults: A systematic review and meta-analysis. *Ageing Research Reviews, 16*, 12–31.

Kenworthy, L., Anthony, L. G., Naiman, D. Q., Cannon, L., Wills, M. C., Luong-Tran, C., . . . Wallace, G. L. (2014). Randomized controlled effectiveness trial of executive function intervention for children on the autism spectrum. *Journal of Child Psychology and Psychiatry, 55*, 374–383.

Kerr, R., & Booth, B. (1978). Specific and varied practice of motor skill. *Perceptual and Motor Skills, 46*(2), 395–401.

Kesler, S., Hadi Hosseini, S. M., Heckler, C., Janelsins, M., Palesh, O., Mustian, K., & Morrow, G. (2013). Cognitive training for improving executive function in chemotherapy-treated breast cancer survivors. *Clinical Breast Cancer, 13*(4), 299–306.

Khan, K., Nelson, K., & Whyte, E. (2014). Children choose their own stories: The impact of choice on children's learning of new narrative skills. *Journal of Child Language, 41,* 949–962.

Khatri, P., Blumenthal, J. A., Babyak, M. A., Craighead, W. E., Herman, S., Baldewicz, T., . . . Krishnan, K. R. (2001). Effects of exercise training on cognitive functioning among depressed older men and women. *Journal of Aging and Physical Activity, 9*(1), 43–57.

Kida, N., Oda, S., & Matsumura, M. (2005). Intensive baseball practice improves the go/nogo reaction time, but not the simple reaction time. *Cognitive Brain Research, 22,* 257–264.

Kielburger, C., & Kielburger, M. (2008). *Me to we: Finding meaning in a material world.* John Wiley & Sons.

Kielburger, C., & Major, K. (1999). *Free the children: A young man fights against child labor and proves that children can change the world.* Harper Collins.

Kim, S. H., Kim, M., Ahn, Y. B., Lim, H. K., Kang, S. G., Cho, J. H., . . . Song, S. W. (2011). Effect of dance exercise on cognitive function in elderly patients with metabolic syndrome: A pilot study. *Journal of Sports Science and Medicine, 10*(4), 671–678.

Kimura, K., Obuchi, S., Arai, T., Nagasawa, H., Yoshitaka, S., Watanabe, S., & Kojima, M. (2010). The influence of short-term strength training on health-related quality of life and executive function. *Journal of Physiological Anthropology, 29,* 95–101.

Kioumourtzoglou, E., Kourtessis, T., Michalopoulou, M., & Derri, V. (1998). Differences in several perceptual abilities between experts and novices in basketball, volleyball, and water-polo. *Perceptual and Motor Skills, 86,* 899–912.

Kit, K. A., Mateer, C. A., Tuokko, H. A., & Spencer-Rodgers, J. (2014). Influence of negative stereotypes and beliefs on neuropsychological test performance in a traumatic brain injury population. *Journal of the International Neuropsychological Society, 20*(2), 157–167.

Klingberg, T. (2010). Training and plasticity of working memory. *Trends in Cognitive Science, 14*(7), 317–324.

Klingberg, T., Fernell, E., Olesen, P., Johnson, M., Gustafsson, P., Dahlstrom, K., . . . Westerberg, H. (2005). Computerized training of working memory in children with ADHD—A randomized, controlled trial. *Journal of American Academy of Child and Adolescent Psychiatry, 44*(2), 177–186.

Klusmann, V., Evers, A., Schwarzer, R., Schlattmann, P., Reischies, F. M., Heuser, I., & Dimeo, F. C. (2010). Complex mental and physical activity in older women and cognitive performance: A 6-month randomized controlled trial. *Journals of Gerontology Series A: Biological Sciences, 65A,* 680–688.

Kochanska, G., Murray, K., & Coy, K. C. (1997). Inhibitory control as a contributor to conscience in childhood: From toddler to early school age. *Child Development, 68*(2), 263–277.

Koutsandréou, F., Wegner, M., Niemann, C., & Budde, H. (2016). Effects of motor versus cardiovascular exercise training on children's working memory. *Medicine & Science in Sports & Exercise, 48,* 1144–1152.

Krafft, C. E., Pierce, J. E., Schwarz, N. F., Chi, L., Weinberger, A. L., Schaeffer, D. J., . . . McDowell, J. E. (2014). An eight month randomized controlled exercise intervention alters resting state synchrony in overweight children. *Neuroscience, 256*, 445–455.

Krafft, C. E., Schaeffer, D. J., Schwarz, N. F., Chi, L. X., Weinberger, A. L., Pierce, J. E., . . . McDowell, J. E. (2014). Improved fronto-parietal white matter integrity in overweight children is associated with attendance in an after-school exercise program. *Developmental Neuroscience, 36*(1), 1–9.

Krafft, C. E., Schwarz, N. F., Chi, L., Weinberger, A. L., Schaeffer, D. J., Pierce, J. E., . . . McDowell, J. E. (2014). An 8-month randomized controlled exercise trial alters brain activation during cognitive tasks in overweight children. *Obesity, 22*(1), 232–242.

Kramer, A. F., & Erickson, K. I. (2007). Capitalizing on cortical plasticity: Influence of physical activity on cognition and brain function. *Trends in Cognitive Sciences, 11*(8), 342–348.

Kramer, A. F., Hahn, S., Cohen, N. J., Banich, M. T., McAuley, E., Harrison, C. R., . . . Colcombe, A. (1999). Ageing, fitness and neurocognitive function. *Nature, 400*(6743), 418–419.

Kray, J., Karbach, J., Haenig, S., & Freitag, C. (2012). Can task-switching training enhance executive control functioning in children with attention deficit/hyperactivity disorder? *Frontiers in Human Neuroscience, 5*, 180.

Kroesbergen, E. H., van't Noordende, J. E., & Kolkman, M. E. (2014). Training working memory in kindergarten children: Effects on working memory and early numeracy. *Child Neuropsychology, 20*(1), 23–37.

Kueider, A. M., Parisi, J. M., Gross, A. L., & Rebok, G. W. (2012). Computerized cognitive training with older adults: A systematic review. *PLoS One, 7*(7), e40588.

Kundu, B., Sutterer, D. W., Emrich, S. M., & Postle, B. R. (2013). Strengthened effective connectivity underlies transfer of working memory training to tests of short-term memory and attention. *The Journal of Neuroscience, 33*(20), 8705–8715.

Kusché, C. A., & Greenberg, M. T. (1994). *The PATHS (Promoting alternative thinking strategies) curriculum*. South Deerfield, MA: Channing-Bete Co.

Kyttälä, M., Kanerva, K., & Kroesbergen, E. (2015). Training counting skills and working memory in preschool. *Scandinavian Journal of Psychology, 56*, 363–370.

Lakes, K. D., Bryars, T., Emmerson, N., Sirisinihal, S., Salim, N., Arastoo, S., . . . Kang, C. J. (2013). The healthy for life Taekwondo pilot study: A preliminary evaluation of effects on executive function and BMI, feasibility, and acceptability. *Mental Health and Physical Activity, 6*, 181–188.

Lakes, K. D., & Hoyt, W. T. (2004). Promoting self-regulation through school-based martial arts training. *Journal of Applied Developmental Psychology, 25*, 283–302.

Landauer, T. K., & Bjork, R. A. (1978). Optimum rehearsal patterns and name learning. In M. M. Gruneberg, P. E. Morris, & R. N. Sykes (Eds.), *Practical aspects of memory* (pp. 625–632). London, UK: Academic Press.

Langlois, F., Vu, T. T. M., Chassé, K., Dupuis, G., Kergoat, M.-J., & Bherer, L. (2013). Benefits of physical exercise training on cognition and quality of life in frail older adults. *Journals of Gerontology Series B-Psychological Sciences and Social Sciences, 68*, 400–404.

Lane, A. M., & Lovejoy, D. J. (2001). The effects of exercise on mood changes: The moderating effect of depressed mood. *Journal of Sports Medicine and Physical Fitness*, 41(4), 539–545.

Larson, R. W. (2000). Toward a psychology of positive youth development. *American Psychologist*, 55(1), 170–183.

Law, L. L. F., Barnett, F., Yau, M. K., & Gray, M. A. (2014). Effects of combined cognitive and exercise interventions on cognition in older adults with and without cognitive impairment: A systematic review. *Ageing Research Reviews*, 15, 61–75.

Leckie, R. L., Oberlin, L. E., Voss, M. W., Prakash, R. S., Szabo-Reed, A., Chaddock-Heyman, L., . . . Erickson, K. I. (2014). BDNF mediates improvements in executive function following a 1-year exercise intervention. *Frontiers in Human Neuroscience*, 8, 985–992.

Lee, Y. Y., Chan, M. F., & Mok, E. (2010). Effectiveness of music intervention on the quality of life of older people. *Journal of Advanced Nursing*, 66, 2677–2687.

Lee, T. D., & Magill, R. A. (1983). The locus of contextual interference in motor-skill acquisition. *Journal of Experimental Psychology: Learning, Memory, and Cognition*, 9, 730–746.

Legault, C., Jennings, J. M., Katula, J. A., Dagenbach, D., Gaussoin, S. A., Sink, K. M., . . . SHARP-P Study Group. (2011). Designing clinical trials for assessing the effects of cognitive training and physical activity interventions on cognitive outcomes: The Seniors Health and Activity Research Program Pilot (SHARP-P) study, a randomized controlled trial. *BMC Geriatrics*, 11(1), 27–37.

Leh, S. E., Petrides, M., & Strafella, A. P. (2010). The neural circuitry of executive functions in healthy subjects and Parkinson's disease. *Neuropsychopharmacology*, 35(1), 70–85.

Lehto, J. E., Juujärvi, P., Kooistra, L., & Pulkkinen, L. (2003). Dimensions of executive functioning: Evidence from children. *British Journal of Developmental Psychology*, 21(1), 59–80.

Leonard, N. R., Jha, A. P., Casarjian, B., Goolsarran, M., Garcia, C., Cleland, C. M., . . . Massey, Z. (2013). Mindfulness training improves attentional task performance in incarcerated youth: A group randomized controlled intervention trial. *Frontiers in Psychology*, 4, 792.

Lerner, R. M., Lerner, J. V., Bowers, E. P., Lewin-Bizan, S., & von Eye, A. (2011). Individual and contextual bases of thriving in adolescence: Findings from the 4-H Study of Positive Youth Development. *Journal of Adolescence*, 34(6), 1107–1114.

Li, A. W., & Goldsmith, C.-A. W. (2012). The effects of yoga on anxiety and stress. *Alternative Medicine Review*, 17(1), 21–35.

Li, S.-C., Schmiedek, F., Huxhold, O., Röcke, C., Smith, J., & Lindenberger, U. (2008). Working memory plasticity in old age: Practice gain, transfer, and maintenance. *Psychology and Aging*, 23(4), 731–742.

Li-Grining, C. P., Raver, C. C., & Pess, R. A. (2011). *Academic impacts of the Chicago School Readiness Project: Testing for evidence in elementary school.* Society for Research in Child Development Biennial Meeting, Montreal, Canada.

Lilienthal, L., Tamez, E., Shelton, J. T., Myerson, J., & Hale, S. (2013). Dual n-back training increases the capacity of the focus of attention. *Psychonomic Bulletin & Review*, *20*(1), 135–141.

Lillard, A., & Else-Quest, N. (2006). The early years: Evaluating Montessori education. *Science*, *313*(5795), 1893–1894.

Ling, D. S., Kelly, M., & Diamond, A. (2016). Human-animal interaction and the development of cognitive control (executive functions). In L. Freund, S. McCune, P. McCardle, L. Esposito, & J. A. Griffin (Eds.), *Social neuroscience of human-animal interaction*. New York, NY: American Psychological Association Press.

Liston, C., McEwen, B. S., & Casey, B. J. (2009). Psychosocial stress reversibly disrupts prefrontal processing and attentional control. *Proceedings of the National Academy of Sciences of the United States of America*, *106*(3), 912–917.

Liu-Ambrose, T., Donaldson, M. G., Ahamed, Y., Graf, P., Cook, W. L., Close, J., . . . Khan, K. M. (2008). Otago home-based strength and balance retraining improves executive functioning in older fallers: A randomized controlled trial. *Journal of American Geriatrics Society*, *56*, 1821–1830.

Liu-Ambrose, T., Nagamatsu, L. S., Graf, P., Beattie, B. L., Ashe, M. C., & Handy, T. C. (2010). Resistance training and executive functions: A 12-month randomized controlled trial. *Archives of Internal Medicine*, *170*(2), 170–178.

Liu-Ambrose, T., Nagamatsu, L. S., Voss, M. W., Khan, K. M., & Handy, T. C. (2012). Resistance training and functional plasticity of the aging brain: A 12-month randomized controlled trial. *Neurobiological Aging*, *33*(8), 1690–1698.

Logue, S. F., & Gould, T. J. (2013). The neural and genetic basis of executive function: Attention, cognitive flexibility, and response inhibition. *Pharmacology, Biochemistry and Behaviour*, *123*, 45–54.

Lohaus, A., & Klein-Hessling, J. (2003) Relaxation in children: Effects of extended and intensified training. *Psychology & Health*, *18*, 237–249.

Loosli, S. V., Buschkuehl, M., Perrig, W. J., & Jaeggi, S. M. (2012). Working memory training improves reading processes in typically developing children. *Child Neuropsychology*, *18*(1), 62–78.

Loprinzi, P. D., & Cardinal, B. J. (2011). Association between objectively measured physical activity and sleep, NHANES 2005–2006. *Mental Health and Physical Activity*, *4*, 65–69.

Lunt, L., Bramham, J., Morris, R. G., Bullock, P. R., Selway, R. P., Xenitidis, K., & David, A. S. (2012). Prefrontal cortex dysfunction and 'jumping to conclusions': Bias or deficit? *Journal of Neuropsychology*, *6*(1), 65–78.

Lustig, C., Shah, P., Seidler, R. S., & Reuter-Lorenz, P. A. (2009). Aging, training and the brain: A review and future directions. *Neuropsychology Review*, *19*, 504–522.

MacCoon, D. G., MacLean, K. A., Davidson, R. J., Saron, C. D., & Lutz, A. (2014). No sustained attention differences in a longitudinal randomized trial comparing mindfulness based stress reduction versus active control. *PLoS One*, *9*(6), e97551.

Mackey, A. P., Hill, S. S., Stone, S. I., & Bunge, S. A. (2011). Differential effects of reasoning and speed training in children. *Developmental Science*, *14*(3), 582–590.

MacLean, K. A., Ferrer, E., Aichele, S. R., Bridwell, D. A., Zanesco, A. P., Jacobs, T. L., ... Saron, C. D. (2010). Intensive meditation training improves perceptual discrimination and sustained attention. *Psychological Science, 21*(6), 829–839.

Mahncke, H. W., Bronstone, A., & Merzenich, M. M. (2006). Brain plasticity and functional losses in the aged: Scientific bases for a novel intervention. *Progress in Brain Research, 157*, 81–109.

Maillot, P., Perrot, A., & Hartley, A. (2012). Effects of interactive physical-activity video-game training on physical and cognitive function in older adults. *Psychology and Aging, 27*, 589–600.

Mak, C., Whittingham, K., Cunnington, R., & Boyd, R. N. (2018). Efficacy of mindfulness-based interventions for attention and executive function in children and adolescents—A systematic review. *Mindfulness, 9*(1), 59–78.

Mallinson, J., & Singleton, M. (2017). *Roots of yoga.* London, UK: Penguin.

Manjunath, N. K., & Telles, S. (2001). Improved performance in the Tower of London test following yoga. *Indian Journal of Physiological Pharmacology, 45*(3), 351–354.

Mao, Z. M., Arnsten, A. F. T., & Li, B. M. (1999). Local infusion of an alpha-1 adrenergic agonist into the prefrontal cortex impairs spatial working memory performance in monkeys. *Biological Psychiatry, 46*, 1259–1265.

Marchetti, R., Forte, R., Borzacchini, M., Vazou, S., Tomporowski, P., & Pesce, C. (2015). Physical and motor fitness, sport skills and executive function in adolescents: A moderated prediction model. *Psychology, 6*, 1915‑1929.

Marek, G. J., & Aghajanian, G. K. (1999). 5-HT2A receptor or alpha1-adrenoceptor activation induces excitatory postsynaptic currents in layer V pyramidal cells of the medial prefrontal cortex. *European Journal of Pharmacology, 367*, 197–206.

Marmeleira, J. F., Godinho, M. B., & Fernandes, O. M. (2009). The effects of an exercise program on several abilities associated with driving performance in older adults. *Accident Analysis and Prevention, 41*(1), 90–97.

Masley, S., Roetzheim, R., & Gualtieri, T. (2009). Aerobic exercise enhances cognitive flexibility. *Journal of Clinical Psychology in Medical Settings, 16*(2), 186–193.

Mawjee, K., Woltering, S., Lai, N., Gotlieb, H., Kronitz, R., & Tannock, R. (2014). Working memory training in ADHD: Controlling for engagement, motivation, and expectancy of improvement (pilot study). *Journal of Attention Disorders.*

Mawjee, K., Woltering, S., & Tannock, R. (2015). Working memory training in postsecondary students with ADHD: A randomized controlled study. *PLoS One, 10*(9), e0137173.

McAuley, E., Mullen, S. P., Szabo, A. N., White, S. M., Wójcicki, T. R., Mailey, E. L., ... Kramer, A. F. (2011). Self-regulatory processes and exercise adherence in older adults: Executive function and self-efficacy effects. *American Journal of Preventive Medicine, 41*, 284–290.

McCarney, R., Warner, J., Iliffe, S., van Haselen, R., Griffin, M., & Fisher, P. (2007). The Hawthorne effect: A randomised, controlled trial. *BMC Medical Research Methodology, 7*, 30.

McClelland, M. M., Cameron, C. E., Connor, C. M., Farris, C. L., Jewkes, A. M., & Morrison, F. J. (2007). Links between behavioral regulation and preschoolers' literacy, vocabulary, and math skills. *Developmental Psychology, 43*(4), 947–959.

McNaughten, D., & Gabbard, C. (1993). Physical exertion and immediate mental performance of sixth-grade children. *Perceptual and Motor Skills, 77*, 1155–1159.

Meland, A., Ishimatsu, K., Pensgaard, A. M., Wagstaff, A., Fonne, V., Garde, A. H., & Harris, A. (2015). Impact of mindfulness training on physiological measures of stress and objective measures of attention control in a military helicopter unit. *The International Journal of Aviation Psychology, 25*(3–4), 191–208.

Melby-Lervåg, M., & Hulme, C. (2012). Is working memory training effective? A meta-analytic review. *Developmental Psychology, 49*(2), 270–291.

Melby-Lervåg, M., Redick, T. S., & Hulme, C. (2016). Working memory training does not improve performance on measures of intelligence or other measures of "far transfer": Evidence from a meta-analytic review. *Perspectives on Psychological Science, 11*, 512–534.

Melhuish, E. (2004). A literature review of the impact of early years provision on young children, with emphasis given to children from disadvantaged backgrounds. London, United Kingdom: National Audit Office.

Melhuish, E., Ereky-Stevens, P., Penderi, A., Rentzo, T., Slot, B., & Leseman, P. (2015). *A review of research on the effects of Early Childhood Education and Care (ECEC) upon child development.* Retrieved from Child Care Canada: http://childcarecanada.org/documents/research-policy-practice/15/10/effects-early-childhood-education-and-care-child-developmen

Menezes, A., Dias, N. M., Trevisan, B. T., Carreiro, L. R. R., & Seabra, A. G. (2015). Intervention for executive functions in attention deficit and hyperactivity disorder. *Arquivos de Neuro-Psiquiatria, 73*, 227–236.

Metzler-Baddeley, C., Cantera, J., Coulthard, E., Rosser, A., Jones, D. K., & Baddeley, R. J. (2014). Improved executive function and callosal white matter microstructure after rhythm exercise in Huntington's disease. *Journal of Huntington's Disease, 3*(3), 278–283.

Michael, J., Sebanz, N., & Knoblich, G. (2016). The sense of commitment: A minimal approach. *Frontiers in Psychology, 6*, 1968.

Miller, G. E., Brody, G. H., Yu, T., & Chen, E. (2014). A family-oriented psychosocial intervention reduces inflammation in low-SES African American youth. *Proceedings of the National Academy of Sciences of the United States of America, 111*(31), 11287–11292. doi:10.1073/pnas.1406578111

Miller, H. V., Barnes, J. C., & Beaver, K. M. (2011). Self-control and health outcomes in a nationally representative sample. *American Journal of Health Behavior, 35*(1), 15–27.

Mischel, W., Ayduk, O., & Mendoza-Denton, R. (2003). Sustaining delay of gratification over time: A hot-cool systems perspective. In G. Loewenstein, D. Read, & R. F. Baumeister (Eds.), *Time and decision: Economic and psychological perspectives on intertemporal choice* (pp. 175–200). New York, NY: Russell Sage Foundation.

Miyake, A., Friedman, N. P., Emerson, M. J., Witzki, A. H., Howerter, A., & Wager, T. D. (2000). The unity and diversity of executive functions and their contributions to complex "frontal lobe" tasks: A latent variable analysis. *Cognitive Psychology*, *41*(1), 49–100.

Moffitt, T. E. (2012). *Childhood self-control predicts adult health, wealth, and crime.* Multi-Disciplinary Symposium Improving the Well-Being of Children and Youth, January 29, 2012, Copenhagen, Denmark.

Moffitt, T. E., Arseneault, L., Belsky, D., Dickson, N., Hancox, R. J., Harrington, H., . . . Caspi, A. (2011). A gradient of childhood self-control predicts health, wealth, and public safety. *Proceedings of the National Academy of Sciences of the United States of America, 108*(7), 2693–2698.

Molloy, D. W., Beerschoten, D. A., Borrie, M. J., Crilly, R. G., & Cape, R. D. T. (1988). Acute effects of exercise on neuropsychological function in elderly subjects. *Journal of the American Geriatrics Society, 36*(1), 29–33.

Monti, J. M., Hillman, C. H., & Cohen, N. J. (2012). Aerobic fitness enhances relational memory in preadolescent children: The FITKids randomized control trial. *Hippocampus, 22,* 1876–1882.

Montessori, M. (1989). *To educate the human potential.* Oxford, UK: ABC-CLIO.

Moody, D. E. (2009). Can intelligence be increased by training on a task of working memory? *Intelligence, 37,* 327–328.

Moreau, D. (2015). Brains and brawn: Complex motor activities to maximize cognitive enhancement. *Educational Psychology Review, 27,* 475–482.

Moreau, D., & Conway, A. R. A. (2013). Cognitive enhancement: A comparative review of computerized and athletic training programs. *International Review of Sport and Exercise Psychology, 6*(1), 155–183.

Moreau, D., & Conway, A. R. A. (2014). The case for an ecological approach to cognitive training. *Trends in Cognitive Science, 18*(7), 334–336.

Moreau, D., Morrison, A. B., & Conway, A. R. A. (2015). An ecological approach to cognitive enhancement: Complex motor training. *Acta Psychologica, 157,* 44–55.

Morgan, C. A., Doran, A., Steffian, G., Hazlett, G., & Southwick, S. M. (2006). Stress-induced deficits in working memory and visuo-constructive abilities in special operations soldiers. *Biological Psychiatry, 60,* 722–729.

Morrison, A. B., & Chein, J. M. (2011). Does working memory training work? The promise and challenges of enhancing cognition by training working memory. *Psychonomic Bulletin & Review, 18*(1), 46–60.

Morrison, A., Goolsarran, M., Rogers, S., & Jha, A. (2014). Taming a wandering attention: Short-form mindfulness training in student cohorts. *Frontiers in Human Neuroscience, 7,* 897–908.

Morrison, F. J., Ponitz, C. C., & McClelland, M. M. (2010). Self-regulation and academic achievement in the transition to school. In S. D. Calkins & M. Bell (Eds.), *Child development at the intersection of emotion and cognition* (pp. 203–224). Washington, DC: American Psychological Association.

Mortimer, J. A., Ding, D., Borenstein, A. R., DeCarli, C., Guo, Q., Wu, Y., . . . Chu, S. (2012). Changes in brain volume and cognition in a randomized trial of exercise and social interaction in a community-based sample of non-demented Chinese elders. *Journal of Alzheimer's Disease, 30,* 757–766.

Moul, J. L., Goldman, B., & Warren, B. (1995). Physical activity and cognitive performance in the older population. *Journal of Aging and Physical Activity, 3*(2), 135–145.

Moynihan, J. A., Chapman, B. P., Klorman, R., Krasner, M. S., Duberstein, P. R., Brown, K. W., & Talbot, N. L. (2013). Mindfulness-based stress reduction for older adults: Effects on executive function, frontal alpha asymmetry and immune function. *Neuropsychobiology, 68*(1), 34–43.

Mrazek, M. D., Franklin, M. S., Phillips, D. T., Baird, B., & Schooler, J. W. (2013). Mindfulness training improves working memory capacity and GRE performance while reducing mind wandering. *Psychological Science, 24*(5), 776–781.

Muraven, M. (2010). Building self-control strength: Practicing self-control leads to improved self-control performance. *Journal of Experimental Social Psychology, 46*(2), 465–468.

Muraven, M., & Baumeister, R. F. (2000). Self-regulation and depletion of limited resources: Does self-control resemble a muscle? *Psychological Bulletin, 126*(2), 247–259.

Murphy, B. L., Arnsten, A. F. T., Goldman-Rakic, P. S., & Roth, R. H. (1996). Increased dopamine turnover in the prefrontal cortex impairs spatial working memory performance in rats and monkeys. *Neurobiology, 93,* 1325–1329.

Murphy, M. C., & Dweck, C. S. (2010). A culture of genius: How an organization's lay theories shape people's cognition, affect, and behavior. *Personality and Social Psychology Bulletin, 36,* 283–296.

Murray, N., Sujan, H., Hirt, E. R., & Sujan, M. (1990). The influence of mood on categorization: A cognitive flexibility interpretation. *Journal of Personality and Social Psychology, 59,* 411–425.

Myer, G. D., Faigenbaum, A. D., Edwards, N. M., Clark, J. F., Best, T. M., & Sallis, R. E. (2015). Sixty minutes of what? A developing brain perspective for activating children with an integrative exercise approach. *British Journal of Sports Medicine, 49,* 1510–1516.

Nagin, D., & Tremblay, R. E. (1999). Trajectories of boys' physical aggression, opposition, and hyperactivity on the path to physically violent and nonviolent juvenile delinquency. *Child Development, 70*(5), 1181–1196.

Napoli, M., Krech, P. R., & Holley, L. C. (2005). Mindfulness training for elementary school students: The Attention Academy. *Journal of Applied School Psychology, 21*(1), 99–125.

National Council on Aging. (2015). *The United States of Aging survey: Full research findings* (pp. 4–7). Retrieved from ncoa.org/wp-content/uploads/USA15-National-Fact-Sheet-Final.pdf

Nepal, B., Brown, L., & Ranmuthugala, G. (2010). Modelling the impact of modifying lifestyle risk factors on dementia prevalence in Australian population aged 45 years and over, 2006–2051. *Australasian Journal on Ageing, 29*(3), 111–116.

Ng, Q. X., Ho, C. Y. X., Chan, H. W., Yong, B. Z. J., & Yeo, W.-S. (2017). Managing childhood and adolescent attention-deficit/hyperactivity disorder (ADHD) with exercise: A systematic review. *Complementary Therapies in Medicine, 34,* 123–128.

Nguyen, M. H., & Kruse, A. (2012). A randomized controlled trial of Tai chi for balance, sleep quality and cognitive performance in elderly Vietnamese. *Clinical Interventions in Aging, 7,* 185–189.

Nicholson, C. (2007, March 26). Beyond IQ: Youngsters who can focus on the task at hand do better in math. *Scientific American.* https://www.scientificamerican.com/article/beyond-iq-kids-who-can-focus-on-task-do-better-math/.

Niendam, T. A., Laird, A. R., Ray, K. L., Dean, Y. M., Glahn, D. C., & Carter, C. S. (2012). Meta-analytic evidence for a superordinate cognitive control network subserving diverse executive functions. *Cognitive, Affective & Behavioral Neuroscience, 12*(2), 241–268.

Nigg, J. T., Wong, M. M., Martel, M. M., Jester, J. M., Puttler, L. I., Glass, J. M., . . . Zucker, R. A. (2006). Poor response inhibition as a predictor of problem drinking and illicit drug use in adolescents at risk for alcoholism and other substance use disorders. *Journal of the American Academy of Child and Adolescent Psychiatry, 45*(4), 468–475.

Noack, H., Lövdén, M., Schmiedek, F., & Lindenberger, U. (2009). Cognitive plasticity in adulthood and old age: Gauging the generality of cognitive intervention effects. *Restorative Neurology and Neuroscience, 27,* 435–453.

Noble, K. G., McCandliss, B. D., & Farah, M. J. (2007). Socioeconomic gradients predict individual differences in neurocognitive abilities. *Developmental Science, 10,* 464–480.

Noice, T., & Noice, H. (1997). *The nature of expertise in professional acting: A cognitive view.* Mahwah, NJ: Erlbaum.

Noice, T., & Noice, H. (2004). A cognitive learning principle derived from the role acquisition strategies of professional actors. *Cognitive Technology, 9,* 1–24.

Noice, H., Noice, T., & Staines, G. (2004). A short-term intervention to enhance cognitive and affective functioning in older adults. *Journal of Aging and Health, 16*(4), 562–585.

Northey, J. M., Cherbuin, N., Pumpa, K. L., Smee, D. J., & Rattray, B. (2017). Exercise interventions for cognitive function in adults older than 50: A systematic review with meta-analysis. *British Journal of Sports Medicine, 52*(3), 154–162.

Oberauer, K. (2005). Binding and inhibition in working memory: Individual and age differences in short-term recognition. *Journal of Experimental Psychology, 134*(3), 268–287.

Okon-Singer, H., Hendler, T., Pessoa, L., & Shackman, A. J. (2015). The neurobiology of emotion-cognition interactions: Fundamental questions and strategies for future research. *Frontiers in Human Neuroscience, 9,* 58.

Olson, D. R. (1964). *Cognitive development: The child's acquisition of diagonality.* New York, NY: Academic Press.

Olson, S. L., Smeroff, A. J., Kerr, D. C., Lopez, N. L., & Wellman, H. M. (2005). Developmental foundations of externalizing problems in young children: The role of effortful control. *Development and Psychopathology, 17*(1), 24–45.

Oken, B. S., Zajdel, D., Kishiyama, S., Flegal, K., Dehen, C., Haas, M., . . . Leyva, J. (2006). Randomized, controlled, six-month trial of yoga in healthy seniors: Effects on cognition and quality of life. *Alternative Therapies in Health and Medicine, 12*(1), 40–47.

Omizo, M. M., Loffredo, D. A., Hammett, V. L. (1982). Relaxation exercises for the learning disabled and family. *Academic Therapy, 17,* 603–608.

O'Shaughnessy, T., Lane, K. L., Gresham, F. M., & Beebe-Frankenberger, M. (2003). Children placed at risk for learning and behavioral difficulties: Implementing a school-wide system of early identification and prevention. *Remedial and Special Education, 24*(1), 27–35.

Oswald, W. D., Gunzelmann, T., Rupprecht, R., & Hagen, B. (2006). Differential effects of single versus combined cognitive and physical training with older adults: The SimA study in a 5-year perspective. *European Journal of Aging, 3,* 179–192.

Owen, A. M., Evans, A. C., & Petrides, M. (1996). Evidence for a two-stage model of spatial working memory processing within lateral frontal cortex: A positron emission tomography study. *Cerebral Cortex, 6*(1), 31–38.

Owen, A. M., Hampshire, A., Grahn, J. A., Stenton, R., Dajani, S., Burns, A. S., . . . Ballard, C. G. (2010). Putting brain training to the test. *Nature, 465*(7299), 775.

Park, D. C., Gutchess, A. H., Meade, M. L., & Stine-Morrow, E. A. L. (2007). Improving cognitive function in older adults: Nontraditional approaches. *Journal of Gerontology, 62,* 45–52.

Park, D. C., Lodi-Smith, J., Drew, L., Haber, S., Hebrank, A., Bischof, G. N., & Aamodt, W. (2014). The impact of sustained engagement on cognitive function in older adults: The synapse project. *Psychological Science, 25*(1), 103–112.

Park, D. C., & Payer, D. (2000). Working memory across the adult lifespan. In F. I. M. Craik & E. Bialystok (Eds.), *Lifespan cognition: Mechanisms of change* (pp. 128–142). New York, NY: Oxford University Press.

Pascoe, M. C., & Bauer, I. E. (2015). A systematic review of randomised control trials on the effects of yoga on stress measures and mood. *Journal of Psychiatric Research, 68* (Supplement C), 270–282.

Pasquinilli, M. (2001). *The child whisperer.* Dayton, OH: Asian Arts Center.

Penedo, F. J., & Dahn, J. R. (2005). Exercise and well-being: A review of mental and physical health benefits associated with physical activity. *Current Opinion in Neurobiology, 18*(2), 189–193.

Penner, I.-K., Vogt, A., Stöcklin, M., Gschwind, L., Opwis, K., & Calabrese, P. (2012). Computerised working memory training in healthy adults: A comparison of two different training schedules. *Neuropsychological Rehabilitation, 22,* 716–733.

Pereg, M., Shahar, N., & Meiran, N. (2013). Task switching training effects are mediated by working-memory management. *Intelligence, 41*(5), 467–478.

Perry, J. L., Joseph, J. E., Jiang, Y., Zimmerman, R. S., Kelly, T. H., Darna, M., . . . Bardo, M. T. (2011). Prefrontal cortex and drug abuse vulnerability: Translation to prevention and treatment interventions. *Brain Research Reviews, 65*(2), 124–149.

Pesce, C. (2012). Shifting the focus from quantitative to qualitative exercise characteristics in exercise and cognition research. *Journal of Sport & Exercise Psychology, 34*(6), 766–786.

Pesce, C., Croce, R., Ben-Soussan, T. D., Vazou, S., McCullick, B., Tomporowski, P. D., & Horvat, M. (2016). Variability of practice as an interface between motor and cognitive development. *International Journal of Sport and Exercise Psychology*, 1–20.

Pesce, C., Crova, C., Marchetti, M., Struzzolino, I., Masci, I., Vannozzi, G., & Forte, R. (2013). Searching for cognitively optimal challenge point in physical activity for children with typical and atypical motor development. *Mental Health and Physical Activity, 6*(3), 172–180.

Pesce, C., Leone, L., Motta, A., Marchetti, R., & Tomporowski, P. D. (2016). From efficacy to effectiveness of a "whole child" initiative of physical activity promotion. *Translational Journal of the American College of Sports Medicine, 1*(3), 18–29.

Pesce, C., Masci, I., Marchetti, R., Vazou, S., Sääkslahti, S., & Tomporowski, P. D. (2016). Deliberate play and preparation jointly benefit motor and cognitive development: Mediated and moderated effects. *Frontiers in Psychology, 7*, 349–366.

Petrides, M. (1994). Frontal lobes and working memory: Evidence from investigations of the effects of cortical excisions in nonhuman primates. In F. Boller & J. Grafman (Eds.), *Handbook of neuropsychology,* Vol. 9 (pp. 59–82). Amsterdam, The Netherlands: Elsevier Science Publishers.

Petrides, M. (1995). Functional organization of the human frontal cortex for mnemonic processing: Evidence from neuroimaging studies. *Annals of the New York Academy of Sciences, 769*, 85–96.

Pfungst, O. (1911). *Clever Hans (the horse of Mr. von Osten): A contribution to experimental animal and human psychology* (C. L. Rahn, Trans.). New York, NY: Henry Holt. (Original work published 1907)

Platt, J., & Spivack, G. (1975). *The MEPS procedure manual*. Philadelphia: Community Mental Health/Mental Retardation Center, Hahnemann Medical College and Hospital.

Plemons, J. K., Willis, S. L., & Baltes, P. B. (1978). Modifiability of fluid intelligence in aging: A short-term longitudinal training approach. *Journal of Gerontology, 33*, 224–231.

Posner, M. I., & DiGirolamo, G. J. (1998). Executive attention: Conflict, target detection, and cognitive control. In R. Parasuraman (Ed.), *The attentive brain* (pp. 401–423). Cambridge, MA: MIT Press.

Postle, B. R., Brush, L. N., & Nick, A. M. (2004). Prefrontal cortex and the mediation of proactive interference in working memory. *Cognitive, Affective, & Behavioral Neuroscience, 4*(4), 600–608.

Powell, R. R. (1974). Psychological effects of exercise therapy upon institutionalized geriatric mental patients. *Journal of Gerontology (Kirkwood), 29*(2), 157–161.

Prakash, R. S., Voss, M. W., Erickson, K. I., & Kramer, A. F. (2015). Physical activity and cognitive vitality. *Annual Review of Psychology, 66,* 769–795.

Predovan, D., Fraser, S. A., Renaud, M., & Bherer, L. (2012). The effect of three months of aerobic training on Stroop performance in older adults. *Journal of Aging Research, 2012,* 269815–269822.

Prins, P. J. M., Dovis, S., Ponsioen, A., ten Brink, E., & van der Oord, S. (2011). Does computerized working memory training with game elements enhance motivation and training efficacy in children with ADHD? *Cyberpsychology, Behavior, and Social Networking, 14*(3), 115–122.

Pugin, F., Metz, A. J., Stauffer, M., Wolf, M., Jenni, O. G., & Huber, R. (2014). Working memory training shows immediate and long-term effects on cognitive performance in children and adolescents. *F1000 Research, 3,* 82–93.

Purohit, S. P., & Pradhan, B. (2017). Effect of yoga program on executive functions of adolescents dwelling in an orphan home: A randomized controlled study. *Journal of Traditional and Complementary Medicine, 7*(1), 99–105.

Putnam, R. D. (2000). *Bowling alone.* NY: Simon & Schuster.

Rabipour, S., & Raz, A. (2012). Training the brain: Fact and fad in cognitive and behavioral remediation. *Brain and Cognition, 79*(2), 159–179.

Raichlen, D. A., Foster, A. D., Gerdeman, G. L., Seillier, A., & Giuffrida, A. (2012). Wired to run: Exercise-induced endocannabinoid signaling in humans and cursorial mammals with implications for the 'runner's high'. *Journal of Experimental Biology, 215*(8), 1331–1336.

Raina, P., Waltner-Toews, D., Bonnet, B., Woodward, C., & Abernathy, T. (1999). Influence of companion animals on the physical and psychological health of older people: An analysis of a one-year longitudinal study. *Journal of the American Geriatrics Society, 47,* 323–329.

Rapport, M. D., Orban, S. A., Kofler, M. J., & Friedman, L. M. (2013). Do programs designed to train working memory, other executive functions, and attention benefit children with ADHD? A meta-analytic review of cognitive, academic, and behavioral outcomes. *Clinical Psychology Review, 33*(8), 1237–1252.

Raver, C. C., Jones, S. M., Li-Grining, C. P., Metzger, M., Champion, K. M., & Sardin, L. (2008). Improving preschool classroom processes: Preliminary findings from a randomized trial implemented in Head Start settings. *Early Childhood Research Quarterly, 23*(1), 10–26.

Raver, C. C., Jones, S. M., Li-Grining, C., Zhai, F., Bub, K., & Pressler, E. (2011). CSRP's impact on low-income preschoolers' preacademic skills: Self-regulation as a mediating mechanism. *Child Development, 82*(1), 362–378.

Rea, C. P., & Modigliani, V. (1985). The effect of expanded versus massed practice on the retention of multiplication facts and spelling lists. *Human Learning: Journal of Practical Research & Applications, 4,* 11–18.

Rebok, G. W., Ball, K., Guey, L. T., Jones, R. N., Kim, H.-Y., King, J. W., . . . Willis, S. L. (2014). Ten-year effects of the advanced cognitive training for independent and vital elderly cognitive training trial on cognition and everyday functioning in older adults. *Journal of the American Geriatrics Society, 62,* 16–24.

Redick, T. S., & Lindsey, D. R. (2013). Complex span and N-back measures of working memory: A meta-analysis. *Psychonomic Bulletin & Review, 20,* 1102–1113.

Redick, T. S., Shipstead, Z., Harrison, T. L., Hicks, K. L., Fried, D. E., Hambrick, D. Z., . . . Engle, R. W. (2013). No evidence of intelligence improvement after working memory training: A randomized, placebo-controlled study. *Journal of Experimental Psychology: General, 142*(2), 359–379.

Redick, T. S., & Webster, S. B. (2014). Videogame interventions and spatial ability interactions. *Frontiers in Human Neuroscience, 8,* 183–190.

Reijnders, J., van Heugten, C., & van Boxtel, M. (2013). Cognitive interventions in healthy older adults and people with mild cognitive impairment: A systematic review. *Ageing Research Reviews, 12*(1), 263–275.

Reed, J. A., Einstein, G., Hahn, E., Hooker, S. P., Gross, V. P., & Kravitz, J. (2010). Examining the impact of integrating physical activity on fluid intelligence and academic performance in an elementary school setting: A preliminary investigation. *Journal of Physical Activity and Health, 7,* 343–351.

Reuter-Lorenz, P. A., Marshuetz, C., Jonides, J., Smith, E. E., Hartley, A., & Koeppe, R. (2001). Neurocognitive ageing of storage and executive processes. *European Journal of Cognitive Psychology, 13,* 257–278.

Riccio, C. A., & Gomes, H. (2013). Interventions for executive function deficits in children and adolescents. *Applied Neuropsychology: Child, 2*(2), 133–140.

Richmond, L. L., Morrison, A. B., Chein, J. M., & Olson, I. R. (2011). Working memory training and transfer in older adults. *Psychology and Aging, 26*(4), 813–822.

Riggs, N. R., Blair, C. B., & Greenberg, M. T. (2003). Concurrent and 2-year longitudinal relations between executive function and the behavior of 1st and 2nd grade children. *Child Neuropsychology, 9*(4), 267–276.

Riggs, N. R., Greenberg, M. T., Kusché, C. A., & Pentz, M. A. (2006). The mediational role of neurocognition in the behavioral outcomes of a social-emotional prevention program in elementary school students: Effects of the PATHS curriculum. *Prevention Science, 7*(1), 91–102.

Riggs, N. R., Spruijt-Metz, D., Sakuma, K. K., Chou, C. P., & Pentz, M. A. (2010). Executive cognitive function and food intake in children. *The Journal of Nutrition Education and Behavior, 42*(9), 398–403.

Roberts, G., Quach, J., Spencer-Smith, M., Anderson, P. J., Gathercole, S., Gold, L., . . . Wake, M. (2016). Academic outcomes 2 years after working memory training for children with low working memory: A randomized clinical trial. *JAMA Pediatrics, 170,* e154568.

Roberts, R., & Gibson, E. (2002). Individual differences in sentence memory. *Journal of Psycholinguistic Research, 31*(6), 573–598.

Roca, M., Manes, F., Cetkovich, M., Bruno, D., Ibáñez, A., Torralva, T., & Duncan, J. (2014). The relationship between executive functions and fluid intelligence in schizophrenia. *Frontiers in Behavioral Neuroscience, 8,* 46.

Rosenbaum, D. A., Carlson, R. A., & Gilmore, R. O. (2001). Acquisition of intellectual and perceptual-motor skills. *Annual Review of Psychology, 52,* 453–470.

Rosenthal, R., & Jacobsen, L. (1968). *Pygmalion in the classroom: Teacher expectation and pupils' intellectual development*. New York, NY: Holt, Rinehart, & Winston.

Röthlisberger, M., Neuenschwander, R., Cimeli, P., Michel, E., & Roebers, C. M. (2012). Improving executive functions in 5- and 6-year-olds: Evaluation of a small group intervention in prekindergarten and kindergarten children. *Infant and Child Development, 21*(4), 411–429.

Rudebeck, S. R., Bor, D., Ormond, A., O'Reilly, J. X., & Lee, A. C. (2012). A potential spatial working memory training task to improve both episodic memory and fluid intelligence. *PLoS One, 7*(11), e50431.

Rueda, M. R., Checa, P., & Cómbita, L. M. (2012). Enhanced efficiency of the executive attention network after training in preschool children: Immediate changes and effects after two months. *Developmental Cognitive Neuroscience, 2*, S192–S204.

Rueda, M. R., Rothbart, M. K., McCandliss, B. D., Saccomanno, L., & Posner, M. I. (2005). Training, maturation, and genetic influences on the development of executive attention. *Proceedings of the National Academy of Sciences of the United States of America, 102*(41), 14931–14935.

Ryan, R. M., & Deci, E. L. (2000). The 'what' and 'why' of goal pursuits: Human needs and the self-determination of behavior. *Psychological Inquiry, 11*, 227–268.

Rydell, R. J., Van Loo, K. J., & Boucher, K. L. (2014). Stereotype threat and executive functions: Which functions mediate different threat-related outcomes? *Personality and Social Psychology Bulletin, 40*(3), 377–390.

Sahdra, B. K., MacLean, K. A., Ferrer, E., Shaver, P. R., Rosenberg, E. L., Jacobs, T. L., . . . Saron, C. D. (2011). Enhanced response inhibition during intensive meditation training predicts improvements in self-reported adaptive socioemotional functioning. *Emotion, 11*(2), 299–312.

Salmon, P. (2001). Effects of physical exercise on anxiety, depression, and sensitivity to stress. *Clinical Psychology Review, 21*(1), 33–61.

Sánchez, M. M., Young, L. J., Plotsky, P. M., & Insel, T. R. (2000). Distribution of corticosteroid receptors in the rhesus brain: Relative absence of glucocorticoid receptors in the hippocampal formation. *Journal of Neuroscience, 20*, 4657–4668.

Savage, R., Cornish, K., Manly, T., & Hollis, C. P. (2006). Cognitive processes in children's reading and attention: The role of working memory, divided attention, and response inhibition. *British Journal of Psychology, 97*(3), 365–385.

Scherder, E. J., Scherder, R., Verburgh, L., Königs, M., Blom, M., Kramer, A. F., & Eggermont, L. (2014). Executive functions of sedentary elderly may benefit from walking: A systematic review and meta-analysis. *American Journal of Geriatric Psychiatry, 22*(8), 782–791.

Scherder, E. J., Van Paasschen, J., Deijen, J.-B., Van Der Knokke, S., Orlebeke, J. F. K., Burgers, I., . . . Sergeant, J. A. (2005). Physical activity and executive functions in the elderly with mild cognitive impairment. *Aging & Mental Health, 9*, 272–280.

Schmidt, M., Jäger, K., Egger, F., Roebers, C. M., & Conzelmann, A. (2015). Cognitively engaging chronic physical activity, but not aerobic exercise, affects executive functions in primary school children: A group-randomized controlled trial. *Journal of Sport & Exercise Psychology, 37*, 575–591.

Schmidt, R. A., & Bjork, R. A. (1992). New conceptualizations of practice: Common principles in three paradigms suggest new concepts for training. *Psychological Science*, *3*, 207–217.

Schmiedek, F., Hildebrandt, A., Lövdén, M., Lindenberger, U., & Wilhelm, O. (2009). Complex span versus updating tasks of working memory: The gap is not that deep. *Journal of Experimental Psychology*, *35*(4), 1089–1096.

Schmiedek, F., Lövdén, M., & Lindenberger, U. (2010). Hundred days of cognitive training enhance broad cognitive abilities in adulthood: Findings from the COGITO study. *Frontiers in Aging Neuroscience*, *2*, 1–10.

Schmiedek, F., Lövdén, M., & Lindenberger, U. (2014). Younger adults show long-term effects of cognitive training on broad cognitive abilities over 2 years. *Developmental Psychology*, *50*(9), 2304–2310.

Schonert-Reichl, K. A., Oberle, E., Lawlor, M. S., Abbott, D., Thomson, K., Oberlander, T., & Diamond, A. (2015). Enhancing cognitive and social-emotional development through a simple-to-administer mindfulness-based school program for elementary school children: A randomized controlled trial. *Developmental Psychology*, *51*(1), 52–66.

Schubert, T., Strobach, T., & Karbach, J. (2014). New directions in cognitive training: On methods, transfer, and application. *Psychological Research*, *78*, 749–755.

Schwaighofer, M., Fischer, F., & Bühner, M. (2015). Does working memory training transfer? A meta-analysis including training conditions as moderators. *Educational Psychologist*, *50*(2), 138–166.

Scudder, M. R., Lambourne, K., Drollette, E. S., Herrmann, S. D., Washburn, R. A., Donnelly, J. E., & Hillman, C. H. (2014). Aerobic capacity and cognitive control in elementary school-age children. *Medicine and Science in Sports and Exercise*, *46*(5), 1025–1035.

Sedlmeier, P., Eberth, J., Schwarz, M., Zimmermann, D., Haarig, F., Jaeger, S., & Kunze, S. (2012). The psychological effects of meditation: A meta-analysis. *Psychological Bulletin*, *138*(6), 1139–1171.

Semrud-Clikeman, M., Nielsen, K. H., Clinton, A., Sylvester, L., Parle, N., & Connor, R. T. (1999). An intervention approach for children with teacher-and parent-identified attentional difficulties. *Journal of Learning Disabilities*, *32*, 581–590.

Shapiro, D. C., & Schmidt, R. A. (1982). The schema theory: Recent evidence and developmental implications. In J. A. S. Kelso & J. E. Clark (Eds.), *The development of movement control and coordination* (pp. 113–50). New York, NY: Wiley.

Sharma, N., Pomeroy, V. M., & Baron, J. C. (2006). Motor imagery: A backdoor to the motor system after stroke? *Stroke: Journal of the American Heart Association*, *37*, 1941–1952.

Shea, J. B., & Morgan, R. L. (1979). Contextual interference effects on the acquisition, retention, and transfer of a motor skill. *Journal of Experimental Psychology: Human Learning & Memory*, *5*, 179–187.

Shipstead, Z., Hicks, K. L., & Engle, R. W. (2012). Cogmed working memory training: Does the evidence support the claims? *Journal of Applied Research in Memory and Cognition*, *1*, 185–193.

Shipstead, Z., Redick, T. S., & Engle, R. W. (2012). Is working memory training effective? *Psychological Bulletin, 138*(4), 628–654.

Sibley, B. A., & Beilock, S. L. (2007). Exercise and working memory: An individual differences investigation. *Journal of Sport and Exercise Psychology, 29*, 783–791.

Sibley, B. A., & Etnier, J. L. (2003). The relationship between physical activity and cognition in children: A meta-analysis. *Pediatric Exercise Science, 15*, 243–256.

Simons, D. J., Boot, W. R., Charness, N., Gathercole, S. E., Chabris, C. F., Hambrick, D. Z., & Stine-Morrow, E. A. L. (2016). Do "brain-training" programs work? *Psychological Science in the Public Interest, 17*, 103–186.

Singh, V. P., Rao, V., Prem, V., Sahoo, R. C., & Keshav, P. K. (2009). Comparison of the effectiveness of music and progressive muscle relaxation for anxiety in COPD—A randomized controlled pilot study. *Chronic Respiratory Disease, 6*(4), 209–216.

Sink, K. M., Reid, K. F., Pahor, M., McDermott, M. M., Lopez, O. L., Katula, J., . . . Williamson, J. D. (2015). Effect of a 24-month physical activity intervention vs health education on cognitive outcomes in sedentary older adults. *JAMA Neurology, 314*(8), 781–790.

Smiley-Oyen, A. L., Lowry, K. A., Francois, S. J., Kohut, M. L., & Ekkekakis, P. (2008). Exercise, fitness, and neurocognitive function in older adults: The "selective improvement" and "cardiovascular fitness" hypotheses. *Annals of Behavioural Medicine, 36*(3), 280–291.

Smith, C., Viljoen, J. T., & McGeachie, L. (2014). African drumming: A holistic approach to reducing stress and improving health? *Journal of Cardiovascular Medicine, 15*, 441–446.

Smith, E. E., & Jonides, J. (1999). Storage and executive processes in the frontal lobes. *Science, 283*(5408), 1657–1661.

Smith, E. E., Jonides, J., Marshuetz, C., & Koeppe, R. A. (1998). Components of verbal working memory: Evidence from neuroimaging. *Proceedings of the National Academy of Sciences of the United States of America, 95*(3), 876–882.

Smith, P. J., Blumenthal, J. A., Hoffman, B. M., Cooper, H., Strauman, T. A., Welsh-Bohmer, K., . . . Sherwood, A. (2010). Aerobic exercise and neurocognitive performance: A meta-analytic review of randomized controlled trials. *Psychosomatic Medicine, 72*(3), 239–252.

Smith, R. E., & Smoll, F. L. (1997). Coaching the coaches: Youth sports as a scientific and applied behavioral setting. *Current Directions in Psychological Science, 6*(1), 16–21.

Smith, R. E., Smoll, F. L., & Barnett, N. P. (1995). Reduction of children's sport performance anxiety through social support and stress reduction training for coaches. *Journal of Applied Developmental Psychology, 16*(1), 125–142.

Smith, R. E., Zane, N. W. S., Smoll, F. L., & Coppel, D. B. (1983). Behavioral assessment in youth sports: Coaching behaviors and children's attitudes. *Medicine and Science in Sports and Exercise, 15*, 208–214.

Smoll, F. L., Smith, R. E., Barnett, N. P., & Everett, J. J. (1993). Enhancement of children's self-esteem through social support training for youth sport coaches. *Journal of Applied Psychology, 78*, 602–610.

Snowden, M., Steinman, L., Mochan, K., Grodstein, F., Prohaska, T. R., Thurman, D. J., . . . Anderson, L. A. (2011). Effect of exercise on cognitive performance in community-dwelling older adults: Review of intervention trials and recommendations for public health practices and research. *Journal of the American Geriatrics Society, 59*(4), 704–716.

Söderqvist, S., Bergman Nutley, S., Ottersen, J., Grill, K. M., & Klingberg, T. (2012). Computerized training of non-verbal reasoning and working memory in children with intellectual disability. *Frontiers in Human Neuroscience, 6*(271), 1–8.

Solomon, T. L. S., Plamondon, A., O'Hara, A., Finch, H., Goco, G., Chaban, P., . . . Tannock, R. (2017). A cluster randomized-controlled trial of the impact of the Tools of the Mind curriculum on self-regulation in Canadian preschoolers. *Frontiers in Psychology, 8*, 18.

Sonuga-Barke, E. J., Brandeis, D., Cortese, S., Daley, D., Ferrin, M., Holtmann, M., . . . European ADHD Guidelines Group. (2013). Nonpharmacological interventions for ADHD: Systematic review and meta-analyses of randomized controlled trials of dietary and psychological treatments. *The American Journal of Psychiatry, 170*, 275–289.

Spencer-Smith, M., & Klingberg, T. (2015). Benefits of a working memory training program for inattention in daily life: A systematic review and meta-analysis. *PLoS One, 10*(3), e0119522.

Spierer, L., Chavan, C. F., & Manuel, A. L. (2013). Training-induced behavioural and brain plasticity in inhibitory control. *Frontiers in Human Neuroscience, 7*(427), 1–9.

St. Clair-Thompson, H. L., & Gathercole, S. E. (2006). Executive functions and achievements in school: Shifting, updating, inhibition, and working memory. *The Quarterly Journal of Experimental Psychology, 59*(4), 745–759.

Staiano, A. E., Abraham, A. A., & Calvert, S. L. (2012). Competitive versus cooperative exergame play for African American adolescents' executive function skills: Short-term effects in a long-term training intervention. *Developmental Psychology, 48*, 337–342.

Stauffer, J. M., Ree, M. J., & Caretta, T. R. (1996). Cognitive-components tests are not much more than g: An extension of Kyllonen's analysis. *Journal of General Psychology, 123*, 193–205.

Steele, C. M., & Aronson, J. (1998). How stereotypes influence the standardized test performance of talented African American students. In C. Jencks & M. Phillips (Eds.), *The Black-White Test score gap* (pp. 401–427). Washington, DC: Brookings Institution.

Stepankova, H., Lukavsky, J., Buschkuehl, M., Kopecek, M., Ripova, D., & Jaeggi, S. M. (2014). The malleability of working memory and visuospatial skills: A randomized controlled study in older adults. *Developmental Psychology, 50*(4), 1049–1059.

Stephenson, C. L., & Halpern, D. F. (2013). Improved matrix reasoning is limited to training on tasks with a visuospatial component. *Intelligence, 41*, 341–357.

Stine-Morrow, E. A. L., & Basak, C. (2011). Cognitive interventions. In K. W. Schaie & S. L. Willis (Eds.), *Handbook of the psychology of aging* (7th ed., pp. 153–171). San Diego, CA: Elsevier.

Stoler-Miller, B. (1998). *Yoga: Discipline of freedom*. New York, NY: Bantam Books.

Streiner, D. L. (2009). The effects of exercise programs on cognition in older adults: A review. *Clinical Journal of Sport Medicine, 19*(5), 438.

Stroth, S., Reinhardt, R. K., Thöne, J., Hille, K., Schneider, M., Härtel, S., Weidemann, W., Bös, K., & Spitzer, M. (2010). Impact of aerobic exercise training on cognitive functions and affect associated to the COMT polymorphism in young adults. *Neurobiology of Learning and Memory, 94*, 364–372.

Subramanian, S. K., Sharma, V. K., Arunachalam, V., Radhakrishnan, K., & Ramamurthy, S. (2015). Effect of structured and unstructured physical activity training on cognitive functions in adolescents—A randomized control trial. *Journal of Clinical and Diagnostic Research, 9*(11), CC04–CC09.

Sullivan, M. B., Erb, M., Schmalzl, L., Moonaz, S., Noggle Taylor, J., & Porges, S. W. (2018). Yoga therapy and polyvagal theory: The convergence of traditional wisdom and contemporary neuroscience for self-regulation and resilience. *Frontiers in Human Neuroscience, 12*, 67–71.

Tamm, L., Epstein, J. N., Peugh, J. L., Nakonezny, P. A., & Hughes, C. W. (2013). Preliminary data suggesting the efficacy of attention training for school-aged children with ADHD. *Developmental Cognitive Neuroscience, 4*, 16–28.

Tang, Y. Y. (2005). *Health from brain, wisdom from brain*. Dalian, China: Dalian University of Technology Press.

Tang, Y. Y. (2009). *Exploring the brain, optimizing the life*. Beijing, China: Science Press.

Tang, Y. Y., Ma, Y., Wang, J., Fan, Y., Feng, S., Lu, Q., . . . Posner, M. I. (2007). Short-term meditation training improves attention and self-regulation. *Proceedings of the National Academy of Sciences of the United States of America, 104*(43), 17152–17156.

Tang, Y. Y., Yang, L., Leve, L. D., & Harold, G. T. (2012). Improving executive function and its neurobiological mechanisms through a mindfulness-based intervention: Advances within the field of developmental neuroscience. *Child Development Perspectives, 6*(4), 361–366.

Tangney, J. P., Baumeister, R. F., & Boone, A. L. (2004). High self-control predicts good adjustment, less pathology, better grades, and interpersonal success. *Journal of Personality, 72*(2), 271–324.

Tardif, S., & Simard, M. (2011). Cognitive stimulation programs in health elderly: A review. *International Journal of Alzheimer's Disease, 2011*(378934), 1–13.

Taylor-Piliae, R. E., Newell, K. A., Cherin, R., Lee, M. J., King, A. C., & Haskell, W. L. (2010). Effects of tai chi and Western exercise on physical and cognitive functioning in healthy community-dwelling older adults. *Journal of Aging and Physical Activity, 18*, 261–279.

Telles, S., Singh, N., Bhardwaj, A. K., Kumar, A., & Balkrishna, A. (2013). Effect of yoga or physical exercise on physical, cognitive and emotional measures in children: A randomized controlled trial. *Child and Adolescent Psychiatry and Mental Health, 7*(1), 37–52.

Theeuwes, J. (1991). Exogenous and endogenous control of attention: The effect of visual onsets and offsets. *Perception and Psychophysics, 49*(1), 83–90.

Thompson, T. W., Waskom, M. L., Garel, K. L. A., Cardenas-Iniguez, C., Reynolds, G. O., Winter, R., . . . Gabrieli, J. D. E. (2013). Failure of working memory training to enhance cognition or intelligence. *PLoS One*, *8*(5), e63614.

Thorell, L. B., Lindqvist, S., Bergman, N., Bohlin, G., & Klingberg, T. (2009). Training and transfer effects of executive functions in preschool children. *Developmental Science*, *12*(1), 106–113.

Tidwell, J. W., Dougherty, M. R., Chrabaszcz, J. R., Thomas, R. P., & Mendoza, J. L. (2014). What counts as evidence for working memory training? Problems with correlated gains and dichotomization. *Psychonomic Bulletin & Review*, *21*(3), 620–628.

Tomporowski, P. D., Davis, C., Miller, P. H., & Naglieri, J. A. (2008). Exercise and children's intelligence, cognition, and academic achievement. *Educational Psychology Review*, *20*, 111–131.

Tomporowski, P. D., Lambourne, K., & Okumura, M. S. (2011). Physical activity interventions and children's mental function: An introduction and overview. *Preventive Medicine*, *52*, S3–S9.

Tomporowski, P. D., McCullick, B., Pendleton, D. M., & Pesce, C. (2015). Exercise and children's cognition: The role of exercise characteristics and a place for metacognition. *Journal of Sport and Health Science*, *4*(1), 47–55.

Traverso, L., Viterbori, P., & Usai, M. C. (2015). Improving executive function in childhood: Evaluation of a training intervention for 5-year-old children. *Frontiers in Psychology*, *6*, 525.

Trulson, M. E. (1986). Martial arts training: A novel "'cure" for juvenile delinquency. *Human Relations*, *39*(12), 1131–1140.

Tseng, C. N., Gau, B. S., & Lou, M. F. (2011). The effectiveness of exercise on improving cognitive function in older people: A systematic review. *The Journal of Nursing Research*, *19*(2), 119–131.

Tucha, O., Tucha, L., Kaumann, G., König, S., Lange, K. M., Stasik, D., . . . Lange, K. W. (2011). Training of attention functions in children with attention deficit hyperactivity disorder. *Attention Deficit and Hyperactivity Disorders*, *3*, 271–283.

Tuckman, B. W., & Hinkle, J. S. (1986). An experimental study of the physical and psychological effects of aerobic exercise on school children. *Health Psychology*, *5*(3), 197–207.

Twenge, J. M., Catanese, K. R., & Baumeister, R. F. (2002). Social exclusion causes self-defeating behavior. *Journal of Personality and Social Psychology*, *83*, 606–615.

Tyagi, A., & Cohen, M. (2016). Yoga and heart rate variability: A comprehensive review of the literature. *International Journal of Yoga*, *9*, 97–113.

Ulrich, R. S. (1983). Aesthetic and affective response to natural environment. In I. Altman & J. F. Wohlwill (Eds.), *Human behavior and environment: Advances in theory and research*, Vol. 6. New York, NY: Plenum.

Unsworth, N., & Engle, R. W. (2007). On the division of short-term and working memory: An examination of simple and complex span and their relation to higher order abilities. *Psychological Bulletin*, *133*(6), 1038–1066.

Trabasso, T., & van den Broek, P. (1985). Causal thinking and the representation of narrative events. *Journal of Memory and Language*, *24*, 612–630.

van der Donk, M., Hiemstra-Beernink, A.-C., Tjeenk-Kalff, A., van der Leij, A., & Lindauer, R. (2015). Cognitive training for children with ADHD: A randomized controlled trial of Cogmed working memory training and 'paying attention in class'. *Frontiers in Psychology, 6*, 1081.

van der Oord, S., Ponsioen, A. J. G. B., Geurts, H. M., ten Brink, E. L., & Prins, P. J. M. (2014). A pilot study of the efficacy of a computerized executive functioning remediation training with game elements for children with ADHD in an outpatient setting: Outcome on parent- and teacher-rated executive functioning and ADHD behavior. *Journal of Attention Disorders, 18*(8), 699–712.

Van Doren, J., Arns, M., Heinrich, H., Vollebregt, M. A., Strehl, U., & Loo, S. K. (2018). Sustained effects of neurofeedback in ADHD: A systematic review and meta-analysis. *European Child & Adolescent Psychiatry*, 1–13.

van Uffelen, J. G. Z., Chinapaw, M. J. M., Hopman-Rock, M., & van Mechelen, W. (2008). The effects of exercise on cognition in older adults with and without cognitive decline: A systematic review. *Clinical Journal of Sport Medicine, 18*, 486–500.

Vazou, S., Pesce, C., Lakes, K., & Smiley-Oyen, A. L. (2019). More than one road leads to Rome: A narrative review and meta-analysis of physical activity intervention effects on cognition in youth. *International Journal of Sport and Exercise Psychology, 17*(2), 153-178.

Vazou, S., & Smiley-Oyen, A. L. (2014). Moving and academic learning are not antagonists: Acute effects on executive function and enjoyment. *Journal of Sport & Exercise Psychology, 36*, 474–485.

Venditti, S., Verdone, L., Pesce, C., Tocci, N., Caserta, M., & Ben-Soussan, T. D. (2015). Creating well-being: Increased creativity and proNGF decrease following Quadrato motor training. *Biomed Research International, 2015*, 275062.

Verburgh, L., Königs, M., Scherder, E. J., & Oosterlaan, J. (2014). Physical exercise and executive functions in preadolescent children, adolescents and young adults: A meta-analysis. *British Journal of Sports Medicine, 48*, 973–979.

Verghese, J., Lipton, R. B., Katz, M. J., Hall, C. B., Derby, C. A., Kuslansky, G., . . . Buschke, H. (2003). Leisure activities and the risk of dementia in the elderly. *New England Journal of Medicine, 348*(25), 2508–2516.

Vitaro, F., Barker, E. D., Brendgen, M., & Tremblay, R. E. (2012). Pathways explaining the reduction of adult criminal behaviour by a randomized preventive intervention for disruptive kindergarten children. *Journal of Child Psychology and Psychiatry, 53*(7), 748–756.

Voelcker-Rehage, C., Godde, B., & Staudinger, U. M. (2011). Cardiovascular and coordination training differentially improve cognitive performance and neural processing in older adults. *Frontiers in Human Neuroscience, 5*, 26–37.

Vollebregt, M. A., van Dongen-Boomsma, M., Buitelaar, J. K., & Slaats-Willemse, D. (2014). Does EEG-neurofeedback improve neurocognitive functioning in children with attention-deficit/hyperactivity disorder? A systematic review and a double-blind placebo-controlled study. *Journal of Child Psychology and Psychiatry, 55*, 954–955.

von Bastian, C. C., & Oberauer, K. (2013). Distinct transfer effects of training different facets of working memory capacity. *Journal of Memory and Language*, 69(1), 36–58.

von Haaren, B., Ottenbacher, J., Muenz, J., Neumann, R., Boes, K., & Ebner-Priemer, U. (2016). Does a 20-week aerobic exercise training programme increase our capabilities to buffer real-life stressors? A randomized, controlled trial using ambulatory assessment. *European Journal of Applied Physiology*, 116(2), 383–394.

von Hecker, U., & Meiser, T. (2005). Defocused attention in depressed mood: Evidence from source monitoring. *Emotion*, 5, 456–463.

Voss, M. W., Nagamatsu, L. S., Liu-Ambrose, T., & Kramer, A. F. (2011). Exercise, brain, and cognition across the lifespan. *Journal of Applied Physiology*, 111(5), 1505–1513.

Vygotsky, L. S. (1978). *Mind in society: The development of higher psychological processes*. Cambridge, MA: Harvard University Press.

Wang, H., & Lupica, C. R. (2014). Release of endogenous cannabinoids from ventral tegmental area dopamine neurons and the modulation of synaptic processes. *Progress in Neuro-Psychopharmacology & Biological Psychiatry*, 52, 24–27.

Wang, H.-X., Jin, Y., Hendrie, H. C., Liang, C., Yang, L., Cheng, Y., . . . Gao, S. (2013). Late leisure activities and risk of cognitive decline. *The Journals of Gerontology Series A: Biological Sciences and Medical Sciences*, 68(2), 205–213.

Wang, J. R., & Hsieh, S. (2013). Neurofeedback training improves attention and working memory performance. *Clinical Neurophysiology*, 124(12), 2406–2020.

Wang, M., Gamo, N. J., Yang, Y., Jin, L. E., Wang, X.-J., Laubach, M., . . . Arnsten, A. F. T. (2011). Neuronal basis of age-related working memory decline. *Nature*, 476, 210–213.

Wang, M. Y., Chang, C. Y., & Su, S.Y. (2011). What's cooking? Cognitive training of executive function in the elderly. *Frontiers in Psychology*, 2, 228-238.

Wass, S., Porayska-Pomsta, K., & Johnson, M. H. (2011). Training attentional control in infancy. *Current Biology*, 21(18), 1543–1547.

Wayne, P. M., Walsh, J. N., Taylor-Piliae, R. E., Wells, R. E., Papp, K. V., Donovan, N. J., & Yeh, G. Y. (2014). The impact of tai chi on cognitive performance in older adults: Systematic review and meta-analysis. *Journal of the American Geriatrics Society*, 62(1), 25–39.

West, J., Otte, C., Geher, K., Johnson, J., & Mohr, D. C. (2004). Effects of Hatha yoga and African dance on perceived stress, affect, and salivary control. *Annals of Behavioral Medicine*, 28(1), 114–118.

Westendorp, M., Houwen, S., Hartman, E., Mombarg, R., Smith, J., & Visscher, C. (2014). Effect of a ball skill intervention on children's ball skills and cognitive functions. *Medicine & Science in Sports & Exercise*, 46, 414–422.

Wilkinson, A. J., & Yang, L. (2012). Plasticity of inhibition in older adults: Retest practice and transfer effects. *Psychology and Aging*, 27, 606–615.

Williams, G. C., Cox, E. M., Kouides, R., & Deci, E. L. (1999). Presenting the facts about smoking to adolescents: The effects of an autonomy supportive style. *Archives of Pediatrics & Adolescent Medicine*, 153, 959–964.

Williams, P., & Lord, S. R. (1997). Effects of group exercise on cognitive functioning and mood in older women. *Australian and New Zealand Journal of Public Health*, *21*(1), 45–52.

Williamson, D., Dewey, A., & Steinberg, H. (2001). Mood change through physical exercise in nine-to ten-year-old children. *Perceptual and Motor Skills*, *93*(1), 311–316.

Williamson, J. D., Espeland, M., Kritchevsky, S. B., Newman, A. B., King, A. C., Pahor, M., ... Miller, M. E. (2009). Changes in cognitive function in a randomized trial of physical activity: Results of the lifestyle interventions and independence for elders pilot study. *Journals of Gerontology Series A: Biological Sciences*, *64A*, 688–694.

Willis, S. L., Tennstedt, S. L., Marsiske, M., Ball, K., Elias, J., Koepke, K. M., ... Wright, E. (2006). Long-term effects of cognitive training on everyday functional outcomes in older adults. *The Journal of the American Medical Association*, *296*(23), 2805–2814.

Winstok, Z. (2009). From self-control capabilities and the need to control others to proactive and reactive aggression among adolescents. *Journal of Adolescence*, *32*(3), 455–466.

Wong, A. S. Y., He, M. Y. Q., & Chan, R. W. S. (2014). Effectiveness of computerized working memory training program in Chinese community settings for children with poor working memory. *Journal of Attention Disorders*, *18*, 318–330.

Wroolie, T. E., Williams, K. E., Keller, J., Zappert, L. N., Shelton, S. D., Kenna, H. A., ... Rasgon, N. L. (2006). Mood and neuropsychological changes in women with midlife depression treated with escitalopram. *Journal of Clinical Psychopharmacology*, *26*, 361–366.

Yang, C. R., & Seamans, J. K. (1996). Dopamine D1 receptor actions in layers V-V1 rat prefrontal cortex neurons in vitro: Modulation of dendritic-somatic signal integration. *Journal of Neuroscience*, *16*, 1922–1935.

Yang, H., Yang, S., & Isen, A. M. (2013). Positive affect improves working memory: Implications for controlled cognitive processing. *Cognition and Emotion*, *27*(3), 474–482.

Ybarra, O., Winkielman, P., Yeh, I., Burnstein, E., & Kavanagh, L. (2011). Friends (and sometimes enemies) with cognitive benefits: What types of social interactions boost cognitive functioning? *Social Psychological and Personality Science*, *2*, 253–261.

Young, J., Angevaren, M., Rusted, J., & Tabet, N. (2015). Aerobic exercise to improve cognitive function in older people without known cognitive impairment. *Cochrane Database of Systematic Reviews*, *4*, CD005381.

Zanesco, A. P., King, B. G., MacLean, K. A., & Saron, C. D. (2013). Executive control and felt concentrative engagement following intensive meditation training. *Frontiers in Human Neuroscience*, *7*(566), 1–13.

Zanto, T. P., Rubens, M. T., Thangavel, A., & Gazzaley, A. (2011). Causal role of the prefrontal cortex in top-down modulation of visual processing and working memory. *Nature Neuroscience*, *14*(5), 656–661.

Zeidan, F., Johnson, S. K., Diamond, B. J., David, Z., & Goolkasian, P. (2010). Mindfulness meditation improves cognition: Evidence of brief mental training. *Consciousness and Cognition, 19*(2), 597–605.

Zenner, C., Herrnleben-Kurz, S., & Walach, H. (2014). Mindfulness-based interventions in schools—A systematic review and meta-analysis. *Frontiers in Psychology, 5,* 603.

Zimmermann, P., & Fimm, B. (2002). A test battery for attentional performance. In M. Leclercq & P. Zimmermann (Eds.), *Applied neuropsychology of attention: Theory, diagnosis and rehabilitation* (pp. 110–151). London, UK: Psychology Press.

Zinke, K., Einert, M., Pfennig, L., & Kliegel, M. (2012). Plasticity of executive control through task switching training in adolescents. *Frontiers in Human Neuroscience, 6,* 41.

Zinke, K., Zeintl, M., Eschen, A., Herzog, C., & Kliegel, M. (2012). Potentials and limits of plasticity induced by working memory training in old-old age. *Gerontology, 58,* 79–87.

Zinke, K., Zeintl, M., Rose, N. S., Putzmann, J., Pydde, A., & Kliegel, M. (2014). Working memory training and transfer in older adults: Effects of age, baseline performance, and training gains. *Developmental Psychology, 50*(1), 304–315.

Zoogman, S., Goldberg, S. B., Hoyt, W. T., & Miller, L. (2015). Mindfulness interventions with youth: A meta-analysis. *Mindfulness, 6*(2), 290–302.

9

Fundamental Questions Surrounding Efforts to Improve Cognitive Function Through Video Game Training

Adam Eichenbaum, Daphne Bavelier, and C. Shawn Green

The Curse of Learning Specificity

Neural plasticity, or the ability of the brain to reorganize its structure and activity, is of critical importance. It underlies not only our ability to acquire knowledge and learn new skills, but also to compensate for damage. For nearly 50 years, the dominant framework in the field of learning and neural plasticity held that the brain was capable of truly large-scale changes only early in life. This framework was buoyed by findings of "critical periods" (Wiesel & Hubel, 1965), times during which the brain was found to be capable of the type of reorganization necessary to recover lost function (e.g., to recover visual function in a deprived eye), but only up until a certain age. However, recent work has demonstrated that the brain retains a significant degree of plasticity well into old age. Emerging evidence even suggests that plasticity that had assumed to be "lost" due to age, injury, or disease may be at least partially re-established via genetic, pharmacological, or behavioral means (Bavelier, Levi, Li, Dan, & Hensch, 2010). These findings have led to a surge of interest in techniques to accelerate the rate of cognitive development, to enhance cognitive skills in young adults, and to slow, to halt, or even to reverse the deterioration of cognitive skill seen with natural aging.

Although humans retain a significant capacity to learn throughout the life span, a second roadblock frequently stands in the way of translating learning gains into practical real-world benefits. This obstacle has been dubbed the "curse of specificity" (Green &Bavelier, 2012). While it is true that, given appropriate training (e.g., a sufficient amount of practice time, practice sessions that are appropriately spaced, and useful feedback), humans will tend to improve on almost any task, the improvements that are observed are often confined to the exact training task, with little to no benefits of the training being observed for seemingly near identical tasks. Such learning specificity has been perhaps most thoroughly explored in the domain of perceptual learning, where, for instance, training to discriminate a grating tilted at 43° from one tilted at 67° leads to no benefit when participants are subsequently asked to discriminate a grating tilted at −40° from one tilted at −30° (Jeter, Dosher, Petrov, & Lu, 2009). In all, perceptual learning specificity has been noted for virtually every low-level visual feature imaginable, including orientation (Fahle, 1997; Spang, Grimsen, Herzog, & Fahle, 2010), retinal position (Skrandies, 2011), contrast (Yu, Klein, & Levi, 2004), motion direction and speed (Ball & Sekuler, 1982), and even the trained eye (Batson, Beer, Seitz, & Watanabe, 2011). Although such specificity of learning has been most fully documented in the perceptual learning literature, it is certainly not limited to it. As Diamond and Ling discuss in Chapter 8, a similar trend toward task-specific learning has been seen in the cognitive training literature as well, where training on one working memory (WM) task does not always lead to improvements in other tasks that engage WM (Zinke et al., 2014).

One intervention that does appear to lead to more generalizable improvements in cognitive performance involves an experience that was not originally designed for practical ends: action video game training. While the focus of this chapter is video game training, music training (Roden et al., 2014; White-Schwoch, Carr, Anderson, Strait, & Kraus, 2013) and meditation training (Lutz et al., 2009; Ramsburg & Youmans, 2014; Slagter, Davidson, & Lutz, 2011), represent other examples of complex forms of experience that lead to enhancements in cognitive performance.

Video Game Experience Affecting Cognitive Function

The earliest research on the effects of playing commercial video games focused strongly on visuomotor coordination and visuospatial cognition. In the early 1980s, home video game consoles had not yet reached a high level of popularity, so initial experiments focused on individuals who frequently played arcade video games. The performance of the frequent gamers was then compared to that of individuals who did not play video games (Kuhlman & Beitel, 1991; Sims & Mayer, 2002; Tirre & Raouf, 1994; Yi & Lee, 1997). For instance, Griffith, Voloschin, Gibb, and Bailey (1983) found that avid video game players outperformed non-video game players on a rotary pursuit task (a common visuomotor coordination task). Similarly, Greenfield, Brannon, and Lohr (1994) found that expert players of the video game *The Empire Strikes Back* outperformed nonplayers in measures of mental rotation ability (i.e., mentally "folding" a two-dimensional figure into a three-dimensional box). Although much of the early work was cross-sectional (i.e., comparing avid video game players with nonplayers), the intervention studies that were performed suggested that the relationship between video game playing and enhanced cognitive skill was causal instead of merely correlational (Drew & Waters, 1986; Dustman, Emmerson, Steinhaus, Shearer, & Dustman, 1992; Subrahmanyam & Greenfield, 1994). For instance, Gagnon assessed participants' mental rotation and hand–eye coordination skills before and after five hours of playing two different spatially demanding video games (Gagnon, 1985). Consistent with a causal relationship, posttraining scores on these measures were significantly improved as compared to pretraining scores (although there are issues related to intervention study designs that do not include an active control group, as discussed below in this chapter).

Taken together, the early work in this domain strongly suggested that, although commercial video games were designed with no purpose other than entertainment, they nonetheless had the capacity to more broadly influence perceptual, cognitive, and motor skills.

Action Video Games

As the research field developed, distinct genres of video games emerged in the commercial sphere, with each genre possessing unique dynamics, mechanics, and content. One particular genre, the "action video game," became the focus of much of the scientific work, as games in this genre appeared to influence perceptual, attentional, and cognitive skills to a significantly larger degree than games from other genres (Green & Bavelier, 2003). The action game genre was differentiated from other genres by several qualitative features, including the need to focus on, and track, numerous fast-moving objects on screen while simultaneously moving one's character through the game environment to avoid enemies and obstacles.

Over the past decade, numerous studies have shown that playing and training on action video games lead to significant improvements on many aspects of human performance, from low-level visual skills (Green & Bavelier, 2007; Li, Polat, Makous, & Bavelier, 2009), to visual attentional skills (Anguera et al., 2013; Awh, Vogel, & Oh, 2006; Dye, Green, & Bavelier, 2009; Feng, Spence, & Pratt, 2007; Green & Bavelier 2006a, 2006b; Li, Polat, Scalzo, & Bavelier, 2010; Mishra, Zinni, Bavelier, & Hillyard, 2011; Spence, Yu, Feng, & Marshman, 2009; for a review, see Eichenbaum, Bavelier, & Green, 2014), to higher-order cognitive skills and executive functions (Anderson, Bavelier, & Green, 2010; Chiappe, Conger, Liao, Caldwell, & Vu, 2013; Colzato, van den Wildenberg, & Hommel, 2014; Green & Bavelier, 2012; Strobach, Frensch, & Schubert, 2012).

Critically, the scope and scale of the effects have been of a magnitude that has led many groups to use off-the-shelf action video games for real-world ends (e.g., for rehabilitative or job-related training). For example, one application can be seen in adults with amblyopia (commonly referred to as "lazy eye")—a condition in which visual information from one eye is essentially ignored by the brain. Although amblyopia has previously been considered to be reasonably irreversible after childhood (Li, Ngo, Nguyen, & Levi, 2011), adults who trained on an action video game using their "bad" eye reported dramatic improvement in their visual ability, with some patients recovering entirely from the affliction (see also Waddingham, Cobb, Eastgate, & Gregson, 2006, and Li et al., 2013, for examples of using video games, although not always pure action games, in the treatment of amblyopia). Similarly, action video game training in a group of Italian-speaking dyslexics resulted in greater reading abilities than did traditional dyslexia-based interventions (Franceschini et al., 2013). In terms of job-related skills, action video game experience has been shown to strongly correlate with laparoscopic and endoscopic surgical ability, with action game experience being a better predictor than previous surgical experience (Ou, McGlone, Camm, & Khan, 2013; Rosser et al., 2007), and specific action video game training leading to substantial benefits in (surgical simulator) performance (Schlickum, Hedman, Enochsson, Kjellin, & Felländer-Tsai, 2009). Finally, an investigation of military drone pilots revealed that those who reported greater amounts of action video game play per week outperformed their counterparts who had not engaged in action video game play (Chiappe et al., 2013; Gopher, Weil, & Bareket, 1994; McKinley, McIntire, & Funke, 2011).

Moderators of Training Improvements

Although the improvements observed following action video game training are quite robust, there may be individual differences in the efficacy of training. For example, training benefits may scale with age, or they may depend on the sex of the participant. So far, there is only research on the effects of age and sex; factors like cultural heritage and various psychosocial factors (e.g., socioeconomic status) have not yet been investigated.

Sex
Few studies have specifically analyzed sex as a potential factor in training benefits. Feng and colleagues (2007) reported that females improved more as a result of training than did males. However, prior to training, females performed worse than males on the measures of spatial cognition being tested. Therefore, it is unclear whether the training effect is related to sex or reflects the reasonably common finding in the learning literature that training leads to larger benefits in individuals who show initially poorer performance. Furthermore, and counter to the findings of Feng and colleagues,

Subrahmanyam and Greenfield (1994) reported that males benefited more from training than females, with both sexes performing equally well at pre-test. Mirroring these conflicting results, meta-analyses performed by Wang and colleagues (2016) and Powers and colleagues (Powers, Brooks, Aldrich, Palladino, & Alfieri, 2013) came to separate conclusions regarding the effect of sex. Whereas Wang and colleagues found no significance of the male–female sex ratio on the effect size of training-related improvements, Powers and colleagues found that males benefited more from action video game experience than did females (although the effect of sex disappeared with removal of cross-sectional studies—which have obvious participant-bias issues—and analysis of training studies only).

Age

Participant age does appear to play a role in the benefits engendered by action video game training. Toril, Reales, and Ballesteros (2013) conducted a meta-analysis on the effects of action video game training specifically for elderly participants. The authors reported that participants between the ages of 71 and 80 years old improved more than did participants between the ages of 60 and 70. This finding is corroborated by the report from Power and colleagues (2013), who also found that the oldest participants improved more than younger groups. Wang and colleagues (2016) reported an effect of age going in the other direction (i.e., younger participants improved more than older); however, the magnitude of this effect, while significant, was very small. An effect of age can most likely be understood as the difference between the improvement of a near-optimized ability in young adults versus a recovery of function in elderly adults. Furthermore, it is worth noting that the types of games employed by studies utilizing child, young adult, and older adult populations often differ dramatically due to issues with content (e.g., studies with children cannot utilize games that feature adult content, such as violence) or motor abilities (e.g., elderly individuals often have difficulty with standard video game controllers). This pattern of covariation between the populations and the methods in turn makes it difficult to confidently attribute differences found among studies to developmental differences.

Longevity of Training Benefits

It is also interesting to note the longevity of the benefits induced by video game training, as there are very few skills or abilities that are perfectly maintained in the absence of any continued practice. To date, few studies in this domain have collected the type of follow-up data that would speak to this point, but so far, the data suggest that the effects persist over a reasonable period of time. Feng and colleagues (2007) trained non-video game players for ten hours and found improvements in both spatial attention and mental rotation compared to controls. At a five-month follow-up, the improvements in both spatial abilities remained (i.e., there was no statistically significant fall off in performance over the intervening period). In addition, Li and colleagues (2009) reported participants had enhanced contrast sensitivity after training for 50 hours. When participants in both the action and control group performed the same experimental task again between 5 and 30 months later, the training-related enhancement in contrast sensitivity remained significant, although it was reduced by some degree. Moreover, participants from both groups reported no change in their video game consumption after cessation of training (i.e., participants did not continue to play video games, including the training games, after the study ended). Last, Anguera and colleagues (2013) reported improvements in multitasking ability after training elderly participants (60–85 years old) for 12 hours on a custom-built racing game. These improvements persisted when participants returned approximately 7.5 months later and performed the same multitasking test. As of yet, no action video game training study has investigated the impact of follow-up "booster" sessions on the maintenance of training-induced improvements. Thus, while improvements resulting from action video game training are certainly not impervious to atrophy, the current evidence suggests that the benefits are quite robust.

Methodological Issues

Although the results in the action video game training literature are clearly promising, issues remain both in terms of methodology (e.g., as is the case in the more general executive function literature, questions apply: What makes for a proper control group? How should training be spaced? How many hours of training are sufficient?) and in terms of theory (e.g., What is the mechanism by which video games produce broad transfer? What aspects of video games are particularly critical in producing generalizable benefits?).

The Need for Intervention/Training Studies

One clear difference between the literature outlining the effects of playing certain types of video games on cognitive function and the literature on the effects of dedicated cognitive training is the presence of pre-existing "expert" video game players (i.e., people who choose to play lots of video games as part of their everyday life; furthermore, the increasing mass popularity of "brain training" regimens may create a similar population of individuals in that domain as well).

In the early stages of video game research, much of the work was cross-sectional—grouping together individuals who did play video games and individuals who did not play video games, then comparing each group's performance on a measure of interest (e.g., hand–eye coordination, mental rotation). Although the studies are certainly informative, it is not possible to establish a causal link between the experience and the outcome via cross-sectional work alone because of the concern about the potential for population bias (i.e., the possibility that individuals with enhanced cognitive skills self-select as avid gamers). Individuals with innately superior cognitive skills may disproportionately choose to become avid gamers, rather than having the gaming itself cause superior cognitive skills. It is thus vital that well-designed training studies be conducted to establish that the act of playing action video games actually causes enhancements in processes of interest. Yet, although this is largely agreed upon, there is significant debate about how the studies should be instantiated.

Issues with Control Task Selection

In many of the earliest intervention studies examining the effects of video game experience, only an experimental group was included. In various studies, participants were required to navigate a maze from a bird's-eye view and shoot enemies (*Targ*; Gagnon, 1985), maneuver a tank through a virtual battle arena (*Battlezone*; Gagnon, 1985), rotate falling geometric objects (*Tetris*; Okagaki & Frensch, 1994), or operate a spaceship in three dimensions (Dorval & Pépin, 1986), with participant performance on a given task (or set of tasks) being measured before and after training. While in each case significant improvements were observed from pretraining to posttraining, the lack of a control group meant that several potential confounds were still possible. The first, and most obvious, would be simple test–retest improvements. As noted in Chapter 8 by Diamond and Ling (2017), individuals tend to improve the second time they perform a task, regardless of any intervening experience. Thus, the changes observed above, which were attributed to the gaming experience, could just as plausibly be explained by repeated exposure to the test.

In an effort to control for test–retest confounds, several later studies included two groups: an experimental group that trained on the video game and a test–retest control group that did not train on the video game and instead only took the pre- and posttest measure (Greenfield, Brannon, et al. 1994). In one study, participants' visuospatial abilities were assessed before and after being randomly assigned to either an experimental group that played *Robotron* for 5 hours or to a control group that did not play the game but instead underwent pre- and posttesting at time intervals similar to those for the experimental group. Those who trained on *Robotron* showed reduced reaction times on the measure of interest compared to the controls, indicating that the difference could not be attributed to simple test–retest effects (Greenfield, DeWinstanley, Kilpatrick, & Kaye, 1994).

However, and again as discussed by Diamond and Ling, many potential issues still remain with simple test–retest control groups. For instance, the individuals in the experimental group interact frequently with the research team, whereas the no-contact test–retest control group do not experience the same engagement. As a result, there may be motivational differences between the groups (i.e., the greater the engagement, the more motivated one may be), which could manifest in performance differences, an effect recognized as the Hawthorne effect (Parsons, 1974). To avoid such participant reactivity effects, the control group must undergo a procedure similar to the procedure administered to the experimental group, but one that critically excludes the components of interest found in the experimental group's training. One of the first studies to include an "active" control group required those in the experimental group to train on *Marble Madness*, while the control group trained on *Conjecture*, an equally engaging computerized word game (Subrahmanyam & Greenfield, 1994). The data acquired from this study much more unequivocally suggested that *Marble Madness* resulted in enhanced visuospatial performance.

While in the field there is now general consensus that proper experimental studies require active control groups, as is true in the executive function literature, there continues to be substantial debate about the ideal active control condition. For instance, the level of task difficulty in most experimental training paradigms changes in response to the participants' skill level (i.e., the demands of the task increase as the participants improve). Thus, many investigators have sought to ensure that the control task has this same quality. In the cognitive function literature, for instance, studies by Redick and colleagues (2013) and Jaeggi, Buschkuehl, Jonides, and Shah (2011) used adaptive psychometric (i.e., visual search task) and quiz-based controls, respectively. In the case of the gaming literature, meanwhile, almost all commercial video games, including those that have been commonly used as controls, become increasingly difficult as the player progresses through the game. For example, in *Tetris* (which has been used as a control game in many studies), the speed at which the blocks fall increases as the player reaches higher levels. There are researchers who believe, though, that the proper control should instead primarily match the experimental condition in terms of visual stimulation and motor response. In the general cognitive training literature, this is seen in experiments by Klingberg and colleagues (2005) and Brehmer, Westerberg, and Bäckman (2012). In these experiments, the control group performed the same task as the experimental group but at an easier level of difficulty that never increased, so as to prevent the possibility of transfer of training. In the gaming literature, a similar strategy was employed by Glass, Maddox, and Love (2013), who used an easier/reduced version of the experimental game (*StarCraft*) as the primary control video game (including a second, different control game as well). While this may fail to control for a sense of interest or engagement, its advantages include matching the experimental conditions almost perfectly (e.g., stimulus characteristics, motor requirements, and task instructions).

Overall, the complexity of modern commercial video games makes control game selection a potentially more difficult endeavor than control selection for studies of general cognitive training. While Redick and colleagues (2013) selected visual search as the control-training task because visual search performance is known—via previous work on individual differences—to be reasonably independent of WM performance (making it an ideal control-training task, as there is little concern that visual search uses the proposed core cognitive processes involved in WM training), no such pairs of commercial video games exist. Instead, most modern studies of video games have attempted to match games along as many dimensions as possible while leaving out the posited critical aspects of action video games. These studies have often used either puzzle games (Green & Bavelier, 2003, 2007; Spence et al., 2009) or life simulation games (Blacker, Curby, Klobusicky, & Chein, 2014; Dye et al., 2009; Glass et al., 2013; Green, Sugarman, Medford, Klobusicky, & Bavelier, 2012; Li et al., 2009; Oei & Patterson, 2013) as the control-training games. These games are selected to match the action (experimental) game for dimensions like arousal, engagement, sense of flow, and identification with character.

Titrating the Level of Difficulty

As discussed above, nearly all experimental executive-function training paradigms use an adaptive level of difficulty. As participants improve, the demands of the task increase. For instance, in N-back training, when participants have mastered the 2-back task, they are moved up to 3-back, and when they master 3-back, they are moved up to 4-back, and so forth. While difficulty levels are not quite so easy to quantify in action video games, similar principles have been followed by most successful video game training studies. For instance, most linear story games, such as *Call of Duty* or *Medal of Honor*, ramp the difficulty level up automatically as the story progresses (and participants cannot progress the story until their level of ability allows them to pass certain checkpoints). In games in which difficulty does not automatically increase, researchers have manually adjusted it. For example, in studies that have used the video game *Unreal Tournament 2004*, researchers tracked participant performance in 20-minute blocks and increased difficulty by one level whenever performance reached a criterion level (i.e., when participants had "killed" twice as many robots as the number of times their own characters had "died").

While most of today's young adults have, to some extent, been raised with some degree of video game familiarity, proper titration of difficulty can be quite problematic in elderly populations. For instance, one study had elderly individuals play a commercial action video game but failed to find any effect of training (Boot et al., 2013). However, because these games are explicitly designed for a young adult audience (specifically, for a young adult *gamer* audience), the action is potentially much too fast for elderly novice game players. This may be why when the video game used for training has been better tailored for an older audience (i.e., of a slower pace, with fewer demands, etc.), those who trained on the video game have shown better task-switching and multitasking performance, as well as enhanced short-term and WM ability (Anguera et al., 2013; Basak, Boot, Voss, & Kramer, 2008; Belchior et al., 2013; Mayas, Parmentier, Andrés, & Ballesteros, 2014; Toril et al., 2014).

Issues Related to Training Schedules

In addition to designing both a proper control group and utilizing an appropriate level of challenge during training, researchers must also be careful when designing training schedules. As any student who is studying for a large exam knows, both the pace at which one studies, as well as the total time spent studying, greatly affect one's ability to perform when ultimately taking the exam.

Massed Versus Distributed Training

For at least three decades, researchers have recognized the schedule of training as one of the most influential factors when measuring the amount of learning. In the seminal paper on this topic, Baddeley and Longman (1978) performed a study to examine the most efficacious schedule for training United States Postal Service workers to type. Several different training schedules were compared, with different numbers of sessions per day and different numbers of hours per session (e.g., 1-hour sessions with one session per day; 2-hour sessions with one session per day; 1-hour sessions with two sessions per day). The results clearly indicated that distributed practice (i.e., shorter training sessions spaced over more days) resulted in by far the most learning. In fact, participants in the most distributed group reached a criterion level of performance (60 correct keystrokes per minute) nearly twice as quickly as participants in the most massed group.

The basic finding that distributed practice is far superior to massed practice has been replicated in a wide variety of domains, including the video game domain (Stafford & Dewar, 2014). However, despite distributed practice being one of the most well-known effects in the field of learning and transfer, there nonetheless continue to be studies that attempt to use highly massed practice schedules. For example, a study by van Ravenzwaaij, Boekel, Forstmann, Ratcliff, and Wagenmakers (2014) failed to find a significant transfer effect after both 10-hour and 20-hour action video game training experiments. Unlike previous work in the field, the authors used a highly massed training schedule (in one case with 4-hour sessions, twice as massed as the most massed training condition used in the formative work by Baddeley & Longman, 1978). Similarly, a study by Shute, Ventura, and Ke (2014) used 8 hours of training divided unevenly across three sessions. Participants trained for 2 hours in the first session and 3 hours in each of the subsequent sessions. While these training schedules are undoubtedly cost- and time-effective for the research team (in that a 20-hour training study, which under normal conditions takes approximately one month, can be completed in only five days), they violate one of the main tenets of learning, and it is unclear what can be gleaned from them.

Total Training Time

An additional factor to consider when conducting a training study is the total amount of training. Obviously, the simplest guideline is to use a length of training where a null result is convincing. For example, showing that 2 hours of training on a "brain trainer" produces no significant benefits is reasonably unconvincing. One could easily argue that benefits would be observed with a longer training duration. The counterargument that if 50 hours of training produced a significant improvement, then some benefits should be seen after 2 hours, is too simplistic. In this vein, much of our recent work has employed 50 hours of training spaced out over several months (Bejjanki et al., 2014; Green, Pouget, & Bavelier, 2010; Li et al., 2009).

It may not be the case, though, that longer training durations always predict more transfer. For example, numerous investigations have shown that training on *Tetris* for only 10 to 20 hours enhances mental rotation and visuospatial skills (De Lisi & Wolford, 2002; Sims & Mayer, 2002; Terlecki, Newcombe, & Little, 2008; Uttal et al., 2013). However, somewhat surprisingly, research has also shown that when visuospatial and mental rotation skills of expert *Tetris* players are compared to the skills of nonexpert players, there is very little difference between the two groups (Destefano, Lindstedt, & Gray, 2011). Together, this pattern of results suggests that there is in fact a "sweet spot" of total training dosage, such that too little would not be sufficient for transfer, but too much would lead to honing of a specific isolated skill (i.e., rotating Tetris blocks) and not of a generalizable ability (i.e., basic visuospatial skills). A computational framework laying out exactly this argument has been put forward by Fulvio, Green, and Schrater (2014), who showed that increasing training time (or manipulating other task parameters, such as stimulus repetitions, stimulus variety, etc.) led to a predictable shift from what is known as a "model-based" system (i.e., one that is flexible, and thus the learning is transferable) to a "model-free" system (i.e., one that uses well-rehearsed heuristics to make choices and is therefore quite rigid). Although the question of whether the shift from more effortful processing to automatization also leads to less transfer in the cognitive domain remains to be fully explored, a more nuanced suggestion for total training time is to use a duration that ensures significant learning has occurred on the training task (otherwise there is nothing to transfer) but not so much that behaviors become automatized.

Test–Retest Effects

While we have discussed the need for groups to act as a control against simple test–retest effects (i.e., we would not want to ascribe changes as being due to our intervention that are due to participants' taking a test a second time), interestingly, in the video game training literature, several authors (e.g., Boot, Blakely, & Simons, 2011; Kristjánsson, 2013) have suggested that significant test–retest improvements must be observed in both the experimental and control groups for the study to be valid. These authors argue that, in cases in which the experimental group shows a significant improvement from pre- to posttest but the control group shows no change, there are two equally plausible interpretations. The first, favored by the majority of the literature, is that the experimental training resulted in significant benefits on the transfer task, while the control training resulted in no change. The second interpretation, favored by the authors above, is that both groups were subject to test–retest improvements, but the control-training experience actually reduced the control participants' performance back to baseline levels. In other words, these authors suggest that, rather than the experimental training having led to benefits in performance, the control training led to decrements in performance. While both interpretations are logically valid, it is difficult to come up with a mechanism by which a control-training paradigm could severely adversely affect participant performance. Indeed, negative transfer effects are reasonably rare in the literature—with the most common examples involving training and transfer tasks that have conflicting stimulus–response contingencies, which is rarely the case in the literature under discussion.

Beyond the question of interpretation, though, a potentially more interesting question is whether test–retest effects are something to aspire to in the first place. While this appears to be the belief of some authors (Boot et al., 2011; Kristjánsson, 2013), several pieces of evidence indicate the contrary, especially when the goal is to establish transfer from a given intervention or training. Indeed, it is well documented that there is no better way to enhance performance on a task than repeatedly performing that very task. The end result is that the practiced task will be learned faster and to a greater extent than any other less direct training intervention. However, as demonstrated by Schmidt and Bjork as early as 1992, training on one given task comes at a price; in particular, there is a reduced ability to transfer the benefits of training to related tasks or contexts. Thus, the conditions that foster fast and efficient learning are not necessarily aligned with those that foster transfer.

The presence of simple test–retest effects would be of less importance if the benefits of repeatedly taking the test were independent of the benefits engendered by the experimental training paradigm. However, this does not appear to be true. Instead, a meta-analysis of training studies in the cognitive domain has shown that designs that use repeated testing consistently report significantly larger improvements in the control group and correspondingly smaller calculated benefits of the experimental group as compared to studies that do not (Uttal et al., 2013). In other words, allowing the control group to improve on the transfer task via repeated testing reduces the maximum difference that can be observed between the control and experimental groups.

Such a finding is consistent with nearly all existing theories of transfer that have at their foundation the idea that transfer will be observed in cases where the training and the transfer task share some critical processing demands (an idea first put forward by Thorndike & Woodworth, 1901; for more contemporary views, see Singley & Anderson, 1989; Taatgen, 2013; Green, Strobach, & Schubert, 2014). This is why investigators use experimental training tasks that they believe share processing demands with their transfer battery and control-training tasks that they believe do not. Through repeated practice on the experimental training, the process common to the transfer battery should be strengthened, and this should, in turn, be manifested by improved performance on the transfer battery that shares this process. Meanwhile, because the control training shares no processing in common with the transfer battery, no improvements should be noted. Yet, because undertaking the same task several times is the most efficient way to improve on a task, improvements due to simple test–retest effects will have the potential to mask transfer effects from intervening training. It may not be surprising then, that the two most prominent failed video game training experiments featured enormous test–retest effects in the control group (Boot, Kramer, Simons, Fabiani, & Gratton, 2008; van Ravenzwaaij et al., 2014).

The point here is not to advocate against control conditions that contain a test–retest design; on the contrary, this is an essential design feature to establish an intervention's efficacy. However, the design should be carefully thought out to choose testing procedures that limit the opportunity for test–retest improvements if the focus of the study is transfer. This is often done in the literature by limiting the amount of exposure participants have to the transfer battery (i.e., by using as few trials as is possible), putting a large delay between the first and second exposure to the transfer battery (which should already be the case given appropriate training time and distribution of practice), eliminating feedback during the transfer battery, and limiting testing protocols to before- and after-intervention tests (repeated testing throughout the intervention is all but guaranteed to result in a null effect). Groups working on validating test batteries to evaluate cognitive or memory skills have been designing matched but different versions of the same tests for this very purpose for years. There are well-accepted ways to minimize test–retest effects, and this is a commendable practice for studies focusing on transfer.

Theoretical Issues

Beyond the specifics of how training interventions should be designed and how control groups should be constructed, the field of action video game training is actively searching for mechanistic explanations of the observed improvements. Critical questions remain, such as what drives the broad transfer of learning seen with action video game training compared to other cognitive training paradigms? What might be the most appropriate way to assess the efficacy of action video game training? Why are action video games relatively unique in their ability to foster certain forms of learning?

Mechanisms Underlying Broad Transfer of Learning From Action Video Games

Most of the literature on training of cognitive function has centered on the topic of transfer of learning—when training on a first task conveys immediate benefits on a second, untrained task. For example, training on one WM task can lead to immediate benefits on a second WM task (Melby-Lervåg & Hulme, 2012) or on a measure of fluid intelligence (Colom et al., 2013; Jaeggi, Buschkuehl, Jonides, & Perrig, 2008; Shipstead, Redick, & Engle, 2012). Although the topic of transfer is of obvious importance, we have recently suggested that the broad benefits seen as a result of action video game play may not in fact reflect transfer of learning, but may instead reflect "learning to learn" (Bavelier, Green, Pouget, & Schrater, 2012). In learning to learn, training on a first task may not necessarily convey any immediate benefits on a second, untrained task, but training on the first task does convey an enhanced ability to learn to perform the second, untrained task. As an example, we have shown that while training with action video games does not result in better performance than training with control video games in the first few trials of a visual detection task, individuals trained with action video games show greatly accelerated learning of the visual detection task. After many blocks of practice with the second task, those who had trained with action video games far outperformed those who had trained with a control video game (Bejjanki et al., 2014). Similar findings were obtained in a visuomotor task (Gozli, Bavelier, & Pratt, 2014).

Studying learning to learn, rather than transfer, necessitates a shift in research methodology. In particular, the pretest → training → posttest framework, while ideal for studying transfer, is inadequate for studying learning to learn. Instead, rather than single, short, identical pre- and posttest tasks, full-learning tasks need to be given before and after training (slightly different tasks at pre- and posttest, counterbalanced across participants). It could be that training paradigms that failed to transfer actually resulted in broader benefits than realized because only immediate transfer, not subsequent learning abilities, was assessed.

Critical Components of Video Games that Engender Learning and Transfer

To some extent, the video game industry has, through selection via mass popularity, evolved to produce games that are finely tuned to be effective learning tools. After all, no one would play a game that requires zero learning to master (i.e., where players are already perfect at the game the first time they try). Neither would one play a game that is impossible to master. Thus, a good game is one that efficiently teaches players to improve their skills. As has been noted by many authors,over the past 30 years, games have essentially "rediscovered" characteristics long known in psychology and education to be critical to generating learning and have instantiated those characteristics in game environments (Gentile & Gentile, 2008). These include both characteristics known to enhance learning in general and characteristics known to promote transferable learning gains.

General Learning

First, unlike standard lab training tasks (or most brain trainers), video games are rewarding. Indeed, the same neural circuits that indicate the presence of primary reinforcers, such as food and water, are also activated by video game play (Koepp et al., 1998). Such reinforcement is important for producing learning in and of itself, as the same neural systems implicated in the coding of reward have also been shown to promote neural plasticity (activity in these reward systems appears to put the brain in a state that is permissive to plasticity; Kilgard & Merzenich, 1998). The reinforcement also engenders motivation to continue playing, as time on task is one of the leading predictors of learning outcomes (Przybylski, Rigby, & Ryan, 2010). Second, video games are physiologically arousing. The well-known Yerkes-Dodson law—first proposed at the turn of the 20th century—states that learning is an inverse U-shaped function of arousal (Yerkes & Dodson, 1908). Too little arousal (i.e., boredom) will produce no learning. Too much arousal (as caused by a severely traumatic event) will also cause poor learning. With a moderate level of arousal, learning is at a maximum. At the neural level, this seems to be reflective of the fact that the system that mediates arousal, like the reward system, acts in a permissive manner toward plasticity (Bao, Chan, & Merzenich, 2001). Video games also get task difficulty right; initial levels of difficulty are typically low, to reduce entrance costs and to avoid immediate quitting. Difficulty is then titrated to ensure that that the demands remain challenging (even as the players' skills progress) but are always eventually doable (i.e., they are in the "zone of proximal development," as discussed by Vygotsky, 1978). And finally, video game designers are experts at developing a player's in-game skill set to the point where the task at hand can always be solved by building upon what has been learned in the previous levels (i.e., scaffolding).

Generalizable Learning

In terms of producing transfer, modern video games are incredibly complex and require something new at every turn from the player. Even when replaying a video game, players must deal with the inherent stochasticity of the game environment (e.g., locations of enemies and allies compared to the last playthrough). In action video games, no specific skills can be developed because no two situations are ever identical. Thus, the best that can be done is to improve task-independent processes (e.g., attentional skills) and to "learn to learn" to perform new tasks. Furthermore, as discussed by Diamond and Ling, the best training tools are those that represent a combination of tasks, a gestalt of responsibilities. It is thus no surprise that the best predictors of cognitive improvement and of sustained abilities in old age are the programs that include more than just a solitary intervention (i.e., aerobic exercise, video game training, and group socialization; Karr, Areshenkoff, Rast, & Garcia-Barrera, 2014).

Conclusions

Unlike many dedicated computerized cognitive training paradigms that produce narrow transfer, action video game play has been repeatedly shown to benefit performance on a wide variety of perceptual, attentional, and cognitive tasks. In fact, the scope and scale of these effects is of such a magnitude that research groups have already started to explore the use of commercial off-the-shelf games in tackling real-world problems ranging from amblyopia rehabilitation to the training of laparoscopic surgeons. However, unanswered questions remain. Some relate to methodology (e.g., how to choose the proper control-training task) and some relate to theory (e.g., exactly what video game characteristics are critical in engendering learning and transfer). Future work will need to break action video games down to a more tractable level in order to effectively test core hypotheses and eventually build evidence-based video game training paradigms.

References

Anderson, A. F., Bavelier, B., & Green, C. S. (2010). Speed–accuracy tradeoffs in cognitive tasks in action game players. *Journal of Vision, 10*(7), 748. doi:10.1167/10.7.748

Anguera, J. A., Boccanfuso, J., Rintoul, J. L., Al-Hashimi, O., Faraji, F., Janowich, J., . . . Gazzaley, A. (2013). Video game training enhances cognitive control in older adults. *Nature, 501*(7465), 97–101. doi:10.1038/nature12486

Awh, E., Vogel, E. K., & Oh, S. H. (2006). Interactions between attention and working memory. *Neuroscience, 139*(1), 201–208. doi:10.1016/j.neuroscience.2005.08.023

Baddeley, A. D., & Longman, D. J. A. (1978). The influence of length and frequency of training sessions on the rate of learning to type. *Ergonomics, 21*(8), 627–635. doi:10.1080/00140137808931764

Ball, K. K., & Sekuler, R. (1982). A specific and enduring improvement in visual motion discrimination. *Science, 218* (4573), 697–698. doi: 10.1126/science.7134968

Bao, S., Chan, V. T., & Merzenich, M. M. (2001). Cortical remodelling induced by activity of ventral tegmental dopamine neurons. *Nature, 412*(6842), 79–83. doi:10.1038/35083586

Basak, C., Boot, W. R., Voss, M. W., & Kramer, A. F. (2008). Can training in a real-time strategy video game attenuate cognitive decline in older adults? *Psychology and Aging, 23*(4), 765–777. doi:10.1037/a0013494

Batson, M. A., Beer, A. L., Seitz, A. R., & Watanabe, T. (2011). Spatial shifts of audio-visual interactions by perceptual learning are specific to the trained orientation and eye. *Seeing and Perceiving, 24*(6), 579–594. doi:10.1163/187847611X603738

Bavelier, D., Green, C. S., Pouget, A., & Schrater, P. (2012). Brain plasticity through the life span: Learning to learn and action video games. *Annual Review of Neuroscience, 35,* 391–416. doi:10.1146/annurev-neuro-060909-152832

Bavelier, D., Levi, D. M., Li, R. W., Dan, Y., & Hensch, T. K. (2010). Removing brakes on adult brain plasticity: From molecular to behavioral interventions. *The Journal of Neuroscience, 30*(45), 14964–14971. doi:10.1523/JNEUROSCI.4812-10.2010

Bejjanki, V. R., Zhang, R., Li, R., Pouget, A., Green, C. S., Lu, Z., & Bavelier, D. (2014). Action video game play facilitates the development of better perceptual templates. *Proceedings of the National Academy of Sciences, 111*(47), 16961–16966. doi:10.1073/pnas.1417056111

Belchior, P., Marsiske, M., Sisco, S. M., Yam, A., Bavelier, D., Ball, K., & Mann, W. C. (2013). Video game training to improve selective visual attention in older adults. *Computers in Human Behavior*, *29*(4), 1318–1324. doi:10.1016/j.chb.2013.01.034

Blacker, K. J., Curby, K. M., Klobusicky, E., & Chein, J. M. (2014). Effects of action video game training on visual working memory. *Journal of Experimental Psychology: Human Perception and Performance*, *40*(5), 1992–2004. doi:10.1037/a0037556

Boot, W. R., Blakely, D. P., & Simons, D. J. (2011). Do action video games improve perception and cognition? *Frontiers in Psychology*, *2*, 226. doi:10.3389/fpsyg.2011.00226

Boot, W. R., Champion, M., Blakely, D. P., Wright, T., Souders, D. J., & Charness, N. (2013). Video games as a means to reduce age-related cognitive decline: Attitudes, compliance, and effectiveness. *Frontiers in Psychology*, *4*, 31. doi:10.3389/fpsyg.2013.00031

Boot, W. R., Kramer, A. F., Simons, D. J., Fabiani, M., & Gratton, G. (2008). The effects of video game playing on attention, memory, and executive control. *Acta Psychologica*, *129*(3), 387–398. doi:10.1016/j.actpsy.2008.09.005

Brehmer, Y., Westerberg, H., & Bäckman, L. (2012). Working-memory training in younger and older adults: Training gains, transfer, and maintenance. *Frontiers in Human Neuroscience*, *6*, 63. doi:10.3389/fnhum.2012.00063

Chiappe, D., Conger, M., Liao, J., Caldwell, J. L., & Vu, K. P. L. (2013). Improving multitasking ability through action videogames. *Applied Ergonomics*, *44*(2), 278–284. doi:10.1016/j.apergo.2012.08.002

Colom, R., Román, F. J., Abad, F. J., Shih, P. C., Privado, J., Froufe, M., . . . Jaeggi, S. M. (2013). Adaptive N-back training does not improve fluid intelligence at the construct level: Gains on individual tests suggest training may enhance visuospatial processing. *Intelligence*, *41*(5), 712–727. doi:10.1016/j.intell.2013.09.002

Colzato, L. S., van den Wildenberg, W. P., & Hommel, B. (2014). Cognitive control and the COMT Val(1)(5)(8)Met polymorphism: Genetic modulation of videogame training and transfer to task-switching efficiency. *Psychological Research*, *78*(5), 670–678. doi:10.1007/s00426-013-0514-8

De Lisi, R., & Wolford, J. L. (2002). Improving children's mental rotation accuracy with computer game playing. *The Journal of Genetic Psychology: Research and Theory on Human Development*, *163*(3), 272–282. doi:10.1080/00221320209598683

Destefano, M., Lindstedt, J. K., & Gray, W. D. (2011). Use of complementary actions decreases with expertise. In L. Carlson, C. Hölscher, & T. Shipley (Eds.), *33rd Annual Conference of the Cognitive Science Society* (pp. 2709–2714). Austin, TX: Cognitive Science Society.

Dorval, M., & Pépin, M. (1986). Effect of playing a video game on a measure of spatial visualization. *Perceptual and Motor Skills*, *62*(1), 159–162. doi:10.2466/pms.1986.62.1.159

Drew, B., & Waters, J. (1986). Video games: Utilization of a novel strategy to improve perceptual motor skills and cognitive functioning in the non-institutionalized elderly. *Cognitive Rehabilitation*, *4*(2), 26–31.

Dustman, R. E., Emmerson, R. Y., Steinhaus, L. A., Shearer, D. E., & Dustman, T. J. (1992). The effects of videogame playing on neuropsychological performance of elderly individuals. *Journal of Gerontology*, *47*(3), 168–171. doi:10.1093/geronj/47.3.P168

Dye, M. G., Green, C., & Bavelier, D. (2009). Increasing speed of processing with action video games. *Current Directions in Psychological Science*, *18*(6), 321–326. doi:10.1111/j.1467-8721.2009.01660.x

Eichenbaum, A., Bavelier, D., & Green, C. S. (2014). Video games: Play that can do serious good. *American Journal of Play*, *7*(1), 50–72.

Fahle, M. (1997). Specificity of learning curvature, orientation, and vernier discriminations. *Vision Research, 37*(14), 1885–1895. doi:10.1016/S0042-6989(96)00308-2

Feng, J., Spence, I., & Pratt, J. (2007). Playing an action video game reduces gender differences in spatial cognition. *Psychological Science, 18*(10), 850–855. doi:10.1111/j.1467-9280.2007.01990.x

Franceschini, S., Gori, S., Ruffino, M., Viola, S., Molteni, M., & Facoetti, A. (2013). Action video games make dyslexic children read better. *Current Biology, 23*(6), 462–466. doi:10.1016/j.cub.2013.01.044

Fulvio, J. M., Green, C. S., & Schrater, P. R. (2014). Task-specific response strategy selection on the basis of recent training experience. *PLoS Computational Biology, 10*(1): e1003425. doi:10.1371/journal.pcbi.1003425

Gagnon, D. (1985). Videogames and spatial skills: An exploratory study. *Educational Communication and Technology Journal, 33*(4), 263–275. doi:10.1007/BF02769363

Gentile, D. A., & Gentile, J. R. (2008). Violent video games as exemplary teachers: A conceptual analysis. *Journal of Youth and Adolescence: A Multidisciplinary Research Publication, 37*(2), 127–141. doi:10.1007/s10964-007-9206-2

Glass, B. D., Maddox, W. T., & Love, B. C. (2013). Real-time strategy game training: Emergence of a cognitive flexibility trait. *PLoS One, 8*(8), e70350. doi:10.1371/journal.pone.0070350

Gopher, D., Weil, M., & Bareket, T. (1994). Transfer of skill from a computer game trainer to flight. *Human Factors, 36*(3), 387–405. doi:10.1177/001872089403600301

Gozli, D. G., Bavelier, D., & Pratt, J. (2014). The effect of action video game playing on sensorimotor learning: Evidence from a movement tracking task. *Human Movement Science, 38C*, 152–162. doi:10.1016/j.humov.2014.09.004

Green, C., & Bavelier, D. (2003). Action video game modifies visual selective attention. *Nature, 423*(6939), 534–537. doi:10.1038/nature01647

Green, C., & Bavelier, D. (2006a). Effect of action video games on the spatial distribution of visuospatial attention. *Journal of Experimental Psychology: Human Perception and Performance, 32*(6), 1465–1478. doi:10.1037/0096-1523.32.6.1465

Green, C. S., & Bavelier, D. (2006b). Enumeration versus multiple object tracking: The case of action video game players. *Cognition, 101*(1), 217–245. doi:10.1016/j.cognition.2005.10.004

Green, C. S., & Bavelier, D. (2007). Action-video-game experience alters the spatial resolution of vision. *Psychological Science, 18*(1), 88–94. doi:10.1111/j.1467-9280.2007.01853.x

Green, C. S., & Bavelier, D. (2012). Learning, attentional control, and action video games. *Current Biology, 22*(6), R197–R206. doi:10.1016/j.cub.2012.02.012

Green, C. S., Pouget, A., & Bavelier, D. (2010). Improved probabilistic inference as a general learning mechanism with action video games. *Current Biology, 20*(17), 1573–1579. doi:10.1016/j.cub.2010.07.040

Green, C. S., Strobach, T., & Schubert, T. (2014). On methodological standards in training and transfer experiments. *Psychological Research, 78*(6), 756–772. doi:10.1007/s00426-013-0535-3

Green, C. S., Sugarman, M. A., Medford, K., Klobusicky, E., & Bavelier, D. (2012). The effect of action video game experience on task-switching. *Computers in Human Behavior, 28*(3), 984–994. doi:10.1016/j.chb.2011.12.020

Greenfield, P. M., Brannon, G., & Lohr, D. (1994). Two-dimensional representation of movement through three-dimensional space: The role of video game expertise. *Journal of Applied Developmental Psychology, 15*(1), 87–103. doi:10.1016/0193-3973(94)90007-8

Greenfield, P. M., DeWinstanley, P., Kilpatrick, H., & Kaye, D. (1994). Action video games and informal education: Effects on strategies for dividing visual attention. *Journal of Applied Developmental Psychology, 15*(1), 105–123. doi:10.1016/0193–3973(94)90008–6

Griffith, J. L., Voloschin, P., Gibb, G. D., & Bailey, J. R. (1983). Differences in eye-hand motor coordination of video-game users and non-users. *Perceptual and Motor Skills, 57*(1), 155–158. doi:10.2466/pms.1983.57.1.155

Jaeggi, S. M., Buschkuehl, M., Jonides, J., & Perrig, W. J. (2008). Improving fluid intelligence with training on working memory. *Proceedings of the National Academy of Sciences 105*(19), 6829–6833. doi:10.1073/pnas.0801268105

Jaeggi, S. M., Buschkuehl, M., Jonides, J., & Shah, P. (2011). Short- and long-term benefits of cognitive training. *Proceedings of the National Academy of Sciences, 108*(25), 10081–10086. doi:10.1073/pnas.1103228108

Jeter, P. E., Dosher, B. A., Petrov, A., & Lu, Z.-L. (2009). Task precision at transfer determines specificity of perceptual learning. *Journal of Vision, 9*(3), 1–13. doi:10.1167/9.3.1

Karr, J. E., Areshenkoff, C. N., Rast, P., & Garcia-Barrera, M. A. (2014). An empirical comparison of the therapeutic benefits of physical exercise and cognitive training on the executive functions of older adults: A meta-analysis of controlled trials. *Neuropsychology, 28*(6), 829–845. doi:10.1037/neu0000101

Kilgard, M. P., & Merzenich, M. M. (1998). Cortical map reorganization enabled by nucleus basalis activity. *Science, 279*(5357), 1714–1718. doi: 10.1126/science.279.5357.1714

Klingberg, T., Fernell, E., Olesen, P. J., Johnson, M., Gustafsson, P., Dahlstrom, K., . . . Westerberg, H. (2005). Computerized training of working memory in children with ADHD—A randomized, controlled trial. *Journal of the American Academy of Child and Adolescent Psychiatry, 44*(2), 177–186. doi:10.1097/00004583-200502000-00010

Koepp, M. J., Gunn, R. N., Lawrence, A. D., Cunningham, V. J., Dagher, A., Jones, T., . . . Grasby, P. M. (1998). Evidence for striatal dopamine release during a video game. *Nature, 393*(6682), 266–268. doi:10.1038/30498

Kristjánsson, Á. (2013). The case for causal influences of action video game play upon vision and attention. *Attention, Perception, and Psychophysics, 75*(4), 667–672. doi:10.3758/s13414-013-0427-z

Kuhlman, J. S., & Beitel, P. A. (1991). Videogame experience: A possible explanation for differences in anticipation of coincidence. *Perceptual and Motor Skills, 72*(2), 483–488. doi:10.2466/pms.1991.72.2.483

Li, R. W., Ngo, C., Nguyen, J., & Levi, D. M. (2011). Video-game play induces plasticity in the visual system of adults with amblyopia. *PloS Biology, 9*(8), e1001135. doi:10.1371/journal.pbio.1001135

Li, R., Polat, U., Makous, W., & Bavelier, D. (2009). Enhancing the contrast sensitivity function through action video game training. *Nature Neuroscience, 12*(5), 549–551. doi:10.1038/nn.2296

Li, R., Polat, U., Scalzo, F., & Bavelier, D. (2010). Reducing backward masking through action game training. *Journal of Vision, 10*(14), 1–13. doi:10.1167/10.14.33

Li, J., Thompson, B., Deng, D., Chan, L. Y., Yu, M., & Hess, R. F. (2013). Dichoptic training enables the adult amblyopic brain to learn. *Current Biology, 23*(8), R308–309. doi:10.1016/j.cub.2013.01.059

Lutz, A., Slagter, H. A., Rawlings, N. B., Francis, A. D., Greischar, L. L., & Davidson, R. J. (2009). Mental training enhances attentional stability: Neural and behavioral evidence. *The Journal of Neuroscience, 29*(42), 13418–13427. doi:10.1523/JNEUROSCI.1614-09.2009

Mayas, J., Parmentier, F. B., Andrés, P., & Ballesteros, S. (2014). Plasticity of attentional functions in older adults after non-action video game training: A randomized controlled trial. *PLoS One, 9*(3), e92269. doi:10.1371/journal.pone.0092269

McKinley, R. A., McIntire, L. K., & Funke, M. A. (2011). Operator selection for unmanned aerial systems: Comparing video game players and pilots. *Aviation, Space, and Environmental Medicine, 82*(6), 635–642.

Melby-Lervåg, M., & Hulme, C. (2012). Is working memory training effective? A meta-analytic review. *Developmental Psychology, 49*(2), 270–291. doi:10.1037/a0028228

Mishra, J., Zinni, M., Bavelier, D., & Hillyard, S. A. (2011). Neural basis of superior performance of action videogame players in an attention-demanding task. *Journal of Neuroscience, 31*(3), 992–998. doi:10.1523/JNEUROSCI.4834-10.2011

Oei, A. C., & Patterson, M. D. (2013). Enhancing cognition with video games: A multiple game training study. *PLoS One 8*(3), e58546. doi:10.1371/journal.pone.0058546

Okagaki, L., & Frensch, P. A. (1994). Effects of video game playing on measures of spatial performance: Gender effects in late adolescence. *Journal of Applied Developmental Psychology, 15*, 33–58. doi:10.1016/0193-3973(94)90005-1

Ou, Y., McGlone, E. R., Camm, C. F., & Khan, O. A. (2013). Does playing video games improve laparoscopic skills? *International Journal of Surgery, 11*(5), 365–369. doi:10.1016/j.ijsu.2013.02.020

Parsons, H. M. (1974). What happened at Hawthorne? New evidence suggests the Hawthorne effect resulted from operant reinforcement contingencies. *Science, 183*(4128), 922–932. doi:10.1126/science.183.4128.922

Powers, K. L., Brooks, P. J., Aldrich, N. J., Palladino, M. A., & Alfieri, L. (2013). Effects of video-game play on information processing: A meta-analytic investigation. *Psychonomic Bulletin & Review, 20*(6), 1055–1079. doi:10.3758/s13423-013-0418-z

Przybylski, A. K., Rigby, C. S., & Ryan, R. M. (2010). A motivational model of videogame engagement. *Review of General Psychology, 14*(2), 154–166. doi:10.1037/a0019440

Ramsburg, J. T., & Youmans, R. J. (2014). Meditation in the higher-education classroom: Meditation training improves student knowledge retention during lectures. *Mindfulness, 5*(4), 431–441. doi:10.1007/s12671-013-0199-5

Redick, T. S., Shipstead, Z., Harrison, T. L., Hicks, K. L., Fried, D. E., Hambrick, D. Z., . . . Engle, R. (2013). No evidence of intelligence improvement after working memory training: A randomized, placebo-controlled study. *Journal of Experimental Psychology: General, 142*(2), 359–379. doi:10.1037/a0029082

Roden, I., Könen, T., Bongard, S., Frankenberg, E., Friedrich, E. K., & Kreutz, G. (2014). Effects of music training on attention, processing speed and cognitive music abilities—Findings from a longitudinal study. *Applied Cognitive Psychology, 28*(4), 545–557. doi:10.1002/acp.3034

Rosser, J. C., Lynch, P. J., Cuddihy, L., Gentile, D. A., Klonsky, J., & Merrell, R. (2007). The impact of video games on training surgeons in the 21st century. *Archives of Surgery, 142*(2), 181–186. doi:10.1001/archsurg.142.2.181

Schlickum, M. K., Hedman, L., Enochsson, L., Kjellin, A., & Felländer-Tsai, L. (2009). Systematic video game training in surgical novices improves performance in virtual reality endoscopic surgical simulators: A prospective randomized study. *World Journal of Surgery, 33*(11), 2360–2367. doi:10.1007/s00268-009-0151-y

Schmidt, R. A., & Bjork, R. A. (1992). New conceptualizations of practice: Common principles in three paradigms suggest new concepts for training. *Psychological Science, 3*, 207–217. doi:10.1111/j.1467-9280.1992.tb00029.x

Shipstead, Z., Redick, T. S., & Engle, R.W. (2012). Is working memory training effective? *Psychological Bulletin, 138*(4), 628–654. doi:10.1037/a0027473

Shute, V., Ventura, M., & Ke, F. (2014). The power of play: The effects of Portal 2 and Lumosity on cognitive and noncognitive skills. *Computers and Education, 80,* 58–67. doi:10.1016/j.compedu.2014.08.013

Sims, V. K., & Mayer, R. E. (2002). Domain specificity of spatial expertise: The case of video game players. *Applied Cognitive Psychology, 16*(1), 97–115. doi:10.1002/acp.759

Singley, M. K., & Anderson, J. R. (1989). *The transfer of cognitive skill.* Cambridge, MA: Harvard University Press.

Skrandies, W. (2011). Perceptual learning induced by short-term training with visual stimuli: Effects of retinal position and stimulus type. *Zeitschrift für Neuropsychologie, 22*(4), 257–262. doi:10.1024/1016-264X/a000055

Slagter, H. A., Davidson, R. J., & Lutz, A. (2011). Mental training as a tool in the neuroscientific study of brain and cognitive plasticity. *Frontiers in Human Neuroscience, 5,* 17. doi:10.3389/fnhum.2011.00017

Spang, K., Grimsen, C., Herzog, M. H., & Fahle, M. (2010). Orientation specificity of learning vernier discriminations.*Vision Research, 50*(4), 479–485. doi:10.1016/j.visres.2009.12.008

Spence, I., Yu, J. J., Feng, J., & Marshman, J. (2009). Women match men when learning a spatial skill. *Journal of Experimental Psychology: Learning, Memory, and Cognition, 35*(4), 1097–1103. doi:10.1037/a0015641

Stafford, T., & Dewar, M. (2014). Tracing the trajectory of skill learning with a very large sample of online game players. *Psychological Science, 25*(2), 511–518. doi:10.1177/0956797613511466

Strobach, T., Frensch, P. A., & Schubert, T. (2012). Video game practice optimizes executive control skills in dual-task and task switching situations. *Acta Psychologica, 140*(1), 13–24. doi:10.1016/j.actpsy.2012.02.001

Subrahmanyam, K., & Greenfield, P. M. (1994). Effect of video game practice on spatial skills in girls and boys. *Journal of Applied Developmental Psychology, 15*(1), 13–32. doi:10.1016/0193-3973(94)90004-3

Taatgen, N. A. (2013). The nature and transfer of cognitive skills. *Psychological Review, 120*(3), 439–471. doi:10.1037/a0033138

Terlecki, M. S., Newcombe, N. S., & Little, M. (2008). Durable and generalized effects of spatial experience on mental rotation: Gender differences in growth patterns. *Applied Cognitive Psychology, 22*(7), 996–1013. doi:10.1002/acp.1420

Thorndike, E. L., & Woodworth, R. S. (1901). The influence of improvement in one mental function upon the efficiency of other functions. *Psychological Review, 8*(3), 247–261. doi:10.1037/h0074898

Tirre, W. C., & Raouf, K. K. (1994). Gender differences in perceptual-motor performance. *Aviation, Space, and Environmental Medicine, 65*(5), A49–A53.

Toril, P., Reales, J. M., & Ballesteros, S. (2014). Video game training enhances cognition of older adults: A meta-analytic study. *Psychology and Aging, 29*(3), 706–716. doi:10.1037/a0037507

Uttal, D. H., Meadow, N. G., Tipton, E., Hand, L. L., Alden, A. R., Warren, C., & Newcombe, N. S. (2013). The malleability of spatial skills: A meta-analysis of training studies. *Psychological Bulletin, 139*(2), 352–402. doi:10.1037/a0028446

van Ravenzwaaij, D., Boekel, W., Forstmann, B. U., Ratcliff, R., & Wagenmakers, E.-J. (2014). Action video games do not improve the speed of information processing in simple perceptual tasks. *Journal of Experimental Psychology: General, 143*(5), 1794–1805. doi:10.1037/a0036923

Vygotsky, L. S. (1978). *Mind in society: The development of higher psychological processes.* Cambridge, MA: Harvard University Press.

Waddingham, P. E., Cobb, S. V., Eastgate, R. M., & Gregson, R. M. (2006). Virtual reality for interactive binocular treatment of amblyopia. *International Journal on Disability and Human Development, 5*(2), 155–161.

Wang, P., Liu, H. H., Zhu, X. T., Meng, T., Li, H. J., & Zuo, X. N. (2016). Action video game training for healthy adults: A meta-analytic study. *Frontiers in Psychology, 7*, 907. doi:10.3389/fpsyg.2016.00907

White-Schwoch, T., Carr, K. W., Anderson, S., Strait, D. L., & Kraus, N. (2013). Older adults benefit from music training early in life: Biological evidence for long-term training-driven plasticity. *The Journal of Neuroscience, 33*(45), 17667–17674. doi:10.1523/JNEUROSCI.2560-13.2013

Wiesel, T. N., & Hubel, D. H. (1965). Comparison of the effects of unilateral and bilateral eye closure on cortical unit responses in kittens. *Journal of Neurophysiology, 28*(6), 1029–1040. doi:10.1152/jn.1965.28.6.1029

Yerkes, R. M., & Dodson, J. D. (1908). The relation of strength of stimulus to rapidity of habit-formation. *Journal of Comparative Neurology and Psychology, 18*(5), 459–482. doi:10.1002/cne.920180503

Yi, S. H., & Lee, S. E. (1997). Video game experience and children's abilities of self-control and visual information processing. *Korean Journal of Child Studies, 18*(2), 105–120.

Yu, C., Klein, S. A., & Levi, D. M. (2004). Perceptual learning in contrast discrimination and the (minimal) role of context. *Journal of Vision, 4*(3), 169–182. doi:10.1167/4.3.4

Zinke, K., Zeintl, M., Rose, N. S., Putzmann, J., Pydde, A., & Kliegel, M. (2014). Working memory training and transfer in older adults: Effects of age, baseline performance, and training gains. *Developmental Psychology, 50*(1), 304–315. doi:10.1037/a0032982

10

Logical and Methodological Considerations in Cognitive Training Research

Benjamin Katz and Priti Shah

Introduction

The evidence that life experiences can affect prefrontal function is substantial and compelling (see Chapter 8 by Diamond and Ling). It is less clear whether existing computer-based cognitive training can improve prefrontal function in a manner relevant to real-life outcomes. We suggest that if the first claim is accepted as true, as is supported by many other chapters in this volume, then an important scientific goal for establishing the veracity of the second claim is to identify the necessary and sufficient experiences that could lead to the creation of efficient and effective cognitive interventions. Researchers who have developed computerized cognitive-training interventions have attempted to do that by identifying the core executive processes of the prefrontal cortex and by targeting and taxing exactly those processes. However, a vast space of possibilities (and potential barriers) remains between the development of targeted and successful cognitive interventions and the natural set of rich early-life experiences that have been shown to be important in the development of prefrontal function. To construct efficient, effective training interventions that may generalize to real-life outcomes, properly mapping out this space is necessary, starting with the consideration of several factors. Thus, we begin by discussing the cognitive skills that are the focus of many cognitive interventions: executive functions, working memory (WM), and functions of the prefrontal cortex.

Working Memory and Executive Functions

Numerous terms—*working memory, executive functions, cognitive* or *effortful control, self-regulation, executive attention*—are used to characterize and taxonomize the functions of the prefrontal cortex. The relationship among the different cognitive functions continues to be an issue of theoretical debate (see Diamond, 2010, for a discussion of these various terms; for other prominent perspectives, see also Diamond, 2013; Kane & Engle, 2002; Mischel & Ayduk, 2002; Miyake et al., 2000; Miyake & Shah, 1999; Nee, Wager, & Jonides, 2007; Rueda, Posner, & Rothbart, 2005). The main point of agreement, however, is that these cognitive abilities control and regulate the mind and behavior. They include the abilities to ignore distracting information (also called "interference resolution"), to inhibit prepotent behaviors (i.e., "response inhibition"), to work actively with task-relevant information held in mind, and to sustain or flexibly shift attention.

Despite its importance, the executive function/prefrontal system is highly limited in capacity (Cowan, 2001), and it shows substantial individual differences (Cowan, 2001; Engle, Kane, & Tuholski, 1999; Rouder et al., 2008). Furthermore, individual differences in executive functions, as assessed by a variety of tasks, play a large role in predicting performance in a wide range of academic and non-academic outcome measures (Clark, Prior, & Kinsella, 2002; Pickering, 2006; Rapport, Scanlan, & Denney, 1999). These outcome measures include relatively complex tasks—such as WM span tasks like the operation span and reading span, the Tower of Hanoi, or the Wisconsin Card Sorting Test—that require multiple executive processes, as well as tasks that attempt to measure an individual executive process, such as the Stroop task (primarily response inhibition). WM and executive functions are highly correlated with vocabulary size (Daneman & Green, 1986), language comprehension (Baddeley, 2003; Gathercole & Baddeley, 2014; Nation, Adams, Bowyer-Crane, & Snowling, 1999; Seigneuric, Ehrlich, Oakhill, & Yuill, 2000), reading ability (Gathercole, Brown, & Pickering, 2003; McVay & Kane, 2012), and mathematics problem-solving skills (Bull & Scerif, 2001; Zheng, Swanson, & Marcoulides, 2011). WM and executive functions, more generally, affect an individual's ability to learn complex skills and to acquire new knowledge (Gathercole, Lamont, & Alloway, 2006) as well as to focus on classroom tasks (Alloway & Alloway, 2010; Barkley, 1997). WM and executive functions are also highly associated with general intelligence (Alloway & Passolunghi, 2011; Brydges, Reid, Fox, & Anderson, 2012; Conway, Kane, & Engle, 2003; Kane & Engle, 2002).

As already pointed out, life experiences have a substantial impact on the development of prefrontal function. In fact, the effects of socioeconomic status (SES) on prefrontal function are larger in magnitude than effects on most other subsystems except language (Evans & Schamberg, 2009; Farah, Noble, & Hurt, 2005; Hackman & Farah, 2009; Hackman, Farah, & Meaney, 2010; Sarsour et al., 2011). Typically, SES is measured in terms of a combination of income (usually income-to-needs ratio), education, and/or occupation (Entwisle & Astone, 1994). In a small but growing number of studies, the reduced cognitive performance that accompanies low SES is associated with neural differences examined with functional magnetic resonance imaging (fMRI) or electroencephalography (EEG; Hackman & Farah, 2009). For example, in one fMRI study, SES differences were found in the amygdala of adults (Gianaros et al., 2008). Another MRI study in children found significant differences in regional brain volume in the hippocampus and amygdala, as well as SES-by-age interactions in the left superior temporal gyrus and the left inferior frontal gyrus (Noble, Houston, Kan, & Sowell, 2012). Finally, in one EEG study accompanied by an attentional task, even when children from high and low SES groups perform similarly on cognitive measures, neural differences on an event-related potential waveform were found between the two groups (D'Anguilli, Herdman, Stapells, & Hertzman, 2008). These findings are complementary with evidence of differences in functional brain activity in extremely deficient environments, such as those present in Romanian orphanages (Chugani et al., 2001) and of structural brain differences in animal models of poor parenting (Barha, Pawluski, & Galea, 2007).

This is not to say that SES in and of itself affects the development of prefrontal function. Rather, the effects are due to factors associated with SES, such as life stress and others that Diamond and Ling also discuss (see Chapter 8 in this volume). Many other factors are proposed to have direct effects on the development of executive functions, and some of these factors are also likely associated with SES. These include parental warmth, physical environment, cognitive stimulation, physical activity and fitness, adiposity, nutritional deprivation, exposure to neurotoxins, anxiety, depression, and trauma (Canfield, Kreher, Cornwall, & Henderson, 2003; DePrince, Weinzierl, & Combs, 2009; Diamond & Lee, 2011; Eysenck, Derakshan, Santos, & Calvo, 2007; Fay-Stammbach, Hawes, & Meredith, 2014; Lawson et al., 2014; Lukowski et al., 2010; Pollak et al., 2010; Rogers et al., 2004; Sarsour et al., 2011; Valiente, Lemery-Chalfant, & Castro, 2007; Wolf et al., 2007). The list of insults that may reduce executive function performance is extensive. It is also possible that these factors may be associated with each other; for example, low SES may be accompanied by life stress and increased exposure to neurotoxins in the environment. It is worth carefully considering the specific cause of impairments in any particular target population when designing a cognitive training program.

Although it may seem obvious that many factors can negatively affect the development of executive functions, some evidence suggests that life experiences can have a positive effect on the development of the executive function and, further, that some interventions may be successful. What is perhaps more surprising is the evidence that some impairments to executive function can be avoided—or ameliorated—through a variety of interventions (for a discussion of some possible approaches, see Diamond, 2014, and Mackey, Raizada, & Bunge, 2011). Although the initial findings are limited and preliminary, researchers have identified beneficial effects of unstructured time and play (Barker et al., 2014), some kinds of curricular experiences, including the Montessori program (Lillard & Else-Quest, 2006), bilingualism (Bialystok, Craik, & Luk, 2008), and responsive parenting that promotes a child's developing autonomy and independent problem-solving (Bernier, Carlson, & Whipple, 2010). However, most of the studies finding a positive impact of early childhood experiences on executive function are more controversial, in part because they almost entirely rely upon correlational data. Additionally, it is easier to identify the underlying mechanisms for certain deleterious factors (e.g., lead poisoning, brain injury) than others (e.g., poor cognitive enrichment). A few randomized, controlled interventions targeting SES do exist, however, and some, like the Abecedarian Program, show considerable promise (Barnett & Masse, 2007). However, even among targeted interventions, the underlying mechanisms that lead to improvements in executive function from these life experiences are much less clear. Determining what element—or combination of elements—led to improvements in executive function is difficult.

One approach to developing interventions has been to identify the factors relevant to healthy development of prefrontal function and to attempt to distill these factors into interventions. Cognitive training focuses primarily on cognitive stimulation, and Diamond and Ling (Chapter 8 of this volume) argue that more complex activities that incorporate cognitive stimulation with physical activity and also have some real-life context may be more effective for yielding transfer. Diamond and Ling propose *Tools of the Mind*, a complex, contextualized intervention for preschool students, as having good potential (Diamond, 2012; Diamond & Ling, Chapter 8 of this volume). In *Tools of the Mind*, for example, some training activities involve structured planning of the tasks to be completed each day; another activity involves reading and listening activities that focus on self-regulation but that use normal, age-appropriate books (Diamond, Barnett, Thomas, & Munro, 2007; Imholz & Petrosino, 2012). Another promising training program designed to improve child attentional capacities incorporated intensive parental involvement (and parental training) to complement child-focused activities (Neville et al., 2013). These more complex interventions show promise, but they remain controversial. Although a careful implementation of *Tools of the Mind* was found to be effective (Diamond et al., 2007), and a more recent multisite implementation suggested a positive impact (Blair & Raver, 2014), one much-publicized study failed to replicate these findings (Wilson & Farran, 2012). It is possible that a good-quality implementation of complex, contextualized interventions is effective, but it is too early to conclude that this type of intervention is better than computerized cognitive-training interventions. In addition to the fact that the studies have not always been replicated, there are very few such studies for each individual intervention or program.

It is difficult to assess the underlying mechanisms of improvements in more complex interventions because they have multiple components. One approach to addressing this issue is by designing studies in which the components of an intervention are evaluated individually and in combination, so that examining the impact of any individual component, as well as the combined effects, is possible. While a small number of individual studies have examined multicomponent interventions, such as transcranial direct current stimulation (tDCS) and cognitive training (Ditye, Jacobson, Walsh, & Lavidor, 2012) and exercise and cognitive training (O'Dwyer, Burton, Pachana, & Brown, 2007), no study as of yet has properly investigated the effect of each component by itself and in conjunction with each other. Additionally, other potentially beneficial components from other domains have not yet been combined with computerized cognitive training. Socioemotional interventions are one surprising example of this, particularly given the success of interventions with significant parental training elements as described by Neville et al. (2013) and within the Perry Preschool Program (Muennig, Schweinhart, Montie, & Neidell, 2009). We also agree with Diamond and Ling (Chapter 8) that the inclusion of a human mentor in *Cogmed* training merits further examination.

As noted above, life experiences have a direct and substantial impact on executive function. That negative life experiences sometimes have dramatic impact is not controversial. While there is some evidence that positive life experiences can also be beneficial, the data are not as clear, and the mechanisms underlying the improvements are usually not fully understood, but with this point, a basic piece of logic applies: If you can make executive function worse through negative factors, you can also make it better by removing them (in the instances where they can, in fact, be removed). Furthermore, it is likely that everyone has at least some limiting factors present during their development. Thus, another piece of logic is that people with "more addressable" factors underlying their difficulties (e.g., lack of cognitive stimulation) will benefit more from intervention than people with less addressable factors (e.g., brain injury). Moreover, some of the most successful programs to alleviate the negative effects of low SES—such as the Abecedarian and the Perry Preschool Programs—include multiple components, some of which, such as physical exercise or socioemotional elements like parental training, may be necessary for creating interventions that generalize to real life. These large-scale, comprehensive programs include elements that potentially address the deleterious effects of factors like life stress and anxiety. While it is unknown whether computerized cognitive training would be more effective if combined with these elements, it is possible that those elements may be found to be necessary in facilitating improvements following any serious intervention. Any conclusions about the efficacy of cognitive training, particularly as an element of a larger course of intervention, are premature until this issue has been more fully investigated.

Rethinking Transfer

While the underlying mechanisms and necessary components of interventions to improve executive function may not be fully understood, it seems that at least some interventions have indeed been successful in improving executive function. In Chapter 8, Diamond and Ling include a frequently made assertion regarding cognitive training, stating, "Participants improve on the skills they practice and that usually transfers to other contexts where those same skills are needed (narrow transfer)—but people *rarely* improve on what they have not practiced." Such narrow transfer is typically regarded as easy to achieve and of somewhat limited utility. The real excitement—and controversy—surrounding cognitive training research is not whether narrow transfer is possible, but if far transfer can be achieved. In the context of cognitive training, far transfer is typically discussed as a practice of some basic cognitive process or skills (WM, sustained attention, or the like) that leads to improvement on some meaningful, complex outcomes, such as performance on intelligence tests (Jaeggi, Buschkuehl, Shah, & Jonides, 2014) or academic achievement (Titz & Karbach, 2014).

The earliest direct discussion of transfer is often attributed to Woodworth and Thorndike (1901), who examined whether practice on one cognitive skill might lead to improved performance on a different, but related, skill, but largely did not find that this practice was generalizable. However, several other types of transfer may take place as well. For example, consider transfer in the context of education. Part of the goal of schooling is that skills taught must also be accessible later than when the task is learned, that is, transfer must occur across time. Additionally, context matters—for cognitive training to be worthwhile as an intervention, it should ideally transfer to different contexts as well as to different tasks. In other words, improved executive function is of limited usefulness if it only works in the lab and not in the workplace or at school. Klahr and Chen (2011) provided a definition of transfer that incorporates these three dimensions—transfer across time, transfer across different tasks, and transfer to different contexts (see Figure 10.1). Transfer in each of these dimensions can be near or far as well. However, just because transfer is not far does not mean it is not worthwhile. Multiplication skills, for example, can be used in different contexts and long after the skill is taught, and the fact that the task itself is very similar in the different contexts—an example of near transfer—does not matter so very much.

Consider what near transfer means in any given example of executive function training. Improvements on untrained tasks that theoretically tapped visuospatial WM capacity following a spatial N-back intervention could be considered near transfer, because the outcome task itself is at least somewhat similar to the training intervention. However, if the training is improving an underlying construct, even by a relatively small amount, this could nonetheless have significant

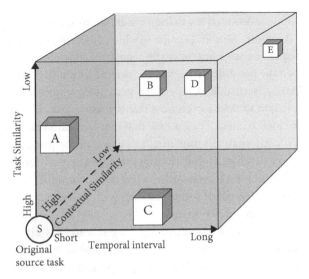

Figure 10.1. Klahr and Chen's (2011) three-dimensional representation of transfer distance space.
(Included with permission from "Finding One's Place in Transfer Space" from the September 2011 issue of *Child Development Perspectives*)

effects on the performance of other tasks that draw on this construct. Therefore, it probably would be more profitable to divide transfer improvements that have been seen thus far in cognitive training studies into the two categories of real transfer and superficial transfer.

Real transfer describes transfer that results in improvement to the underlying mechanism (e.g., WM in the example of WM training). Superficial transfer, however, indicates improvements on a transfer task that is not the result of an improvement in the underlying mechanism. Superficial transfer could result from a variety of factors. For example, consider a training study where participants improved performance not because some underlying ability was improved, but because they had learned to use a certain strategy or had become more cautious in answering during the tests. These results would still be potentially interesting, and the intervention perhaps still useful in a practical sense, but from a theoretical standpoint, the results would not provide as much information about the underlying mechanisms of executive function. In other words, superficial transfer is transfer that does not occur as a result of improvement of the central underlying cognitive mechanism.

It is important to be confident that near-transfer effects are not merely superficial. In fact, most cognitive training paradigms have taken seriously the idea that training tasks should be designed, and transfer tasks selected, so that people cannot develop very narrow task-specific strategies (Schmidt & Bjork, 1992). As an example, suppose one is trying to improve children's spatial WM by having the children play the game "Simon" (a game in which different points on a circle light up, and participants must remember the sequence of locations that lit up). This task leaves open the possible strategy of developing locations on a clock, making this training task susceptible to superficial strategy development. In this case, it would be important to assess even near transfer using tasks that do not permit the use of the clock strategy (e.g., a task that requires individuals to remember verbal material). Alternatively, one might design training protocols that are resistant to strategy development, including using a variety of training tasks (e.g., intermittently training on Simon and a verbal task; Schmidt & Bjork, 1992) and purposely hindering the use of some strategies (e.g., by having participants recite the alphabet while playing Simon).

Unfortunately, although cognitive training researchers attempt to follow the broad outlines of the advice above, it is impossible to completely remove the possibility that strategy development is responsible for transfer. In fact, we know that practice on executive control tasks typically involves strategy development (Rubinstein, Meyer, & Evans, 2001). Furthermore, one recent study found that participants reported developing grouping strategies during an adaptive WM training task that they also applied to near-transfer tests of visuospatial and verbal WM (Dunning & Holmes, 2014). Von Bastian & Oberauer (2013) offer a variety of strategies that could explain improvements on both near- and far-transfer tasks, such as optimizing the speed and accuracy trade-off, and they suggest that transfer tasks must be selected with these strategies in mind if one is to exclude strategy development as a mechanism for transfer. While some of the strategies cannot absolutely be ruled out (for example, the speed/accuracy trade-off will apply to any timed task), some measures may indeed be more resistant to strategy development than others. Ultimately, whether a near-transfer task is more or less susceptible to strategy development depends also on how expansive one's definition of near transfer is. While tests like Dot Matrix, Corsi block and digit span may be prone to basic strategies like chunking, others, like the Surface Development and Form Board tests (Jaeggi et al., 2014) are likely different enough to defy basic strategy use, even though they target visual WM.

A final issue related to determining whether transfer is superficial or real is that it is often difficult to draw the connection between improvement on the underlying mechanism during training on a task and improvements on a measure of far transfer. As Melby-Lervåg and Hulme point out in their meta-analysis (2013), if far transfer, but not near transfer, is observed in a study, the theoretical underpinnings of the far-transfer improvements become difficult to discern. An individual might, through training, improve their motivation or some other dispositional factors that could improve performance on another task or in another context. Merely observing far transfer in this situation might suggest, incorrectly, that the individual had improved due to gains on a pure cognitive function.

Many researchers have concluded that near transfer can be found reliably, and, presumably, it's not merely superficial (Karbach & Verhaeghen, 2014; Melby-Lervåg & Hulme, 2013). That is, transfer occurs from some measures of executive function or WM to other measures—such as a variety of visuospatial WM tasks—that are unlikely to be accounted for by mere superficial factors. There seems to be broad consensus, as stated by Diamond and Ling in Chapter 8, that near transfer exists and is not superficial. In Melby-Lervåg and Hulmes's meta-analysis, near transfer to the visuospatial WM domain persisted even at long-term follow-up. In Karbach and Verhaeghen's considerably more expansive meta-analysis, near (and even far) transfer occurred reliably for both younger and older adults.

If near transfer is not superficial, then it should also yield some far-transfer effects, but the far-transfer effects may not always be detectable. To clarify, if near transfer is real, in that a specific cognitive skill similar to that which is trained is improved at the level of the latent construct, by definition potential should exist for far transfer to tasks for which that specific cognitive skill is a rate-limiting factor to performance (see, for example, Richey, Phillips, Schunn, & Schneider, 2014).[1] In the case of WM and executive functions, there is strong evidence that these abilities are the rate-limiting factors to performance for many complex cognitive tasks. The correlation between WM, executive functions, and performance on complex cognitive tasks likely arises because these basic skills serve as rate-limiting factors to performance. Consider, for example, the classic inductive-reasoning task often used as a measure of general or fluid intelligence, Raven's Progressive Matrices. In addition to knowing that executive function and WM skills correlate with Raven's performance, we know that test-item difficulty is largely explained by the WM and executive function demands of an item. For example, Just and Carpenter (1979) found that people did worse on items that had more rules that had to be actively maintained and tested than on items that had fewer rules to be maintained and tested. Furthermore, in a nice theoretical demonstration, a computational model with more WM capacity performed better on Raven's test items than one that had less (Just & Carpenter, 1992; see also Duncan et al., 2008, 2012). Gray, Chabris, and Braver (2003) also found that people do worse on items that include lures that are similar to the correct response, suggesting that inhibitory processes also limit performance on Raven's Matrices. Likewise, Jarosz and Wiley (2012) found that the test items that explain

---

[1] We note that Diamond does not consider this to be far transfer, however (personal communication).

the relationship between WM capacity and fluid intelligence are those that contain the most highly distracting lures.

If we accept that (a) there is some non-strategy-related near transfer of executive function training to other WM and executive function tasks, and (b) these skills are likely rate-limiting factors to performance on more complex cognitive tasks, such as the Raven's test, then it is somewhat puzzling that evidence for far transfer remains elusive. There are two possible explanations. One possibility is that what we assume to be real near transfer is actually superficial (Shipstead, Hicks, & Engle, 2012). "Intelligence" as measured by the Weschler Abbreviated Scale of Intelligence and that measured by Raven's Progressive Matrices correlate at .56 (Kane et al., 2004). Perhaps people develop a strategy that allows them to get better at some training task(s) as well as the Raven's Matrices, and this strategy is not relevant for the construct of intelligence or cognitive control. A second possibility is that something like a "rate-determining step" in chemistry could occur. Rate-determining steps are the factors that may influence the rate at which a particular chemical reaction takes place (Chang, 2004). Something similar could be taking place in cognitive training, specifically in relation to transfer. If a person's WM improves, but that person still has low vocabulary skills, then performance on a reading comprehension measure may not show improvement, at least in the short run. Both WM and vocabulary may have been limiting factors, and only one has improved. Although good cognitive assessments primarily target a single skill, they rarely, or perhaps even never, involve only one single cognitive capacity. The limiting factors may be expansive and may include some of the life factors related to SES—such as emotional responsivity or anxiety in response to stress—discussed earlier.

Two additional explanations for improvements on cognitive constructs via training but lack of far-transfer effects relate to use of the two-time-point repeated-measures study design, wherein participants receive a pretest before training and a posttest after training. This design does not allow for the detection of sleeper effects that might arise at some point following training. For example, participants might require a certain amount of time after the training for consolidation (similar to the idea of distributed practice in verbal recall tasks; for a quantitative review, see Cepeda, Pashler, Vul, Wixted, & Rohrer, 2006). While many cognitive training studies do include a long-term follow-up assessment one to three months after the posttest assessment (Holmes, Gathercole, & Dunning, 2009; Jaeggi, Buschkuehl, Jonides, & Shah, 2011; Klingberg et al., 2005), the spacing of assessments used in many cognitive training studies may not be ideal for capturing transfer to untrained tasks. In the rate-determining factor example above, improvements in WM capacity may, in time, lead to improvements in vocabulary that could then improve reading comprehension. Also, while the amount of training people receive in studies often varies, most computerized cognitive-training studies include less than 10 hours of total training. The training dosage used is often based on previous research and the limited resources available for running a time-intensive training study. We agree with Diamond and Ling (Chapter 8) that this is problematic. Without a full understanding of dose–response effects in training, there is little reason to expect transfer at any given point. Recent meta-analyses have done little to clarify this matter, with neither the meta-analysis by Au et al. (2015), which focused on N-back training, nor the more expansive meta-analysis of WM and executive functioning training by Karbach and Verhaeghen (2014) finding that time training was a significant moderator of transfer gain. However, there is little variability in the training time in different studies, and it is unclear whether longer training times than are typically used may be necessary to find consistent transfer effects.

Another potential barrier to observing transfer effects relates specifically to the design of the assessment. Many of the tests used to observe far transfer have been manipulated beyond the original, validated version, due to the use of multiple time points of testing. For example, Redick et al. (2013) found no transfer as a function of dual N-back training after dividing many of the assessments used into multiple versions. However, the act of dividing assessments by two—and sometimes three or four—can be problematic. Shortened versions may be significantly less reliable than the full tests. Additionally, it is possible that, for many tasks, the difference in scores between higher- and lower-performing individuals may be due to responses to only one or two problems. This is perhaps the case for the Raven's Advanced Progressive Matrices. If such an assessment is divided into multiple versions, the shortened test may not have enough granularity to detect changes in performance from one instance of the assessment to the next. Thus, even if improvements are not observed, it may not necessarily be because the participants are not improving in some underlying ability. Indeed, if improvements on near-transfer tasks are easier to detect than improvements on far-transfer tasks because of the noise in transmission from near to far transfer; it is quite likely that far-transfer effects will have smaller effect size than near transfer and require far more participants for effects to be detected.

A potential solution to this problem would be to include "medium" transfer measures—that is, an assessment that provides a theoretical bridge between near- and far-transfer measures. Well-designed medium-transfer measures may also be more resistant to strategy development (e.g., the Heads, Toes, Knees, Shoulder task, which has been used successfully as a measure of self-regulation in school-children; Ponitz et al., 2008; Ponitz, McClelland, Matthews, & Morrison, 2009). This task involves a more ecologically relevant activity than many computerized executive-function tasks and is also more closely connected to response inhibition, WM, and attention than classroom achievement measures. Combined with near- and far-transfer measures, medium-transfer tasks would help explain the underlying mechanisms of cognitive training and provide a firmer theoretical basis for real transfer when it is observed.

There are a variety of strategies that might improve confidence in transfer effects following a cognitive training program, including, but not limited to, extending the amount of time spent training, including multiple time points of testing, and ensuring that assessments are long and challenging enough to capture potential improvements. The reality of cognitive training studies, however, is such that certain solutions (such as offering assessments at multiple time points) may not be feasible without the painstaking work of creating new test batteries. While researchers will doubtlessly make significant progress toward improved training/testing paradigms in the future, re-examining the existing evidence for transfer effects following training is still worthwhile.

Evidence of Transfer After Training

There is conflicting evidence about whether training executive functions can successfully lead to transfer to fluid intelligence or other relevant outcomes, such as academic achievement. For example, several studies that use identical WM interventions have yielded dramatically different results (Chooi & Thompson, 2012; Heinzel et al., 2014; Jaeggi, Buschkuehl, Jonides, & Perrig, 2008; Kundu, Sutterer, Emrich, & Postle, 2013; Redick et al., 2013; Stephenson & Halpern, 2013; Thompson et al., 2013). Several researchers have attempted to address this disparity through reviews and meta-analyses. Since 2010 no less than eight significant meta-analyses and many reviews have focused specifically on training executive function and WM. Meta-analyses offer great promise in furthering cognitive training research.

One issue with cognitive training studies is that the sample size is often small, as recruiting subjects and encouraging them to complete a complicated and challenging intervention is often difficult. Meta-analyses combine multiple studies and partially address the issues of small samples sizes. Since Melby-Lervåg and Hulme conducted their highly cited meta-analysis of 23 WM training studies (2013), Au et al. (2015) released a meta-analysis focusing solely on N-back WM training, Karbach and Verhaeghen (2014) published a more expansive meta-analysis of both WM and executive function training in both young and old adults, and Karr, Areshenkoff, Rast, and Garcia-Barrera (2014) compared the effects of both cognitive training and physical exercise on executive function. The number of reviews is too expansive to list here (we found more than two dozen since 2010; see Table 8.3 in Chapter 8, which lists various reviews on different methods of improving executive functions), but they generally focus on a specific aspect of cognitive training or on training a single demographic group. This includes reviews by Noack, Lövdén, and Schmiedek (2014), Karch, Albers, Renner, Lichtenauer, and von Kries (2013), Shipstead, Redick, and Engle (2012), Morrison and Chein (2011), and Buschkuehl and Jáeggi (2010). We note that Chapter 8 by Diamond and Ling in this volume is an exception.

For the purposes of this chapter, we will instead focus on the more limited subset of recent meta-analyses featuring systematic and empirical evidence for or against cognitive training. The conclusions of these meta-analyses differ somewhat dramatically—while Au et al., Karr et al., and Karbach and Verhaeghen conclude that training executive functions like WM may be effective in improving capacities such as fluid intelligence, Melby-Lervåg and Hulme suggest that transfer gains are nonsignificant or minor at best. In particular, Melby-Lervåg and Hulme do not find evidence of longer-term or non-near transfer and conclude that their analyses "show clearly that these training programs give only near-transfer effects, and that there is no convincing evidence that even such near-transfer effects are durable" (2013, p. 283). However, besides including a somewhat different group of studies (in part because new research is constantly being generated), key differences between these meta-analyses likely lead to these vastly different conclusions. For example, the populations included in each meta-analysis differ. Melby-Lervåg and Hulme included both typically developing children and adults as well as clinical populations in their analysis, Karr et al. included both impaired and healthy older adults, Au et al. included only studies with healthy young adults, and Karbach and Verhaeghen include studies with healthy older adults (but many of those also included samples of younger adults).

Additionally, the interventions included in each meta-analysis differed significantly. Au et al. included only N-back working-memory training interventions. Karbach and Verhaeghen included both working-memory and executive-function interventions more generally. Additionally, Karbach & Verhaeghen included studies that examined acute effects from a single training session, while the others only included studies of longer-term practice; an important distinction given that it is uncertain how long the effects of a single session may persist. Melby-Lervåg and Hulme included a variety of computerized working-memory interventions (as well as interventions that target executive functions more generally, although these are included under the banner of WM training). And Karr et al. included studies that used a variety of computerized and non-computerized training programs. Thus, it is unsurprising that these four meta-analyses arrive at vastly different conclusions.

Each has its own strengths and weaknesses for answering broad questions about the efficacy of cognitive training. The analysis by Au et al., with a focus only on N-back training, is perhaps too narrow to answer the broader questions of effectiveness of cognitive training. Melby-Lervåg and Hulme include a variety of interventions, populations, and age groups, such that any individual demographic group may only have a few studies included. Karbach and Verhaeghen only include studies with older adults (although younger groups are included, studies with only younger adults or children were not). And Karr et al. include an expansive set of interventions under the banner of executive function training.

We do not seek to discuss the strengths and weaknesses of meta-analyses more generally, but rather to point out some of the significant differences across meta-analyses of WM and executive function training. The differences in conclusions across the meta-analyses included here should illustrate that even a systematic and quantitative analyses of multiple studies cannot definitively answer the outstanding questions regarding cognitive training; even the influence of different moderating factors, such as age, may matter more or less across meta-analyses— while Karbach and Verhaeghen did not find an effect of age on transfer, Melby-Lervåg suggest that such an effect may exist, for example.

Small differences in keyword choice for these meta-analyses, or a single year of time difference, can result in dramatically different groups of studies included—and dramatically different outcomes as well. Thus, while meta-analyses are certainly a useful tool in addressing some of the issues related to cognitive training research, such as small sample size, they are by no means conclusive. Review articles, particularly those where there are no outlined criteria for which studies to include and which to exclude, are also problematic in this regard. For example, in the review by Noack et al. (2014) of cognitive training research, key words included *training, transfer, latent, intelligence, working memory, executive control,* and *executive functions,* whereas the meta-analysis by Melby-Lervåg and Hulme (2013) used simply *working memory training.* Simple differences in keyword choice can result in different sets of papers available for inclusion. Consider, for example, the Noack et al. (2014) review; even though they used an expansive list of key terms, some notable examples of WM training during the 5 years covered by their review were not included, such as Jaeggi et al. (2011) and Stephenson and Halpern (2013). Some of these choices reflect differences in how researchers conceptualize "cognitive training." Consider two different chapters within this volume: while Hicks and Engle in Chapter 1 considers only computerized WM training programs, Chapter 8 by Diamond and Ling includes a variety of interventions ranging from exercise to meditation to computerized training.

The issue with the choice of keywords, or how one operationalizes cognitive training, is that the studies one then includes for examination may differ quite widely—and the conclusions reached may differ significantly as well. To illustrate this point, we examined four recent reviews or meta-analyses completed prior to the time of writing in addition to the Diamond and Ling chapter simply to identify how much overlap each of these papers had in the list of studies included. These other papers included reviews from Morrison and Chein (2011), von Bastian and Oberauer (2013), and Noack et al. (2014), as well as the meta-analysis from Melby-Lervåg and Hulme (2013). While the differences in publication date mean that it is difficult to directly compare the studies included in each of these papers, well over 100 interventions were included across all five papers. The Morrison and Chein piece included 21, the Melby-Lervåg meta-analysis included 22, the von Bastian and Oberauer review included 46, the Noack et al. review included 28, and the Diamond and Ling chapter in this volume included 103. Only a handful of studies, including the original Jaeggi et al. (2008) WM training paper and the COGITO Study (Schmiedek, Lövdén, & Lindenberger, 2010), were included in all five. As noted, the Diamond and Ling chapter includes a variety of interventions that go beyond the scope of what most researchers would include as cognitive training; however, even if one does not consider this chapter, the number of studies included across each of the four other pieces only total six. While in some cases the exclusions are quite understandable (such as in the case of the meta-analysis, where certain compromises must be made for the sake of analysis), these disparities mean that the range of works covered in each of these reviews or meta-analyses is significantly different.

One feature that is shared by this group of reviews and meta-analyses, however, is the initial cut-off date for cognitive training studies included for discussion: the earliest paper across any of the five studies is from 1997. This reveals yet another issue with the use of key words as the sole means of identifying papers for a meta-analysis or review. It is possible that there are other studies, or perhaps even entire branches of research, that are being missed due to shifts in psychological terminology over time.

Historical Examples and Instrumental Enrichment

Although the cognitive revolution didn't happen until the mid-1950s, interventions that would even today probably be categorized as cognitive training, such as the Pelmanism program popular in Great Britain, have existed for more than 100 years (Katz, 2014). While many of these early interventions (like most psychological work at the time) lacked controlled trials, by the middle part of the twentieth century, interventions were being developed and evaluated through modern empirical methodology. Feuerstein's experience with Holocaust survivors led him to believe that experience could impair one's movement through Piagetian stages, and thus he developed an initial intervention targeted at "deeply deprived" and "significantly deprived" individuals (Feuerstein & Hamburger, 1965). The program itself was based heavily on the sorts of tasks

Piaget himself used in his experiments (worth considering when the discussion turns to more modern interventions), beginning with simple conservation tasks, led by a facilitator and customized individually for each student, and gradually increasing in complexity and difficulty to include a variety of pen-and-paper tasks, or, as Feuerstein referred to them, *instruments* (Feuerstein & Jensen, 1980). The Instrumental Enrichment program was customized based on age and Feuerstein's initial assessment of the student through a variety of memory and attention tests. Each instrument was repeatable and could be adaptively made more difficult (the initial version of his program included several hundred pages of tasks). Fourteen instruments made up the building blocks of the program, including comparisons, spatial orientation, analytic perception, categorization, numerical progressions, temporal relations, and syllogisms, among others. The program was designed for classroom use, for approximately 60 minutes a day, up to five times a week, for as many as three years. Many of these tasks involve executive function.

The results of initial evaluations of the intervention were promising. One study with low-SES Israeli adolescents found improved numeracy skills, visuospatial reasoning, and reading comprehension (Rand, Tannenbaum, & Feuerstein, 1979). One implementation of the program in Detroit was found to improve school attendance for some participants and to improve teacher and parent ratings of study behavior, although the program's impact on abstract thinking was unclear (Stavros, 1989). A study with high-performing adolescents demonstrated generalization to both verbal and nonverbal reasoning tasks as well as improvement in academic achievement measures (Kirschenbaum, 1998). While one large-scale review of an implementation in England found that after one year of the program (instead of two), improvements in academic achievement and standardized tests remained elusive (Blagg, 1991), more recent meta-analytic work from Romney and Samuels (2001) identified an effect size of 0.43 on reasoning ability after completing Feuerstein's program.

Importantly, despite the evidence base supporting this early cognitive training program's utility, no Instrumental Enrichment study can be identified using keywords like *executive functions*. This is not surprising, given that the Piagetian framework used by Feuerstein existed well before the term *executive function* had been coined. Changes in terminology can also potentially explain why other early examinations of how video game training might influence cognitive performance, such as Patricia Greenfield's early thoughts on games such as Pac-Man in *Mind and Media*, are left out of contemporary discussions of computerized cognitive training (Greenfield, 1984).

Instrumental Enrichment is only a single example, but there are others. Although *Space Fortress*, an action computer game designed as a psychological research tool, it has been used as an intervention to improve cognitive abilities

(Fabiani et al., 1989; Stern et al., 2011). Even though the *Space Fortress* intervention itself is in many ways indistinguishable from current computerized cognitive training, it is possible that this program (part of the "Learning Strategies Program" effort) might be left out of a conversation because it focused more on examining strategy development than the improvement of executive function itself (Donchin, 1995). In this context, the strategy development, which occurs naturally through computerized game play, is somewhat different than previous cognitive training programs (Douglas, Parry, Marton, & Garson, 1976), which only focused on strategy development. Without attempting to include an exhaustive list, other examples of cognitive training interventions that have been left out of the conversation include *Logo*, a computer program which has been used successfully since the 1980s to promote the development of a variety of cognitive skills (Clements, 1986), the concentration programs developed by Kossow and Vehreschild in East Germany in the early 1980s (Kossow & Vehreschild, 1983; Vehreschild, Kossow, & Schulz-Wulf, 1984), *Project Intelligence*, which was widely used in Venezeula in the 1980s (Nickerson, 1985), *WFF N' PROOF* and *Equations*, two problem-solving games developed by Layman Allen in the 1960s (Allen, Allen, & Miller, 1966; Allen, 1970), and inductive-reasoning training (Klauer & Phye, 2008). In addition, there are programs used with older adults such as *Senior Odyssey* (Stine-Morrow, 2007) and programs that use activities such as practicing a musical instrument (Costa-Giomi, 1999; Kaviani, Mirbaha, Pournaseh, & Sagan, 2014; Moreno et al., 2009; Schellenberg, 2004), dancing (Kimura & Hozumi, 2012), or learning a second language (Krizman, Marian, Shook, Skoe, & Kraus, 2012). We include these examples to emphasize the extent to which most contemporary reviews of cognitive training are somewhat limited in scope.

As statistical techniques and research methodology has improved since Feuerstein's intervention was developed in the middle of the twentieth century and there are a variety of reasonable justifications for excluding older studies, we hope that researchers will eventually embark on a larger-scale meta-analysis that encompasses this significant body of work, particularly when researchers are including a larger variety of interventions in their discussion of executive function training, as Diamond and Ling do. Table 10.1 shows several interventions completed prior to 1997 that we believe are worth review but that will likely not be identified using modern keyword searches.

Moderating Factors of Cognitive Training

Even after including a greater variety of training studies, however, it quickly becomes obvious that, for any individual intervention, the outcome of cognitive

Table 10.1. Historical Examples of Cognitive Training

Paper	Year	Type of Intervention
Allen, Allen, & Miller	1966	Game-based logic training
Baltes, Dittmann-Kohli, & Kliegl	1986	Inductive-reasoning training
Boger	1952	Perceptual training
Stankov & Chen	1988	Problem-solving training
Kossow & Vehreschild	1983	Concentration training
Rand et al.	1979	Instrumental Enrichment
Nickerson	1985	Project Intelligence
de Bono	1985	CoRT thinking program

training can often be summed up in one word: inconsistent. In the example of dual N-back training, this issue arises if one compares the findings of Jaeggi et al. (2008) and Stephenson and Halpern (2013), where far transfer is found, to Redick et al. (2013) and Thompson et al. (2013), where no far transfer is observed. Von Bastian and Oberauer (2013) point out in their review of WM training that there are a variety of possible moderators of transfer improvements following cognitive training that must be fully explored. These include individual difference factors (e.g., age, baseline ability, disposition and personality factors, country of origin; Au et al., 2015), as well as the specific makeup of the cognitive training program (von Bastian & Oberauer, 2013). We group these potential moderators into three categories: outcome measures, participant characteristics, and design of interventions.

Outcome Measures

Significant variation exists in the variety of transfer tasks used across studies. Much recent cognitive training work promisingly includes a wider variety of tasks (Jaeggi et al., 2014; Klingberg et al., 2005; Redick et al., 2013; Thompson et al., 2013). However, greater variety is only helpful if these studies include overlapping tasks, and, unfortunately, this has not always been the case. For example, the type of transfer tasks used across the studies often varies depending on the focus of the experimenter—studies aimed at determining if cognitive training may improve intelligence may include matrices-based assessments, while executive-function training may include near-transfer measures related to WM, attention, and response inhibition. In our own preliminary examination of transfer measures used in the cognitive training studies included in Chapter 8 (see Tables 8.1b and 8.2b), three previous reviews (Morrison & Chein, 2011; Noack et al., 2014; von Bastian & Oberauer, 2013), and one meta-analysis (Melby-Lervåg & Hulme, 2013) discussed earlier, we identified 269 different outcome measures. Although some, like Raven's Advanced Progressive Matrices, were used in more than 10 studies, the vast majority (217 assessments) were only used in one study. Only 24 assessments were used in at least three studies. This issue proves especially problematic for meta-analytic work. In Melby-Lervåg and Hulme's meta-analysis of WM training, even the Stroop test, which was used in more of the studies examined than any other individual assessment, was only used in ten of the included comparisons. One potential means of addressing this issue could be to develop an easy-to-administer and freely available outcome battery with the goal of using it across a wide variety of WM or executive function studies.

There may also be significant differences in how tests are administered. Whether testing sessions are broken up into multiple days and whether participants are given breaks and tutorial instruction may significantly impact performance, but such factors are not always discussed in descriptions of cognitive-training work. For some studies with significant dropout rates (Jaeggi et al., 2014; Redick et al., 2013), it is possible that the approach to the administration of assessments (for example, including too many assessments without breaks in between) may contribute to the participants' decision to withdraw.

Participant Characteristics

Participant characteristics may be the most problematic moderator from the perspective of the meta-analyses described above. Within Melby-Lervåg's (2013) meta-analysis, which included 23 studies, four different age groups (young children, children, healthy adults, older adults) were included. From a general population perspective, it should not necessarily be surprising that using children versus neurotypical adults versus ADHD individuals versus older adults might impact the results. For example, it is unclear whether ADHD individuals or healthy children stand to benefit more from cognitive training (Rutledge, van den Bos, McClure, & Schweitzer, 2012). Also, some studies suggest that older adults may improve differently from younger individuals during a training intervention, even if they both experience substantial gains (Brehmer, Westerberg, & Bäckman, 2012). Although one recent meta-analysis did not identify differences between older and younger adults on training outcomes (Karbach & Verhaeghen, 2014), some individual studies seem to indicate that younger adults may experience larger transfer effects than older adults (Richmond, Morrison, Chein, & Olson, 2011).

Other specific demographic factors may also influence improvements from training. As one interesting example, Hackman and Farah (2009) suggest that SES may significantly impact the development of executive function, particularly in young children. With as few as 20 participants included in certain studies, it is possible differences in SES may factor into the outcome of these studies. In two comparable studies (Jaeggi et al., 2014; Redick et al., 2013) that arrived at somewhat different outcomes, there are likely some differences related to SES (based on regional disparities and differences in subject pools) that could potentially influence transfer. A recent study as well as research from our own laboratory suggests that differences in SES could contribute to significant differences in performance on transfer tasks (Blair & Raver, 2014; Katz & Shah, 2013).

Another personal characteristic factor often left out of the discussion is how motivation—both to complete the study but also to improve cognition—may influence training improvements and outcome measures. While some studies inform participants that they may improve their intelligence or cognitive function during the study (Jaeggi et al., 2008; Klingberg et al., 2005), others only mention they will be practicing computerized tasks (Redick et al., 2013). Additionally, the use of payment as a means incentive in training studies varies (Au et al., 2015) and may also support—or undermine—outcomes following the intervention. Ideally, future meta-analytic and review work would take these factors into account.

Design of Interventions

Even when transfer tasks and participants are carefully considered, a surprising amount of variation exists across the design of interventions, and it is not always explicitly discussed. For example, the spacing of training sessions is generally not manipulated with training studies, but evidence suggests that spacing may have an impact on how much participants benefit from training (Wang, Zhou, & Shah, 2014). Additionally, although Karbach and Verhaeghen (2014) did not find an impact of dosage in their meta-analysis of WM and executive function training, the dose–response relationship for different training interventions may require some threshold of training time (and possibly performance) be met to lead to transfer effects following training. This factor is also intrinsically linked to participant characteristics; the rate at which an individual improves on a cognitive performance task (which may also be related to the design and implementation of the training program as well as personal characteristics) may also contribute to improvements on untrained tasks (Bürki, Ludwig, Chicherio, & de Ribarpierre, 2014).

Another significant question is whether multimodal interventions are more effective than interventions featuring a single modality of training. For example, interventions from Alloway and Alloway (2008) and Klingberg et al. (2005) use training that include multiple games targeted to improve different abilities, or subsets of abilities, while others (Jaeggi et al., 2008) only include a single repeated task that targets a specific cognitive domain. There are potential tradeoffs with each of these approaches. While it is possible that multimodal training may be more likely to be effective, it has been proposed that it may be more difficult to determine exactly what is leading to transfer effects in this sort of training (Karbach & Verhaeghen, 2014). Using single tasks may not be the solution, though, because even simple tasks that involve more than one cognitive process and potential strategy leave open the possibility of different underlying mechanisms.

Also, some interventions are gamelike in nature (Jaeggi et al., 2011; Klingberg et al., 2005) while others do not include gamelike features, such as scoring and theming. However, there is some evidence that game elements may influence performance on the tasks involved (Katz, Jaeggi, Buschkuehl, Stegman, & Shah, 2014). Interestingly, this difference has appeared most sharply in training targeted toward children versus training targeted toward adults, as though gamelike features are more appropriate for younger individuals than older ones. Even when the core training task is kept the same in terms of cognitive demands, the protocol of studies may differ in various ways.

Conclusion

Developmental cognitive scientists generally agree on many premises that, if true, should fulfill the conditions for cognitive training to be effective and meaningful. A primary assumption is that skills do not merely unfold with brain development, but rather that the brain requires specific experiential inputs for normal development to take place. Presumably, the nature and quality of those inputs will have an impact on the quality of outcomes. Implicit in these assumptions is the idea that if a child requires certain experiences to develop a specific skill or set of knowledge, then there is also a set of experiences that may be provided to a child who did not have an optimized set of experiences. Depending on when in development they are applied, they may alter the child's outcomes. A second important assumption is that rate-limiting factors in performance, across a wide range of complex cognitive tasks (e.g., reading comprehension, inductive reasoning, and numeracy) may ultimately impact capacity for benefiting from these enriching experiences (cognitive training or otherwise).

We feel that researchers working in the field of cognitive training would do well to refocus their energies from the simple question of whether a particular intervention is able to improve untrained tasks, particularly in a small population. Rather, the question of underlying mechanisms is of far greater relevance at present. What can cognitive training tell us about the potential for cognitive plasticity? To what extent may we be able to improve the cognitive capacities that underlie learning and success in intellectual pursuits in life? What domains may be improved, and what domains may be resistant to improvement? Under some conditions, cognitive training clearly leads to transfer, and in other conditions it does not; however, these conditions have not been fully explored. We have provided several examples of possible methodological issues or moderators of improvement in cognitive training work. Before we as a research community make strong claims regarding the effectiveness of cognitive training or the lack thereof, this high-quality research needs to be completed. However, few of these are identified or discussed in widely cited meta-analyses. For example, while Melby-Lervåg and Hulme's (2013) meta-analysis does recognize the fact that people of different ages may respond differently to cognitive training, they do not consider personality characteristics or motivation.

Until these moderating factors have been fully addressed, the conclusions reached from existing work must be tempered. We have aimed to demonstrate the need to re-examine moderating factors and potential issues that could limit the conclusions drawn from any individual training study. As Richard Feynman noted in his 1974 Caltech address:

> If you're doing an experiment, you should report everything that you think might make it invalid—not only what you think is right about it. Other causes that could possibly explain your results and things you thought of that you've eliminated by some other experiment and how they worked—to make sure the other fellow can tell they have been eliminated. Details that could throw doubt on your interpretation must be given, if you know them. You must do the best you can—if you know anything at all wrong or possibly wrong—to explain it. (Feynman, 1974)

Advertisements for brain-training products are now pervasive on the radio, the internet, and television—even during the Superbowl (Hunt & Jaeggi, 2013). They are often couched in the language of neuroscience, no doubt to make scientific claims seem more credible, a phenomenon that itself has been the focus of study (Rhodes, Rodriguez, & Shah, 2014; Weisberg, Keil, Goodstein, Rawson, & Gray, 2008). The frustration many must feel at the proliferation of these products is likely compounded by the numerous psychologists and neuroscientists who have become involved with commercialization of cognitive training products based on their own research.

However, this tendency to focus on the commercial side of cognitive training—rather than the large body of legitimate research—may have unfortunate implications. The consumer interest in brain training likely reflects a widespread interest in another question that has not been settled: how much our cognitive capacities might be improved through experience beyond what may be predicted by genetics? This interest predates modern psychological science by many hundreds of years, but some researchers might be surprised to learn that, as early as the turn of the twentieth century, one brain training program—with language and marketing not unlike that used today—rose to international prominence, ultimately counting over 750,000 individuals as members (Thomson, 2006). The *Pelmanism* program mentioned earlier—once endorsed by British luminaries—promised the public that they could improve their ability to concentrate, remember, and reason beyond their innate abilities (Katz, 2014).

While psychological science has changed dramatically in the past 100 years, the desire to improve cognition has not. This should not be surprising, as our capacity for cognitive plasticity—particularly in the context of important prefrontal executive functions—has significant ramifications for education, work, and our general ability to succeed in an increasingly competitive global environment. But the outcome of any individual study, of any individual intervention, and, as we have illustrated, any individual meta-analysis—cannot be construed as a conclusive answer to the question of how much cognitive function might be improved through experience or intervention. This point is often missed in press releases and media interviews, potentially at great cost, as scientists may not realize the extent to which policy, from funding decisions to educational curricula to decisions about whether to implement programs to address early childhood deprivation, may be affected. Rather than being frustrated and overwhelmed by the popularity of brain training, researchers might do better to consider the implications of their research—and redouble their efforts to improve their methodology. Whatever their results, they should recognize that the public, educators, and policy-makers—in addition to fellow scientists—will be listening.

References

Allen, L. E. (1970). *WFF'N PROOF EQUATIONS: The game of creative mathematics.* Autoelic Instructional Materials Publisher: New Haven, CT.

Allen, L. E., Allen, R. W., & Miller, J. C. (1966). Programmed games and the learning of problem solving skills: The WFF'N PROOF example. *The Journal of Educational Research, 60*(1), 22–26. doi:10.1080/00220671.1966.10883424

Alloway, T. P., & Alloway, R. G. (2008). *Jungle memory training program.* Memosyne: Edinburgh, UK.

Alloway, T. P., & Alloway, R. G. (2010). Investigating the predictive roles of working memory and IQ in academic attainment. *Journal of Experimental Child Psychology, 106*(1), 20–29. doi:10.1016/j.jecp.2009.11.003

Alloway, T. P., & Passolunghi, M. C. (2011). The relationship between working memory, IQ, and mathematical skills in children. *Learning and Individual Differences, 21*(1), 133–137. doi:10.1016/j.lindif.2010.09.013

Au, J., Sheehan, E., Tsai, N., Duncan, G. J., Buschkuehl, M., & Jaeggi, S. M. (2015). Improving fluid intelligence with training on working memory: A meta-analysis. *Psychonomic Bulletin & Review, 22*(2), 366–377. doi:10.3758/s13423-014-0699-x

Baddeley, A. (2003). Working memory: Looking back and looking forward. *Nature Reviews Neuroscience, 4*(10), 829–839. doi:10.1038/nrn1201

Barha, C. K., Pawluski, J. L., & Galea, L. A. (2007). Maternal care affects male and female offspring working memory and stress reactivity. *Physiology & Behavior, 92*(5), 939–950. doi:10.1016/j.physbeh.2007.06.022

Barker, J. E., Semenov, A. D., Michaelson, L., Provan, L. S., Snyder, H. R., & Munakata, Y. (2014). Less-structured time in children's daily lives predicts self-directed executive functioning. *Frontiers in Psychology, 5*, 593. doi:10.3389/fpsyg.2014.00593

Barkley, R. A. (1997). Behavioral inhibition, sustained attention, and executive functions: Constructing a unifying theory of ADHD. *Psychological Bulletin, 121*(1), 65–94. doi:10.1037/0033-2909.121.1.65

Barnett, W. S., & Masse, L. N. (2007). Comparative benefit–cost analysis of the Abecedarian program and its policy implications. *Economics of Education Review, 26*(1), 113–125. doi:10.1016/j.econedurev.2005.10.007

Bernier, A., Carlson, S. M., & Whipple, N. (2010). From external regulation to self-regulation: Early parenting precursors of young children's executive functioning. *Child Development, 81*, 326–339. doi:10.1111/j.1467-8624.2009.01397.x

Bialystok, E., Craik, F., & Luk, G. (2008). Cognitive control and lexical access in younger and older bilinguals. *Journal of Experimental Psychology: Learning, Memory, and Cognition, 34*(4), 859–873. doi:10.1037/0278-7393.34.4.859

Blagg, N. (1991). *Can we teach intelligence? A comprehensive evaluation of Feuerstein's Instrumental Enrichment program.* New York, NY: Psychology Press.

Blair, C., & Raver, C. C. (2014). Closing the achievement gap through modification of neurocognitive and neuroendocrine function: Results from a cluster randomized controlled trial of an innovative approach to the education of children in kindergarten. *PloS One, 9*(11), e112393. doi:10.1371/journal.pone.0112393

Brehmer, Y., Westerberg, H., & Bäckman, L. (2012). Working-memory training in younger and older adults: Training gains, transfer, and maintenance. *Frontiers in Human Neuroscience, 6*, 63. doi:10.3389/fnhum.2012.00063

Brydges, C. R., Reid, C. L., Fox, A. M., & Anderson, M. (2012). A unitary executive function predicts intelligence in children. *Intelligence, 40*(5), 458–469. doi:10.1016/j.intell.2012.05.006

Bull, R., & Scerif, G. (2001). Executive functioning as a predictor of children's mathematics ability: Inhibition, switching, and working memory. *Developmental Neuropsychology, 19*(3), 273–293.

Bürki, C. N., Ludwig, C., Chicherio, C., & de Ribaupierre, A. (2014). Individual differences in cognitive plasticity: An investigation of training curves in younger and older adults. *Psychological Research, 78*(6), 821–835. doi:10.1007/s00426-014-0559-3

Buschkuehl, M., & Jaeggi, S. M. (2010). Improving intelligence: A literature review. *Swiss Medical Weekly, 140*(19–20), 266–272. doi:smw-12852

Canfield, R. L., Kreher, D. A., Cornwell, C., & Henderson, C. R. (2003). Low-level lead exposure, executive functioning, and learning in early childhood. *Child Neuropsychology, 9*(1), 35–53. doi:10.1076/chin.9.1.35.14496

Cepeda, N. J., Pashler, H., Vul, E., Wixted, J. T., & Rohrer, D. (2006). Distributed practice in verbal recall tasks: A review and quantitative synthesis. *Psychological Bulletin, 132*(3), 354. doi:10.1037/0033-2909.132.3.354

Chang, R. (2004). *Physical chemistry for the biosciences.* Sausalito, CA: University Science Books.

Chooi, W.-T., & Thompson, L. A. (2012). Working memory training does not improve intelligence in healthy young adults. *Intelligence, 40*(6), 531–542. doi:10.1016/j.intell.2012.07.004

Chugani, H. T., Behen, M. E., Muzik, O., Juhász, C., Nagy, F., & Chugani, D. C. (2001). Local brain functional activity following early deprivation: A study of postinstitutionalized Romanian orphans. *Neuroimage, 14*(6), 1290–1301. doi:10.1006/nimg.2001.0917

Clark, C., Prior, M., & Kinsella, G. (2002). The relationship between executive function abilities, adaptive behaviour, and academic achievement in children with externalising behaviour problems. *Journal of Child Psychology and Psychiatry, 43*(6), 785–796. doi:10.1111/1469-7610.00084

Clements, D. H. (1986). Effects of Logo and CAI environments on cognition and creativity. *Journal of Educational Psychology, 78*(4), 309. doi:10.1037//0022-0663.78.4.309

Conway, A. R., Kane, M. J., & Engle, R. W. (2003). Working memory capacity and its relation to general intelligence. *Trends in Cognitive Sciences, 7*(12), 547–552. doi:10.1016/j.tics.2003.10.005

Costa-Giomi, E. (1999). The effects of three years of piano instruction on children's cognitive development. *Journal of Research in Music Education, 47*(3), 198–212. doi:10.2307/3345779

Cowan, N. (2001). Metatheory of storage capacity limits. *Behavioral and Brain Sciences, 24*(1), 154–176. doi:10.1017/S0140525X0161392X

D'Angiulli A., Herdman, A., Stapells, D., & Hertzman, C. (2008). Children's event-related potentials of auditory selective attention vary with their socioeconomic status. *Neuropsychology. 22*, 293–300. doi:10.1037/0894-4105.22.3.293

Daneman, M., & Green, I. (1986). Individual differences in comprehending and producing words in context. *Journal of Memory and Language, 25*(1), 1–18. doi:10.1016/0749-596X(86)90018-5

DePrince, A. P., Weinzierl, K. M., & Combs, M. D. (2009). Executive function performance and trauma exposure in a community sample of children. *Child Abuse & Neglect, 33*(6), 353–361. doi:10.1016/j.chiabu.2008.08.002

Diamond, A. (2010). The evidence base for improving school outcomes by addressing the whole child and by addressing skills and attitudes, not just content. *Early Education and Development, 21*(5), 780–793. doi:10.1080/10409289.2010.514522

Diamond, A. (2012). Activities and programs that improve children's executive functions. *Current Directions in Psychological Science, 21*, 335–341. doi:10.1177/0963721412453722

Diamond, A. (2013). Executive functions. *Annual Review of Psychology, 64*, 135–168. doi:10.1146/annurev-psych-113011-143750

Diamond, A. (2014). Want to optimize executive functions and academic outcomes? Simple, just nourish the human spirit. *Minnesota Symposia on Child Psychology, 37*, 203–230. doi:10.1002/9781118732373.ch7

Diamond, A., Barnett, W. S., Thomas, J., & Munro, S. (2007). Preschool program improves cognitive control. *Science, 318*(5855), 1387–1388. doi:10.1126/science.1151148

Diamond, A., & Lee, K. (2011). Interventions shown to aid executive function development in children 4–12 years old. *Science, 333*(6045), 959–964. doi:10.1126/science.1204529

Ditye, T., Jacobson, L., Walsh, V., & Lavidor, M. (2012). Modulating behavioral inhibition by tDCS combined with cognitive training. *Experimental Brain Research*, *219*(3), 363–368. doi:10.1007/s00221-012-3098-4

Donchin, E. (1995). Video games as research tools: The Space Fortress game. *Behavior Research Methods, Instruments, & Computers*, *27*(2), 217–223. doi:10.3758/BF03204735

Douglas, V. I., Parry, P., Marton, P., & Garson, C. (1976). Assessment of a cognitive training program for hyperactive children. *Journal of Abnormal Child Psychology*, *4*(4), 389–410. doi:10.1007/BF00922535

Duncan, J., Parr, A., Woolgar, A., Thompson, R., Bright, P., Cox, S., . . . Nimmo-Smith, I. (2008). Goal neglect and Spearman's g: Competing parts of a complex task. *Journal of Experimental Psychology: General*, *137*, 131–148. doi:10.1037/0096-3445.137.1.131

Duncan, J., Schramm, M., Thompson, R., & Dumontheil, I. (2012). Task rules, working memory, and fluid intelligence. *Psychonomic Bulletin & Review*, *19*(5), 864–870. doi:10.3758/s13423-012-0225-y

Dunning, D. L., & Holmes, J. (2014). Does working memory training promote the use of strategies on untrained working memory tasks? *Memory & Cognition*, *42*(6), 854–862. doi:10.3758/s13421-014-0410-5

Engle, R. W., Kane, M. J., & Tuholski, S. W. (1999). Individual differences in working memory capacity and what they tell us about controlled attention, general fluid intelligence, and functions of the prefrontal cortex. In A. Miyake, & P. Shah (Eds.), *Models of working memory: Mechanisms of active maintenance and executive control* (pp. 102–134). New York, NY: Cambridge University Press.

Entwisle, D. R., & Astone, N. M. (1994). Some practical guidelines for measuring youth's race/ethnicity and socioeconomic status. *Child Development*, *65*(6), 1521–1540. doi:10.2307/1131278

Evans, G. W., & Schamberg, M. A. (2009). Childhood poverty, chronic stress, and adult working memory. *Proceedings of the National Academy of Sciences*, *106*(16), 6545–6549. doi:10.1073/pnas.0811910106

Eysenck, M. W., Derakshan, N., Santos, R., & Calvo, M. G. (2007). Anxiety and cognitive performance: Attentional control theory. *Emotion*, *7*(2), 336–353. doi:10.1037/1528-3542.7.2.336

Fabiani, M., Buckley, J., Gratton, G., Coles, M. G. H., Donchin, E., & Logie, R. (1989). The training of complex task performance. *Acta Psychologica*, *71*(1), 259–299. doi:10.1016/0001-6918(89)90012-7

Farah, M. J., Noble, K. G., & Hurt, H. (2005). Poverty, privilege, and brain development: Empirical findings and ethical implications. In J. Illes (Ed.), *Neuroethics: Defining the issues in theory, practice, and policy*. New York, NY: Oxford University Press.

Fay-Stammbach, T., Hawes, D. J., & Meredith, P. (2014). Parenting influences on executive function in early childhood: A review. *Child Development Perspectives*, *8*(4), 258–264. doi:10.1111/cdep.12095

Feuerstein, R., & Hamburger, M. (1965). *A proposal to study the process of redevelopment in several groups of deprived early adolescents in both residential and nonresidential settings*. The Youth-Aliyah Department of the Jewish Agency, Jerusalem. Unpublished report.

Feuerstein, R., & Jensen, M. R. (1980). Instrumental Enrichment: Theoretical basis, goals, and instruments. *The Educational Forum*, *44*(4), 401–423. doi:10.1080/00131728009336184

Feynman, R. (1974). Commencement address at the California Institute of Technology, Pasadena, CA.

Gathercole, S. E., & Baddeley, A. D. (2014). *Working memory and language processing.* New York, NY: Psychology Press. doi:10.4324/9781315804682

Gathercole, S. E., Brown, L., & Pickering, S. J. (2003). Working memory assessments at school entry as longitudinal predictors of National Curriculum attainment levels. *Educational and Child Psychology, 20*(3), 109–122.

Gathercole, S. E., Lamont, E., & Alloway, T. P. (2006). Working memory in the classroom. In S. J. Pickering (Ed.), *Working memory and education* (pp. 219–240). Burlington, MA: Academic Press.

Gianaros, P. J., Horenstein, J. A., Hariri, A. R., Sheu, L. K., Manuck, S. B., Matthews, K. A., & Cohen, S. (2008). Potential neural embedding of parental social standing. *Social Cognitive and Affective Neuroscience, 3*(2), 91–96. doi:10.1093/scan/nsn003

Gray, J. R., Chabris, C. F., & Braver, T. S. (2003). Neural mechanisms of general fluid intelligence. *Nature Reviews Neuroscience, 6*(3), 316–322. doi:10.1038/nn1014

Greenfield, P. M. (1984). *Mind and media.* Cambridge, MA: Harvard University Press.

Hackman, D. A., & Farah, M. J. (2009). Socioeconomic status and the developing brain. *Trends in Cognitive Sciences, 13*(2), 65–73.

Hackman, D. A., Farah, M. J., & Meaney, M. J. (2010). Socioeconomic status and the brain: Mechanistic insights from human and animal research. *Nature Reviews Neuroscience, 11*(9), 651–659. doi:10.1016/j.tics.2008.11.003

Heinzel, S., Schulte, S., Onken, J., Duong, Q. L., Riemer, T. G., Heinz, A., . . . Rapp, M. A. (2014). Working memory training improvements and gains in non-trained cognitive tasks in young and older adults. *Neuropsychology, Development, and Cognition, Section B, Aging Neuropsychology and Cognition, 21*(2), 146–173. doi:10.1080/13825585.2013.790338

Holmes, J., Gathercole, S. E., & Dunning, D. L. (2009). Adaptive training leads to sustained enhancement of poor working memory in children. *Developmental Science, 12*(4), F9–F15. doi:10.1111/j.1467-7687.2009.00848.x

Hunt, E., & Jaeggi, S. M. (2013). Challenges for research on intelligence. *Journal of Intelligence, 1*(1), 36–54. doi:10.3390/jintelligence1010036

Imholz, S., & Petrosino, A. (2012). Teacher observations on the implementation of the tools of the mind curriculum in the classroom: Analysis of interviews conducted over a one-year period. *Creative Education, 3*(2), 185–192. doi:10.4236/ce.2012.32029

Jaeggi, S. M., Buschkuehl, M., Jonides, J., & Perrig, W. J. (2008). Improving fluid intelligence with training on working memory. *Proceedings of the National Academy of Sciences, 105*(19), 6829–6833. doi:10.1073/pnas.0801268105

Jaeggi, S. M., Buschkuehl, M., Jonides, J., & Shah, P. (2011). Short-and long-term benefits of cognitive training. *Proceedings of the National Academy of Sciences, 108*(25), 10081–10086. doi:10.1073/pnas.1103228108

Jaeggi, S. M., Buschkuehl, M., Shah, P., & Jonides, J. (2014). The role of individual differences in cognitive training and transfer. *Memory & Cognition,42*(3), 464–480. doi:10.3758/s13421-013-0364-z

Jarosz, A. F., & Wiley, J. (2012). Why does working memory capacity predict RAPM performance? A possible role of distraction. *Intelligence, 40*(5), 427–438. doi:10.1016/j.intell.2012.06.001

Just, M. A., & Carpenter, P. (1979). The computer and eye processing pictures: The implementation of a raster graphics device. *Behavior Research Methods & Instrumentation, 11*(2), 172–176. doi:10.3758/BF03205642

Just, M. A., & Carpenter, P. A. (1992). A capacity theory of comprehension: Individual differences in working memory. *Psychological Review*, *99*(1), 122–149. doi:10.1037//0033-295X.99.1.122

Kane, M. J., & Engle, R. W. (2002). The role of prefrontal cortex in working-memory capacity, executive attention, and general fluid intelligence: An individual-differences perspective. *Psychonomic Bulletin & Review*, *9*(4), 637–671. doi:10.3758/BF03196323

Kane, M. J., Hambrick, D. Z., Tuholski, S. W., Wilhelm, O., Payne, T. W., & Engle, R. W. (2004). The generality of working memory capacity: A latent-variable approach to verbal and visuospatial memory span and reasoning. *Journal of Experimental Psychology: General*, *133*(2), 189–217. doi:10.1037/0096-3445.133.2.189

Karbach, J., & Verhaeghen, P. (2014). Making working memory work: A meta-analysis of executive-control and working memory training in older adults. *Psychological Science*, *25*(11), 2027–2037. doi:10.1177/0956797614548725

Karch, D., Albers, L., Renner, G., Lichtenauer, N., & von Kries, R. (2013). The efficacy of cognitive training programs in children and adolescents: A meta-analysis. *Deutsches Ärzteblatt International*, *110*(39), 643–652. doi:10.3238/arztebl.2013.0643

Karr, J. E., Areshenkoff, C. N., Rast, P., & Garcia-Barrera, M. A. (2014). An empirical comparison of the therapeutic benefits of physical exercise and cognitive training on the executive functions of older adults: A meta-analysis of controlled trials. *Neuropsychology*, *28*(6), 829. doi:10.1037/neu0000101

Katz, B. (2014). Brain-training isn't just a modern phenomenon, the Edwardians were also fans. *The Conversation*. Retrieved from http://theconversation.com/brain-training-isnt-just-a-modern-phenomenon-the-edwardians-were-also-fans-29515.

Katz, B., Jaeggi, S., Buschkuehl, M., Stegman, A., & Shah, P. (2014). Differential effect of motivational features on training improvements in school-based cognitive training. *Frontiers in Human Neuroscience*, *8*, 242. doi:10.3389/fnhum.2014.00242

Katz B, & Shah P. (2013, January). *Socioeconomic status as a moderator of improvements in executive function following cognitive training in adolescents*. Poster session presented at Determinants of Executive Function and Dysfunction Conference, Boulder, CO.

Kaviani, H., Mirbaha, H., Pournaseh, M., & Sagan, O. (2014). Can music lessons increase the performance of preschool children in IQ tests? *Cognitive Processing*, *15*(1), 77–84. doi:10.1007/s10339-013-0574-0

Kimura, K., & Hozumi, N. (2012). Investigating the acute effect of an aerobic dance exercise program on neuro-cognitive function in the elderly. *Psychology of Sport and Exercise*, *13*(5), 623–629. doi:10.1016/j.psychsport.2012.04.001

Kirschenbaum, R. J. (1998). Dynamic assessment and its use with under-served gifted and talented populations. *Gifted Child Quarterly*, *42*(3), 140–147. doi:10.1177/001698629804200302

Klahr, D., & Chen, Z. (2011). Finding one's place in transfer space. *Child Development Perspectives*, *5*(3), 196–204. doi:10.1111/j.1750-8606.2011.00171.x

Klauer, K. J., & Phye, G. D. (2008). Inductive reasoning: A training approach. *Review of Educational Research*, *78*(1), 85–123. doi:10.3102/0034654307313402

Klingberg, T., Fernell, E., Olesen, P. J., Johnson, M., Gustafsson, P., Dahlström, K., . . . Westerberg, H. (2005). Computerized training of working memory in children with ADHD—A randomized, controlled trial. *Journal of the American Academy of Child & Adolescent Psychiatry*, *44*(2), 177–186. doi:10.1097/00004583-200502000-00010

Kossow, H.-J., & Vehreschild, T. (1983). [Formation of a concentration training program for concentration disordered children]. *Psychiatrie, Neurologie, und Medizinische Psychologie, 35*(1), 31–36.

Krizman, J., Marian, V., Shook, A., Skoe, E., & Kraus, N. (2012). Subcortical encoding of sound is enhanced in bilinguals and relates to executive function advantages. *Proceedings of the National Academy of Sciences, 109*(20), 7877–7881. doi:10.1073/pnas.1201575109

Kundu, B., Sutterer, D. W., Emrich, S. M., & Postle, B. R. (2013). Strengthened effective connectivity underlies transfer of working memory training to tests of short-term memory and attention. *The Journal of Neuroscience, 33*(20), 8705–8715. doi:10.1523/JNEUROSCI.5565-12.2013

Lawson, G. M., Hook, C. J., Hackman, D. A., Farah, M. J., Griffin, J. A., Freund, L. S., & McCardle, P. (2014). Socioeconomic status and neurocognitive development: Executive function. In J. A. Griffin, L. S. Freund, & P. McCardle (Eds.), *Executive function in preschool children: Integrating measurement, neurodevelopment, and translational research.* Washington, DC: American Psychological Association Press.

Lillard, A., & Else-Quest, N. (2006). The early years: Evaluating Montessori education. *Science, 313*(5795), 1893–1894. doi:10.1126/science.1132362

Lukowski, A. F., Koss, M., Burden, M. J., Jonides, J., Nelson, C. A., Kaciroti, N., . . . Lozoff, B. (2010). Iron deficiency in infancy and neurocognitive functioning at 19 years: Evidence of long-term deficits in executive function and recognition memory. *Nutritional Neuroscience, 13*(2), 54–70. doi:10.1179/147683010X12611460763689

Mackey, A., Raizada, R., & Bunge, S. A. (2011). Environmental influences on prefrontal development. In D. Stuss & R. Knight (Eds.), *Oxford handbook of frontal lobe functions* (2nd ed., pp. 145–164). New York, NY: Oxford University Press.

McVay, J. C., & Kane, M. J. (2012). Why does working memory capacity predict variation in reading comprehension? On the influence of mind wandering and executive attention. *Journal of Experimental Psychology: General, 141*(2), 302–320. doi:10.1037/a0025250

Melby-Lervåg, M., & Hulme, C. (2013). Is working memory training effective? A meta-analytic review. *Developmental Psychology, 49*(2), 270–291. doi:10.1037/a0028228

Mischel, W., & Ayduk, O. (2002). Self-regulation in a cognitive–affective personality system: Attentional control in the service of the self. *Self and Identity, 1*(2), 113–120. doi:10.1080/152988602317319285

Miyake, A., Friedman, N. P., Emerson, M. J., Witzki, A. H., Howerter, A., & Wager, T. D. (2000). The unity and diversity of executive functions and their contributions to complex "frontal lobe" tasks: A latent variable analysis. *Cognitive Psychology, 41*(1), 49–100. doi:10.1006/cogp.1999.0734

Miyake, A., & Shah, P. (Eds.). (1999). *Models of working memory: Mechanisms of active maintenance and executive control.* New York, NY: Cambridge University Press.

Moreno, S., Marques, C., Santos, A., Santos, M., Castro, S. L., & Besson, M. (2009). Musical training influences linguistic abilities in 8-year-old children: More evidence for brain plasticity. *Cerebral Cortex, 19*(3), 712–723. doi:10.1093/cercor/bhn120

Morrison, A. B., & Chein, J. M. (2011). Does working memory training work? The promise and challenges of enhancing cognition by training working memory. *Psychonomic Bulletin & Review, 18*(1), 46–60. doi:10.3758/s13423-010-0034-0

Muennig, P., Schweinhart, L., Montie, J., & Neidell, M. (2009). Effects of a prekindergarten educational intervention on adult health: 37-year follow-up results of a randomized controlled trial. *American Journal of Public Health, 99*(8), 1431–1437. doi:10.2105/AJPH.2008.148353

Nation, K., Adams, J. W., Bowyer-Crane, C. A., & Snowling, M. J. (1999). Working memory deficits in poor comprehenders reflect underlying language impairments. *Journal of Experimental Child Psychology, 73*(2), 139–158. doi:10.1006/jecp.1999.2498

Nee, D. E., Wager, T. D., & Jonides, J. (2007). Interference resolution: Insights from a meta-analysis of neuroimaging tasks. *Cognitive, Affective, & Behavioral Neuroscience, 7*(1), 1–17. doi:10.3758/CABN.7.1.1

Neville, H. J., Stevens, C., Pakulak, E., Bell, T. A., Fanning, J., Klein, S., & Isbell, E. (2013). Family-based training program improves brain function, cognition, and behavior in lower socioeconomic status preschoolers. *Proceedings of the National Academy of Sciences, 110*(29), 12138–12143. doi:10.1073/pnas.1304437110

Nickerson, R. S. (1985). Project intelligence: An account and some reflections. *Special Services in the Schools, 3*(1–2), 83–102. doi:10.1300/J008v03n01_06

Noack, H., Lövdén, M., & Schmiedek, F. (2014). On the validity and generality of transfer effects in cognitive training research. *Psychological Research, 78*(6), 773–789. doi:10.1007/s00426-014-0564-6

Noble, K. G., Houston, S. M., Kan, E., & Sowell, E. R. (2012). Neural correlates of socioeconomic status in the developing human brain. *Developmental Science, 15*(4), 516–527. doi:10.1111/j.1467-7687.2012.01147.x

O'Dwyer, S., Burton, N. W., Pachana, N. A., & Brown, W. J. (2007). Protocol for "Fit Bodies, Fine Minds": A randomized controlled trial on the effect of exercise and cognitive training on cognitive functioning in older adults. *BMC Geriatrics, 7*(1), 23. doi:10.1186/1471-2318-7-23

Pickering, S. J. (Ed.). (2006). *Working memory and education.* Burlington, MA: Academic Press.

Pollak, S. D., Nelson, C. A., Schlaak, M. F., Roeber, B. J., Wewerka, S. S., Wiik, K. L., . . . Gunnar, M. R. (2010). Neurodevelopmental effects of early deprivation in post-institutionalized children. *Child Development, 81*(1), 224–236. doi:10.1111/j.1467-8624.2009.01391.x

Ponitz, C. E. C., McClelland, M. M., Jewkes, A. M., Connor, C. M., Farris, C. L., & Morrison, F. J. (2008). Touch your toes! Developing a direct measure of behavioral regulation in early childhood. *Early Childhood Research Quarterly, 23*(2), 141–158. doi:10.1016/j.ecresq.2007.01.004

Ponitz, C. C., McClelland, M. M., Matthews, J. S., & Morrison, F. J. (2009). A structured observation of behavioral self-regulation and its contribution to kindergarten outcomes. *Developmental Psychology, 45*(3), 605–619. doi:10.1037/a0015365

Rand, Y. A., Tannenbaum, A. J., & Feuerstein, R. (1979). Effects of Instrumental Enrichment on the psychoeducational development of low-functioning adolescents. *Journal of Educational Psychology, 71*(6), 751–763. doi:10.1037//0022-0663.71.6.751

Rapport, M. D., Scanlan, S. W., & Denney, C. B. (1999). Attention-deficit/hyperactivity disorder and scholastic achievement: A model of dual developmental pathways. *Journal of Child Psychology and Psychiatry, 40*(8), 1169–1183. doi:10.1017/S0021963099004618

Redick, T. S., Shipstead, Z., Harrison, T. L., Hicks, K. L., Fried, D. E., Hambrick, D. Z., . . . Engle, R. W. (2013). No evidence of intelligence improvement after working memory training: A randomized, placebo-controlled study. *Journal of Experimental Psychology: General, 142*(2), 359–379. doi:10.1037/a0029082

Rhodes, R., Rodriguez, F., & Shah, P. (2014). Explaining the alluring influence of neuroscience in scientific reasoning. *Journal of Experimental Psychology: Learning, Memory, & Cognition, 40*(5), 1432–1440. doi:10.1037/a0036844

Richey, J. E., Phillips, J. S., Schunn, C. D., & Schneider, W. (2014). Is the link from working memory to analogy causal? No analogy improvements following working memory training gains. *PloS One, 9*(9), e106616. doi:10.1371/journal.pone.0106616

Richmond, L. L., Morrison, A. B., Chein, J. M., & Olson, I. R. (2011). Working memory training and transfer in older adults. *Psychology and Aging, 26*(4), 813–822. doi:10.1037/a0023631

Rogers, M. A., Kasai, K., Koji, M., Fukuda, R., Iwanami, A., Nakagome, K., . . . Kato, N. (2004). Executive and prefrontal dysfunction in unipolar depression: A review of neuropsychological and imaging evidence. *Neuroscience Research, 50*(1), 1–11. doi:10.1016/j.neures.2004.05.003

Romney, D. M., & Samuels, M. T. (2001). A meta-analytic evaluation of Feuerstein's Instrumental Enrichment program. *Educational and Child Psychology, 18*(4), 19–34.

Rouder, J. N., Morey, R. D., Cowan, N., Zwilling, C. E., Morey, C. C., & Pratte, M. S. (2008). An assessment of fixed-capacity models of visual working memory. *Proceedings of the National Academy of Sciences, 105*(16), 5975–5979. doi:10.1073/pnas.0711295105

Rubinstein, J. S., Meyer, D. E., & Evans, J. E. (2001). Executive control of cognitive processes in task switching. *Journal of Experimental Psychology: Human Perception and Performance, 27*(4), 763–779. doi:10.1037//0096-1523.27.4.763

Rueda, M. R., Posner, M. I., & Rothbart, M. K. (2005). The development of executive attention: Contributions to the emergence of self-regulation. *Developmental Neuropsychology, 28*(2), 573–594. doi:10.1207/s15326942dn2802_2

Rutledge, K. J., van den Bos, W., McClure, S. M., & Schweitzer, J. B. (2012). Training cognition in ADHD: Current findings, borrowed concepts, and future directions. *Neurotherapeutics, 9*(3), 542–558. doi:10.1007/s13311-012-0134-9

Sarsour, K., Sheridan, M., Jutte, D., Nuru-Jeter, A., Hinshaw, S., & Boyce, W. T. (2011). Family socioeconomic status and child executive functions: The roles of language, home environment, and single parenthood. *Journal of the International Neuropsychological Society, 17*(1), 120–132. doi:10.1017/S1355617710001335

Schellenberg, E. G. (2004). Music lessons enhance IQ. *Psychological Science, 15*(8), 511–514. doi:10.1111/j.0956-7976.2004.00711.x

Schmidt, R. A., & Bjork, R. A. (1992). New conceptualizations of practice: Common principles in three paradigms suggest new concepts for training. *Psychological Science, 3*(4), 207–217. doi:10.1111/j.1467-9280.1992.tb00029.x

Schmiedek, F., Lövdén, M., & Lindenberger, U. (2010). Hundred days of cognitive training enhance broad cognitive abilities in adulthood: Findings from the COGITO study. *Frontiers in Aging Neuroscience, 2*, 27. doi:10.3389/fnagi.2010.00027

Seigneuric, A., Ehrlich, M.-F., Oakhill, J. V., & Yuill, N. M. (2000). Working memory resources and children's reading comprehension. *Reading and Writing, 13*(1–2), 81–103. doi:10.1023/A:1008088230941

Shipstead, Z., Hicks, K. L., & Engle, R. W. (2012). Cogmed working memory training: Does the evidence support the claims?. *Journal of Applied Research in Memory and Cognition, 1*(3), 185–193. doi:10.1016/j.jarmac.2012.06.003

Shipstead, Z., Redick, T. S., & Engle, R. W. (2012). Is working memory training effective? *Psychological Bulletin, 138*(4), 628–654. doi:10.1037/a0027473

Stavros, D. (1989). *Evaluation of the Instrumental Enrichment project*. Detroit, MI: Research and Evaluation Department Office of Instructional Improvement: Detroit Public Schools.

Stephenson, C. L., & Halpern, D. F. (2013). Improved matrix reasoning is limited to training on tasks with a visuospatial component. *Intelligence, 41*(5), 341–357. doi:10.1016/j.intell.2013.05.006

Stern, Y., Blumen, H. M., Rich, L. W., Richards, A., Herzberg, G., & Gopher, D. (2011). Space Fortress game training and executive control in older adults: A pilot intervention. *Aging, Neuropsychology, and Cognition, 18*(6), 653–677. doi:10.1080/13825585.2011.613450

Stine-Morrow, E. A. (2007). The Dumbledore hypothesis of cognitive aging. *Current Directions in Psychological Science, 16*(6), 295–299. doi:10.1111/j.1467-8721.2007.00524.x

Thompson, T. W., Waskom, M. L., Garel, K. L. A., Cardenas-Iniguez, C., Reynolds, G. O., Winter, R., . . . Gabrieli, J. D. (2013). Failure of working memory training to enhance cognition or intelligence. *PloS One, 8*(5), e63614. doi:10.1371/journal.pone.0063614

Thomson, M. (2006). *Psychological subjects: Identity, culture, and health in twentieth-century Britain*. Oxford, UK: Oxford University Press.

Titz, C., & Karbach, J. (2014). Working memory and executive functions: Effects of training on academic achievement. *Psychological Research, 78*(6), 852–868. doi:10.1007/s00426-013-0537-1

Valiente, C., Lemery-Chalfant, K., & Castro, K. S. (2007). Children's effortful control and academic competence: Mediation through school liking. *Merrill-Palmer Quarterly, 53*(1), 1–25. doi:10.1353/mpq.2007.0006

Vehreschild, T., Kossow, H.-J., & Schulz-Wulf, G. (1984). [Results of concentration training in normally intelligent students with poor concentration]. *Psychiatrie, Neurologie, und Medizinische Psychologie, 36*(3), 152–160.

von Bastian, C. C., & Oberauer, K. (2013). Effects and mechanisms of working memory training: A review. *Psychological Research, 78*(6), 803–820. doi:10.1007/s00426-013-0524-6

Wang, Z., Zhou, R., & Shah, P. (2014). Spaced cognitive training promotes training transfer. *Frontiers in Human Neuroscience, 8*, 217. doi:10.3389/fnhum.2014.00217

Weisberg, D. S., Keil, F. C., Goodstein, J., Rawson, E., & Gray, J. R. (2008). The seductive allure of neuroscience explanations. *Journal of Cognitive Neuroscience, 20*(3), 470–477. doi:10.1162/jocn.2008.20.3.470

Wilson, S. J., & Farran, D. C. (2012, March). Experimental evaluation of the Tools of the Mind preschool curriculum. Paper presented at the Society for Research on Educational Effectiveness, Washington, DC.

Wolf, P. A., Beiser, A., Elias, M. F., Au, R., Vasan, R. S., & Seshadri, S. (2007). Relation of obesity to cognitive function: Importance of central obesity and synergistic influence of concomitant hypertension. The Framingham Heart Study. *Current Alzheimer Research, 4*(2), 111–116. doi:10.2174/156720507780362263

Woodworth, R. S., & Thorndike, E. L. (1901). The influence of improvement in one mental function upon the efficiency of other functions. (I). *Psychological Review, 8*(3), 247. doi:10.1037/h0074898

Zheng, X., Swanson, H. L., & Marcoulides, G. A. (2011). Working memory components as predictors of children's mathematical word problem solving. *Journal of Experimental Child Psychology, 110*(4), 481–498. doi:10.1016/j.jecp.2011.06.001

11

Music Training

Contributions to Executive Function

Brooke M. Okada and L. Robert Slevc

Introduction

Imagine you are the first violinist in a string quartet. Each time you and your colleagues rehearse, you engage multiple cognitive processes. While playing, you must continuously listen and attend to multiple auditory streams of music (i.e., your own playing as well as the music from your violinist, violist, and cellist) to coordinate your group's sound. You must also constantly adjust your tempo, volume, and timbre to match that of your group to highlight certain melodies and coordinate nuances within the music. Simultaneously, you are also reading your music filled with notes and extra markings, and these items must be held in mind until played. These processes likely require a variety of cognitive processes, such as the cognitive flexibility to switch between auditory streams and the ability to update working memory when reading from the score. Both are part of a larger set of cognitive processes termed *executive functions* (EFs).

EFs, also known as cognitive control, are a set of top-down processes involved in the planning and regulation of cognition and behavior (e.g., Diamond, 2013; Diamond & Ling, 2019; Miyake & Friedman, 2012). Most models of EF posit three core related but separable components: inhibitory control, cognitive flexibility (also known as switching or shifting), and working memory updating (Diamond, 2013; Lehto, Juujärvi, Kooistra, & Pulkkinen, 2003; Logue & Gould, 2014; Miyake et al., 2000; Miyake & Friedman, 2012). Inhibitory control requires the control of one's thoughts or behavior to override a prepotent response. Cognitive flexibility requires one to change perspectives or switch between task demands. Lastly, working memory updating requires one to maintain, add, and delete items from memory. These EFs develop through adolescence and can be improved with practice (Diamond, Barnett, Thomas, & Munro, 2007; for a review, see Diamond, 2013).

Because EFs are such a critical aspect of cognition (e.g., EFs are important predictors of success in school and mental health; Diamond, 2013), there is considerable interest in predictors of EF abilities and how one might improve EFs (see Diamond & Ling, 2019). Accordingly, a variety of EF training paradigms have been developed (e.g., Chein & Morrison, 2010; Jaeggi, Buschkuehl, Jonides, & Perrig, 2008), although it is debatable whether these programs lead to improvements extending beyond the specific EF skill trained (particularly regarding fluid intelligence; e.g., Au et al., 2015; Dougherty, Hamovitz, & Tidwell, 2015; Owen et al., 2010). However, it is clear that specific EF processes can be improved from even relatively short experiences (Harrison et al., 2013). That is, there is good evidence for process-specific "near transfer," where improvements reflect the specific abilities taxed by a training task. Thus, regular engagement in tasks that rely on specific EF abilities could benefit those specific aspects of EF.

While carefully controlled EF training paradigms provide essential tools to understand cognitive training and transfer effects, one limitation with many of these paradigms is that the training tasks are not particularly interesting or rewarding on their own. Even if intense training on, for example, a dual N-back task can lead to improvements in EFs or other cognitive abilities (Jaeggi et al., 2008), it seems unlikely that many people would like to spend hours practicing their N-back skills. Some informal evidence for this comes from the fact that most cognitive training studies pay participants relatively well, yet still have relatively high rates of attrition. For example, Redick et al. (2013) reported an approximately 20% attrition rate despite paying $352 to participants who completed all parts of the study. Many commercial (and noncommercial) cognitive training programs address this problem of engagement by making training tasks game-like (e.g., the web-based packages *Cogmed* and *Lumosity*). However, an alternative possibility is to target other tasks that draw heavily on specific EF abilities but that have their own intrinsic rewards. One such complex, yet commonly pursued activity is music training (e.g., Hannon & Trainor, 2007; Moreno & Farzan, 2015; Slevc & Okada, 2015).

The idea that musical experience might lead to benefits to cognitive processes (as well as social outcomes) is both longstanding and widely held. For example, Jose Antonio Abreu, the founder of *El Sistema* (a music program targeting children in need) has said:

> *Children engaged in the programme [El Sistema] attain above-average results in school and show a tremendous capacity for collective community action. The orchestra and the choirs, the heart of the programme, help create a sense of solidarity. Involvement becomes a weapon against poverty and inequality, violence and drug abuse. (Burton-Hill, 2012, p. 1)*

Note that Abreu's suggestion is that *active* engagement in a music program, not *passive* exposure to music, will affect cognitive processes. Although passive music listening can affect a range of behaviors, including shopping behavior (Andersson, Kristensson, Wästlund, & Gustafsson, 2012; Milliman, 1982), worker productivity and attention (Shih, Huang, & Chiang, 2012), and exercise (Karageorghis & Priest, 2012), there is only limited evidence for effects of passive musical exposure on cognitive functioning (i.e., there is little evidence for the "Mozart Effect"; Pietschnig, Voracek, & Formann, 2010). Instead, cognitive effects of passive musical exposure likely reflect underlying effects on emotion and arousal rather than transfer effects per se (Thompson, Schellenberg, & Husain, 2001; Slevc & Okada, 2014).

In contrast, there is growing evidence that active music training is associated with a variety of nonmusical cognitive abilities (e.g., the special edition of *Music Perception* on "Music Training and Nonmusical Abilities" [Schellenberg & Winner, 2011]). As mentioned above, the idea that music training might produce collateral benefits for EFs is particularly interesting in part because many people find learning to play an instrument (or sing) intrinsically enjoyable in contrast to computerized EF training paradigms, for example. Another benefit of music training as an EF training paradigm is that it naturally involves multiple cognitive processes since it is a multifaceted activity. Playing an instrument requires that one learn to read music, translate printed notes into planned motor sequences, acquire fine motor coordination, and hold many notes and pieces in memory (Peretz & Zatorre, 2005). Indeed, Diamond (2013; Diamond & Ling, 2019) has posited that activities requiring progressively challenging EF skills, and in which individuals find enjoyment and emotional investment, should be the best at improving EFs.

In addition, there is good reason to think that music training can lead to both neural and cognitive changes. Much of the evidence for this focuses on consequences for motor and auditory processing. For example, compared to a matched control group, those who took 15 months of private keyboard lessons showed structural brain changes (based on deformation-based morphometry) in motor and auditory areas involved in learning music (Hyde et al., 2009). Similarly, professional keyboard players have greater gray matter density in motor areas (left Heschl's gyrus and left inferior frontal gyrus) in comparison to amateur musicians and nonmusicians (Gaser & Schlaug, 2003). In addition to structural brain changes, music training has also been associated with behavioral advantages in a wide range of cognitive processes. Music lessons have been found to correlate with multiple cognitive processes including, but not limited to: verbal memory (Chan, Ho, & Cheung, 1998; Ho, Cheung, & Chan, 2003; Jakobson, Cuddy, & Kilgour, 2003; Jakobson, Lewycky, Kilgour, & Stoesz, 2008); reading ability (Lamb & Gregory, 1993; Butzlaff, 2000); mathematical ability (Vaughn, 2000); spatial skills (Bilhartz, Bruhn, & Olson, 2000); IQ and academic ability (Schellenberg, 2006, but see Sala & Gobet, 2017; Schellenberg & Moreno,

2009); selective auditory attention (Strait & Kraus, 2011); auditory working memory (Pallesen et al., 2010); and processing speed (Bugos & Mostafa, 2011).

Of particular interest here is whether musical experience can specifically affect EF abilities. There is at least some evidence for EF-relevant neural effects of music training. For example, playing an instrument is related to an increased rate of cortical thickness maturation not only in motor areas recruited for planning and producing movement, but also in the dorsolateral prefrontal cortex and orbitofrontal cortex, areas often involved in executive functioning (Hudziak et al., 2014). However, effects of musical experience on EF abilities are only predicted for those aspects of EF on which music training relies (see Principle #2 in Diamond & Ling, 2019). Thus, a first step to understand the relationship of music training and EFs is to consider how music training draws on EF abilities. Below, we detail evidence that music processing involves and taxes different aspects of EF as well as correlational evidence that music training is associated with advantages in EF abilities. We then describe the small, but growing, literature examining the effects of longitudinal music training programs on EF abilities.

Music Training and Cognitive Flexibility

Music is often played in ensembles (e.g., symphonic orchestras, string quartets, or choirs), which requires that musicians play in coordination with each other (Palmer, 2013). Such coordination requires performers to switch flexibly between multiple auditory streams (Loehr, Kourtis, Vesper, Sebanz, & Knoblich, 2013) and to adjust to other members of the group (e.g., Loehr & Palmer, 2011; Moore & Chen, 2010). This may result in relatively general advantages in cognitive flexibility; indeed, musicians outperform nonmusicians on the Trail-Making Test (a standard switching task; Hanna-Pladdy & MacKay, 2011; see also Bugos, Perlstein, McCrae, Brophy, & Bedenbaugh, 2007) and in cued task-switching (Moradzadeh, Blumenthal, & Wiseheart, 2015). Musically trained children also show greater BOLD (blood oxygenation level dependent) responses in a task-switching paradigm (to bivalent versus univalent stimuli) in bilateral ventrolateral prefrontal cortex and supplementary motor area (Zuk, Benjamin, Kenyon, & Gaab, 2014), which are regions linked to executive functioning (Nachev, Kennard, & Husain, 2008; Nee et al., 2013).

Music Training and Inhibitory Control

Adjusting one's own playing to that of other performers not only requires the shifting of attention, but also exercising inhibitory control to monitor for conflict and control one's own performance (Jentzsch, Mkrtchian, & Kansal, 2014; Palmer, 2013). In addition, listening to or tapping complex polyrhythms (e.g., tapping a main meter with one hand and a counter meter with the other) also requires inhibitory control; for example, tapping "4 against 3" is associated with activation in Broadmann Area 47 and the anterior cingulate cortex (ACC) (Vuust, Roepstorff, Wallentin, Mouridsen, & Østergaard, 2006; Vuust, Wallentin, Mouridsen, Østergaard, & Roepstorff, 2011). Because the ACC is associated with aspects of inhibitory control (particularly with conflict monitoring; Botvinick, Braver, Barch, Carter, & Cohen, 2001), polyrhythmic processing likely involves inhibitory control processes. Thus, music training may lead to more general advantages in inhibitory control. Indeed, adult musicians show smaller interference effects in both a pitch-based auditory Stroop task (where the words "high" or "low" are sung in either a high or low pitch) and in a visual Simon Arrows task (where left- or right-pointing arrows appear on the left or right side of the screen) (Bialystok & DePape, 2009). Musicians have also been shown to outperform nonmusicians on a stop-signal task (Strait, Kraus, Parbery-Clark, Ashley, 2010; see also Moreno, Wodniecka, Tays, Alain, & Bialystok, 2014), suggesting an advantage in motor, as well as cognitive, inhibition.

Music Training and Working Memory Updating

Sight-reading music, or playing unpracticed from a score, requires looking ahead in the music to prepare one's performance, and good sight-readers typically look about four notes ahead of where they are playing (Furneaux & Land, 1999; cf. Drake & Palmer, 2000; Goolsby, 1994). Thus, sight-readers are constantly updating the contents of working memory (WM): they must keep in mind which notes are being played and which are yet-to-be played. Correspondingly, sight-reading ability is related to nonmusical measures of WM capacity, as indexed by an average of scores for operation span, reading span, rotation span, and matrix span (Meinz & Hambrick, 2010). In addition, musicians outperform nonmusicians on N-back tasks (Oechslin, Van De Ville, Lazeyras, Hauert, & James, 2013; Pallesen et al., 2010; Slevc, Davey, Buschkuehl, & Jaeggi, 2016) and show enhanced BOLD responses in the prefrontal cortex for N-back difficulty (Pallesen et al., 2010). Finally, music training predicts overall performance on measures of WM updating (i.e., combined performance on spatial N-back, letter memory, and keep track tasks, Okada, 2016; Okada & Slevc, 2018), and months of music lessons in children correlate with performance on tasks tapping inhibition, updating, and shifting (Degé, Wehrum, Stark, & Schwarzer, 2011; but see Schellenberg, 2011).

Randomized Controlled Trials: Music Training and EF

There is considerable evidence that music training is associated with advantages in all three domains of EF. However, the studies described above are correlational in nature, and have contrasted musicians and nonmusicians by recruiting participants who had already undergone various amounts and types of music training. Consequently, it is entirely possible that musicians' advantages in EF tasks do not result from training per se, but instead reflect pre-existing differences that lead some people with relatively high EF abilities to pursue music training (cf. Schellenberg, 2015).

Although the possibility of pre-existing differences and selection bias preclude claims of causation from these studies, promising results have still been found while trying to control for these factors. Therefore, these correlational findings are best interpreted as suggesting appropriate avenues to explore in experimental studies that (ideally) randomly assign participants to music training or a well-matched control training regimen. Although the few longitudinal, randomized studies that have been conducted used different tasks that span a wide range of cognitive abilities (e.g., WM, reading ability, and verbal intelligence; Moreno et al., 2011) and employed music training regimens of varying types, duration, and intensity, intriguing results have been found suggesting that taking music lessons may indeed enhance EF performance. The small body of experimental studies of music training and EFs are summarized in Table 11.1.

In the first training study to specially evaluate EFs, Bugos et al. (2007) randomly assigned 31 older adults (aged 60 to 85 years) to receive individualized piano lessons or to a no-treatment control group. The weekly piano lessons included lessons about music theory, practice with etudes, and songs from a lesson book. After six months of lessons, those who received music training showed improved cognitive functioning on the Trail-Making Test, Part B, in which subjects connect dots alternating between sequential numbers and letters in alphabetical order (i.e., A, 1, B, 2, C, 3, etc.). Given that the comparison performance was with a no-treatment control group, these results must be taken with caution (Shadish, Cook, & Campbell, 2002). Nonetheless, these data still suggest that music lessons may improve performance on nonmusical aspects of cognitive flexibility.

While it is difficult to draw conclusions from studies with a no-treatment control group given the possibility of cognitive "placebo" effects (cf. Boot, Simons, Stothart, & Stutts, 2013), more definitive conclusions can be drawn when there is a comparable active control group. One such randomized study looked at the effect of a computerized music program on the EF component of inhibitory control. Moreno et al. (2011) assigned four- to six-year-old children to either a computerized music training program, in which they learned about basic musical

Table 11.1. Summary of Experimental Studies of Music Training and EFs

Study	Bugos et al., 2007	Moreno et al., 2011	Mehr et al., 2013 Exp 1	Mehr et al., 2013; Exp 2	Roden et al., 2014	Schellenberg, 2004
Notes on Main/ Ancillary Effects	Musicians showed improved performance on Trail-Making Test Part B / Digit Symbol	Musicians had larger peak P2 amplitudes on go/no-go task than control group	No overall results when analyzing all tasks, but when looking at two spatial tasks, musicians were better at map use/ navigation test	No significant differences	Musicians were better at central executive measures than control group	Music group had larger increases in full-scale IQ than control
Citation Count*	167	280	41	41	32	751
EF Effect? Effect Size?	Yes (d = 0.38)	Yes (partial η^2 = 0.21)	Yes (d = 0.65)	No	Yes (Counting Span, partial η^2 = 0.1; Complex Span, partial η^2 = 0.12)	Yes (d = 0.35)
EF Component Tested	Cognitive Flexibility	Inhibitory Control	Working Memory	Working Memory	Working Memory	Reasoning/ Problem-Solving
Measures Assessed	Trail-Making Test - Part B	ERP - P2 on Go/no-go task	Visual form analysis Map Use/ Navigation	Visual Form Analysis Map Use/ Navigation	Counting Span Complex Span	WISC-III: Full-scale IQ
Additional Measures	Digit Symbol	Vocabulary and Block Design Subtests from WPPSI-III	PPVT-III Numerical Discrimination	PPVT-III Numerical Discrimination	One-syllable Word Span Non-Word Recall Test Corsi Block TestMatrix Span Counting Span Color Span Backwards	K-TEA BASC

(continued)

Table 11.1. Continued

Study	Bugos et al., 2007	Moreno et al., 2011	Mehr et al., 2013 Exp 1	Mehr et al., 2013; Exp 2	Roden et al., 2014	Schellenberg, 2004
Training Length (total hours/ duration)	13 hours/6 months	40 hours/4 weeks	4.5 hours/6 weeks	4.5 hours/6 weeks	58.5 hours/1.5 years	28 hours/36 weeks
Music Curriculum	Weekly individualized piano instruction (Theory, Etudes, Lesson); "broad-based music education program"	Computerized music program: primarily listening activities	"Parent-child play" modeled after Eastman Community Music School's Early Childhood Music Program	"Parent-child play" modeled after Eastman Community Music School's Early Childhood Music Program	Weekly lessons on musical instrument of their choice; different instruction based on age	Either standard keyboard lessons or Kodály voice lessons
Control Group	No treatment control	Computerized visual arts training: development of visuospatial skills	Visual arts training; "parent-child play" in the style of Emilio Reggalia	No treatment control	Natural science training program	Either drama lessons or no lessons
Age of Subjects (SD) Music Group / Control Group	69.6 (4.7) years/71.4 (6.4)	63.8 months/ 63.7 months 5.32 years/ 0.31 years 4-6 years	4.86 (0.307) years/4.64 (.268)	4.71 (0.26) years/4.72 (0.353)	7.36 (0.57) years/7.72 (0.68)	Keyboard: 6.20 (0.21) years Voice: 6.28 (.21) Drama: 6.20 (0.23) No Lessons: 6.31 (0.22)
Initial/Final Sample Size	39 / 31	71 / 48	32 / 29	46 / 45	50	144 / 132

* Google citation count as of February 22, 2017

concepts like pitch and rhythm, or a computerized visual arts training program, in which they learned about concepts like shape or color. (Note that, in contrast to most studies investigating transfer from music training, the tasks used by Moreno et al. were all perceptual, and children did not learn how to play an instrument.) After four weeks of daily training, the children who received music training showed greater gains in inhibitory control than the children who received visual arts training, as shown in larger P2 peak amplitudes in no-go trials in a go/no-go task (see Moreno & Farzan, 2015, for discussion).

Two other studies have focused on the EF component of working memory, albeit using different tasks to measure this construct. Roden, Grube, Bongard, and Kreutz (2014) assigned 7–8-year-old children to either 18 months of weekly music lessons on an instrument of their choosing or a natural science training program. They found that those in the music group performed better on a counting span test and a complex span test (Roden et al., 2014). More equivocal evidence comes from Mehr, Schachner, Katz, and Spelke (2013), who randomly assigned 4-year-old children to either music or visual arts training, both of which were designed to cover many content standards in the National Standards for Arts Education and to encourage "parent-child play" (Mehr et al., 2013, Exp. 1). After six weeks of weekly lessons, those who had music lessons performed better on a map use/navigation task, which involves holding a 2-D map with landmarks (i.e., circles that represented barrels from an aerial view) in working memory and manipulating it to navigate a 3-D world (i.e., real barrels in the room). Both tasks necessitate the storing and processing of information, and can be considered tasks tapping working memory ability. However, Mehr et al. (2013) did not find differences on an omnibus test of all four tasks assessed between the music and visual arts group but found this difference only when analyzing data from the two spatial tasks described here. Moreover, in a follow-up study comparing a new group of children randomly assigned to either music lessons or to a no-treatment control, this effect was not found (Mehr et al., 2013, Exp. 2).

Lastly, Schellenberg (2004) reported evidence that music lessons can impact measures of general intelligence, which likely involve multiple aspects of EFs (e.g., reasoning and problem solving). Six-year-old children were randomly assigned to standard keyboard lessons, Kodály voice lessons, drama lessons, or no lessons. After 36 weeks of training, those who received keyboard and voice lessons had greater increases in full-scale IQ than those taking drama lessons or no lessons. There is some debate as to whether these effects on general intelligence reflect underlying EF advantages. In a subsequent correlational study, Schellenberg (2011) found no link between music lessons and most measures of EF assessed (Phonological Fluency, Sun-Moon Stroop, Tower of Hanoi, and Wisconsin Card Sorting Test). However, these tasks may not have recruited EFs given the age group (9-12 year olds, Bialystok, 2011), and other evidence from Degé, Kubicek, and Schwarzer (2011) suggests that EFs do indeed mediate the relation between music lessons and intelligence when using different EF tasks (from the NEPSY-II; Korkman, Kirk, & Kemp, 2007). Given that these studies use different categorizations of musicians versus nonmusicians, as well as different EF tasks, it remains unclear whether increases in IQ are mediated by improvements of lower-level EFs.

Despite these promising results, other evidence suggests that music lessons may not benefit cognitive processing. As mentioned earlier, results from Mehr et al.'s (2013) randomized control experiments were equivocal overall. Furthermore, Rickard, Bambrick, and Gill (2012) found no differences between 10–13-year-old children who took 6.5 months of classroom-based music lessons and children who instead took 6.5 months of drama or art lessons on tests of verbal memory as well as IQ (Rickard et al., 2012). However, these studies examined only classroom-based group music lessons, which may place different demands on EFs than private music lessons.

In sum, the small body of extant randomized control studies on EFs and music training have found somewhat mixed results. Because of the disparity in the EF measures used, EF components studied, and the length, intensity, and type of training administered, it is difficult to compare and draw general conclusions about these results. Additional work is thus clearly needed. Nevertheless, these few studies combined with a larger body of correlational evidence (see above) suggests that music training might indeed show transfer to tasks of EFs. And because EFs are predictive of multiple positive outcomes (e.g., success in school, quality of life, and mental health; see Diamond, 2013), it would be tremendously valuable to know if, and how, musical experiences could function as a kind of EF training. Below, we consider several relevant questions posed by Diamond and Ling's challenge chapter (2019) from the perspective of music training and EFs. It is our hope that these considerations can help guide future research on if and how music training might serve not only as an intrinsically valuable pursuit, but also as an enjoyable "naturalistic" EF training program.

Who benefits most from what? Does the answer differ by age, gender, or cultural group of the participants, or any other variable? For example, are different methods for improving EFs more beneficial at different ages?

Given the limited populations investigated in previous randomized studies of music training, it is not yet clear if certain age groups benefit more from music training. All but one of the randomized studies listed in Table 11.1 looked at young children (age range of 4-7 years) because EFs are thought to be more malleable during childhood and adolescence. However, one study on older adults aged 60 to 85 did find EF improvements after piano lessons as well (albeit without an active control group for comparison; Bugos et al., 2007). There is also correlational evidence that music training may help prevent perceptual cognitive decline in the aging brain (Kraus & White-Schwoch, 2014), which suggests that music training may be beneficial for the older adult population as well. A likely, but so far untested, possibility is that beneficial effects of musical experience will be more pronounced in populations with relatively lower EF abilities, such as children, elderly adults, or neuropsychological patients, as has been found regarding "bilingual advantages" in inhibitory control (cf. Bialystok, Craik, & Luk, 2012).

Differences related to gender and culture are yet unknown: no study examining music training has reported differential EF benefits by gender, and all studies to date of EF and music training teach music grounded in the Western tonal system. Presumably, different cultures would benefit more from training in their own musical system, but it is not necessarily the case that different musical systems place the same demands on EFs.

If a real-life activity improves EFs (be it theater, martial arts, soccer, a school curriculum, or something else), what aspect(s) of the program are responsible for that? Why did the program improve EFs?

As stated above, future studies should begin to investigate the effect of different aspects of music training (e.g., sight-reading and ensemble playing) on EFs. By using well-controlled, randomized studies to examine various components of music training, we will be able to determine which components are responsible for improving EFs. Thus, we will have a better understanding of how music training, as a whole, taxes and affects EFs.

While it is of utmost importance to perform well-controlled longitudinal studies in the lab (to determine causal relations), it is equally vital to perform quasi-experimental or observational studies of music training in the field. Even though these types of studies will suffer from selection bias, it is still important to understand the types of people who choose to study music and how this training affects them. Specifically, these more "naturalistic" music training programs should also produce insights into the effects of real-life music training on EFs. Furthermore, because taking music lessons is associated not only with cognitive benefits, but also with various social effects (e.g., music lessons are associated with increased quality of school life [Eerola & Eerola, 2014] and joint music making leads to more prosocial behavior [Kirschner & Tomasello, 2010]), examination of a range of noncognitive abilities affected by music training is also needed.

What makes the best control condition?

As has been discussed in other domains of training and transfer (e.g., video games: Boot et al., 2013), a proper control condition needs to be equally engaging as music training (i.e., a convincing placebo control). Additionally, if the control condition does not account for all differential expectations, it will confound the experiment, and no causal claims can be made about the music training group (Boot et al., 2013). Therefore, a proper control condition should also incorporate as many nonmusical aspects of the music training program as possible (e.g., motor training, engaging lessons, self-efficacy, any social aspects, etc.) so that any differences in the outcome variables can be attributed solely to the music lessons. (Of course, potential EF benefits from these other aspects of music training are also worthy of investigation, even if not specific to the *musical* aspect of training.) Early work on music training did not use active control groups and so offers only suggestive evidence that music training enhances aspects of cognitive ability in children (Bilhartz et al., 2000; Costa-Giomi, 1999; 2004) and older adults (Bugos et al., 2007). However, more recent studies have used drama lessons, natural science lessons, and visual arts lessons for active control groups (Schellenberg, 2006; Roden et al., 2014 Moreno et al., 2011; Mehr et al., 2013), and, in an ongoing large longitudinal study, sports training (Habibi et al., 2014; Habibi, Cahn, Damasio, & Damasio, 2016). The variety in these training programs is advantageous as they are likely to be effective controls for different nonmusical aspects of music training.

What are the best doses, frequencies, and durations for different programs intended to improve EFs?
Does the answer differ by type of program, EF component (e.g., WM or response inhibition), the age, gender, or cultural group of the participants, or any other variable? For example, does the optimal dose or frequency differ by age?

Because of the shortage of longitudinal, randomized studies with music training programs, there is a lack of converging evidence for effects of different types of music training on each of the three EF components. Furthermore, the length and duration of training regimens greatly differs in the few randomized studies that have been conducted (see Table 11.1). Extant studies have used training programs ranging from only 4.5 hours over six weeks (Mehr et al., 2013) to 45 minutes weekly over 1.5 years (Roden et al., 2014). Improved performance on a switching task was found after six months of lessons (Bugos et al. 2007), and increased performance on an inhibition task was found after only 28 days of more intense daily training (Moreno et al., 2011). Furthermore, effects such as increased full-scale IQ have been seen in musicians after as little as a total of 28 hours (over 36 weeks) of music lessons (Schellenberg, 2004). However, Mehr et al. (2013) failed to find any effects on tasks of receptive vocabulary, numerical quantity discrimination, and spatial reasoning after 4.5 hours of lessons over six weeks (see Table 11.1 for summary), which suggests that music training programs may need to exceed this low dosage to see transfer effects.

Taken together, these findings suggest that training studies should employ at least 13 hours of lessons (as in Bugos et al., 2007) and should ideally allow for practice daily (Moreno et al., 2011) or weekly (Bugos et al., 2007). Because many correlations have been found between cognitive processes and real-world music lessons, which typically occur once a week, and because it is important to maximize the external validity of these studies, it seems reasonable to recommend that a music training program should administer lessons at least once per week.

> What factors affect how long benefits last? Does it matter which method is used to try to sustain benefits (e.g., booster sessions or embed in daily activities)? If refresher or booster sessions are used, at what intervals should they be given and for what duration? Do the answers to these questions differ by type of program, EF component (e.g., WM or response inhibition), the age, gender, or cultural group of the participants, or other variables?

Unfortunately, very little evidence exists regarding the degree to which musically-induced EF benefits persist (Costa-Giomi, 2014). To our knowledge, the only longitudinal music training study using delayed testing was Bugos et al. (2007), who found that only improvements on the digit symbol task, a test of processing speed, remained three months after training had stopped (without a booster session). However, improvements on other tasks improved by training (specifically, the Trail-Making Test, an EF task) did not persist after the three-month delay. Other evidence from a correlational study showed that adults who had taken music lessons during childhood had more robust brain stem responses to speech in noise than adults who had not taken music lessons as children (Skoe & Kraus, 2012), suggesting that neural changes may persist for several years after music training has been discontinued. However, more work is clearly necessary to examine the longevity of these effects.

What about training people in strategies to minimize the need for EFs, so that people do not have to expend so much effort trying to exercise EFs?

Although previous work has not explicitly investigated this topic, evidence that music training leads to improvements in EF task performance could also be interpreted as music training minimizing the need for EFs in those tasks. For example, developing skill in reading and playing music may allow musicians to do these tasks with less cognitive effort (e.g., via expertise-dependent strategies for storing and processing musical material; e.g., Barry & Hallam, 2003; Chaffin & Imreh, 1997; 2002; Nielsen, 2001), which is interesting from the perspective of musical skill acquisition. In addition, this raises the possibility that developing strategies that minimize reliance on EFs in musical contexts might transfer to some other sorts of EF-demanding tasks, which would be a quite different explanation for transfer from the EF-enhancement account discussed here.

Will the type of program end up mattering more or the way it is done?

The type of program and the way it is enacted are likely equally important. To see benefits, any training program should be done in a way that constantly taxes and strengthens EFs and continues to increase in difficulty (Diamond, 2013). In terms of a music training program, then, music lessons should incorporate skills that build upon one another, with gradual increases in complexity (Bugos et al., 2007). In addition, one presumed advantage of music training interventions is the intrinsic value of musical experiences. That is, people engage in playing instruments or singing because it is an enjoyable activity. For this reason, it is important that music training programs teach self-efficacy and motivate participants to practice (Bugos et al., 2007).

If we expect to see near-transfer from music lessons to specific components of EF, then a music training program should be one that contains lessons that tax each specific underlying component of EF. For example, to produce transfer from music training to working memory updating, music lessons might usefully include practice reading music and sight-reading. As discussed above, these processes require one to visually scan ahead in the musical score (Goolsby, 1994) and hold the representations of these notes in mind while simultaneously updating the notes that have already been and are yet to be played. If, indeed, practice reading music and sight-reading can strengthen working memory updating, one should expect to see transfer effects to tasks of working memory updating.

Similarly, to investigate transfer effects to cognitive flexibility, a music training program might include practice playing in an ensemble, which may tap cognitive flexibility (e.g., Loehr & Palmer, 2011; Moore & Chen, 2010). Lastly, a music training program targeting inhibitory control might usefully include practice with complex polyrhythms (e.g., Vuust et al., 2006; 2011).

Conclusion

Because many aspects of music training (e.g., reading music, practicing, playing in ensembles) are likely to engage EFs, and because many people enjoy music lessons solely for their intrinsic value, music training might be considered a relatively easy to implement "naturalistic" EF training program. Given the pervasive role of EFs that may underlie learning to play an instrument, future research should consider the many different types of processes engaged in music training programs as well as the various EF components each process might affect. To use music training as a "naturalistic" EF training program, we must first understand which specific aspects of music lessons are causing the benefits in specific components of EF. Fortunately, there is growing interest in the cognitive underpinnings of musical tasks and in the potential for nonmusical effects of music training. This blossoming area of research will likely paint a clearer picture of the exact associations between music training and EFs.

References

Andersson, P. K., Kristensson, P., Wästlund, E., & Gustafsson, A. (2012). Let the music play or not: The influence of background music on consumer behavior. *Journal of Retailing and Consumer Services, 19*(6), 553–560. doi:10.1016/j.jretconser.2012.06.010

Au, J., Sheehan, E., Tsai, N., Duncan, G. J., Buschkuehl, M., & Jaeggi, S. M. (2015). Improving fluid intelligence with training on working memory: A meta-analysis. *Psychonomic Bulletin & Review, 22*(2), 366–377. doi:10.3758/s13423-014-0699-x

Barry, N., & Hallam, S. (2002) Practising. In R. Parncutt and G. McPherson (Eds.), *The Science and Psychology of Music Performance: Creative Strategies for Teaching and Learning* (pp. 151–166). Oxford University Press: Oxford.

Bialystok, E. (2011). Commentary: How does experience change cognition? Evaluating the evidence. *British Journal of Psychology, 102*, 303–305. doi:10.1111/j.2044-8295.2011.02008.x

Bialystok, E., Craik, F. I. M., & Luk, G. (2012). Bilingualism: Consequences for mind and brain. *Trends in Cognitive Sciences, 16*(4), 240–250. doi:10.1016/j.tics.2012.03.001

502   DEVELOPMENTAL PERSPECTIVE

Bialystok, E., & Depape, A.-M. (2009). Musical expertise, bilingualism, and executive functioning. *Journal of Experimental Psychology: Human Perception and Performance*, 35(2), 565–574. doi:10.1037/a0012735

Bilhartz, T. D., Bruhn, R. A., & Olson, J. E. (2000). The effect of early music training on child cognitive development. *Journal of Applied Developmental Psychology*, 20(4), 615–636. doi:10.1016/S0193-3973(99)00033-7

Boot, W. R., Simons, D. J., Stothart, C., & Stutts, C. (2013). The pervasive problem with placebos in psychology: Why active control groups are not sufficient to rule out placebo effects. *Perspectives on Psychological Science*, 8(4), 445–454. doi:10.1177/1745691613491271

Botvinick, M. M., Braver, T. S., Barch, D. M., Carter, C. S., & Cohen, J. D. (2001). Conflict monitoring and cognitive control. *Psychological Review*, 108(3), 624–652. doi:10.1037//0033-295X.108.3.624

Bugos, J. A., Perlstein, W. M., McCrae, C. S., Brophy, T. S., & Bedenbaugh, P. H. (2007). Individualized piano instruction enhances executive functioning and working memory in older adults. *Aging & Mental Health*, 11(4), 464–471. doi:10.1080/13607860601086504

Bugos, J., & Mostafa, W. (2011). Musical training enhances information processing speed. *Bulletin of the Council for Research in Music Education*, 187, 7–18. doi:10.2307/41162320

Burton-Hill, A. (2012, June 14). José Antonio Abreu on Venezuela's El Sistema miracle. *The Guardian*. http://www.theguardian.com/music/2012/jun/14/abreu-el-sistema-venezuela-interview-clemency-burton-hill

Butzlaff, R. (2000). Can music be used to teach reading? *Journal of Aesthetic Education*, 34(3–4), 167–178. doi:10.2307/3333642

Chaffin, R., & Imreh, G. (1997). "Pulling Teeth and Torture": Musical Memory and Problem Solving. *Thinking & Reasoning*, 3(4), 315–336. doi:10.1080/135467897394310

Chaffin, R., & Imreh, G. (2002). Practicing perfection: Piano performance as expert memory. *Psychological Science*, 13(4), 342–349.

Chan, A. S., Ho, Y. C., & Cheung, M. C. (1998). Music training improves verbal memory. *Nature*, 396(6707), 128. doi:10.1038/24075

Chein, J. M., & Morrison, A. B. (2010). Expanding the mind's workspace: Training and transfer effects with a complex working memory span task. *Psychonomic Bulletin & Review*, 17(2), 193–199. doi:10.3758/PBR.17.2.193

Costa-Giomi, E. (1999). The effects of three years of piano instruction on children's cognitive development. *Journal of Research in Music Education*, 47(3), 198–212. doi:10.2307/3345779

Costa-Giomi, E. (2004). Effects of three years of piano instruction on children's academic achievement, school performance and self-esteem. *Psychology of Music*, 32(2), 139–152. doi:10.1177/0305735604041491

Costa-Giomi, E. (2014). The long-term effects of childhood music instruction on intelligence and general cognitive abilities. *Update: Applications of Research in Music Education*, 33(2), 20–26. doi:10.1177/8755123314540661

Degé, F., Kubicek, C., & Schwarzer, G. (2011). Music lessons and intelligence: A relation mediated by executive functions. *Music Perception*, 29(2), 195–201. doi:10.1525/mp.2011.29.2.195

Degé, F., Wehrum, S., Stark, R., & Schwarzer, G. (2011). The influence of two years of school music training in secondary school on visual and auditory memory.

European Journal of Developmental Psychology, 8(5), 608–623. doi:10.1080/17405 629.2011.590668

Diamond, A. (2013). Executive functions. *Annual Review of Psychology*, 64(1), 135–168. doi:10.1146/annurev-psych-113011-143750

Diamond, A., Barnett, W. S., Thomas, J., & Munro, S. (2007). Preschool program improves cognitive control. *Science, 318*(5855), 1387–1388. doi:10.1126/science.1151148

Diamond, A., & Ling, D. (2019). Review of the evidence on fundamental questions surrounding and efforts to improve executive functions (including working memory). In J. Novick, M. Bunting, M. Dougherty, & R. W. Engle (Eds.), *Cognitive and working memory training: Perspectives from psychology, neuroscience, and human development*. New York, NY: Oxford University Press.

Dougherty, M. R., Hamovitz, T., & Tidwell, J. W. (2015). Reevaluating the effectiveness of n-back training on transfer through the Bayesian lens: Support for the null. *Psychonomic Bulletin & Review, 22*(3), 1–11. doi:10.3758/s13423-015-0865-9

Drake, C., & Palmer, C. (2000). Skill acquisition in music performance: Relations between planning and temporal control. *Cognition, 74*(1), 1–32. doi:10.1016/S0010-0277(99)00061-X

Eerola, P.-S., & Eerola, T. (2014). Extended music education enhances the quality of school life. *Music Education Research, 16*(1), 88–104. doi:10.1080/14613808.2013.829428

Furneaux, S., & Land, M. F. (1999). The effects of skill on the eye-hand span during musical sight-reading. *Proceedings of the Royal Society B: Biological Sciences, 266*(1436), 2435–2440. doi:10.1098/rspb.1999.0943

Gaser, C., & Schlaug, G. (2003). Brain structures differ between musicians and non-musicians. *The Journal of Neuroscience, 23*(27), 9240–9245. doi:10.1523/JNEUROSCI.23-27-09240.2003

Goolsby, T. W. (1994). Eye movement in music reading: Effects of reading ability, notational complexity, and encounters. *Music Perception, 12*(1), 77–96. doi:10.2307/40285756

Habibi, A., Cahn, B. R., Damasio, A., & Damasio, H. (2016). Neural correlates of accelerated auditory processing in children engaged in music training. *Developmental Cognitive Neuroscience, 21*, 1–14. doi:10.1016/j.dcn.2016.04.003

Habibi, A., Ilari, B., Crimi, K., Metke, M., Kaplan, J. T., Joshi, A. A., . . . Damasio, H. (2014). An equal start: Absence of group differences in cognitive, social, and neural measures prior to music or sports training in children. *Frontiers in Human Neuroscience, 8.* doi:10.3389/fnhum.2014.00690

Hanna-Pladdy, B., & MacKay, A. (2011). The relation between instrumental musical activity and cognitive aging. *Neuropsychology, 25*(3), 378–386. doi:10.1037/a0021895

Hannon, E. E., & Trainor, L. J. (2007). Music acquisition: Effects of enculturation and formal training on development. *Trends in Cognitive Sciences, 11*(11), 466–472. doi:10.1016/j.tics.2007.08.008

Harrison, T. L., Shipstead, Z., Hicks, K. L., Hambrick, D. Z., Redick, T. S., & Engle, R. W. (2013). Working memory training may increase working memory capacity but not fluid intelligence. *Psychological Science, 24*(12), 2409–2419. doi:10.1177/0956797613492984

Ho, Y.-C., Cheung, M.-C., & Chan, A. S. (2003). Music training improves verbal but not visual memory: Cross-sectional and longitudinal explorations in children. *Neuropsychology, 17*(3), 439–450. doi:10.1037/0894-4105.17.3.439

Hudziak, J. J., Albaugh, M. D., Ducharme, S., Karama, S., Spottswood, M., Crehan, E., . . . Botteron, K. N. (2014). Cortical thickness maturation and duration of music training: Health-promoting activities shape brain development. *Journal of the*

American Academy of Child & Adolescent Psychiatry, 53(11), 1153–1161. doi:10.1016/j. jaac.2014.06.015

Hyde, K. L., Lerch, J., Norton, A., Forgeard, M., Winner, E., Evans, A. C., & Schlaug, G. (2009). Musical training shapes structural brain development. *The Journal of Neuroscience, 29*(10), 3019–3025. doi:10.1523/JNEUROSCI.5118-08.2009

Jaeggi, S. M., Buschkuehl, M., Jonides, J., & Perrig, W. J. (2008). Improving fluid intelligence with training on working memory. *Proceedings of the National Academy of Sciences of the United States of America, 105*(19), 6829–6833. doi:10.1073/pnas.0801268105

Jakobson, L. S., Cuddy, L. L., & Kilgour, A. R. (2003). Time tagging: A key to musicians' superior memory. *Music Perception, 20*(3), 307–313. doi:10.1525/mp.2003.20.3.307

Jakobson, L. S., Lewycky, S. T., Kilgour, A. R., & Stoesz, B. M. (2008). Memory for verbal and visual material in highly trained musicians. *Music Perception, 26*(1), 41–55. doi:10.1525/mp.2008.26.1.41

Jentzsch, I., Mkrtchian, A., & Kansal, N. (2014). Improved effectiveness of performance monitoring in amateur instrumental musicians. *Neuropsychologia, 52*(100), 117–124. doi:10.1016/j.neuropsychologia.2013.09.025

Karageorghis, C. I., & Priest, D.-L. (2012). Music in the exercise domain: A review and synthesis (part I). *International Review of Sport and Exercise Psychology, 5*(1), 44–66. doi:10.1080/1750984X.2011.631026

Kirschner, S., & Tomasello, M. (2010). Joint music making promotes prosocial behavior in 4-year-old children. *Evolution and Human Behavior, 31*(5), 354–364. doi:10.1016/j. evolhumbehav.2010.04.004

Korkman, M., Kirk, J., & Kemp, S. (2007). *A developmental neuropsychological assessment; NEPSY-II* (2nd ed.). San Antonio, TX: Harcourt Assessment, Inc.

Kraus, N., & White-Schwoch, T. (2014). Music training: Lifelong investment to protect the brain from aging and hearing loss. *Acoustics Australia, 42*(2), 117–123.

Lamb, S. J., & Gregory, A. H. (1993). The relationship between music and reading in beginning readers. *Educational Psychology, 13*(1), 19–27. doi:10.1080/0144341930130103

Lehto, J. E., Juujärvi, P., Kooistra, L., & Pulkkinen, L. (2003). Dimensions of executive functioning: Evidence from children. *British Journal of Developmental Psychology, 21*(1), 59–80. doi:10.1348/026151003321164627

Loehr, J. D., Kourtis, D., Vesper, C., Sebanz, N., & Knoblich, G. (2013). Monitoring individual and joint action outcomes in duet music performance. *Journal of Cognitive Neuroscience, 25*(7), 1049–1061. doi:10.1162/jocn_a_00388

Loehr, J. D., & Palmer, C. (2011). Temporal coordination between performing musicians. *The Quarterly Journal of Experimental Psychology, 64*(11), 2153–2167. doi:10.1080/17 470218.2011.603427

Logue, S. F., & Gould, T. J. (2014). The neural and genetic basis of executive function: Attention, cognitive flexibility, and response inhibition. *Pharmacology Biochemistry and Behavior, 123*, 45–54. doi:10.1016/j.pbb.2013.08.007

Mehr, S. A., Schachner, A., Katz, R. C., & Spelke, E. S. (2013). Two randomized trials provide no consistent evidence for nonmusical cognitive benefits of brief preschool music enrichment. *PLoS One, 8*(12), e82007. doi:10.1371/journal.pone.0082007

Meinz, E. J., & Hambrick, D. Z. (2010). Deliberate practice is necessary but not sufficient to explain individual differences in piano sight-reading skill: The role of working memory capacity. *Psychological Science, 21*(7), 914–919. doi:10.1177/0956797610373933

Milliman, R. E. (1982). Using background music to affect the behavior of supermarket shoppers. *Journal of Marketing, 46*(3), 86–91. doi:10.2307/1251706

Miyake, A., & Friedman, N. P. (2012). The nature and organization of individual differences in executive functions: Four general conclusions. *Current Directions in Psychological Science*, *21*(1), 8–14. doi:10.1177/0963721411429458

Miyake, A., Friedman, N. P., Emerson, M. J., Witzki, A. H., Howerter, A., & Wager, T. D. (2000). The unity and diversity of executive functions and their contributions to complex "frontal lobe" tasks: A latent variable analysis. *Cognitive Psychology*, *41*(1), 49–100. doi:10.1006/cogp.1999.0734

Moore, G. P., & Chen, J. (2010). Timings and interactions of skilled musicians. *Biological Cybernetics*, *103*(5), 401–414. doi:10.1007/s00422-010-0407-5

Moradzadeh, L., Blumenthal, G., & Wiseheart, M. (2015). Musical training, bilingualism, and executive function: A closer look at task switching and dual-task performance. *Cognitive Science*, *39*(5), 992–1020. doi:10.1111/cogs.12183

Moreno, S., Bialystok, E., Barac, R., Schellenberg, E. G., Cepeda, N. J., & Chau, T. (2011). Short-term music training enhances verbal intelligence and executive function. *Psychological Science*, *22*(11), 1425–1433. doi:10.1177/0956797611416999

Moreno, S., & Farzan F. (2015). Music training and inhibitory control: A multidimensional model. *Annals of the New York Academy of Sciences*, *1337*(1), 147–152. doi:10.1111/nyas.12674

Moreno, S., Wodniecka, Z., Tays, W., Alain, C., & Bialystok, E. (2014). Inhibitory control in bilinguals and musicians: Event related potential (ERP) evidence for experience-specific effects. *PLoS One*, *9*(4), e94169. doi:10.1371/journal.pone.0094169

Nachev, P., Kennard, C., & Husain, M. (2008). Functional role of the supplementary and pre-supplementary motor areas. *Nature Reviews: Neuroscience*, *9*(11), 856–869. doi:10.1038/nrn2478

Nee, D. E., Brown, J. W., Askren, M. K., Berman, M. G., Demiralp, E., Krawitz, A., & Jonides, J. (2013). A meta-analysis of executive components of working memory. *Cerebral Cortex*, *23*(2), 264–282. doi:10.1093/cercor/bhs007

Nielsen, S. (2001). Self-regulating learning strategies in instrumental music practice. *Music Education Research*, *3*(2), 155–167. doi:10.1080/14613800120089223

Oechslin, M. S., Van De Ville, D., Lazeyras, F., Hauert, C. A., & James, C. E. (2013). Degree of musical expertise modulates higher order brain functioning. *Cerebral Cortex*, *23*(9), 2213–2224. doi:10.1093/cercor/bhs206

Okada, B. M. (2016). *Musical training and executive functions*. (Master's thesis). Retrieved from the digital repository at the University of Maryland. http://hdl.handle.net/1903/18933

Okada, B. M., & Slevc, L. R. (2018). Individual differences in musical training and Executive Functions: A latent variable approach. *Memory & Cognition, 46*, 1076-1092. doi:10.3758/s13421-018-0822-8

Owen, A. M., Hampshire, A., Grahn, J. A., Stenton, R., Dajani, S., Burns, A. S., . . . Ballard, C. G. (2010). Putting brain training to the test. *Nature*, *465*(7299), 775–778. doi:10.1038/nature09042

Pallesen, K. J., Brattico, E., Bailey, C. J., Korvenoja, A., Koivisto, J., Gjedde, A., & Carlson, S. (2010). Cognitive control in auditory working memory is enhanced in musicians. *PLoS One*, *5*(6), e11120. doi:10.1371/journal.pone.0011120

Palmer, C. (2013). Music performance: Movement and coordination. In D. Deutsch (Ed.), *The psychology of music* (3rd ed., pp. 405–422). San Diego, CA: Academic Press.

Peretz, I., & Zatorre, R. J. (2005). Brain organization for music processing. *Annual Review of Psychology*, *56*, 89–114. doi:10.1146/annurev.psych.56.091103.070225

Pietschnig, J., Voracek, M., & Formann, A. K. (2010). Mozart effect-Shmozart effect: A meta-analysis. *Intelligence, 38*(3), 314–323. doi:10.1016/j.intell.2010.03.001

Redick, T. S., Shipstead, Z., Harrison, T. L., Hicks, K. L., Fried, D. E., Hambrick, D. Z., . . . & Engle, R. W. (2013). No evidence of intelligence improvement after working memory training: A randomized, placebo-controlled study. *Journal of Experimental Psychology: General, 142*(2), 359–379. doi:10.1037/a0029082

Rickard, N. S., Bambrick, C. J., & Gill, A. (2012). Absence of widespread psychosocial and cognitive effects of school-based music instruction in 10–13-year-old students. *International Journal of Music Education, 30*(1), 57–78. doi:10.1177/0255761411431399

Roden, I., Grube, D., Bongard, S., & Kreutz, G. (2014). Does music training enhance working memory performance? Findings from a quasi-experimental longitudinal study. *Psychology of Music, 42*(2), 284–298. doi:10.1177/0305735612471239

Sala, G., & Gobet, F. (2017). When the music's over. Does music skill transfer to children's and young adolescents' cognitive and academic skills? A meta-analysis. *Educational Research Review, 20*, 55–67. doi:10.1016/j.edurev.2016.11.005

Schellenberg, E. G. (2004). Music lessons enhance IQ. *Psychological Science, 15*(8), 511–514. doi:10.1111/j.0956-7976.2004.00711.x

Schellenberg, E. G. (2006). Long-term positive associations between music lessons and IQ. *Journal of Educational Psychology, 98*(2), 457–468. doi:10.1037/0022-0663.98.2.457

Schellenberg, E. G. (2011). Examining the association between music lessons and intelligence. *British Journal of Psychology, 102*(3), 283–302. doi:10.1111/j.2044-8295.2010.02000.x

Schellenberg, E. G. (2015). Music training and speech perception: A gene-environment interaction. *Annals of the New York Academy of Sciences, 1337*(1), 170–177. doi:10.1111/nyas.12627

Schellenberg, E. G., & Moreno, S. (2009). Music lessons, pitch processing, and *g*. *Psychology of Music, 38*(2) 209–221. doi:10.1177/0305735609339473

Schellenberg, E. G., & Winner, E. (Eds.) (2011). Music training and nonmusical abilities: Introduction. *Music Perception, 29*(2), 129–132. doi:10.1525/mp.2011.29.2.129

Shadish, W. R., Cook, T. D., & Campbell, D. T. (2002). *Experimental and quasi-experimental designs for generalized causal inference*. Boston, MA: Houghton Mifflin Company.

Shih, Y. N., Huang, R. H., & Chiang, H. Y. (2012). Background music: Effects on attention performance. *Work, 42*(4), 573–578. doi:10.3233/WOR-2012-1410

Skoe, E., & Kraus, N. (2012). A little goes a long way: How the adult brain is shaped by musical training in childhood. *Journal of Neuroscience, 32*(34), 11507–11510. doi:10.1523/JNEUROSCI.1949-12.2012

Slevc, L. R., Davey, N. S., Buschkuehl, M., & Jaeggi, S. M. (2016). Tuning the mind: Exploring the connections between musical ability and executive functions. *Cognition, 152*, 199–211. doi:10.1016/j.cognition.2016.03.017

Slevc, L. R., & Okada, B. M. (2014). Nonmusical abilities. In W. Thompson (Ed.), *Music in the social and behavioral sciences: An encyclopedia, Vol. 12* (pp. 815–817). Thousand Oaks, CA: SAGE Publications.

Slevc, L. R., & Okada, B. M. (2015). Processing structure in language and music: A case for shared reliance on cognitive control. *Psychonomic Bulletin & Review, 22*(3), 637–652. doi:10.3758/s13423-014-0712-4

Strait, D., & Kraus, N. (2011). Playing music for a smarter ear: Cognitive, perceptual and neurobiological evidence. *Music Perception, 29*(2), 133–146. doi:10.1525/mp.2011.29.2.133

Strait, D. L., Kraus, N., Parbery-Clark, A., & Ashley, R. (2010). Musical experience shapes top-down auditory mechanisms: Evidence from masking and auditory attention performance. *Hearing Research, 261,* 22–29. doi:10.1016/j.heares.2009.12.021

Thompson, W. F., Schellenberg, E. G., & Husain, G. (2001). Arousal, mood, and the Mozart effect. *Psychological Science, 12*(3), 248–251. doi:10.1111/1467-9280.00345

Vaughn, K. (2000). Music and mathematics: Modest support for the oft-claimed relationship. *Journal of Aesthetic Education, 34*(3–4), 149–166. doi:10.2307/3333641

Vuust, P., Roepstorff, A., Wallentin, M., Mouridsen, K., & Østergaard, L. (2006). It don't mean a thing . . . Keeping the rhythm during polyrhythmic tension, activates language areas (BA47). *NeuroImage, 31*(2), 832–841. doi:10.1016/j.neuroimage.2005.12.037

Vuust, P., Wallentin, M., Mouridsen, K., Østergaard, L., & Roepstorff, A. (2011). Tapping polyrhythms in music activates language areas. *Neuroscience Letters, 494*(3), 211–216. doi:10.1016/j.neulet.2011.03.015

Zuk, J., Benjamin, C., Kenyon, A., & Gaab, N. (2014). Behavioral and neural correlates of executive functioning in musicians and non-musicians. *PLoS One, 9*(6), 1–14. doi:10.1371/journal.pone.0099868

12

The Effectiveness of Training
in Task Switching

New Insights and Open Issues From a Life-Span View

Jutta Kray and Sandra Dörrenbächer

Life-Span Changes in Task Switching

As Diamond and Ling outline in Chapter 8, the ability to flexibly switch between task rules and mental sets is seen as a fundamental component of cognitive control (Diamond, 2013; Miyake et al., 2000) and is known to be associated with fluid intelligence (Duncan, Burgess, & Emslie, 1995). While researchers focusing on developmental changes of cognitive flexibility in early and middle childhood have mainly applied a specific variant of task switching, the dimensional change card sort (DCCS) task (for a review, see Zelazo, 2006), life-span researchers have mainly used variants of the task-switching paradigm (for recent reviews on childhood development, see Kray & Ferdinand, 2013, and on adult development, see Kray & Ferdinand, 2014).

The advantage of the task-switching paradigm is that it allows researchers to measure separately different cognitive control processes, such as task preparation, interference, and switching processes within the same experimental paradigm (for a review, see Kiesel et al., 2010). Although there now exist quite a number of different variants of the task-switching paradigm, a majority of studies apply an experimental design in which participants have to switch between two tasks, A and B, within the same block of trials (i.e., mixed-task blocks) as well as to perform only one of the two tasks, A or B, within a block (i.e., single-task blocks). This allows researchers to determine two types of costs: mixing costs and switching costs. Mixing costs are usually defined as differences between performance in mixed-task blocks versus single-task blocks (also termed *global* or *general switch costs*; cf. Kray & Lindenberger, 2000; Mayr, 2001),[1] and switching

---

[1] It should be noted that there are different definitions of mixing costs (Cragg & Chevalier, 2012; Marí-Beffa & Kirkham, 2014): mixing costs referred to as "general," or "global," are measured as the difference in performance on single-block trials and mixed-block trials (i.e., the latter encompassing both switch as well as nonswitch trials); mixing costs referred to as "nonswitch-specific," however, are measured as the difference between single-block trials and only nonswitch trials from mixed blocks.

costs are defined as differences between performance on switch trials (a change of task A to B or B to A) and on nonswitch trials (a repetition of the same task A) within mixed-task blocks (also termed local or specific switch costs). Mixing costs are assumed to reflect control processes that are required for maintaining multiple task sets and for selecting between them (i.e., being in a switching situation), while switching costs are associated with the reconfiguration of tasks itself (i.e., performing a task switch).

Regarding life-span changes in task switching, it is now well documented that age differences in mixing costs show a clear U-shaped developmental curve across the life span that is less pronounced for switching costs. During childhood, nearly all studies investigating older children from middle childhood on find larger age-related changes in mixing costs than in switching costs, suggesting the ability to switch between task and rules develops earlier than the ability to maintain and select task sets (Cepeda, Kramer, & Gonzalez de Sather, 2001; Crone, Ridderinkhof, Worm, Somsen, & van der Molen, 2004; Dibbets & Jolles, 2006; Huizinga & van der Molen, 2007; Karbach & Kray, 2007; Kray, Eber, & Karbach, 2008; Kray, Eber, & Lindenberger, 2004; Kray, Karbach, & Blaye, 2012; Manzi, Nessler, Czernochowski, & Friedman, 2011; Reimers & Maylor, 2005). During adulthood, the majority of task-switching studies report substantial age differences in mixing costs, suggesting age-related impairments in maintaining and selecting between task sets (Buchler, Hoyer, & Cerella, 2008; Cepeda et al., 2001; Koch, 2012; Kray, 2006; Kray & Lindenberger, 2000; Kray et al., 2004, 2008; Lawo, Philipp, Schuch, & Lien, Ruthruff, & Kuhns, 2008; Mayr, 2001; Meiran, Gotler, & Perlman, 2001; Reimers & Maylor, 2005; van Asselen & Ridderinkhof, 2000; but see Kray, Li, & Lindenberger, 2002). Although some studies also found age differences in switching costs (e.g., Meiran et al., 2001), most of the studies showed smaller age differences in switching costs than in mixing costs (e.g., Karayanidis, Jamadar, & Sanday, 2013; Karayanidis, Whitson, Heathcote, & Michie, 2011; Kray & Lindenberger, 2000; Kray et al., 2004, 2008; Kray, Eppinger, & Mecklinger, 2005; Lien et al., 2008; Mayr, 2001; Mayr & Liebscher, 2001; Reimers & Maylor, 2005; Whitson, Karayanidis, & Michie, 2011; Whitson et al., 2014), and often age differences in switching costs failed to reach significance (e.g., Hahn, Anderson, & Kramer, 2004; Kray & Lindenberger, 2000; Salthouse, Fristoe, McGuthry, & Hambrick, 1998). This indicates that older adults show fewer process-specific limitations in executing a task switch, but they show problems in being in a switch situation requiring the selection between tasks and their maintenance (for a meta-analysis, see Wasylyshyn, Verhaeghen, & Sliwinski, 2011). Hence, in light of these findings, it seems particularly promising to foster cognitive processes required for being in a switching situation to reduce age-related differences in task switching.

As is outlined in detail in the overview in Chapter 8 by Diamond and Ling, given the importance of cognitive control for early academic and school success as well as for cognitive functioning in old age, we should really care to find effective ways to improve cognitive control by optimal cognitive intervention programs.

Flexibility and Plasticity of Task-Switching Performance Across the Life Span

The possible range of practice-induced improvements in task-switching abilities strongly depends on an individual's potential for cognitive flexibility as well as for plasticity (cf. Lövdén, Bäckman, Lindenberger, Schaefer, & Schmiedek, 2010), both varying across the life span (Mackey, Raizada, & Bunge, 2013). Per a theoretical framework of cognitive plasticity recently suggested by Lövdén et al. (2010), the premise for the brain to work efficiently is, on the one hand, to maintain a certain level of stability that preserves computational resources (referred to as the brain's dynamic equilibrium). On the other hand, the brain demonstrates the capability to adapt to changing environmental demands. *Plasticity* refers to the actual structural constraints of the dynamic equilibrium on brain function and performance. *Flexibility*, by contrast, refers to the possible range of changes within these structural (plastic) constraints, that is, the brain's capacity to optimize cognitive performance therein. If the environmental demands cause an imbalance with the actual brain supply, the brain will react to this supply-demand mismatch with functional (flexible) or structural (plastic) changes.

While flexibility leads to an immediate response of the behavioral system by recruitment of given cognitive functions (i.e., a primary reaction to altered demands within the pre-existing range of supply), cognitive plasticity requires a more prolonged supply-demand mismatch to overcome the equalizing status quo (i.e., a sluggish, secondary reaction to prolonged altered demands by changing the pre-existing range of functional supply). Within this framework, cognitive interventions may induce changes at the level of flexibility or plasticity depending on a prolonged exposure to demand-supply mismatches. However, such mismatches of supply and demands are a necessary, but not sufficient, condition to prompt cognitive plasticity—the individual intrinsic potential that is shaped differently throughout the life span (by both individual genetics as well as life experiences) determines the ultimate amount of possible plastic changes. Hence, cognitive functioning is considered to be adaptive in nature while at the same time to be restricted to an individual's ultimate range of cognitive performance.

Life-Span Changes in Task-Switching Supply

The developmental changes on the flexibility to switch between task demands have been shown to be mainly related to the brain supply of the frontal lobes and associated cortical and subcortical structures (Bunge & Wright, 2007; Bunge & Zelazo, 2006; Casey, Tottenham, Liston, & Durston, 2005; Enriquez-Geppert, Huster, & Herrmann, 2013; Luna et al., 2001; Luna, Padmanabhan, & O'Hearn, 2010; for recent reviews on the neural development in childhood and adolescence, see Anderson & Spencer-Smith, 2013; Giedd, Raznahan, & Lenroot, 2013; Karbach & Unger, 2014; and in old age, see Cabeza & Dennis, 2013).

The prefrontal cortex (PFC) shows an inverse U-shaped maturational trajectory over the life span, encompassing a steep structural growth in childhood, a maturational peak in young adulthood, and a steady decline in volume in older age (Blakemore, Burnett, & Dahl, 2010; Hedden & Gabrieli, 2004). Based on these neural underpinnings, Bunge and Zelazo (2006) proposed a brain-based framework to account for the functional development of rule processing depending on the maturational change of the frontal lobe. They suggest that successive changes on rule use in childhood (from using a single rule to switching between compatible rules to switching between incompatible sets of rules) emerge from age-related progress in the ability to represent increasingly complex rule hierarchies. Distinct subregions of the PFC may subserve these different complexity levels of rule representations due to a maturational structural differentiation that is paralleled by a functional specialization of frontal regions. Specifically, it has been demonstrated that the orbitofrontal cortex develops very early in childhood and codes simple stimulus-reward contingencies.

The lateral part of the PFC continues to develop into young adulthood, whereas different lateral networks represent, according their own specific trajectories, rules of increasingly higher complexity levels (i.e., the ventrolateral and the dorsolateral PFC both code conditional rules of medium complexity, and the rostrolateral PFC, being the last to develop, codes complex higher-order rules). Hence, the conceptual understanding of the hierarchical rule system that may underlie complex task-switching abilities evolves along the (structural and functional) maturation of distinct networks of the forebrain in childhood (Bunge & Zelazo, 2006). At the opposite end of the life span, however, even the PFC seems to be first and strongest compromised by age-related decline (Moscovitch & Winocur, 1992; West, 1996; for a recent review, see Cabeza & Dennis, 2013). This is expressed both in volume shrinkage as well as in a functional de-differentiation (Braver et al., 2001; Double et al., 1996; Raz, 2004). The neural regression may hamper the ability of complex rule switching, thus reducing the brain supply for task switching in the elderly.

According to these life-span changes, even childhood and older age appear to be highly sensitive periods to incoming demands, as the immature or over-mature frontal supplies may lead to large mismatches. Interventions aiming at improvement of PFC functioning within these developmental trajectories thus might accelerate its maturation in childhood development and might decelerate its decline in older age (Wass, Porayska-Pomsta, & Johnson, 2011; Watson, Lambert, Miller, & Strayer, 2011). Hence, the identification of the most effective training programs for various age ranges is essential to maximize the range of plastic modification by practice in task switching.

Life-Span Changes in Flexible Adaptations to Task-Switching Demands

Life-span changes in the flexible adaptation to immediate alterations of environmental demands, that is, when individuals are required to switch back and forth between two or more tasks, are only briefly summarized, because extensive reviews have been provided elsewhere (e.g., Kray & Ferdinand, 2014; Peters & Crone, 2013). Age-related differences in task-switching costs are shown to be reduced with lower demands on PFC functioning, such as (a) with the presence of external task cues that support the maintaining and retrieving of the currently relevant task (Cepeda et al., 2001; Hahn et al., 2004; Kray, 2006), (b) with the use of internal task prompts such as task-supporting verbalizations (Kray et al., 2008; see also Kray & Ferdinand, 2014), (c) with an increase of preparation time (Crone, Bunge, van der Molen, & Ridderinkhof, 2006; De Jong, 2001; Kray, 2006), (d) with lower interference between the involved tasks by nonoverlapping task-set representations (Mayr, 2001; see also Eppinger, Kray, Mecklinger, & John, 2007), as well as by (e) task practice (Cepeda et al., 2001; Karbach & Kray, 2009). Although age-related differences can be substantially reduced by these manipulations, age differences in mixing costs are still reliable, supporting the view of age differences in structural constraints of brain supply.

Life-Span Changes in the Plasticity of Task-Switching Performance

What are the preconditions to induce plastic changes in task switching across the life span? Considering again the framework suggested by Lövdén et al. (2010), the training should create a mismatch between functional supply and experienced demands. Prior to a cognitive intervention, the dynamic equilibrium is determined by a relative balance between supply and demand, that is, the demand curve employs about equal proportions of negative mismatch (demand > supply) and positive mismatch (supply > demand). The training intervention ideally produces a strong mismatch between the demands of the new training condition and the current brain supply. With increasing practice (a prolonged training interval), this mismatch slowly reduces as now the brain supply adapts to the high task demands. This is the manifestation of training-induced cognitive plasticity, and, as a result, the system rebalances after training with an enlarged (plastic) range for flexible functioning. The training thus stimulates a broadening of the cognitive range so that the individual will act on a higher equilibrium level (cf. Lövdén et al., 2010).

In the previous section, we identified age-sensitive challenging demands of a task-switching situation that can create an initial mismatch between brain supply and task demands, such as the absence of task cues, limited preparation time, the presence of task interference, and less task practice. However, critical from a life span perspective is that the cognitive training intervention not only needs to evoke cognitive activity that is as far away as possible from the routine demand—the mismatch also needs to be launched *relative* to the respective life period and to the intrinsic potential that presumably varies in childhood and older adulthood. Hence, assuming constraints of brain supply in children and elderly people as compared to young adult age, we rather need a moderate mismatch to induce an optimum level of requirement that is activating and challenging but not overtaxing. To apply cognitive interventions and to induce such a relative, or optimal, mismatch for different age ranges across the life span as well as to keep the training intervention challenging throughout a prolonged training interval, researchers have often used an adaptive training procedure that adjusts the task difficulty to individual abilities throughout the training until individuals reach an asymptotic performance level (testing the limits; Kliegl, Smith, & Baltes, 1989; see also Klingberg, 2010; Shipstead, Reddick, & Engle, 2012).

Some researchers have indeed argued and provided empirical evidence that such adaptive testing procedures lead to larger plastic changes as compared to nonadaptive testing, especially in the context of working memory (WM) training (Brehmer, Westerberg, & Bäckman, 2012; Dunning, Holmes, & Gathercole, 2013; Holmes, Gathercole, & Dunning, 2009; Klingberg et al., 2005; but see Karbach & Verhaeghen, 2014, who found no differences between benefits associated with adaptive and nonadaptive testing in a meta-analysis of executive control and WM training in older adults). For example, Brehmer et al. (2012) investigated whether an adaptive WM training regimen would affect the training situation and promote the transfer to domains different from the trained ones, both in younger and older adults. They contrasted a group receiving an adaptive training on various spatial and verbal WM tasks, spaced over five weeks, against an active control sample completing a nonadaptive form of training (i.e., working on the same material but with low-level task demands). Results of this study indicated that an adaptive training procedure led to larger training and transfer gains as compared to low-level practice. These gains were, partly, greater for younger than for older adults and, partly, comparable across age groups. The latter age effects are, however, surprising with respect to a life-span interpretation of the framework by Lövdén et al. (2010), which gave reason to expect pronounced room for cognitive plasticity in older age.

Age Differences in the Effectiveness of Task-Switching Training Interventions: Training, Transfer, and Maintenance Effects

As is outlined in Chapter 8, the principles for designing a good training intervention are often violated. Typically, a training intervention is designed as a longitudinal pretest → training → posttest experiment best with a follow-up session. The efficiency of the intervention can be measured by direct training benefits (i.e., improvements in task performance during the practice sessions) and especially by transfer and long-term maintenance effects, and the intervention should optimally include an active control group. Transfer refers to indirect training gains in similar or different task domains from those performed in the training sessions, whereby long-term maintenance relates to the lastingness or stability of the attained effects. Transfer is often assessed by various tasks that differ from the trained ones and are performed prior to and after extensive intervention (i.e., at pre- and posttest) to determine whether the training had tapped generalizable processes. We distinguish between two transfer scopes: by *near transfer*, we mean the generalization of training-induced improvements to a new but structurally very similar task to the trained one, while by *far transfer*, we mean a broader generalization of training-induced benefits to dissimilar task domains or theoretical constructs (cf. Karbach & Kray, 2009).

Process-Based Task-Switching Interventions

The general idea of process-based interventions is that the intensive practice of switching between tasks and their underlying cognitive-control processes will result in a positive transfer to those cognitive tasks that also partly rely on the same cognitive-control processes. To prove the effectiveness of a task-switching training regimen, researchers have used as an active control a group that performs exactly the same tasks as the treatment group but practices the tasks in single-task block conditions instead of mixed-block conditions (see Minear & Shah, 2008).

Training Gains

Ample evidence suggests that task-switching trainings produce robust gains on training-task performance at different ages (Karbach & Kray, 2009; Kray, Karbach, Haenig, & Freitag, 2012). For example, Karbach and Kray (2009) showed that specific switch costs were substantially reduced as a function of practice in switching—for the different task-switching conditions, net effects ranged from 0.85 standard deviations (SD) to 1.88 SD. The amount of this reduction was similar across age groups but was modulated by the specific training condition (see the section "The Role of Training Variability"). Zinke, Einert, Pfennig, and Kliegel (2012) trained a sample of adolescents (10–14 years) with a similar training setting by further considering the moderating influence of physical exercise. They compared three conditions: one group performing task switching in combination with physical exercise, one group performing task switching without physical exercise, and one passive control group. They found a reduction of latencies for switch trials of approximately 25% and latencies for nonswitch trials of about 18% after task-switching training. Thus, improvements were larger for switch than for nonswitch trials, indicating an increase in the ability to meet the requirements of the training task (i.e., improvements in switch costs).

In addition, Strobach, Liepelt, Schubert, and Kiesel (2012) investigated practice-induced changes of an alternating-run paradigm on mixing and switching costs by varying stimulus and response interference in young adults between 21 and 30 years old. Results of their study revealed larger reductions on mixing than on switching costs. Importantly, mixing costs were eliminated after training, while switching costs were still existent. Training gains were independent of stimulus or response valence (see also Kray & Fehér, 2014). Hence, it seems that the overlap between stimulus and response valence during task-switching training is not sufficient to create the necessary mismatch between environmental demands and brain supply in younger adults.

A recent meta-analysis of Karbach and Verhaeghen (2014) examined the effects of executive-function trainings including task switching in older adults. Results on improvement in the trained task indicated clear benefits in old age, with raw gains of about 0.9 SD and net gains (after subtracting the effects of active controls) of about 0.5 SD. These robust effect sizes point to consistent gains from a task-switching training on latency switch costs even though it should be noted that on error costs, effects were rather mixed (Karbach, 2008; Zinke et al., 2012).

Near Transfer

Numerous studies have shown that practicing task switching indeed induces near transfer to similar switching settings in different age groups. For instance, Kray, Karbach, Haenig, and Freitag (2012) showed that a task-switching training (with alternating runs and no-task cues) led to substantial reduction of task-switching costs (mixing and switching costs) in children 7 to 12 years old with attention deficit/hyperactivity disorder (ADHD), yielding a net effect on mixing costs of about 1.3 SD. The results of Zinke et al. (2012) on adolescents indicated near transfer for both switching groups to a similar switching task, but this transfer was limited to global mixing costs, that is, the ability to maintain and to select between different task sets.

However, this exclusive change of the maintenance abilities might be related to the higher developmental plasticity (and thus the wider room for improvement) for mixing costs than for switching costs between ages 11 and 15 (cf. Huizinga & van der Molen, 2007).

Minear and Shah (2008) investigated undergraduates comparing transfer gains in an alternating-run task-switching setting with and without tasks cues. Interestingly, in line with Zinke et al. (2012), they obtained near transfer gains only on mixing costs and only in the uncued training condition. These findings support the theoretical considerations regarding a considerable supply-demand mismatch to induce plasticity in task-switching performance in adolescents and young adults.

In the life-span study by Karbach and Kray (2009), younger adults also attained near-transfer gains on mixing and switching costs by using an uncued alternating-run task-switching setting. Karbach and Kray compared benefits in healthy children (8–10 years) with the ones obtained in young adults (18–26 years) and older adults (62–79 years) and found even larger near transfer gains for children and older adults than for younger adults on switching tasks that were different from the trained ones. (In all age groups, effects on the reduction of mixing costs were substantially larger for task-switching training groups, ranging from 0.98 SD to 2.15 SD, than for single-task training groups, ranging from 0.11 SD to –55 SD.) Notably, the task-switching training was not adaptive in this study, so the larger mismatch induced by the training task in children and older adults resulted in larger transfer effects, although practice gains were comparable across the three age groups. In the meta-analysis of Karbach and Verhaeghen (2014) clear near-transfer effects of executive-functions training were shown for older adults (net gain score after subtracting controls of about 0.5 SD), supporting the effectiveness of interventions aimed at improving latent cognitive control.

Far Transfer

Results on far transfer effects due to a task-switching training are somewhat heterogeneous: In one of our first task-switching training studies (Karbach & Kray, 2009), we found substantial far transfer to measures of inhibition, WM, and even to fluid intelligence for all age groups (with most effect sizes after task-switching training > 0.70 SD for children, > 0.60 for younger adults, and > 0.40 for older adults as compared to low or negative effects after single-task training). Children with ADHD showed far transfer benefit from an alternating-run task-switching training on inhibition (effect sizes up to 1.6 SD after switching training) and verbal WM (effects up to 0.9 SD after switching training), but not on fluid intelligence (Kray et al., 2012). In contrast, adolescents did not show any clear far transfer to other task domains (Zinke et al., 2012). They indeed benefitted from switching training on their reaction dynamics in a choice re-action time task and (by tendency) in an updating task but not on inhibitory ability. However, the training dose in this study was somewhat smaller than in the one of Karbach and Kray (2009), and the duration of the training regimen may play a crucial role for training effects. Even the optimal spacing of training is still an open question (see the section "The Role of Training Intensity and Duration").

Far-transfer effects were also scarce in a study by von Bastian and Oberauer (2013). They trained three groups of young adults (18–36 years old) in different facets of WM, whereby they defined the ability of task-set switching to be a sub-ordinated facet (termed *supervision*) of a three-componential functional WM model (see also Oberauer, Süß, Wilhelm, & Wittmann, 2003). The supervision group trained on a switching paradigm while the other groups performed training tasks with regard to other functional categories of WM (storage and processing, relational integration). The switching training was adapted from Karbach and Kray (2009) with an alternating-run setting and ambiguous stimuli. Performance of each training group was compared against the one of an active control group that practiced unrelated perceptual-matching tasks. Results indicated, similar to the study of Karbach and Kray, that the supervision training led to substantial near transfer to a similar switching task as well as to far transfer effects to reasoning (fluid intelligence). It did not, however, generalize to inhibition or WM.

Pereg, Shahar, and Meiran (2013) tried to directly replicate the results of the study of Karbach and Kray (2009) by investigating a group of undergraduate students of about 24 years of age. Yet, they only found near transfer to a very similar switching task and found no far transfer to WM or inhibition. Similarly, in one of our recent studies we also tried to replicate our previous results and to further examine which kind of cognitive-control processes (i.e., switching, interference control, updating), contributed to the training success of task-switching training in younger and older adults (Kray & Fehér, 2014). To this end, we systematically varied demands on switching (single vs. mixed task blocks), interference control (univalent or bivalent stimuli), and updating (memory) demands (with or without task cues) across the training groups. Interestingly, for younger adults training and transfer gains did not vary across the training groups (see also Strobach et al., 2012), but training and transfer gains did for the elderly.

Groups trained in resolving task interference (bivalent groups) showed larger (near) transfer gains than the other two groups, though these gains did not differ across groups trained under different memory demands during task switching (with or without task cues). However, we did not find evidence for broad far transfer to untrained cognitive control tasks in this study. Hence, the type of stimuli and interference induced by the involved training tasks may be more important than we first thought (cf. Karbach & Kray, 2009).

In line with these findings are results of a recent study of Anguera et al. (2013), who applied a multitasking training to determine cognitive plasticity in older age.[2] They found considerable near-transfer gains of the training and argued that especially the practice in resolving task interference required in multitasking situations is critical for inducing cognitive plasticity in the elderly. Moreover, they also reported far transfer of the dual-tasking training to sustained attention as well as WM tasks, and this with medium to large effect sizes for far transfer (all effect sizes > 0.5–1.0 SD). Importantly, this study also revealed first robust correlations between behavioral improvement and changes on neural signatures of cognitive control, the latter being indicated by enhanced midline frontal theta power and long-range coherence (e.g., neural-behavioral correlations of the change in spectral power with the multitasking behavioral gain preservation in the training task after six months, $r = 0.76$, and with the behavioral improvement on a vigilance transfer task, $r = 0.56$). In contrast, in their recent meta-analysis of executive-control training in older adults, Karbach and Verhaeghen (2014) found clear but rather small far-transfer gains across training studies (the net gain after subtracting the effects of control treatments remaining being approximately 0.2 SD).

---

[2] Dual-tasking is not to be confused with task switching, because dual-tasking is considered a separate component of the executive functions cluster with idiosyncratic properties (Strobach, Salminen, Karbach, & Schubert, 2014). However, in respect of the similarity of performance costs that result from coordinating two task sets in both paradigms, we present the very promising study of Anguera et al. (2013) on dual-tasking as well.

Training Variability

Very few studies examined whether the variability of training conditions influence practice and transfer gains of (task-switching) training. In light of theoretical considerations of the supply-demand mismatch model, the variation of task-specific processes practiced through the training should lead to reduced practice gains, because in each of the training sessions individuals are confronted with a new task situation so that the mismatch is constantly high throughout the intervention. For instance, in the study by Karbach and Kray (2009), participants received new sets of stimuli and tasks in each of the four training sessions. Indeed, in contrast to the other task-switching training groups that received the same sets of stimuli and tasks in all four sessions, practice gains were nearly absent in this group (cf. Karbach, 2008). Most interestingly from a life-span view is that transfer gains to a new switching task were substantially larger for younger and older adults after the variable training but strongly reduced in children.[3] Hence, the prolonged mismatch was promoting the plasticity of task-switching performance in adults while it hampered it in children. This further underlines that only an optimal level of mismatch in each age group results in positive plasticity.

Training Intensity and Duration

The question remains as to how much practice is necessary to obtain far-reaching effects of task-switching training. To date, no study engaged in systematic measurement of the optimum session length, the optimum session number, and the optimum spacing of an intervention. Former training regimens vary considerably in the amount of training according to their protocols—three weekly sessions of 25 minutes (the total amount being 75 minutes; Zinke et al., 2012) led to satisfying results on training benefit or near transfer, and four weekly sessions of 35 minutes each (the total amount being 140 minutes; Karbach & Kray, 2009) showed large effects on far transfer as well (see above). In general, short-time interventions dominate the area since they are of high importance from an applied perspective (i.e., they are likely to be easier to implement than longer-duration interventions in real-life contexts; cf. Zinke, 2012). The optimum intensity has not been identified yet, as previous short-time results still vary in their effectiveness (see also von Bastian & Oberauer, 2013; Pereg et al., 2013). An important extension should be to adapt the total duration and the spacing

---

[3] It should be noted that, in this training condition, participants were also required to verbalize the task goals, so it is not fully clear whether the variability or the increased dual-task demands reduced the transfer gains in the children group.

of trainings to the life age, as different life-span periods might require different amounts of practice to receive the optimum impetus for change.

Long-Term Maintenance Effects

Training effectiveness is especially measured by the stability of the revealed effects. To date, there are very few studies that specifically examined long-term effects of a task-switching training regimen. For instance, van der Oord, Ponsioen, Geurts, Ten Brink, and Prins (2012) investigated in a pilot study the short- and long-term effects of an executive-function remediation training for children with ADHD (age range: 8–12 years) with a follow-up session after nine weeks. The training battery consisted of a combined WM, inhibition, and cognitive-flexibility training, whereas the latter was again adapted from Karbach and Kray (2009). Long-term effects on executive functioning and on symptom relief were, however, only assessed by parent- and teacher-rated questionnaire data, which were indeed positive for the total executive-functioning score.

Nonetheless, we cannot draw clear conclusions from this result as, first, it was derived from mere subjective ratings, and second, we have no reliable index of whether it was in fact the task-switching training that led to long-term improvement or whether it was the specific WM, the inhibition training, or even the joint latent share of all three interventions.

Strategy-Based Interventions

Another way to improve task-switching performance by cognitive interventions is to instruct individuals to apply specific strategies how to best handle the switching situation and to train them in using these strategies in the hope that they transfer these strategies to similar task situations. One prominent intervention that has been investigated a lot in task-switching studies is the potential beneficial role of verbal processes by instructing individuals to apply task-supporting verbalizations (for reviews, see Cragg & Nation, 2010; Kray & Ferdinand, 2013, 2014). A number of studies have shown that individuals use inner speech processes to maintain and update the currently relevant tasks, especially when they have to internally keep track of the task sequence and external task cues are missing (as in the alternating-run task-switching setting; Bryck & Mayr, 2005; Cragg & Nation, 2010).

To demonstrate the use of inner speech processes, researchers have used either articulatory suppression, in which subjects were instructed to verbalize aloud an overlearned sequence of words while switching between tasks (Miyake,

Emerson, Padilla, & Ahn, 2004; Saeki & Saito, 2004; Saeki, Saito, & Kawaguchi, 2006; Weywadt & Butler, 2013), or by verbalizing task-irrelevant words during task preparation (Karbach, Kray, & Hommel, 2011; Kray et al., 2008) in a switching situation. Results indicated that, compared to a secondary nonverbal task, mixing costs were substantially increased under articulatory suppression conditions (Baddeley, Chincotta, & Adlam, 2001; Emerson & Miyake, 2003). Moreover, developmental researchers found evidence that children and older adults profit from such verbal self-instruction strategies (e.g., Kirkham, Cruess, & Diamond, 2003; Kray et al., 2008; Kray, Gaspard, Karbach & Blaye, 2013; Lucenet, Blaye, Chevalier, & Kray, 2014; for reviews, see Cragg & Nation, 2010; Kray & Ferdinand, 2014). For instance, Kirkham and colleagues (2003) investigated means to help preschoolers (3 and 4 years old) improve their behavior on a dimension-switching task (Kirkham et al., 2003). Importantly, results revealed that in a condition where the children had to label the relevant sorting dimension on each trial, most 3-year-olds succeeded. The authors suggest that children's labeling would help them refocus their attention, overcoming "attentional inertia" (i.e., the pull to continue attending to a previously relevant dimension). Taking a life-span perspective, Kray et al. (2008) compared groups of children (age range: 7–13 years), younger and older adults (20–27 years and 66–77 years, respectively) in three different switching conditions. One of the conditions was with task-relevant verbalizing, one was with task-irrelevant verbalizing, and one was without verbalizing (control condition). Results revealed that mixing costs substantially decreased in groups with task-relevant verbalizing and increased in groups with irrelevant verbalizing. Furthermore, age differences therein were also reduced in the task-relevant verbalizing condition, pointing to promising reductions of action-control deficits in children and older adults by verbal labeling. Hence, verbal labeling the task goals seems to be a valuable tool to optimize task-switching performance during life-span development, as it supports the engagement in task maintenance and task preparation in task-switching situations (e.g., Kray, Blay, & Lucenet, 2010). Moreover, it seems that older adults and children use such verbal strategies less spontaneously than younger adults but benefit when they apply them after instruction to do so (cf. Kray & Ferdinand, 2013). Therefore, the critical question here is to what extent such verbal strategies are useful after practice and whether they can be transferred to new task-switching situations.

Training Effects

Empirical evidence so far suggests verbalization benefits on mixing costs are strongly reduced after three sessions of practice, especially in younger children and older adults, but age differences in task-switching performance are still reliable after practice and use of task-supporting verbalization (Kray et al., 2008). This finding further emphasizes that verbal processes are not only supportive during the implementation of the task representations but are still needed for the engagement in task preparation. Second, for verbal labeling to be effective, children as well as adults, need some practice in the primary task alone (here task switching) before verbal labeling becomes beneficial (see also Kray et al., 2010).

Transfer Effects

Given that verbal labeling the next task is such a beneficial strategy, especially for children and older adults, we were also interested whether training such a strategy in combination with task switching lead to stronger transfer effects than training in task switching alone, assuming that a similar strategy is also applied in a new switching situation. Therefore, we compared transfer effects of a task-switching training group with a group that also applied the verbal labeling (Karbach & Kray, 2009). Although the reduction of mixing costs was larger as compared to an active control group, we found no differences in transfer gains between these two training conditions. However, subjects were not instructed to use verbal labeling at post-test, which may explain the lack of transfer to a new switching situation. As such, we ran a further study in which we compared four training conditions in a group of elderly adults between ages 56–78 to determine whether transfer occurs when the training and the transfer situation are more similar to each other and involve overlapping strategy use (Karbach, Mang, & Kray, 2010). In a pretest → training → posttest design in which all groups were trained in an alternating-run switching setting, we systematically varied the use of verbal task labeling during training and at posttest. One group did not verbalize the task goals in the training session or at posttest; the second group used verbal labeling only in the training sessions; the third group used verbal labeling in the training and at posttest; the fourth group only verbalized at posttest. Results were in line with our assumption that transfer effects would be larger in the third group than in the first two groups. However, the largest transfer effects (reduction in mixing costs) were found in the fourth group, which had received no practice in verbal labeling only in task switching.

These findings again suggest that verbal labeling is a powerful cognitive intervention in a given situation, but is not easily transferred to new situations. Moreover, practice in a specific task-labeling strategy can even result in negative (i.e., smaller) transfer (cf. Karbach et al., 2010).

The Role of Interindividual Differences on Training and Transfer of Task Switching

As is outlined in the beginning of the chapter, the maximum range for plasticity in task-switching performance will not only depend on the optimal mismatch between environmental demands and brain supply but also on the individual intrinsic potential.

Interindividual Differences in Baseline Performance

There are two major models of interindividual differences in training gains depending on baseline performance, namely the amplification and the compensation models. Both models assume interactions between the initial aptitude and the range of practice-induced cognitive plasticity. The amplification model posits that those individuals who are already performing well at the beginning gain the most from an intervention. This relation, also referred to as the Matthew effect (as in the allegory "the rich get richer, and the poor get poorer"), which describes an accumulation of the initial advantages or disadvantages. In the context of life span, we would predict from this model that children and older adults with initially poorer performances would profit less from training, and therefore, age differences would be magnified as a result of practice. In contrast, the compensation model assumes that a cognitive intervention would specifically enhance the abilities of those who are performing low at the beginning, while the well-performing individuals would not record noticeable profit. Hence, children and older adults would gain substantially more than younger adults from a training intervention, and, consequently, age differences would be reduced as a result of practice.

Empirical evidence from task-switching studies suggest, for comparisons at the group level, larger training and transfer gains occur after a task-switching training for children and elderly people as compared to younger adults, supporting the compensation account (e.g., Cepeda et al., 2001; Karbach & Kray, 2009; Kray & Fehér, 2014; for a meta-analysis, see Karbach & Verhaeghen, 2014). However, training studies on age differences in episodic memory show that age differences between younger and older adults increase after practice in applying a mnemonic strategy (Lindenberger, Kliegl, & Baltes, 1992), while younger adults benefitted more from the strategy training in line with an amplification model (see also Lövdén, Brehmer, Li, & Lindenberger, 2012; Verhaeghen & Marcoen, 1996).

Studies that examined the individual status-benefit correlations on task switching found strong negative correlations between initial baseline performance and transfer gains. For example, Karbach (2008) found that the pretest performance (status) seemed to be a reliable predictor for training and transfer benefits in a task-switching training: poorer performance at the beginning was associated with larger training gains (correlations ranging from $r = -0.66$ to $r = -0.83$) and greater transfer benefit (correlations ranging from $r = -0.19$ to $r = -0.72$), pointing to compensatory rather than Matthew effects. However, it is a question for future research as to whether those correlations are mediated by initial brain supply, other intrinsic factors like motivation and personality, or another currently unknown interaction.

Motivational Influences

Motivation is assumed to be integrally linked to cognitive control (evidence comes from the field of cognitive neuroscience; e.g., Kouneiher, Charron, & Koechlin, 2009; Zelazo, Qu, & Kesek, 2010) and seems to affect specifically the outcomes of a task-switching setting (Kleinsorge & Rinkenauer, 2012). Motivated behavior becomes apparent in the interplay of personal traits and environmental determinants. We first provide insights into an influential trait concept and, second, describe an important environment variable that may tap into the cognitive training success.

Powerful motivational variables at the trait level that might affect intervention outcomes are self-regulatory beliefs, such as self-efficacy (Bandura, 1993, 1997). Self-efficacy beliefs describe judgments of the personal capability to perform successfully on a given domain, for example, on memory tasks (Bandura, 1986). Schunk (1984) demonstrated a particular sensitivity of self-efficacy to instructional interventions (see also Zimmerman, 2000). In the educational context, efficacy beliefs have been proven to mediate achievement outcomes (i.e., to directly link learning environments to outcomes; Liem, Lau, & Nie, 2008; Moriarty, Douglas, Punch, & Hattie, 1995; Schunk, 1984; Zimmerman, 2000). Of interest from a life-span perspective, elderly people seem to dispose of poorer self-efficacy resources than younger adults, leading to less effort spent on demanding cognitive tasks (Bruce, Coyne, & Botwinick, 1982; Lachman & Jelalian, 1984; Murphy, Sanders, Gabriesheski, & Schmitt, 1981; Valentijn et al., 2006; Wells & Esopenko, 2008; West, Bagwell, & Dark-Freudemann, 2008).

The influence of self-efficacy on cognitive trainings (and age effects therein) has been investigated especially for strategy trainings on episodic memory (West et al., 2008). Several studies showed that initial memory efficacy beliefs explained a considerable amount of the variability in training outcomes, pointing to self-efficacy and cognition being interrelated at the latent level (Carretti, Borella, Zavagnin, & De Beni, 2011; Valentijn et al., 2006; West et al., 2008; West &

Hastings, 2011). Compelling evidence suggests that a combined training of memory and efficacy-enhancing strategies led to higher gains on cognitive memorizing than an isolated memory training (West et al., 2008). Moreover, a recent study first addressed the role of memory self-efficacy for process-specific cognitive interventions (Payne et al., 2012); the researchers examined the relation of efficacy beliefs with individual differences on an inductive reasoning training in older adults. They found that efficacy beliefs seem to be positively related to the degree to which elderly people can gain from training on fluid abilities. Yet, to the best of our knowledge, no published study has investigated the impact of such self-efficacy beliefs on training in a task-switching setting, where this impact may also vary across the life span. Given their importance for explaining individual differences in intellectual abilities in older age, we are currently investigating and including such self-regulation traits in our ongoing training studies.

A growing body of research has used another approach to keep the training motivation and willingness high through the provision of a motivational training setting with video-game elements (for reviews on video-game playing, see Bavelier, Green, Pouget, & Schrater, 2012; Green & Bavelier, 2006; see also the commentary of Karbach, 2014). Ryan, Rigby, and Przybylski (2006) highlight a motivational pull emerging from video games that satisfies the need for self-determination (Ryan & Deci, 2000) along with relatedness, autonomy, and competency. In the narrow context of computerized games, they describe relatedness as presence (i.e., the sense of being completely taken in the game or being a part thereof). Autonomy refers to provisions of choice, informational feedback by means of rewards, and noncontrolling instructions. Competency nurtures a need for challenge in terms of opportunities to acquire new skills for being successful in game-playing. Relatedness, autonomy, and competency in sum might enhance the intrinsic interest in the game content.

Although these different factors have not yet been systematically investigated, some evidence suggests that cognitive control functioning and task-switching performance can be enhanced by training with video games in younger adults (Glass, Maddox, & Love, 2013) as well as in older adults (Basak, Boot, Voss, & Kramer, 2008). For example, Anguera et al. (2013) investigated the trainability of multitasking in older adults by embedding the training in a custom-designed game simulation, called *NeuroRacer*, in which elderly participants had, in one task, to adapt the driving behavior of a vehicle on a winding road (drive task) and, in another task, to respond to road signs (sign task). They determined training and transfer gains between three training conditions; that is, participants either were to perform both tasks simultaneously (multitasking condition), to complete each task separately (single-task condition/active control group), or to perform none of the tasks (passive control group). Of most importance from our life-span focus, they compared the training benefits from the multitasking condition between a group of younger adults at about 20 years old and a group of older adults between 60 and 85 years old. Results indicated that older adults in the multitasking condition substantially reduced their multitasking costs as compared to the active or passive control condition, and the elderly reached performance levels that were even beyond the ones of untrained younger adults. Hence, embedding the training in a game setting is a promising approach to maintain motivation and training willingness throughout the cognitive intervention and this may also promote transfer to untrained tasks.

However, the direct effects of a game setting have rarely been investigated (e.g., in the context of WM, see Prins, Dovis, Ponsioen, Ten Brink, & van der Oord, 2011). In one of our recent training studies, we investigated systematically the impact of a motivating game setting on a task-switching training in healthy middle-age children (8–11 years old; Dörrenbächer, Müller, Tröger, & Kray, 2014). In a pretest → training → posttest design, we included two active control groups (single-task training) and two task-switching groups that varied in the training setting and either contained game elements (high-motivational training setting) or no game elements (low-motivational training setting). More specifically, in the high-motivational conditions, we embedded the training procedure in the frame of an adventure game, presenting challenging monster battles on a foreign planet. Accordingly, the stimuli, the task labels, the goal instructions, and the feedback were presented in a manner reflecting the game universe.

The low-motivational conditions instead received the standard single- or task-switching setting, containing no elements of a game world. That is to say, children here were confronted with scrambled stimuli lacking the sensation of animated characters, with neutral task labels and simple feedback texts. As a manipulation check of our motivational variation, five times per session, we asked children for their willingness to perform voluntarily additional training blocks. Training motivation was then scored by the total number of additional blocks per training session. Importantly, we ensured that this score was not confounded with the training experience by determining that all children performed in fact the same number of training blocks, irrespectively of their actual choice to play additional ones. This was covered by presenting the willingness questions at random positions during the training session. Results of this study showed that (a) a high-motivational setting indeed enhanced children's training willingness; (b) children benefitted on their task and switching performance on latencies after a task-switching training as compared to a single-task training, and especially when this switching training was embedded in a high-motivational environment; (c) near transfer was obtainable for all training groups, while the group that had received a high-motivational switching training achieved largest effect sizes on switching and mixing costs at near-transfer assessment; and (d) training gains did not consistently propagate to far transfer measures.

In general, the motivational setting primarily influenced the processing speed but did not generalize to specific control processes required by the untrained cognitive-control tasks. This means that, in children, video-game elements seem to facilitate specifically the response energy invested in task processing. As far as higher-order control is concerned, the presence of such high-motivational settings appears to interfere with the problem-solving behavior, maybe by eliciting approach tendencies that overshadow goal-directed control (see also Zelazo et al., 2010). This may account for the lack of far transfer in our game study. The control of individual differences at the trait level or of age differences in the sensitivity to different variations of the setting should, however, expand current insights into the nature of the interplay between motivation and cognition.

Summary and Conclusions

Empirical evidence on life-span changes in task switching has been shown that cognitive-control processes required for being in a switching situation (e.g., task maintenance and selection processes, though not the switching process itself) are highly age sensitive. These cognitive processes mature relatively late during childhood and decline relatively early in older age. Life-span changes in these processes have been attributed to life-span changes in brain supply, in particular in the frontal cortex. Researchers have also identified several environmental demands in task-switching situations that induce a stronger mismatch with the brain supply, especially in childhood and old age, such as the absence of environmental prompts (task cues), less preparation time and task practice, and interference between the involved tasks by overlapping task representations (stimulus and response ambiguity).

Hence, these environmental demands should be challenging and, considering the theoretical framework by Lövdén et al. (2010), be suitable to induce plasticity in task-switching performance across the life span. The findings on the effectiveness of training in task switching are generally in line with these theoretical considerations. Training conditions that produce a larger mismatch indeed result in larger transfer gains: (1) Under identical training conditions (no adaptive training), those age groups show larger transfer gains who showed the largest mixing costs at the beginning of the training, probably because the relative mismatch between demands and brain supply was larger in children and older adults than in younger adults; (2) transfer gains (reduction in mixing costs) are larger for uncued than cued task-switching conditions and (3) larger for variable training conditions in which task demands change in every training session than for constant training conditions at least in adulthood; and (4) for older adults, transfer gains are larger under training conditions that practice the resolving of task interference. Moreover, depending on the brain supply in specific age ranges, (5) training conditions can be too challenging (e.g., children did not show transfer gains to a new switching situation when they received a variable training in which they had to verbalize the task goals in addition), and (6) training conditions that result in a reduced mismatch as they support task-switching performance, such as applying verbal strategies as well as motivational influences like a game setting, seem to be effective in the training situation but are less suitable to produce transfer gains to a new switching situation.

In contrast, the boundary conditions for the generalization of training in task switching are less clear and more heterogeneous. Training studies either found relative broad transfer of the task-switching training or nearly no transfer, relatively independent of the training conditions and age groups, which may also be because of a lack in statistical power due to small sample sizes in the training groups. The occurrence of far transfer is not only dependent on the amount of overlapping cognitive-control processes but also on the overlap of task domains, reliability of measurement, and initial baseline task performance.

Open Issues for Designing Cognitive Intervention Across the Life Span

Although research has identified some age-sensitive task-switching training conditions that lead to larger transfer gains in various age ranges, several unsolved issues remain to be addressed in future training studies.

Finding the Optimal Balance Among Intensity, Duration, and Variability of the Training

Most training studies on the plasticity of task switching only practice a relatively short time, between three and four sessions on a weekly basis with an intensity of less than one hour. Although nearly all the training studies obtained near transfer effects to an untrained new switching task, results on far transfer gains vary considerably. Hence, it is an open question as to how much practice is necessary to obtain the largest effects of task-switching training. An important extension should be to adapt the total duration and weekly distribution of trainings to the life age, as different life-span periods might require different amounts of practice to receive the adequate impetus for change. Furthermore, the optimal balance of training duration, weekly intensity, and variability might also strongly depend on the demand-supply mismatch induced at the beginning of training.

The Optimal Balance Between Intervention Generality and Specificity

While most training studies so far used different types of a specific task-switching training, other studies trained this ability in combination with other cognitive control processes. The advantage of using more specific task-switching interventions is that it allows better explanation regarding what kind of processes have contributed to the transfer gains at the expense of generality of training effects. By contrast, the advantage of using combined cognitive control training interventions is that they may produce more general effects at the expense of understanding which component has caused these effects. However, published training studies do not indicate a clear pattern to support these considerations: more specific task-switching training studies reported relatively broad transfer, and more general cognitive-control training interventions reported relatively narrow transfer to other cognitive tasks. Finding the optimal balance between generality and specificity is thus an important issue to consider in future training studies.

Process- Versus Strategy-Based Interventions

Only a few studies examined a potential positive transfer of instructed verbal strategies to untrained new switching situations and found the transfer of those strategies to be limited in contrast to process-based training interventions. However, so far only the effectiveness of verbal strategies has been investigated in training studies, so whether other strategies are better suited for transfer to new switching situations remains unknown.

Motivational and Emotional Influences

Several training researchers have argued that motivational factors play a crucial role in training effectiveness. However, the role of individual differences in motivation at the trait level (e.g., self-efficacy beliefs), especially for different age brackets, is fully neglected in interpreting current training results and age effects therein. Future research should include such motivational determinants to achieve a broader rationale of cognitive plasticity. Further, the moderating impact of the instructional training context on cognitive performance changes should be considered in detail for different life ages; a large body of evidence shows that cognitive-control trainings with video-game elements lead to training and transfer effects in various age ranges. However, only recently have researchers tried to show whether a video-game training setting indeed results in larger training and transfer benefits compared with a training setting without game elements. At least in children, it seems that video-game elements primarily have an impact on the training willingness and task performance in the training situation but less influence in the generalization of training effects. Whether age groups vary in their sensitivity to different kinds of training settings and motivational aspects (e.g., autonomy, competency, and rewards of various types and magnitudes) is unknown but it would be an inspiring avenue to create new training studies and interventions based on known insights about the effectiveness of task-switching trainings.

References

Anderson, V., & Spencer-Smith, M. (2013). Children's frontal lobes: No longer silent? In D. T. Stuss & R. T. Knight (Eds.), *Principles of frontal lobe function* (2nd ed., pp. 118–135). New York, NY: Oxford University Press.

Anguera, J. A., Boccanfuso, J., Rintoul, J. L., Al-Hashimi, O., Faraji, F., Janowich, J., . . . Gazzaley, A. (2013). Video game training enhances cognitive control in older adults. *Nature, 501*(7465), 97–101. doi:10.1038/nature12486

Baddeley, A., Chincotta, D., & Adlam, A. (2001). Working memory and the control of action: Evidence from task switching. *Journal of Experimental Psychology: General, 130*(4), 641–657. doi:10.1037//0096-3445.130.4.641

Bandura, A. (1986). *Social foundations of thought and action: A social cognitive theory.* Englewood Cliffs, NJ: Prentice-Hall.

Bandura, A. (1993). Perceived self-efficacy in cognitive development and functioning. *Educational Psychologist, 28*(2), 117–148. doi:10.1207/s15326985ep2802_3

Bandura, A. (1997). *Self-efficacy: The exercise of control.* New York, NY: W. H. Freeman.

Basak, C., Boot, W. R., Voss, M. W., & Kramer, A. F. (2008). Can training in a real-time strategy video game attenuate cognitive decline in older adults? *Psychology and Aging, 23*(4), 765–777. doi:10.1037/a0013494

Bavelier, D., Green, C. S., Pouget, A., & Schrater, P. (2012). Brain plasticity through the life span: Learning to learn and action video games. *Annual Review of Neuroscience, 35*, 391–416. doi:10.1146/annurev-neuro-060909-152832

Blakemore, S.-J., Burnett, S., & Dahl, R. E. (2010). The role of puberty in the developing adolescent brain. *Human Brain Mapping, 31*(6), 926–933. doi:10.1002/hbm.21052

Braver, T. S., Barch, D. M., Keys, B. A., Carter, C. S., Cohen, J. D., Kaye, J. A., . . . Reed, B. R. (2001). Context processing in older adults: Evidence for a theory relating cognitive control to neurobiology in healthy aging. *Journal of Experimental Psychology: General, 130*(4), 746–763. doi:10.1037//0096-3445.130.4.746

Brehmer, Y., Westerberg, H., & Bäckman, L. (2012). Working memory training in younger and older adults: Training gains, transfer, and maintenance. *Frontiers in Human Neuroscience, 6*, 63. doi:10.3389/fnhum.2012.00063

Bruce, P. R., Coyne, A. C, & Botwinick, J. (1982). Adult age differences in meta-memory. *Journal of Gerontology, 37*, 354–357. doi:10.1093/geronj/37.3.354

Bryck, R. L., & Mayr, U. (2005). On the role of verbalization during task set selection: Switching or serial order control? *Memory & Cognition, 33*(4), 611–623. doi:10.3758/BF03195328

Buchler, N. G., Hoyer, W. J., & Cerella, J. (2008). Rules and more rules: The effects of multiple tasks, extensive training, and aging on task-switching performance. *Memory & Cognition, 36*(4), 735–748. doi:10.3758/MC.36.4.735

Bunge, S. A., & Wright, S. B. (2007). Neurodevelopmental changes in working memory and cognitive control. *Current Opinion in Neurobiology, 17*(2), 243–250. doi:10.1016/j.conb.2007.02.005

Bunge, S. A., & Zelazo, P. D. (2006). A brain-based account of the development of rule use in childhood. *Current Directions in Psychological Science, 15*(3), 118–121. doi:10.1111/j.0963-7214.2006.00419.x

Cabeza, R., & Dennis, N. A. (2013). Frontal lobes and aging: Deterioration and compensation. In D. T. Stuss & R. T. Knight (Eds.), *Principles of frontal lobe function* (2nd ed., pp. 628–653). New York, NY: Oxford University Press.

Carretti, B., Borella, E., Zavagnin, M., & De Beni, R. (2011). Impact of metacognition and motivation on the efficacy of strategic memory training in older adults: Analysis of specific, transfer and maintenance effects. *Archives of Gerontology and Geriatrics, 52*(3), e192–e197. doi:10.1016/j.archger.2010.11.004

Casey, B. J., Tottenham, N., Liston, C., & Durston, S. (2005). Imaging the developing brain: What have we learned about cognitive development? *Trends in Cognitive Sciences, 9*(3), 104–110. doi:10.1016/j.tics.2005.01.011

Cepeda, N. J., Kramer, A. F., & Gonzalez de Sather, J. C. M. (2001). Changes in executive control across the life span: Examination of task-switching performance. *Developmental Psychology, 37*(5), 715–730. doi:10.1037//0012-1649.37.5.715

Cragg, L., & Chevalier, N. (2012). The processes underlying flexibility in childhood. *Quarterly Journal of Experimental Psychology, 65*(2), 209–232. doi:10.1080/17470210903204618

Cragg, L., & Nation, K. (2010). Language and the development of cognitive control. *Topics in Cognitive Science, 2*(4), 631–642. doi:10.1111/j.1756-8765.2009.01080.x

Crone, E. A., Bunge, S. A., van der Molen, M. W., & Ridderinkhof, K. R. (2006). Switching between tasks and responses: A developmental study. *Developmental Science, 9*(3), 278–287. doi:10.1111/j.1467-7687.2006.00490.x

Crone, E. A., Ridderinkhof, K. R., Worm, M., Somsen, R. J. M., & van der Molen, M. W. (2004). Switching between spatial stimulus-response mappings: A developmental study of cognitive flexibility. *Developmental Science, 7*(4), 443–455. doi:10.1111/j.1467-7687.2004.00365.x

De Jong, R. (2001). Adult age differences in goal activation and goal maintenance. *European Journal of Cognitive Psychology, 13*(1–2), 71–89. doi:10.1080/09541440042000223

Diamond, A. (2013). Executive functions. *Annual Review of Psychology, 64*(1), 135–168. doi:10.1146/annurev-psych-113011-143750

Diamond, A., & Ling, D. S. (2017). Review of the evidence on, and fundamental questions surrounding, efforts to improve executive functions, including working memory. In M. Bunting & J. Novick (Eds.), *Cognitive training.* Oxford, UK: Oxford University Press.

Dibbets, P., & Jolles, J. (2006). The switch task for children: Measuring mental flexibility in young children. *Cognitive Development, 21*(1), 60–71. doi:10.1016/j.cogdev.2005.09.004

Dörrenbächer, S., Müller, P., Tröger, J., & Kray, J. (2014). Dissociable effects of game elements on motivation and cognition in a task-switching training in middle childhood. *Frontiers in Psychology*, 5, 1275. doi:10.3389/fpsyg.2014.01275

Double, K. L., Haliday, G. M., Kril, J. J., Harasty, J. A., Cullen, K., Brooks, W. S., . . . Broe, G. A. (1996). Topography of brain atrophy during normal aging and Alzheimer's disease. *Neurobiology of Aging*, 17(4), 513–521. doi:10.1016/0197-4580(96)00005-X

Duncan, J., Burgess, P., & Emslie, H. (1995). Fluid intelligence after frontal lobe lesions. *Neuropsychologia*, 33(3), 261–268. doi:10.1016/0028-3932(94)00124-8

Dunning, D. L., Holmes, J., & Gathercole, S. E. (2013). Does working memory training lead to generalized improvements in children with low working memory? A randomized controlled trial. *Developmental Science*, 16(6), 915–925. doi:10.1111/desc.12068

Emerson, M. J., & Miyake, A. (2003). The role of inner speech in task switching: A dual-task investigation. *Journal of Memory and Language*, 48(1), 148–168. doi:10.1016/S0749-596X(02)00511-9

Enriquez-Geppert, S., Huster, R. J., & Herrmann, C. S. (2013). Boosting brain functions: Improving executive functions with behavioral training, neurostimulation, and neurofeedback. *International Journal of Psychophysiology*, 88(1), 1–16. doi:10.1016/j.ijpsycho.2013.02.001

Eppinger, B., Kray, J., Mecklinger, A., & John, O. (2007). Age differences in task switching and response monitoring: Evidence from ERPs. *Biological Psychology*, 75(1), 52–67. doi:10.1016/j.biopsycho.2006.12.001

Giedd, J. N., Raznahan, A., & Lenroot, R. K. (2013). Adolescent frontal lobes: Under construction. In D. T. Stuss & R. T. Knight (Eds.), *Principles of frontal lobe function* (2nd ed., pp. 135–145). New York, NY: Oxford University Press.

Glass, B. D., Maddox, W. T., & Love, B. C. (2013). Real-time strategy game training: Emergence of a cognitive flexibility trait. *PLoS One*, 8(8), e70350. doi:10.1371/journal.pone.0070350

Green, C. S., & Bavelier, D. (2006). The cognitive neuroscience of video games. In P. Messaris & L. Humphreys (Eds.), *Digital media: Transformations in human communication* (pp. 211–225). New York, NY: Peter Lang Publishing.

Hahn, S., Andersen, G. J., & Kramer, A. F. (2004). Age influences on multi-dimensional set switching. *Aging, Neuropsychology, and Cognition*, 11, 25–36. https://doi.org/10.1076/anec.11.1.25.29360

Hedden, T., & Gabrieli, J. D. (2004). Insights into the aging mind: A view from cognitive neuroscience. *Nature Reviews, Neuroscience*, 5(2), 87–96. doi:10.1038/nrn1323

Holmes, J., Gathercole, S. E., & Dunning, D. L. (2009). Adaptive training leads to sustained enhancement of poor working memory in children. *Developmental Science*, 12(4), F9–F15. doi:10.1111/j.1467-7687.2009.00848.x

Huizinga, M., & van der Molen, M. W. (2007). Age-group differences in set-switching and set-maintenance on the Wisconsin Card Sorting Task. *Developmental Neuropsychology*, 31(2), 193–215. doi:10.1080/87565640701190817

Karayanidis, F., Jamadar, S., & Sanday, D. (2013). Stimulus-level interference disrupts repetition benefit during task switching in middle childhood. *Frontiers in Human Neuroscience*, 7, 841. doi:10.3389/fnhum.2013.00841

Karayanidis, F., Whitson, L. R., Heathcote, A., & Michie, P. T. (2011). Variability in proactive and reactive cognitive control processes across the adult lifespan. *Frontiers in Psychology*, 2, 318. doi:10.3389/fpsyg.2011.00318

Karbach, J. (2008). *Potential and limits of executive control training: Age differences in the near and far transfer of task-switching training* (Doctoral dissertation). Retrieved from http://scidok.sulb.uni-saarland.de/volltexte/2008/1772/

Karbach, J. (2014). Game-based cognitive training for the aging brain. *Frontiers in Human Neuroscience, 5,* 1100. doi:10.3389/fpsyg.2014.01100

Karbach, J., & Kray, J. (2007). Developmental changes in switching between mental task sets: The influence of verbal labeling in childhood. *Journal of Cognition and Development, 8*(2), 205–236. doi:10.1080/15248370701202430

Karbach, J., & Kray, J. (2009). How useful is executive control training? Age differences in near and far transfer of task-switching training. *Developmental Science, 12*(6), 978–990. doi:10.1111/j.1467-7687.2009.00846.x

Karbach, J., Kray, J., & Hommel, B. (2011). Action-effect learning in early childhood: Does language matter? *Psychological Research, 75*(4), 334–340. doi:10.1007/s00426-010-0308-1

Karbach, J., Mang, S., & Kray, J. (2010). Transfer of verbal self-instruction training in older age. *Psychology and Aging, 25*(3), 677–683. doi:10.1037/a0019845

Karbach, J., & Unger, K. (2014). Executive control training from middle childhood to adolescence. *Frontiers in Developmental Psychology, 5,* 390. doi:10.3389/fpsyg.2014.00390

Karbach, J., & Verhaeghen, P. (2014). Making working memory work: A meta-analysis of executive control and working memory training in younger and older adults. *Psychological Science, 25*(11), 2027–2037. doi:10.1177/0956797614548725

Kiesel, A., Steinhauser, M., Wendt, M., Falkenstein, M., Jost, K., Philipp, A. M., & Koch, I. (2010). Control and interference in task switching–A review. *Psychological Bulletin, 136*(5), 849–874. doi:10.1037/a0019842

Kirkham, N. Z., Cruess, L., & Diamond, A. (2003). Helping children apply their knowledge to their behavior on a dimension-switching task. *Developmental Science, 6*(5), 449–467. doi:10.1111/1467-7687.00300

Kleinsorge, T., & Rinkenauer, G. (2012). Effects of monetary incentives on task switching. *Experimental Psychology, 59*(4), 216–226. doi:10.1027/1618-3169/a000146

Kliegl, R., Smith, J., & Baltes, P. B. (1989). Testing-the-limits and the study of adult age differences in cognitive plasticity of a mnemonic skill. *Developmental Psychology, 25*(2), 247–256. doi:10.1037//0012-1649.25.2.247

Klingberg, T. (2010). Training and plasticity of working memory. *Trends in Cognitive Sciences, 14*(7), 317–324. doi:10.1016/j.tics.2010.05.002

Klingberg, T., Fernell, E., Olesen, P. J., Johnson, M., Gustafsson, P., Dahlström, K., . . . Westerberg, H. (2005). Computerized training of working memory in children with ADHD—A randomized, controlled trial. *Journal of the American Academy of Child & Adolescent Psychiatry, 44*(2), 177–186. doi:10.1097/00004583-200502000-00010

Kouneiher, F., Charron, S., & Koechlin, E. (2009). Motivation and cognitive control in the human prefrontal cortex. *Nature Neuroscience, 12*(7), 939–945. doi:10.1038/nn.2321

Kray, J. (2006). Task-set switching under cue-based versus memory-based switching conditions in younger and older adults. *Brain Research, 1105*(1), 83–92. doi:10.1016/j.brainres.2005.11.016

Kray, J., Blaye, A., & Lucenet, J. (2010). Can older adults enhance task-switching performance by using verbal self-instructions? The influence of working-memory load and early learning. *Frontiers in Aging Neuroscience, 2,* 147. doi:10.3389/fnagi.2010.00147

Kray, J., Eber, J., & Karbach, J. (2008). Verbal self-instructions in task switching: A compensatory tool for action-control deficits in childhood and old age? *Developmental Science*, *11*(2), 223–236. doi:10.1111/j.1467-7687.2008.00673.x

Kray, J., Eber, J., & Lindenberger, U. (2004). Age differences in executive functioning across the lifespan: The role of verbalization in task preparation. *Acta Psychologica*, *115*(2–3), 143–165. doi:10.1016/j.actpsy.2003.12.001

Kray, J., Eppinger, B., & Mecklinger, A. (2005). Age differences in attentional control: An event-related potential approach. *Psychophysiology*, *42*(4), 407–416. doi:10.1111/j.1469-8986.2005.00298.x

Kray, J., & Fehér, B. (2014). *Age differences in the transfer of task-switching training: The impact of working memory and inhibition control demands.* Invited symposium paper presented at the 12th International Conference on Cognitive Neuroscience (ICON), Brisbane, Australia.

Kray, J., & Ferdinand, N. K. (2013). How to improve cognitive control in development during childhood: Potentials and limits of cognitive interventions. *Child Development Perspectives*, *7*(2), 121–125. doi:10.1111/cdep.12027

Kray, J., & Ferdinand, N. K. (2014). Task switching and aging. In J. Grange & G. Houghton (Eds.), *Task switching and cognitive control* (pp. 350–373). New York, NY: Oxford University Press.

Kray, J., Gaspard, H., Karbach, J., & Blaye, A. (2013). Developmental changes in using verbal self-cueing in task-switching situations: The impact of task practice and task sequencing demands. *Frontiers in Psychology*, *4*, 940. doi:10.3389/fpsyg.2013.00940

Kray, J., Karbach, J., & Blaye, A. (2012). The influence of stimulus-set size on developmental changes in cognitive control and conflict adaptation. *Acta Psychologica*, *140*(2), 119–128. doi:10.1016/j.actpsy.2012.03.005

Kray, J., Karbach, J., Haenig, S., & Freitag, C. (2012). Can task-switching training enhance executive control functioning in children with attention deficit/hyperactivity disorder? *Frontiers in Human Neuroscience*, *5*, 180. doi:10.3389/fnhum.2011.00180

Kray, J., Li, K. Z. H., & Lindenberger, U. (2002). Age-related changes in task-switching components: The role of task uncertainty. *Brain and Cognition*, *49*(3), 363–381. doi:10.1006/brcg.2001.1505

Kray, J., & Lindenberger, U. (2000). Adult age differences in task switching. *Psychology and Aging*, *15*(1), 126–147. doi:10.1037//0882-7974.15.1.126

Lachman, M. E., & Jelalian, E. (1984). Self-efficacy and attributions for intellectual performance in young and elderly adults. *Journal of Gerontology*, *39*(5), 577–582. doi:10.1093/geronj/39.5.577

Lawo, V., Philipp, A. M., Schuch, S., & Koch, I. (2012). The role of task preparation and task inhibition in age-related task-switching deficits. *Psychology and Aging*, *27*(4), 1130–1137. doi:10.1037/a0027455

Liem, A. D., Lau, S., & Nie, Y. (2008). The role of self-efficacy, task value, and achievement goals in predicting cognitive engagement, task disengagement, peer relationship, and achievement outcome. *Contemporary Educational Psychology*, *33*(4), 486–512. doi:10.1016/j.sbspro.2010.07.214

Lien, M., Ruthruff, E., & Kuhns, D. (2008). Age-related differences in switching between cognitive tasks: Does internal control ability decline with age? *Psychology and Aging*, *23*(2), 330–341. doi:10.1037/0882-7974.23.2.330

Lindenberger, U., Kliegl, R., & Baltes, P. B. (1992). Professional expertise does not eliminate negative age differences in imagery-based memory performance during adulthood. *Psychology and Aging, 7*(4), 585–593. doi:10.1037/0882-7974.7.4.585

Lövdén, M., Bäckman, L., Lindenberger, U., Schaefer, S., & Schmiedek, F. (2010). A theoretical framework for the study of adult cognitive plasticity. *Psychological Bulletin, 136*(4), 659–676. doi:10.1037/a0020080

Lövdén, M., Brehmer, Y., Li, S.-C., & Lindenberger, U. (2012). Training-induced compensation versus magnification of individual differences in memory performance. *Frontiers in Human Neuroscience, 6*, 141. doi:10.3389/fnhum.2012.00141

Lucenet, J., Blaye, A., Chevalier, N., & Kray, J. (2014). Cognitive control and language across the life span: Does labeling improve reactive control? *Developmental Psychology, 50*(5), 1620–1627. doi:10.1037/a0035867

Luna, B., Padmanabhan, A., & O'Hearn, K. (2010). What has fMRI told us about the development of cognitive control through adolescence? *Brain and Cognition, 72*(1), 101–113. doi:10.1016/j.bandc.2009.08.005

Luna, B., Thulborn, K. R., Munoz, D. P., Merriam, E. P., Garver, K. E., Minshew, N. J., . . . Sweeney, J. A. (2001). Maturation of widely distributed brain function subserves cognitive development. *NeuroImage, 13*(5), 768–793. doi:10.1006/nimg.2000.0743

Mackey, A. P., Raizada, R. D. S., & Bunge, S. A. (2013). Environmental influences on prefrontal development. In D. T. Stuss & R. T. Knight (Eds.), *Principles of frontal lobe function* (2nd ed., pp. 145–164). New York, NY: Oxford University Press.

Manzi, A., Nessler, D., Czernochowski, D., & Friedman, D. (2011). The development of anticipatory cognitive control processes in task-switching: An ERP study in children, adolescents, and young adults. *Psychophysiology, 48*(9), 1258–1275. doi:10.1111/j.1469-8986.2011.01192.x

Marí-Beffa, P., & Kirkham, A. (2014). The mixing cost as a measure of cognitive control. In J. Grange & G. Houghton (Eds.), *Task switching and cognitive control* (pp. 74–101). New York, NY: Oxford University Press.

Mayr, U. (2001). Age differences in the selection of mental sets: The role of inhibition, stimulus ambiguity, and response set overlap. *Psychology and Aging, 16*(1), 96–109. doi:10.1037//0882-7974.16.1.96

Mayr, U., & Liebscher, T. (2001). Is there an age deficit in the selection of mental sets? *European Journal of Cognitive Psychology, 13*(1–2), 47–69. doi:10.1080/09541440042000188

Meiran, N., Gotler, A., & Perlman, A. (2001). Old age is associated with a pattern of relatively intact and relatively impaired task-set switching abilities. *The Journal of Gerontology Psychological Sciences & Social Sciences, 56*(2), 88–102. doi:10.1093/geronb/56.2.P88

Minear, M., & Shah, P. (2008). Training and transfer effects in task switching. *Memory & Cognition, 36*(8), 1470–1483. doi:10.3758/MC.336.8.1470

Miyake, A., Emerson, M. J., Padilla, F., & Ahn, J.-C. (2004). Inner speech as a retrieval aid for task goals: The effects of cue type and articulatory suppression in the random task cuing paradigm. *Acta Psychologica, 115*(2–3), 123–142. doi:10.1016/j.actpsy.2003.12.004

Miyake, A., Friedman, N. P., Emerson, M. J., Witzki, A. H., Howerter, A., & Wager, T. D. (2000). The unity and diversity of executive functions and their contributions to complex "frontal lobe" tasks: A latent variable analysis. *Cognitive Psychology, 41*(1), 49–100. doi:10.1006/cogp.1999.0734

Moriarty, B., Douglas, G., Punch, K., & Hattie, J. (1995). The importance of self-efficacy as a mediating variable between learning environments and achievement. *British Journal of Educational Psychology, 65*(1), 73–84. doi:10.1111/j.2044-8279.1995.tb01132.x

Moscovitch, M., & Winocur, G. (1992). The neuropsychology of memory and aging. In F. I. M. Craik & T. A. Salthouse (Eds.), *The handbook of aging and cognition* (pp. 315–372). Hillsdale, NJ: Erlbaum.

Murphy, M. D., Sanders, R. E., Gabriesheski, A. S., & Schmitt, F. A. (1981). Metamemory in the aged. *Journal of Gerontology, 36*(2), 185–193. doi:10.1093/geronj/36.2.185

Oberauer, K., Süß, H.-M., Wilhelm, O., & Wittman, W. W. (2003). The multiple faces of working memory: Storage, processing, supervision, and coordination. *Intelligence, 31*, 167–193. doi:10.1016/S0160-2896(02)00115-0

Payne, B. R., Jackson, J. J., Hill, P. L., Gao, X., Roberts, B. W., & Stine-Morrow, E. A. (2012). Memory self-efficacy predicts responsiveness to inductive reasoning training in older adults. *The Journal of Gerontology, Series B, Psychological Sciences and Social Sciences, 67B*(1), 27–35. doi:10.1093/geronb/gbr073

Peters, S., & Crone, E. A. (2013). Cognitive flexibility in childhood and adolescence. In J. Grange & G. Houghton (Eds.), *Task switching and cognitive control* (pp. 332–350). New York, NY: Oxford University Press.

Pereg, M., Shahar, N., & Meiran, N. (2013). Task switching training effects are mediated by working-memory management. *Intelligence, 41*(5), 467–478. doi:10.1016/j.intell.2013.06.009

Prins, P. J. M., Dovis, S., Ponsioen, A., Ten Brink, E., & van der Oord, S. (2011). Does computerized working memory training with game elements enhance motivation and training efficacy in children with ADHD? *Cyberpsychology, Behavior, and Social Networking, 14*(3), 115–122. doi:10.1089/cyber.2009.0206

Raz, N. (2004). The aging brain observed in vivo: Differential changes and their modifiers. In R. Cabeza, L. Nyberg, & D. C. Park (Eds.), *Cognitive neuroscience of aging: Linking cognitive and cerebral aging* (pp. 17–55). New York, NY: Oxford University Press.

Reimers, S., & Maylor, E. A. (2005). Task switching across the life span: Effects of age on general and specific switch costs. *Developmental Psychology, 41*(4), 661–671. doi:10.1037/0012-1649.41.4.661

Ryan, R. M., & Deci, E. L. (2000). Intrinsic and extrinsic motivations: Classic definitions and new directions. *Contemporary Educational Psychology, 25*(1), 54–67. doi:10.1006/ceps.1999.1020

Ryan, R. M., Rigby, C. S., & Przybylski, A. (2006). The motivational pull of video games: A self-determination theory approach. *Motivation and Emotion, 30*, 347–363. doi:10.1007/s11031-006-9051-8

Saeki, E., & Saito, S. (2004). Effect of articulatory suppression on task-switching performance: Implications for models of working memory. *Memory, 12*(3), 257–271. doi:10.1080/09658210244000649

Saeki, E., Saito, S., & Kawaguchi, J. (2006). Effects of response-stimulus interval manipulation and articulatory suppression on task switching. *Memory, 14*(8), 965–976. doi:10.1080/09658210601008973

Salthouse, T. A., Fristoe, N., McGuthry, K. E., & Hambrick, D. Z. (1998). Relation of task switching to speed, age, and fluid intelligence. *Psychology and Aging, 13*(3), 445–461. doi:10.1037//0882-7974.13.3.445

Schunk, D. H. (1984). Self-efficacy perspective on achievement behavior. *Educational Psychologist, 19*(1), 48–58. doi:10.1080/00461528409529281

Shipstead, Z., Redick, T. S., & Engle, R. W. (2012). Is working memory training effective? *Psychological Bulletin, 138*(4), 628–654. doi:10.1037/a0027473

Strobach, T., Liepelt, R., Schubert, T., & Kiesel, A. (2012). Task switching: Effects of practice on switch and mixing costs. *Psychological Research, 76*(1), 74–83. doi:10.1007/s00426-011-0323-x

Strobach, T., Salminen, T., Karbach, J., & Schubert, T. (2014). Practice-related optimization and transfer of executive functions: A general review and a specific realization of their mechanisms in dual tasks. *Psychological Research 78*(6), 836–851. doi:10.1007/s00426-014-0563-7

Valentijn, S. A., Hill, R. D., Van Hooren, S. A., Bosma, H., Van Boxtel, M. P., Jolles, J., & Ponds, R. W. (2006). Memory self-efficacy predicts memory performance: Results from a 6-year follow-up study. *Psychology and Aging, 21*(1), 165–172. doi:10.1037/0882-7974.21.2.165

van Asselen, M., & Ridderinkhof, K. R. (2000). Shift costs of predictable and unexpected set shifting in young and older adults. *Psychologica Belgica, 40*(4), 259–273.

van der Oord, S., Ponsioen, A. J. G. B., Geurts, H. M., Ten Brink, E. L., & Prins, P. J. M. (2012). A pilot study of the efficacy of a computerized executive functioning remediation training with game elements for children with ADHD in an outpatient setting: Outcome on parent- and teacher-rated executive functioning and ADHD behavior. *Journal of Attention Disorders, 20*(8), 1–14. doi:10.1177/1087054712453167

Verhaeghen, P., & Marcoen, A. (1996). On the mechanisms of plasticity in young and older adults after instruction in the method of loci: Evidence for an amplification model. *Psychology and Aging, 11*(1), 164–178. doi:10.1037//0882-7974.11.1.164

von Bastian, C. C., & Oberauer, K. (2013). Distinct transfer effects of training different facets of working memory capacity. *Journal of Memory and Language, 69*, 36–58. doi:10.1016/j.jml.2013.02.002

Wass, S. V., Porayska-Pomsta, K., & Johnson, M. H. (2011). Training attentional control in infancy. *Current Biology, 21*(18), 1543–1547. doi:10.1016/j.cub.2011.08.004

Wasylyshyn, C., Verhaeghen, P., & Sliwinski, M. J. (2011). Aging and task switching: A meta-analysis. *Psychology and Aging, 26*(1), 15–20. doi:10.1037/a0020912

Watson, J., Lambert, A. E., Miller, A. E., & Strayer, D. L. (2011). The magical letters P, F, C, and sometimes U: The rise and fall of executive attention with the development of prefrontal cortex. In K. L. Fingerman, C. A. Berg, J. Smith, & T. C. Antonucci (Eds.), *Handbook of life-span development* (pp. 409–438). New York, NY: Springer.

Wells, G. D., & Esopenko, C. (2008). Memory self-efficacy, aging, and memory performance: The roles of effort and persistence. *Educational Gerontology, 34*(6), 520–530. doi:10.1080/03601270701869386

West, R. L. (1996). An application of prefrontal cortex function theory to cognitive aging. *Psychological Bulletin, 120*(2), 272–292. doi:10.1037/0033-2909.120.2.272

West, R. L., Bagwell, D. K., & Dark-Freudeman, A. (2008). Self-efficacy and memory aging: The impact of a memory intervention based on self-efficacy. *Aging, Neuropsychology & Cognition, 15*(3), 302–329. doi:10.1080/13825580701440510

West, R. L., & Hastings, E. C. (2011). Self-regulation and recall: Growth curve modelling of intervention outcomes for older adults. *Psychology & Aging, 26*(4), 803–812. doi:10.1037/a0023784

Weywadt, C., & Butler, K. M. (2013). The role of verbal short term memory in task selections: How articulatory suppression influences task choice in voluntary task switching. *Psychonomic Bulletin & Review, 20*(2), 334–340. doi:10.3758/s13423-012-0349-0

Whitson, L. R., Karayanidis, F., Fulham, R., Provost, A., Michie, P. T., Heathcote, A., & Hsieh, S. (2014). Reactive control processes contributing to residual switch cost and mixing cost across the adult lifespan. *Frontiers in Psychology, 5, 383.* doi:10.3389/fpsyg.2014.00383

Whitson, L. R., Karayanidis, F., & Michie, P. T. (2011). Task practice differentially modulates task-switching performance across the adult lifespan. *Acta Psychologica, 139*(1), 124–136. doi:10.1016/j.actpsy.2011.09.004

Zelazo, P. D. (2006). The dimensional change card sort (DCCS): A method of assessing executive function in children. *Nature Protocols, 1*(1), 297–301. doi:10.1038/nprot.2006.46

Zelazo, P. D., Qu, L., & Kesek, A. C. (2010). Hot executive function: Emotion and the development of cognitive control. In S. D. Calkins & M. A. Bell (Eds.), *Child development at the intersection of emotion and cognition* (pp. 97–111). Washington, DC: American Psychological Association.

Zimmerman, B. J. (2000). Attaining self-regulation: A social cognitive perspective. In M. Boekaerts, P. R. Pintrich, & M. Zeidner (Eds.), *Handbook of self-regulation* (pp. 13–41). San Diego, CA: Academic Press.

Zinke, K., Einert, M., Pfennig, L., & Kliegel, M. (2012). Plasticity of executive control through task switching training in adolescents. *Frontiers in Human Neuroscience, 6,* 41. doi:10.3389/fnhum.2012.00041

Epilogue

Don't Buy the Snake Oil

Michael R. Dougherty and Randall W. Engle

This volume aims to summarize the state of the art in the working memory (WM) training literature. The idea for the book was conceived in 2011 and at the time the notion that general cognitive abilities could be improved through training was exciting yet controversial. The early evidence, although imperfect, was tantalizing: Several studies had seemingly illustrated that people's performance on untrained measures of intellectual abilities could be improved after just a few hours of training on cognitively demanding tasks (Chein & Morrison, 2010; Jaeggi, Buschkuel, Jonides, & Perrig, 2008; Klingberg et al., 2005; Mahncke et al., 2006). Proponents of training pointed to these early successes as reason for optimism and funding agencies released a flood of money to support the work, while critics pointed to shortcomings in experimental design as reason to remain skeptical.

In most areas of science, progress in resolving open scientific questions is slow, sometimes requiring decades of careful research. However, this has not been the case for research on WM training, which has exploded over the past decade. The primary open question addressed by the explosion of research was whether WM training is indeed effective in improving intellectual abilities. Researchers tackled the question by addressing several methodological issues: Do the effects persist when placebo (i.e., active) control conditions are used as comparisons? Are the effects present when the sample size is adequate? Are the effects dependent on randomization to condition (training versus control)? Are the observed transfer effects dependent on the overlap between training and transfer tasks? How long do training effects persist?

Getting to the bottom of these questions has meant diving into the nitty-gritty details to discern when, where, and why training effects transfer. Such work lacks the "pizzazz" and excitement that comes with grandiose claims that IQ can be increased, but it's a necessary step in our understanding of human nature. Unfortunately, the outcome of the systematic study of WM training has not been favorable to the proponents: We now know that many of the most promising findings in the WM training literature were likely false starts. The imperfections cited by the critics of some of the seminal studies have proven to be crucial to their outcome (Simons et al., 2016).

A complete review of the literature is well beyond the scope of this brief epilogue, but it may be useful to point to a few recent meta-analyses to address some of the above questions. Meta-analyses are useful because they address one of the main methodological challenges that has plagued the WM training literature, namely, the issue of small samples.

Are the Effects of WM Training Dependent on the Type of Control Condition Used in the Study Design?

The motivation for this question stems from the long history of research documenting that people's behavior often changes in response to placebos and/or expectations implied or otherwise communicated to participants (Boot, Simons, Stothart & Stutts, 2013; Kirsch, 1985). Controlling for participants' expectation is important for differentiating between causal mechanisms: Do transfer effects arise because the cognitive mechanisms improve with training, or do such effects arise due to expectations? This question has been addressed in a number of recent meta-analyses of WM training. For instance, in a comparison of N-back training studies using active versus passive controls, Dougherty, Hamovitz, and Tidwell (2016) observed that the meta-analytic effect for improvements on measures of fluid intelligence was nearly half a standard deviation for passive control designs but approximately zero for active control designs. Analysis of active control studies using Bayesian methods revealed strong support for the null hypothesis. An analysis by von Bastian, Guye, and De Simoni (see Chapter 4 of this volume) offers a similar conclusion, as do several other recent meta-analyses (see Melby-Lervag, Redick, & Hulme, 2016; Sala & Gobet, 2017, among others). The issue of whether transfer of training depends on type of control group is settled. Effects quite clearly manifest when training is compared to a passive control but there is little to no effect when active controls are used.

Does Random Assignment to Condition Matter?

Randomization is a bedrock principle of sound methodology. Without it, any statistical differences (or lack thereof, for that matter) observed between groups can be attributable to pre-existing group differences, self-selection into groups, or biases in how experimenters assign participants to groups. Randomization is the one tool that researchers have at their disposal to ensure that the participants assigned to one group are, on average, identical to those in another group on all possible dimensions. Of course, randomization works best when sample sizes are large, an issue that we've already noted as problematic for many WM training studies. So, does randomization matter? In a meta-analysis of children age 3 to 16, Sala and Gobet (2017) observed that the meta-analytic effect size for far transfer dropped from $G = 0.27$ for nonrandomized samples to $G = 0.07$ for randomized samples. The effect size for studies using both active controls and randomization was 0.03. Clearly, these results indicate that randomization does indeed matter and that transfer effects are negligible when sizable randomized samples are tested.

Are Training Effects Dependent on the Overlap Between Training and Transfer Task Operations or Stimuli?

The allure of WM training as a cure for intellectual deficits and cognitive decline is the promise that training on one task transfers to an ostensibly different task that bears little resemblance to the trained task. There are two orthogonal dimensions to consider in this regard. One dimension concerns the relation between the underlying theoretical constructs. Unsurprisingly, there is clear evidence that transfer effects are larger for tasks that measure a construct closely related to the task that is trained (see Melby-Verlag et al., 2016; Simons et al., 2016). The second dimension concerns the types of stimuli or operations included in the tasks that measure the construct. For instance, an N-back training task that uses letters as stimuli and an episodic memory task that uses letters as stimuli may ostensibly reflect different theoretical constructs, yet share task variability due to the overlap between stimuli. As far as we know, research addressing this distinction is relatively underdeveloped, but it may account for some observed transfer effects (see Soveri, Antfolk, Salo, & Laine, 2017; Sprenger et al., 2013). One thing is for sure, however: the ideal approach would involve an assessment of transfer using a latent variable approach, which addresses the issues of measurement error and task-specific variability (see Chapter 3 of this volume, by Schmiedek, Lövdén, & Lindenberger).

What Might a Convincing Study Look Like?

Taking the above considerations into account, the ideal study is one that uses a double-blind procedure with an active control group, collects multiple measures of cognitive abilities pre- and posttest, randomly assigns a large sample, collects measures of prior expectations, and uses statistical tests that allow one to quantify evidence for or against the null hypothesis. De Simoni and von Bastian (2018) conducted a study that closely approximates this idealized standard, and once again the results are no friend of the proponents. In their study, which included roughly 60 participants per condition, De Simoni and von Bastian's data showed evidence for the null hypothesis of no transfer of training across all measured constructs, with most Bayes factors for far transfer ranging from 5.76 to 11.36 in favor of the null.

But What About the Brain?

Of course, this volume is about more than transfer. Various chapters discuss the neural underpinnings of WM training and postulate ways in which neuroscience methods may allow identification of the neural markers associated with training as well as potential methods for modulating training benefits (e.g., tCDS). As interesting as these chapters may be, the usefulness of these methods as they pertain to transfer always comes back to the issue of behavioral change. It is trivial to note correlations between brain and behavior, or to observe changes in brain activation, structure, or synchrony as a function of training. Naturally, one would expect changes in brain to accompany any practiced activity. Even placebos elicit changes in brain activity (see Zunhammer, Bingel, & Wager, 2018). However, in the absence of sustained behavioral improvements on untrained measures of cognitive ability, there is not much for neuroscience to explain. That's not to say that the neuroscience data aren't valuable in other ways, but rather that neuroscience data cannot and do not substantiate claims of transfer that aren't observed behaviorally.

Summary

There's a saying in investigative research: "Follow the evidence." Unfortunately, the evidence now strongly, if not decisively, indicates that WM training does not transfer to performance on nontrained tasks. The critics were right: WM training does not lead to long-lasting generalizable improvements in cognitive functioning. We're sure that this conclusion will be disappointing to many readers of this book, as well as to the many individuals who have purchased commercial brain-training products. We're sorry to offer such a gloomy outlook on the state of the art, but sometimes science leads us where it leads us. Every parent of a child struggling in school will tell you that they would do almost anything to see their child succeed and love school. This has led parents of children classified as "reading disabled" to purchase glasses with colored lenses (Irlen lenses) because some optometrists told them the problem was in their child's vision. Likewise, promoters of WM and brain training have franchised sales of their products through psychiatrists and psychologists with the lure to distraught parents that their child could be "fixed" by several hours of playing video games. Military agencies that need their enlistees to learn incredibly complex and dangerous jobs, such as demolition experts for explosive devices (as depicted in the film *The Hurt Locker*), are desperate to have recruits who are able to read, to study, and to learn the vast volumes necessary to successfully perform those duties. The prospect that WM and brain training might do that has led to enormous efforts to find the effects promised by the proponents of WM and brain training. With that vast amount of money and effort, if the effects were there, they would have been found, would have been shown to be robust to random assignment and the use of placebo controls, and would have been replicated. They were not.

References

Boot, W. R., Simons, D. J., Stothart, C., & Stutts, C. (2013). The pervasive problem with placebos in psychology: Why active control groups are not sufficient to rule out placebo effects. *Perspectives on Psychological Science, 8*, 445–545.

Chein, J. M., & Morrison, A. B. (2010). Expanding the mind's workspace: Training and transfer effects with a complex working memory span task. *Psychonomic Bulletin & Review, 17*, 193–199.

De Simoni, C., & von Bastian, C. C. (2018). Working memory updating and binding training: Bayesian evidence supporting the absence of transfer. *Journal of Experimental Psychology: General, 147*(6), 829–858. doi:10.1037/xge0000453

Dougherty, M. R., Hamovitz, T., & Tidwell, J. W. (2016). Re-evaluating the effectiveness of n-back training on transfer through the Bayesian lens: Support for the null. *Psychonomic Bulletin and Review, 23*, 306–316.

Jaeggi, S. M., Buschkuehl, M., Jonides, J., & Perrig, W. J. (2008). Improving fluid intelligence with training on working memory. *Proceedings of the National Academy of Sciences of the United States of America, 105*, 6829–6833.

Kirsch, I. (1985). Response expectancy as a determinant of experience and behavior. *American Psychologist, 40*, 1189–1202.

Klingberg, T., Fernell, E., Olesen, P. J., Johnson, M., Gustafsson, P., Dahlström, K., . . . Westerberg, H. (2005). Computerized training of working memory in children with ADHD—A randomized, controlled trial. *Journal of the American Academy of Child and Adolescent Psychiatry, 44*, 177–186.

Mahncke, H. W., Connor, B. B., Appelman, J., Ahsanuddin, O. N., Hardy, J. L., Wood, R. A., . . . Merzenich, M. M. (2006). Memory enhancement in healthy older adults using a brain plasticity-based training program: A randomized, controlled study. *Proceedings of the National Academy of Sciences of the United States of America, 103*, 12523–12528.

Melby-Lervåg, M., Redick, T. S., & Hulme, C. (2016). Working memory training does not improve performance on measures of intelligence or other measures of "far transfer": Evidence from a meta-analytic review. *Perspectives on Psychological Science, 11*, 512–534.

Sala, G., & Gobet, F. (2017). Working memory training in typically developing children: A meta-analysis of the available evidence. *Developmental Psychology, 53*, 671–685.

Simons, D. J., Boot, W. R., Charness, N., Gathercole, S. E., Chabris, C. F., Hambrick, D. Z., & Stine-Morrow, E. A. (2016). Do "brain-training" programs work? *Psychological Science in the Public Interest, 17*(3), 103–186.

Soveri, A., Antfolk, J., Karlsson, L., Salo, B., & Laine, M. (2017). Working memory training revisited: A multi-level meta-analysis of n-back training studies. *Psychonomic Bulletin & Review, 24*, 1077–1096.

Sprenger, A. M., Atkins, S. M., Bolger, D. J., Harbison, J. I., Novick, J. M., Weems, S. A., ... Dougherty, M. R. (2013). Training working memory: Limits of transfer. *Intelligence, 41*, 638–663. doi.org/10.1016/j.intell.2013.07.013

Zunhammer, M., Bingel, U., & Wager, T. D. (2018). Placebo effects on the neurologic pain signature: A meta-analysis of individual participant functional magnetic resonance imaging data. *JAMA Neurology*. doi:10.1001/jamaneurol.2018.2017

Index

Page numbers followed by *f* and *t* indicate figures and tables, respectively. Numbers followed by n indicate footnotes.

For the benefit of digital users, indexed terms that span two pages (e.g., 52–53) may, on occasion, appear on only one of those pages.